# INTRODUCTION
# TO
# AVIATION SCIENCE

PRENTICE-HALL INTERNATIONAL, INC., *London*
PRENTICE-HALL OF AUSTRALIA, PTY. LTD., *Sydney*
PRENTICE-HALL OF JAPAN, INC., *Tokyo*
PRENTICE-HALL OF CANADA, LTD., *Toronto*
PRENTICE-HALL OF INDIA PRIVATE LTD., *New Delhi*

# INTRODUCTION TO AVIATION SCIENCE

**Joseph W. Benkert**

*Chairman, Aviation Department*
*Bryant & Stratton School*
*(A Junior College of Business)*
*Boston, Massachusetts*

Prentice-Hall Inc., *Englewood Cliffs, New Jersey*

Current printing (last digit):

10 9 8 7 6 5 4 3 2 1

13-478222-4

Library of Congress Catalog Card Number: 71-135754
Printed in the United States of America

***To Mother and Dad . . .***
*without whom my aviation career*
*would have been short-lived.*

# PREFACE

Introduction to Aviation Science is a textbook for flight. It is a study of the theory of flight; the mechanics of flight; and how flight is controlled aloft.

The atmospheric phenomena through which we fly, methods of navigation, and the "rules of the road" are discussed in detail, as are the techniques of two-way radio communications, instrument interpretation, and emergency procedures.

During the writing of this text, two simple rules were followed. The first rule determined what material was to be included in the book; the "need to know" was the primary criteria. The information which we need to know is included in detail.

The second rule determined the language of the text. No word or phrase is used where a simpler, more easily understood word or phrase will suffice.

These two rules make this volume ideal for the fledgling pilot or interested layman who wishes to build his aviation knowledge on a solid foundation.

Each subject area is segmented into short, precise, easily assimilated capsules which are designed to guide the reader through the diverse fields of the knowledge required of today's student pilot.

Questions following each chapter assist the reader in reviewing the important points discussed. He should re-read those sections with which he remains unfamiliar.

Answers to the review questions are contained in the appendix and are written in such a way that they provide a concise review of the entire text.

Federal Aviation Administration Exam-O-Grams are used liberally after many chapters. Since these publications are provided by the FAA to fill the need for written examination preparation material, the reader will find them to be invaluable aids.

The text is extremely comprehensive in that all material required of a student pilot is included in its pages.

The body of knowledge concerning flight has constantly expanded. It will continue to expand. We stand on the threshold of a phenomenal growth in the aviation industry. We have not yet tapped the unseen reservoir of new challenges which the aviation industry will one day master.

Welcome to the realm of flight. You will enjoy the adventures and rewards ahead.

*Joseph W. Benkert*

# ACKNOWLEDGEMENTS

To all who helped in the writing of this text, a sincere "Thank you."

Special notes of thanks must go to my wife, Janet, and children, Gretchen and Joseph, who exhibited great patience and understanding during the twenty-four months spent in production of the manuscript.

During the endless hours of proofreading, many people came to my aid, including Mr. and Mrs. Dennis Doucette, Miss Judith Shaia, Miss Linda Marie Connors, Miss Jackie Welch, and Miss Charlene Ernesti.

I am especially greatful to Ann and Joseph Todesca for their continuous encouragement and help.

The use of Federal Aviation Administration publications and the navigation computer by Felsenthal Instrument Company is greatfully acknowledged. I also extend thanks to Anthony's Studio, Lawrence, Massachusetts for the photographs of the navigation computer used in this book.

# NOTE TO THE READER

One of the interesting facts concerning Aviation Science is that it is a constantly changing subject. New theories are being advanced and old theories are being proven or dis-proven. If you will fill in your name and address on a post card, I will, by return mail, inform you of any important changes or developments in the study of Aviation Science.

In addition, you will receive a final examination covering all of the pertinent information in this text which will prove helpful to you in the preparation for the written Federal Aviation Administration pilot license examinations.

I would also enjoy hearing your comments or constructive criticism of this text.

Fill out a post card as illustrated below so that I may mail you any pertinent revisions and your final examination.

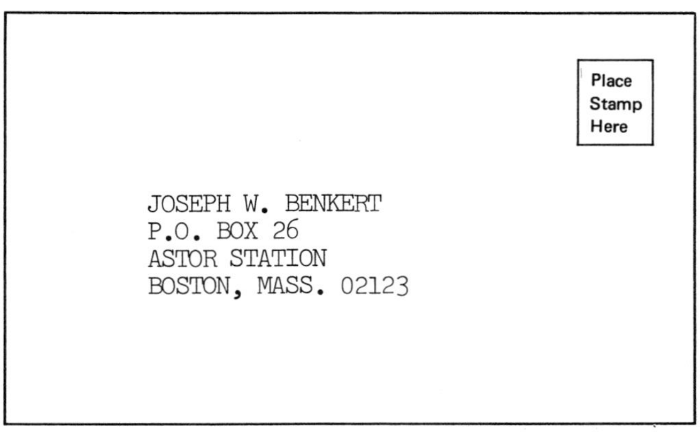

# CONTENTS

# PART I

## AIRCRAFT OPERATIONS

# 1

# THE AIRPLANE

The airplane is a fascinating machine. It is efficiently designed and incorporates safety devices and engineering techniques which make it a unique machine.

As we build with experience on the foundation laid in this text, the wonder of the airplane will be increasingly apparent.

Before we begin with the bricks and mortar of our foundation, we must look at the airplane and become familiar with its major components.

Figure 1 shows a typical light aircraft, and lists the major parts with which we must immediately become familiar.

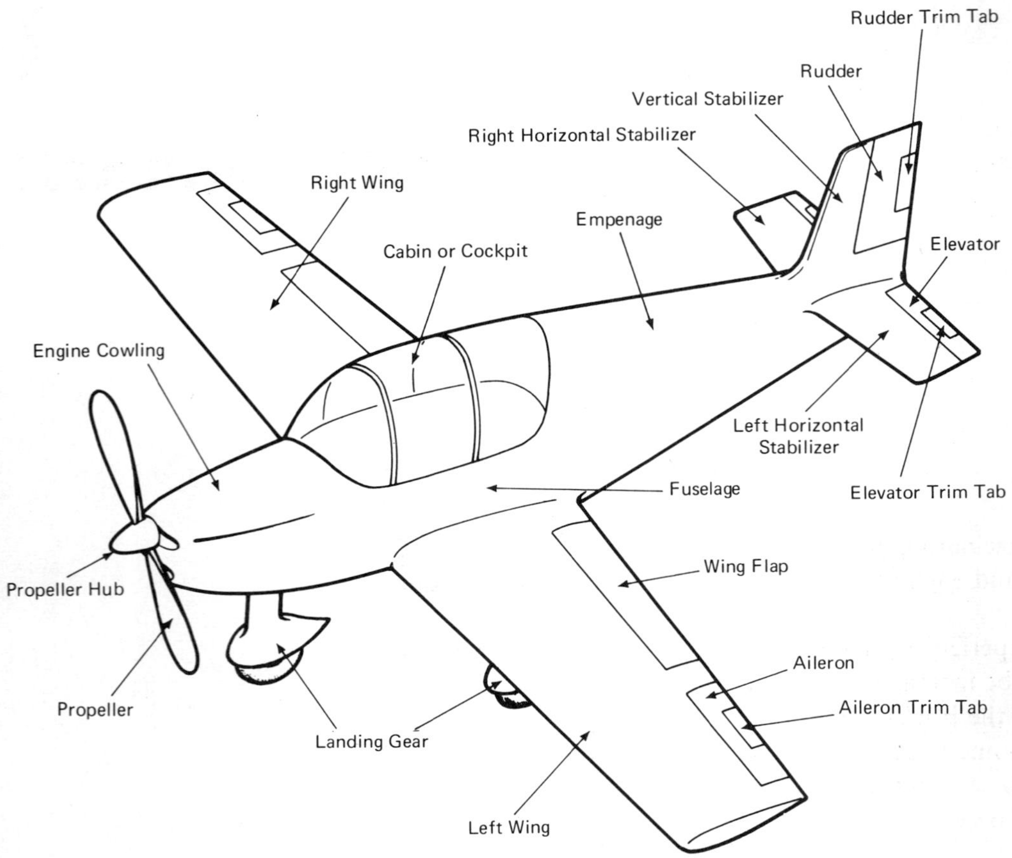


Fig. 1. Aircraft nomenclature.

# 2

# THE FOUR FORCES OF FLIGHT

When in flight, every aircraft is subjected to four forces. Two of them are natural forces, and Mother Nature imposes them on all matter on or near the earth. Man had to produce two counteracting forces in order to achieve flight.

## The Natural Forces

*Gravity* is the force of attraction by which terrestrial bodies tend to fall toward the center of the earth.

*Drag* is the force exerted on a moving body which tends to reduce its forward motion. Air is a fluid. It has weight and substance, which can be seen by watching it in motion, bending the limbs of a strong oak tree. When we attempt to move through this thick fluid mass we feel resistance, called "drag." The amount of resistance depends on the size and shape of the penetrating body, the speed at which penetration is attempted (drag increases as speed increases), the denseness of the air or fluid being penetrated, and the coefficient of friction of the penetrating material.

## The Man-Made Forces

*Lift* is the man-made force which counteracts gravity, allowing man and his machine to rise above the earth. Particular shapes, or airfoils, produce the phenomenon called "lift" when moved through the air.

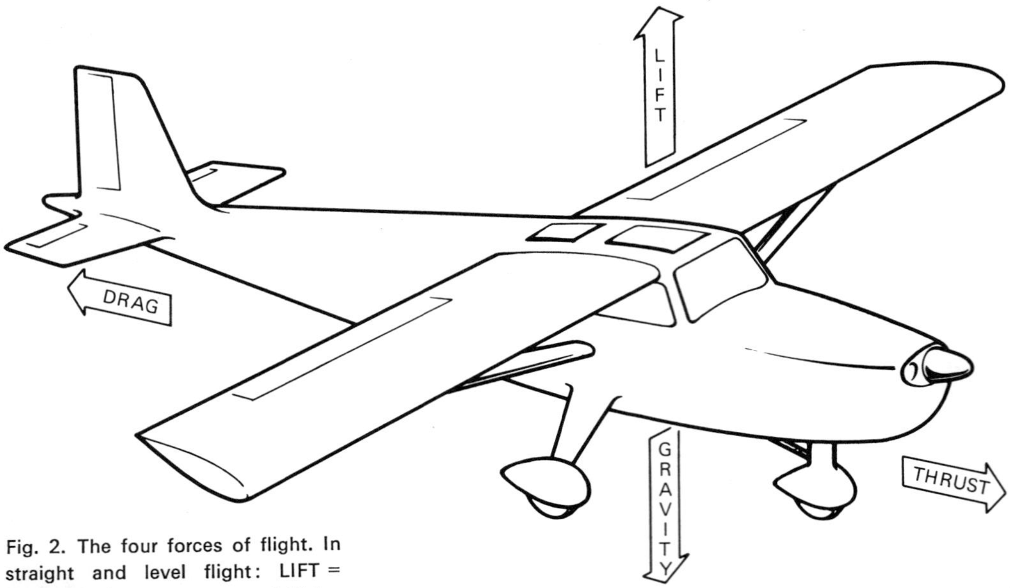


Fig. 2. The four forces of flight. In straight and level flight: LIFT = GRAVITY, in constant speed flight: THRUST = DRAG.

*Thrust* is the man-made force which counteracts drag. Actually, thrust, which is developed by the whirling propeller, is also lift acting in a different direction to pull the aircraft forward.

## The Balance of Forces

In straight and level flight, with a constant airspeed, lift exactly equals gravity and thrust exactly equals drag. By changing the value and/or the direction of the two man-made forces of flight we control the aircraft. If lift becomes greater than gravity, the aircraft will climb. Conversely, if lift is less than gravity, the aircraft will descend. (See Fig. 2.)

When thrust is increased and becomes greater than drag the aircraft will accelerate, but only to a speed at which drag again equals thrust. At that point, the aircraft speed will stabilize. If thrust is decreased, drag will cause the aircraft to decelerate until drag again equals thrust.

## The Production of Lift

Obviously, any object moving through the air will not produce lift. Only an object of a particular shape has the capability of producing lift when air passes by it. This shape is called an "airfoil."

Referring to Fig. 3 we see a view of the stub end of a typical airfoil.

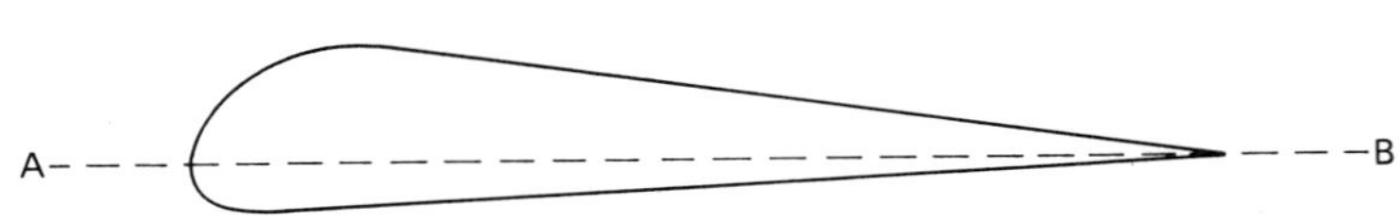


Fig. 3. An airfoil.

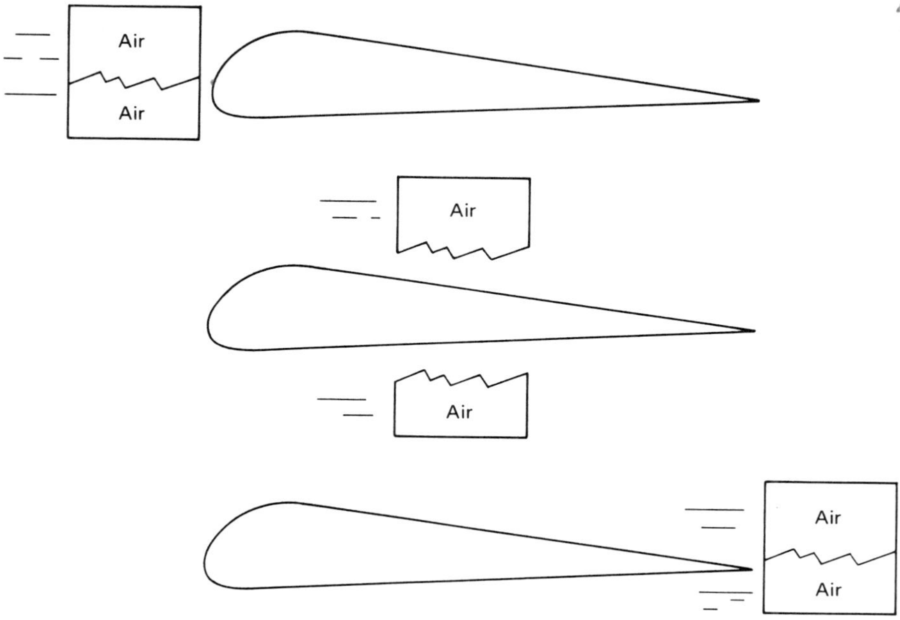


Fig. 4. Lift production.

This same shape would be found if we sliced through a propeller blade, for it is a standard airfoil shape.

The curved upper surface of the airfoil is called the "camber." Line AB is an imaginary straight line from the leading edge to the trailing edge of the airfoil, and is called the "chord" line.

Picture a cubic foot block of air striking the leading edge of an airfoil as shown in Fig. 4. The parcel of air breaks in half, with one piece going over the top or camber of the airfoil, and the other piece passing along the bottom of the airfoil.

The two parcels of air must meet and rejoin at or beyond the trailing edge of the airfoil. The camber is a curved line. The bottom of the airfoil is nearly a straight line. Therefore, the parcel of air passing over the camber has farther to go than the air passing under the airfoil, so it must travel faster to get to the trailing edge at the same time as the parcel which took the more direct route.

Bernoulli's theorem of physics states that the faster a fluid (and air *is* a fluid) travels, the lower will be the pressure surrounding it. The area of fast moving air, over the camber of the airfoil, is an area of lower barometric pressure.

Air, like water, tends to flow from high to low. The higher pressure air under the airfoil tries to flow toward the lower pressure area above the camber. It naturally cannot go through the airfoil, but pushes it up into the low pressure area, creating lift.

The difference in barometric pressure between the air under the airfoil and the air above the camber does not have to be great. A small differential in pounds per square foot of pressure, when multiplied by the many square feet of lifting surface, will create many hundreds of pounds of lift.

Not all of the lift created by an airfoil is derived from the Bernoulli effect. The exact percentage depends on the shape of the airfoil. Normally, at least one-half of the total lift is credited to Bernoulli. The remaining lift comes from the high-speed air ramming under the airfoil. We can demonstrate this powerful force by holding our extended palm from the window of a moving auto. As we tilt our palm from the horizontal, we will feel ram air pushing our hand upward.

The total amount of lift produced by an airfoil depends on many factors including the shape and area of the airfoil, the speed of the air moving by it, the density of that air, and the angle with which the air strikes the airfoil.

*The Shape of the Airfoil*

Naturally, the thicker the airfoil, the more lift will be produced, because the air passing over the camber will have to travel much farther and faster than the air under the airfoil. This creates a larger pressure differential above and below the lifting surface. But airfoil shape is limited by two important factors. If it is too thick it will create a huge amount of drag. This kind of drag is called *parasite drag*. A more streamlined shape will cause less parasite drag.

If the air is asked to travel too far along the camber it will "burble" or tear away from the smooth flow, as shown in Fig. 5, filling in the low pressure area and destroying lift.

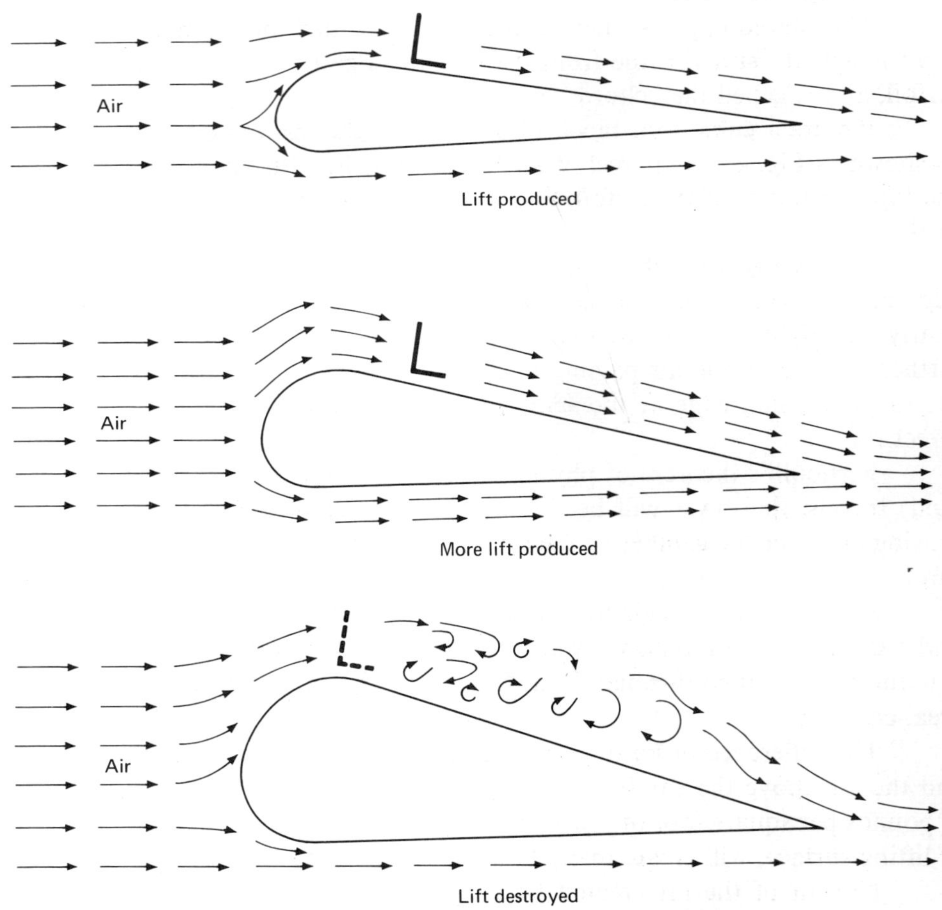


Fig. 5. Airfoil shape determines the amount of lift produced.

**The Speed of the Air**

If the speed of the air passing the airfoil is increased, the amount of lift produced will be increased. This is because more ram effect will be present, and the pressure differential between the air above and below the airfoil will be greater.

**The Density of the Air**

"Density" is a measure of the number of molecules of air present in a given volume. Every molecule of air adds to the lift produced by the airfoil. Hence, the more molecules passing the airfoil, the more lift will be produced.

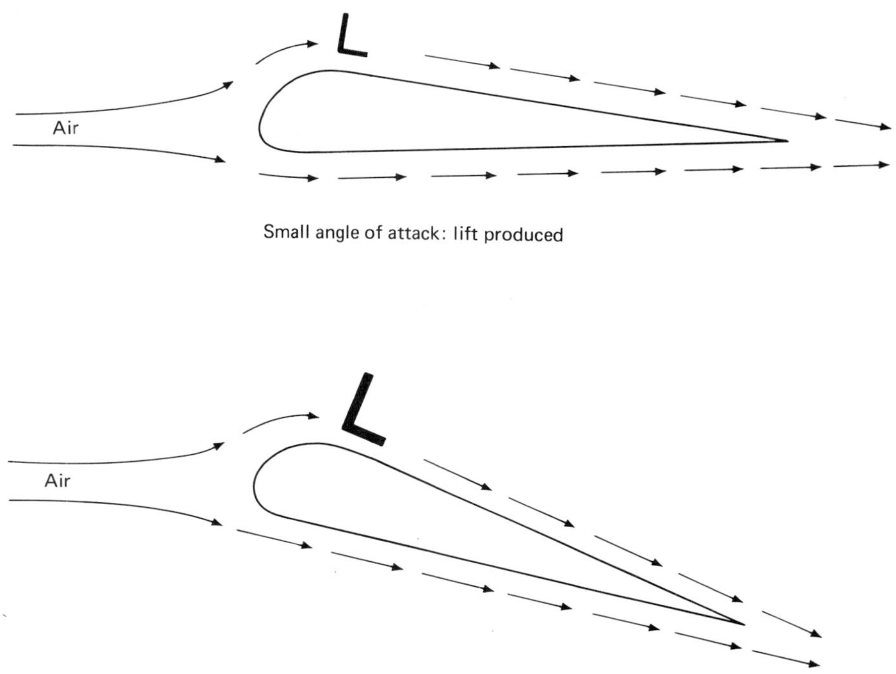


Fig. 6. The angle of attack determines the amount of lift produced.

**The Angle of Attack**

The "angle of attack" is the angle at which the passing air strikes the airfoil. As this angle is increased more lift will be generated.

Figure 6 shows an airfoil with a small and a large angle of attack. Notice that the air passing over the camber of the airfoil with the higher angle of attack must travel still farther (and faster) thus increasing the differential pressure between the top and bottom of the airfoil. Also, more ram effect is produced.

## Relative Wind

The air which passes the airfoil in flight is called the "relative wind." It is wind which is present ***only*** because of the ***relative motion*** between the airfoil and the air. Relative wind has nothing whatever to do with the natural wind that stirs the leaves of trees.

Relative motion can be demonstrated on the surface of the earth by observing a power boat in a stream. Figure 7 shows a boat moving through the water at a speed of 5 miles per hour.

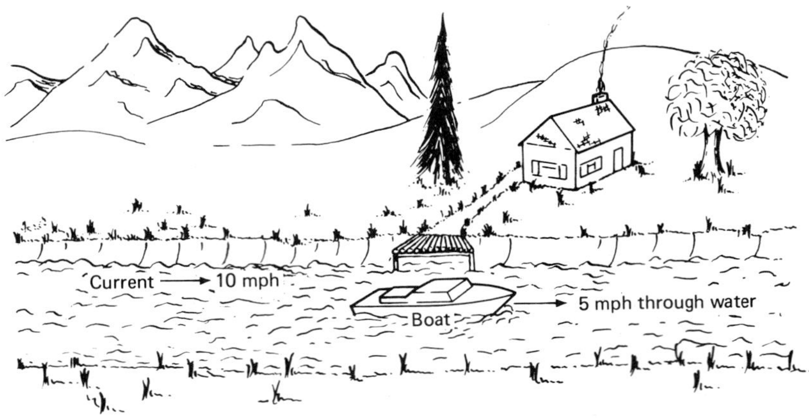


Fig. 7. Relative speed. A boat in a stream has two speeds: (1) speed relative to the water (5 mph), and (2) speed relative to the shore (15 mph).

The boat moves at a speed of 5 miles per hour because the engine produces a certain amount of power.

However, the water is moving downstream at a speed of 10 miles per hour. Therefore, the boat has a relative speed in the water of 5 miles per hour, but a relative speed to the shore of 15 miles per hour (10 mph current speed plus 5 mph boat speed).

If we turned the boat around, its engine would still produce the same amount of power and hence would still drive it through the water at 5 miles per hour. But relative to the shore, it would be moving backward at 5 miles per hour (10 mph current speed minus 5 mph boat speed). The boat does not handle any differently however. It is still plowing through the water at a speed of 5 miles per hour.

For an aircraft in flight, ***only the relative wind is important to aircraft performance***. Just as the engine on the boat drove it through the water at a certain speed, either upstream or downstream, so will the aircraft engine drive the aircraft through the air at a certain speed, either into the natural wind or with the natural wind. The aircraft will perform identically when flying into the natural wind, across the natural wind, or with the natural wind, because it does not "know" that the air through which it is flying is moving across the surface of the earth.

Figure 8 shows an aircraft flying through a parcel of air at a speed of 100 miles per hour. The parcel of air is moving across the ground at a speed of 20 miles per hour. The aircraft has two speeds. First, it has a speed relative to the air through which it is flying of 100 miles per hour, because the engine is developing sufficient power to drive it through the air at that speed. Second, it has a speed relative to the ground of 80 miles per hour (100 mph airspeed minus 20 mph natural wind speed).

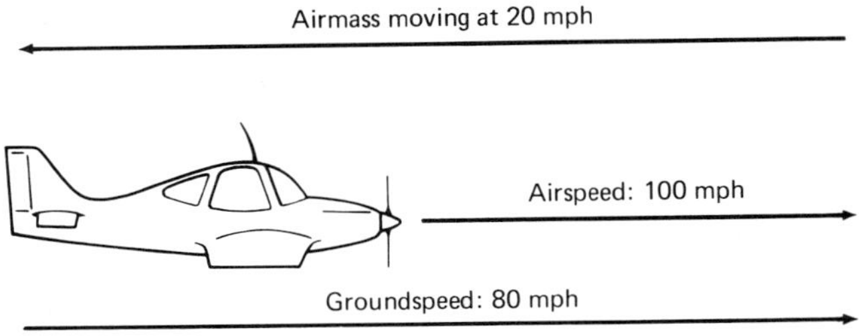


Fig. 8. Airspeed versus groundspeed. An airplane in flight has two speeds: (1) speed relative to the air (airspeed), and (2) speed relative to the ground (groundspeed).

*But the aircraft does not perform according to ground speed.* It performs *only* in relation to the speed and direction of the relative wind passing the aircraft structure.

Now let us turn the aircraft around. Since the power output of the engine has not changed, the speed of the aircraft through the air will not change. The relative wind will remain at 100 miles per hour. The speed over the ground will now be 120 miles per hour, but the aircraft does not perform any differently.

The relative wind is considered to be from a direction opposite to the flight path of the aircraft. That is, if the flight path is forward and upward, the relative wind will be rearward and downward. (See Fig. 9.)

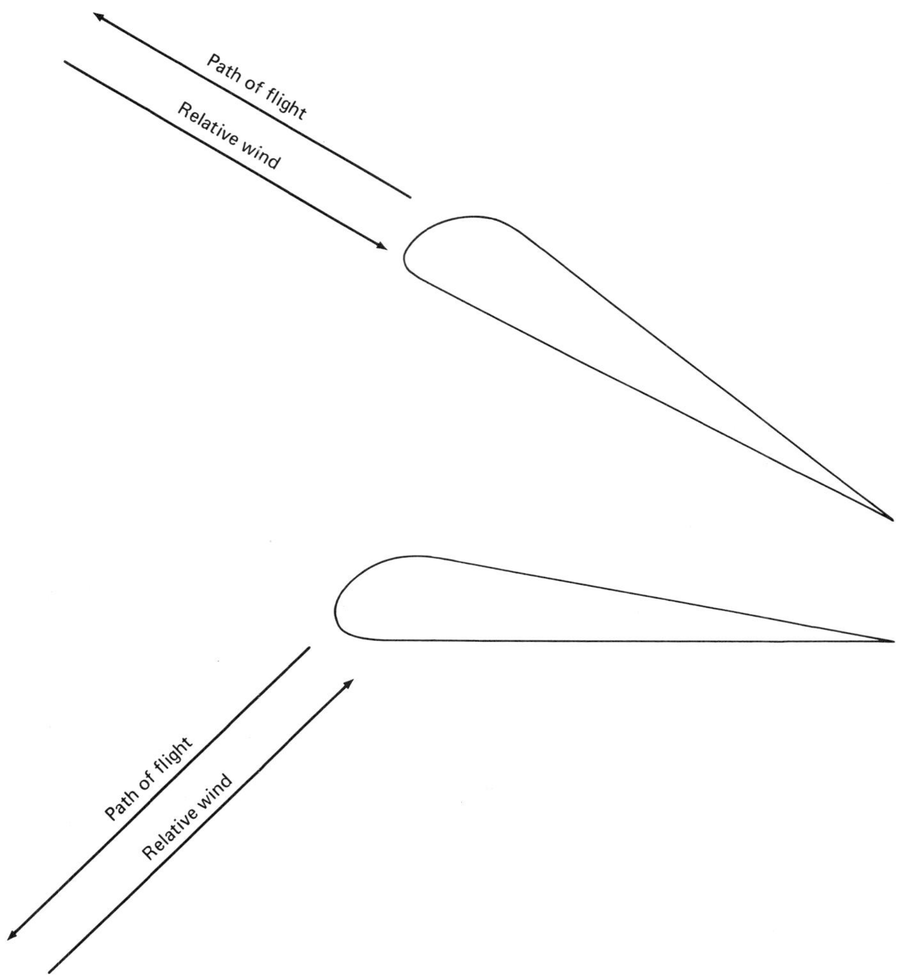


Fig. 9. The relative wind is opposite to the path of flight.

*The Path of Flight*

The path of flight is the path that the aircraft is following through the air. It is *not* necessarily the same direction as the nose points. Figure 10 shows an aircraft in varying conditions, some of which show the path of flight in approximately the same direction that the nose is pointing, and other conditions in which the path of flight is far from the direction the nose is pointing.

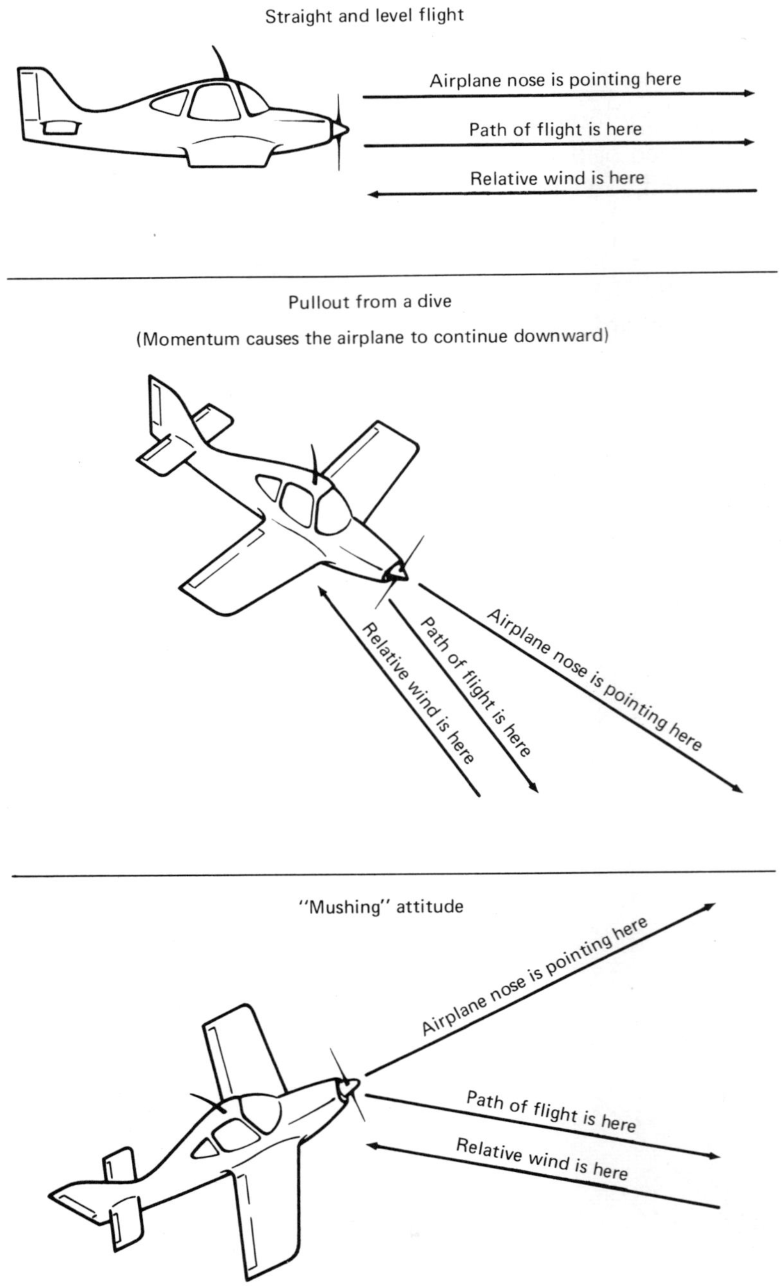


Fig. 10. The path of flight.

## Angle of Attack—Technical Definition

As we have already learned, the angle of attack is the angle with which the passing air strikes the airfoil. That air is the relative wind. To ensure that we all use the same reference point on the airfoil to measure angle of attack, the chord line has been designated as the reference line. Hence, the technical definition of angle of attack is *the angle between the relative wind and the chord of the airfoil*. (See Fig. 11.)

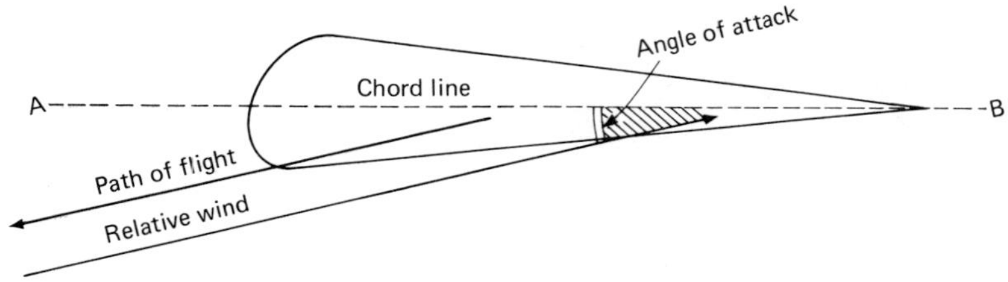


Fig. 11. The angle of attack.

## Angle of Incidence

When the wings are attached to the aircraft they are mounted so that the chord line is *not* parallel to the nose-to-tail axis of the aircraft. This slight angle (shown exaggerated in Fig. 12) between the chord of the wing and the aircraft axis is called the "angle of incidence." This produces a condition where some ram effect is present even though the aircraft axis is parallel to the horizon (in straight and level flight) or parallel to the path of flight.

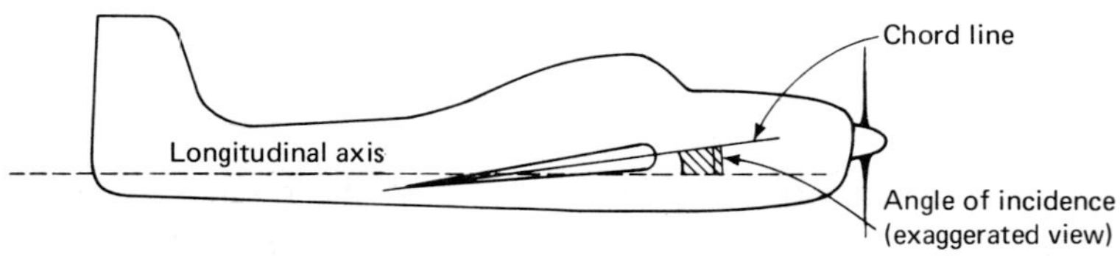


Fig. 12. The angle of incidence (exaggerated view).

## The Stall

We have seen that the amount of lift produced by an airfoil can be increased by increasing the angle of attack. However, there is a limit to which that angle may be increased before the relative wind passing over the camber will tear away or burble. When the angle of attack exceeds 18–20 degrees, the relative wind will burble and fill in the low pressure area above the camber, thus destroying lift. This condition is called a "stall" and is shown in Fig. 13.

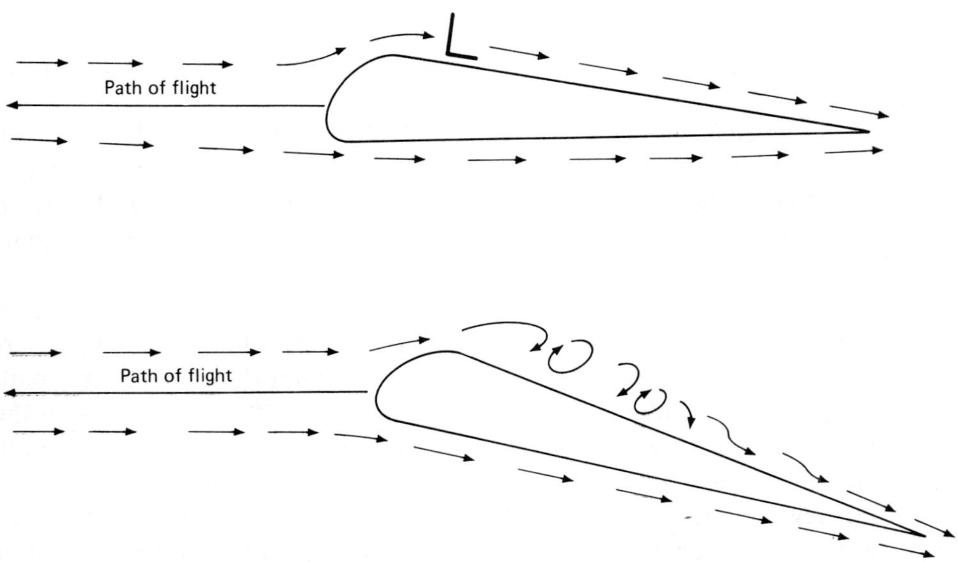


Fig. 13. The stall.

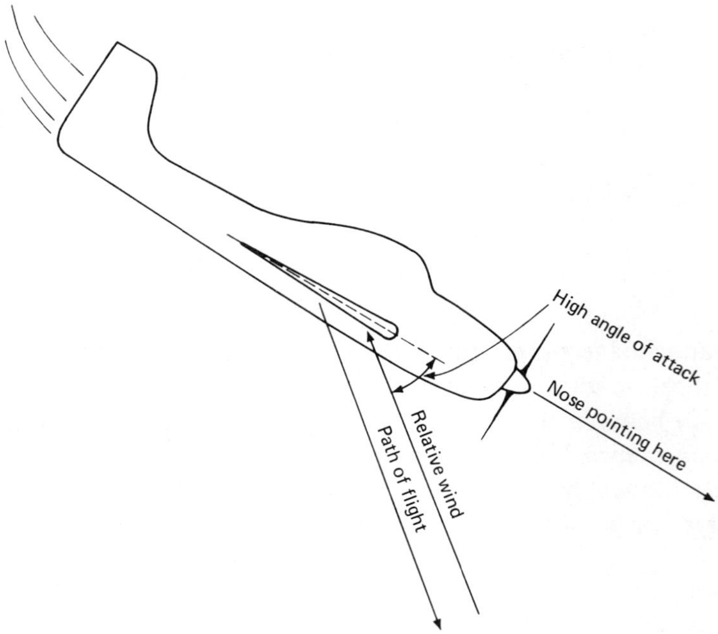


Fig. 14. Momentum. A high angle of attack may be developed during a pullout from a dive.

It is important to remember that the relative wind is considered to be opposite to the flight path of the aircraft. With this in mind, we will learn that an excessive angle of attack may be produced with the aircraft in any attitude in relation to the horizon. Figure 14 shows an aircraft in a pullout from a dive.

Even though the nose is starting upward, momentum continues to carry the aircraft downward. Notice the high angle of attack, causing a stall and loss of lift. An aircraft is a heavy mass. Time is required to change its direction. We must keep this in mind when changing the path of flight so as to avoid high angles of attack and unexpected stalls.

*Causes of the Stall*

There are three conditions which will cause an angle of attack large enough to produce a stall.

1. Insufficient airspeed: If the speed of the relative wind is not fast enough to produce enough lift to sustain the desired flight path, the aircraft will descend or "mush," changing the path of flight to a downward and forward direction. The relative wind will be upward and rearward, increasing the angle of attack.

2. Excessively violent flight maneuvers: When we cause a change in the direction in which the nose is pointing, the aircraft does not immediately take up a new path of flight. Momentum keeps the aircraft on its original path for a time. If we force the nose of the aircraft to point 18–20 degrees from the flight path the angle of attack will become excessive, again resulting in a stall. *This may occur at any airspeed, and from any aircraft attitude.*

3. Severe vertical gusts: A vertical gust is a shaft of air moving vertically, either toward or away from the earth's surface. When we fly an aircraft

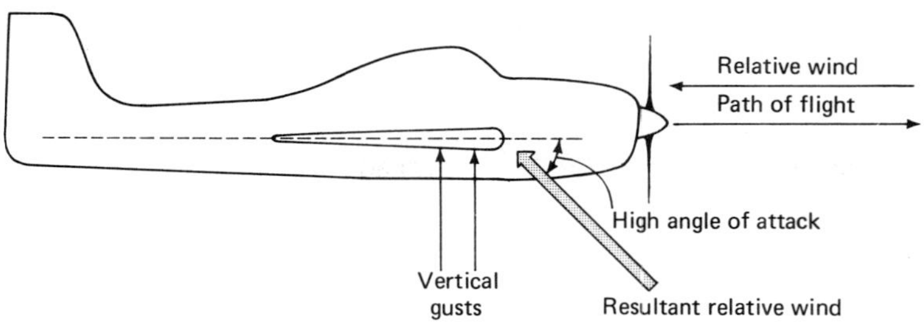


Fig. 15. Resultant relative wind. Vertical gusts may increase the angle of attack.

through vertical gusts, the *resultant relative wind* may strike the chord of the airfoil at an excessive angle of attack, causing a stall.

*Effect of the Stall*

When the aircraft approaches the stalled condition, we will feel the controls becoming sluggish, and the air burbling across the wing. When the stall occurs a large portion of the lift will be destroyed. Gravity will be stronger than the remaining lift, causing the aircraft to fall. Built-in stability, inherent to the aircraft design will aid us in restoring lift. (See Chapter 9.)

*Stall Recovery*

A stall is the result of an excessive angle of attack. Stall recovery is accomplished by simply returning to a lesser angle of attack. We must use the appropriate control action to make the nose of the aircraft point within 18–20 degrees of the path of flight. As soon as we do this, the relative wind will again flow smoothly over the airfoil, producing lift.

This recovery procedure from a stall must be practiced constantly because the stall may occur due to careless pilot technique at *any airspeed* and *from any attitude*, and rapid recovery with little altitude loss is essential.

## CHECK YOUR READING

1. List the four forces of flight.
2. What force prevents the aircraft from accelerating to an infinite speed when thrust is increased?
3. Briefly describe the process by which lift is produced by an airfoil.
4. State Bernoulli's theorem.
5. List the factors which determine the amount of lift which an airfoil will produce.
6. What is relative wind?
7. What effect does the natural wind have on an aircraft in flight?
8. Define angle of attack.
9. Define a stall.
10. What causes an airfoil to stall?
11. What is the angle of incidence? What is its purpose?
12. At what airspeed will an aircraft stall?
13. From what attitude will an aircraft stall?

# 3

# AIRCRAFT CONTROL

An aircraft has three axes, and a flight control to cause movement about each axis. We must have a clear picture of these axes and controls so that we can learn how the aircraft is made to respond to our command.

## The Longitudinal Axis

The longitudinal axis is a straight line from the nose to the tail of the aircraft. Movement about this axis is called *roll.*

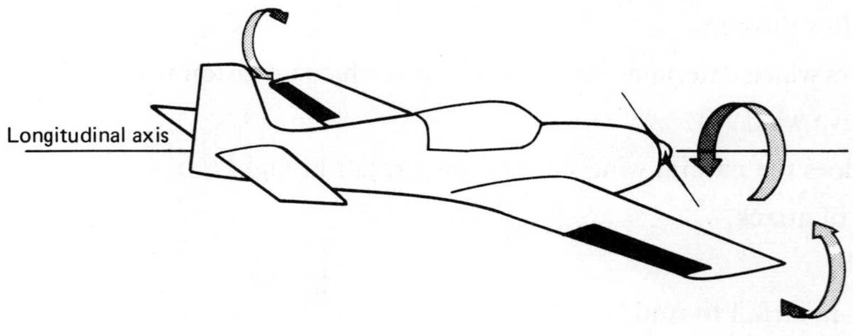


Fig. 16. The longitudinal axis. The aircraft rolls about the longitudinal axis.

*The Aileron* The control surfaces which cause the aircraft to roll about its longitudinal axis are the ailerons. These moveable panels on the trailing edge of each wing are connected by cables or pushrods to the control column in the cockpit. The ailerons are rigged to move in an opposite direction to each other. That is, when one moves upward, above the trailing edge of the wing, the other moves downward.

To understand how the aileron causes the aircraft to roll, we must return to our discussion of lift. Remember that lift is produced because air flowing over the camber of an airfoil is forced to travel farther and faster than the air under the airfoil. We can produce more lift by increasing the camber of the airfoil. The aileron does just that.

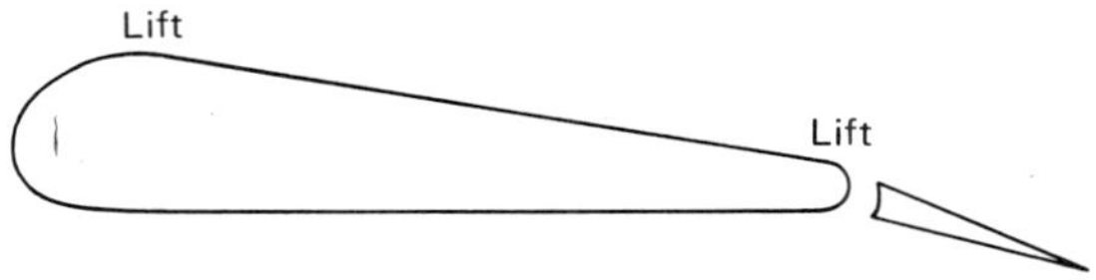


Fig. 17. The aileron. The "down" aileron produces lift above the wing.

Figure 17 shows a cross section of a wing with the aileron in the down position. Notice how the camber of the wing has increased, causing a new low pressure area above it, which produces additional lift.

At the same time, the aileron on the other wing is deflected upward as shown in the cross sectional view in Fig. 18. The up aileron has formed a small camber on the bottom of this wing, with an associated low pressure area. Hence, this aileron is producing lift in a direction opposite to the main airfoil lift. The resultant lift produced in an upward direction by this wing will be less.

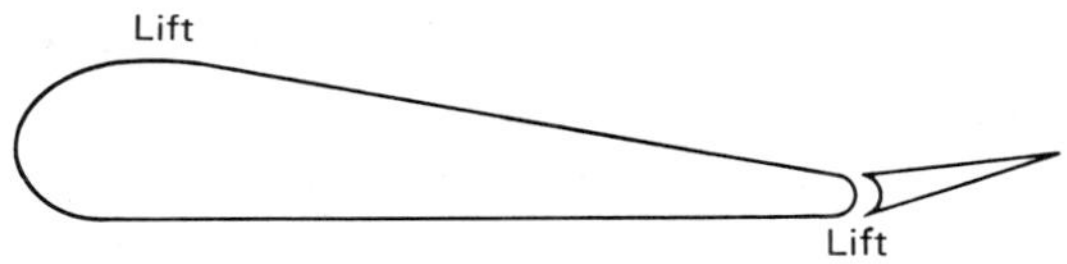


Fig. 18. The aileron. The" up" aileron produces lift under the wing.

The difference in net lift produced by the *up aileron* wing and the *down aileron* wing causes the aircraft to roll (about the longitudinal axis) toward the up aileron.

*Bank and Turn.* When an aircraft rolls, the wings are said to be *banked.* Why do we wish to bank the wings?

Lift produced by the wing is considered to act perpendicular to the relative wind and to the wing span of the aircraft. If we bank the aircraft to the right as shown in Fig. 19, the lift will be diverted to the right. It is this diversion of lift which causes the aircraft to move in that direction, or turn.

Therefore, we can see that the aircraft is turned by banking the wings. The aileron, being the control to produce bank, is the control which produces the turn. If the wings are level the aircraft cannot turn. If the wings are banked the aircraft *must* turn. (Assume no other control is used.)

We also notice in Fig. 19 that the total lift produced must be greater in a level turn than in straight and level flight. If the aircraft weighs 2,000 pounds, the lift produced in level flight must be 2,000 pounds, in order to exactly counteract gravity. But some lift is also used to pull the aircraft around the turn.

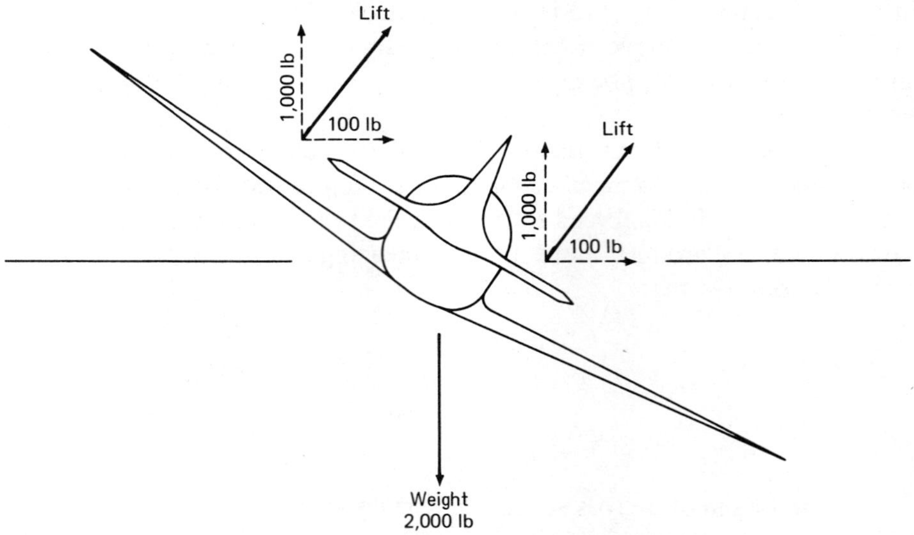


Fig. 19. The diversion of lift makes the aircraft turn.

Hence, more than 2,000 pounds of lift must be produced in a turn to counteract gravity *plus* cause the aircraft to turn. Normally, this extra lift will be produced by increasing the angle of attack while turning. We will see later that the amount of extra lift required in a turn varies with the amount of bank. (See Chapter 5.)

## The Lateral Axis

The lateral axis is a straight line drawn from wing tip to wing tip, perpendicular to the longitudinal axis.

Aircraft movement about the lateral axis is called *pitch*.

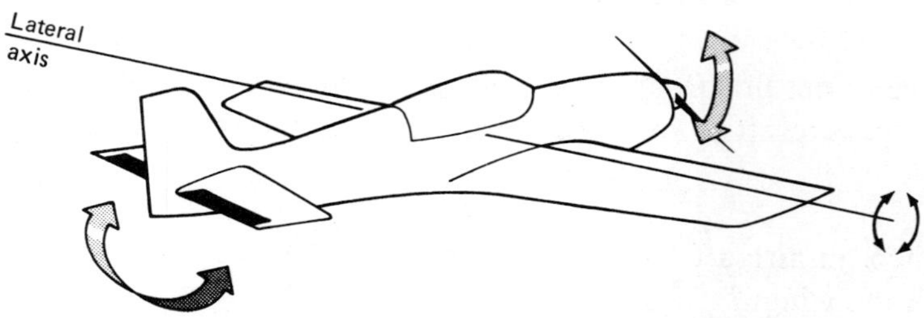


Fig. 20. The lateral axis. The aircraft pitches about the lateral axis.

### *The Elevator*

Pitch is controlled by the elevator, which is a moveable panel attached to the trailing edge of the horizontal stabilizer.

Pressure applied to the control column is transmitted by cable or push-rod from the cockpit to the elevator. When forward pressure is applied to the control column, the elevator is deflected downward, increasing the camber and the amount of lift produced by the horizontal tail surface, as shown in Fig. 21.

The increased lift produced causes the tail to move upward and the nose to pitch downward, with rotation occuring about the lateral axis. When rearward pressure is applied to the control column the elevator is deflected upward, reversing the camber of the tail surface which produces downward lift. The tail

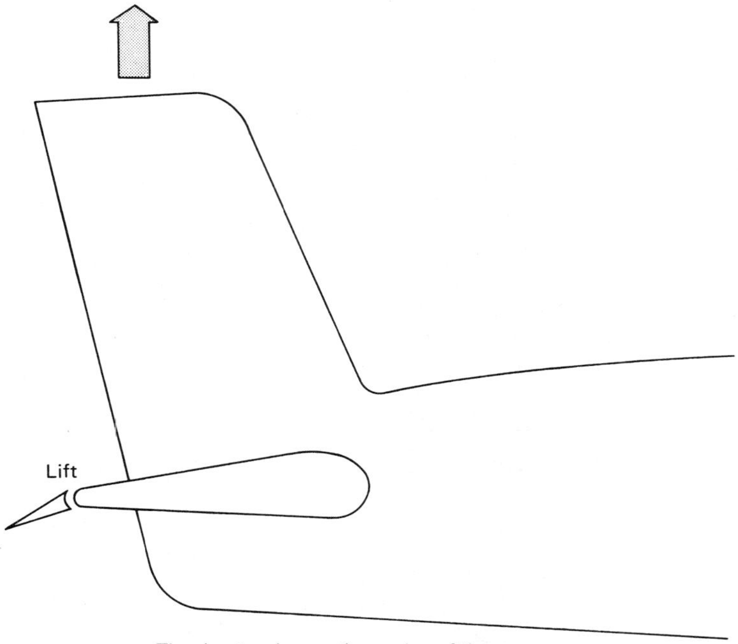


The elevator changes the camber of the horizontal stabilizer, thereby changing aircraft pitch.

Fig. 21. The elevator.

will move downward into that low pressure area and the nose will pitch upward.

*Why Pitch Control*? We must control pitch for two reasons. Pitch may control aircraft speed. With a constant power setting, the aircraft will increase speed with nose-low pitch, and will decrease speed with nose-high pitch.

Lift production may be controlled with pitch also. The angle of attack changes when pitch changes. Again, with constant power, high angles of attack will produce high amounts of lift; low angles of attack produce less lift.

## The Vertical Axis

The vertical axis is a straight line from the top to the bottom of the aircraft, passing through the point of intersection of the longitudinal axis and the lateral axis, and perpendicular to both axes.

Movement about the vertical axis is called *yaw*.

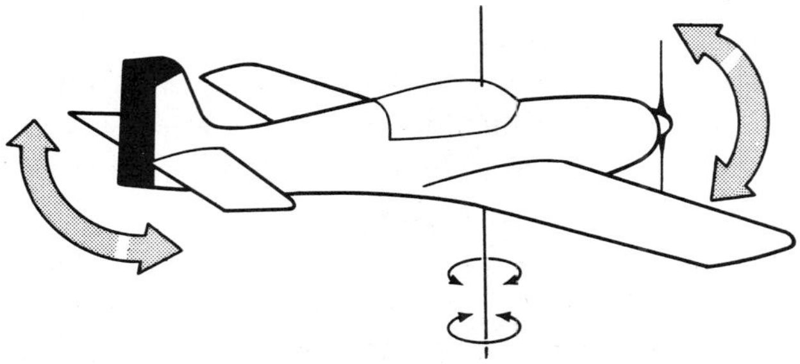


Fig. 22. The vertical axis. The aircraft yaws about the vertical axis.

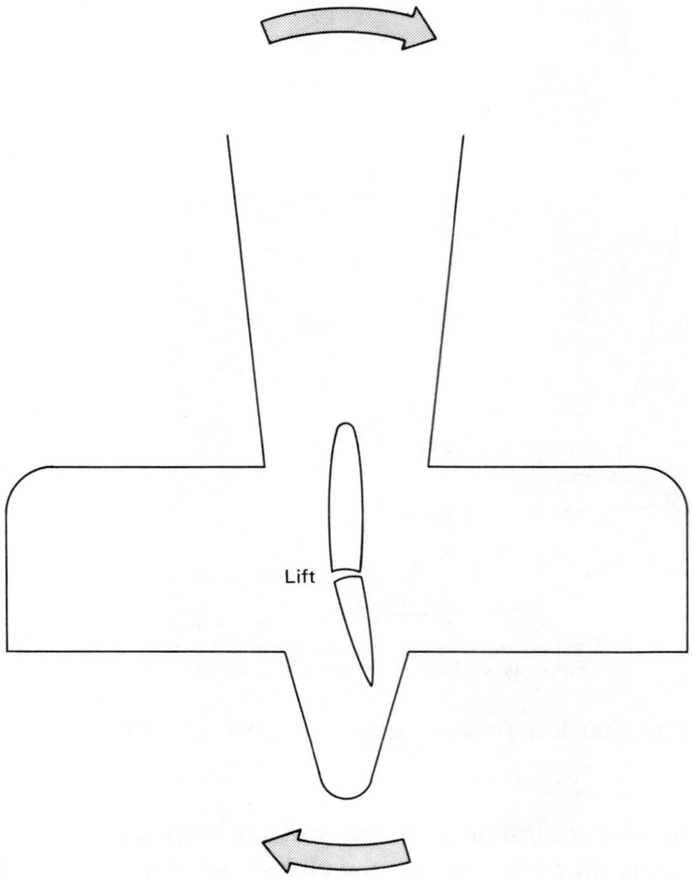


Fig. 23. The rudder.

The rudder changes the camber of the vertical stabilizer, thereby producing yaw.

*The Rudder*

Yaw is controlled by the rudder, which is a moveable panel attached to the trailing edge of the vertical stabilizer. Rudder pedals in the cockpit, operated by foot, transmit pressure by cable or pushrod to deflect the rudder from dead center.

When the rudder is deflected it forms a camber on the vertical stabilizer assembly and an associated low pressure area. Lift is thus created, and the tail of the aircraft moves toward the low pressure. When the tail moves in one direction, the nose yaws in the opposite direction, with rotation of the aircraft occuring about the vertical axis.

*Rudder Use.* Contrary to popular belief, the rudder *does not* cause the aircraft to turn. However, it does serve many other purposes in flight.

Do you remember our discussion of aileron movement and the resulting action? We learned that one aileron moves upward when the other moves downward. The downward moving aileron is displaced into higher pressure air, due to the *packing* of ram air under the wing. The upward moving aileron is displaced into an area of lower pressure above the wing. Higher pressure air produces more drag, which slows the speed of the down aileron wing, causing the aircraft to yaw in the direction opposite to the desired turn. Rudder pressure is applied when the turn is initiated in order to prevent the yaw.

The second use for the rudder concerns "torque." Torque is a twisting force. When we open a faucet or turn a door knob we apply torque. It is common practice to consider aircraft torque as one force, but it is really a combination of several forces.

The spinning propeller produces drag as well as lift. The drag tends to slow the propeller rotation. Picture a strongman in front of the aircraft, holding the once spinning propeller. This herculean task would result in the stoppage of the propeller, but the aircraft would rotate in the opposite direction. Drag is a strongman. It does not stop the propeller but it does try to stop it. Therefore, the aircraft tries to rotate in the opposite direction.

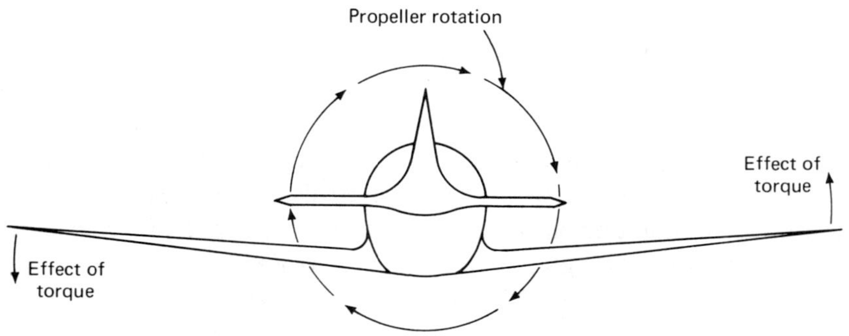


Fig. 24. Torque.

When we look at the propeller from the pilot's view, rotation is clockwise. Torque effect on the aircraft is counterclockwise, causing the aircraft to bank left.

In addition to this force, the slipstream, or disturbed air swirling rearward from the propeller, strikes the vertical stabilizer on the left side, forcing the tail to the right and the nose to the left.

The aircraft manufacturer attaches the vertical stabilizer at a small angle to the longitudinal axis to aid in counteracting the torque effect. However, the amount of torque effect changes as the power applied to the propeller changes. The manufacturer adjusts the attach points of the vertical stabilizer for proper torque correction with regard to *cruise power and speed*, because that is the most common flight condition.

If we direct more than cruise power to the propeller and cause additional torque, we must assist the offset vertical fin in preventing left yaw by applying right rudder pressure. Conversely, with less than cruise power, less torque will be developed. The effect of the offset vertical fin must be diminished by applying left rudder pressure.

The effectiveness of the offset vertical stabilizer is also controlled by the speed of the relative wind. As we increase our speed through the air the stabilizer will become over-effective. A decrease in effectiveness will prevail with low speed through the air. Therefore, we must constantly adjust rudder pressure to prevent yaw to the left with high torque or low speeds, and yaw to the right with low torque or high speeds.

*Angle of Incidence*. Torque, and its effect on the aircraft is so important that the manufacturer builds in another correction for it. The left wing is attached to the aircraft with a slightly greater angle of incidence than is the right wing, counteracting the left banking tendency caused by torque.

## Trim Tabs

Naturally the aircraft will not be loaded identically for every flight. We may be flying alone one day and have many passengers on the next. As the load distribution changes, so will the balance of the aircraft.

Picture a flight with a great amount of weight in the rear of the aircraft. This will cause the aircraft to fly tail low, pitching the nose upward. We must maintain desired pitch so we will be forced to exert constant forward pressure on the control column. It is humanly impossible to hold constant pressure against anything. The muscles of the body become fatigued, and many small corrections will result. This action will cause a series of pitch changes. To aid us in our efforts to provide a comfortable flight, and to prevent fatigue, the manufacturer provides tiny control panels attached to the trailing edge of the aileron, rudder, and elevator. These controls are *trim tabs*. We can adjust these small tabs in flight or on the ground so that each control is constantly displaced without pressure being applied on the control column or rudder pedals.

For example, Fig. 25 shows an aircraft which is loaded heavily rearward. The trim tab is displaced upward, causing a chain reaction. A tiny camber is formed on the elevator, displacing it downward. The down elevator forces the tail up, pitching the nose down.

Experience in flight will aid us in "fine tuning" the trim tabs so that the aircraft will fly in any predetermined attitude for prolonged periods with no control use on our part. This adds safety to the flight by insuring a constant flight attitude if our attention should be momentarily diverted to another cockpit duty.

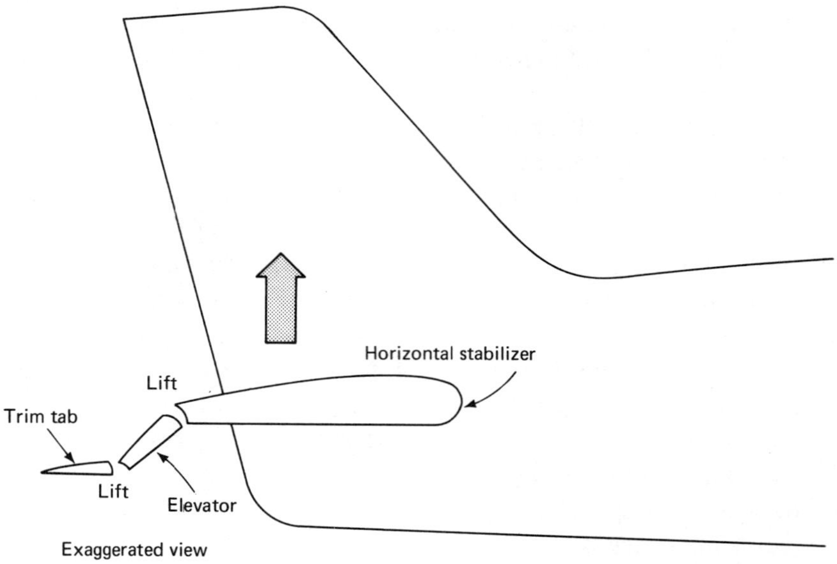


Fig. 25. The elevator trim tab.

## CHECK YOUR READING

1. List the three axes of the aircraft, and the movement about each axis.
2. What is the primary purpose of the aileron?
3. Describe the force which makes the aircraft turn.
4. Why must more lift be produced in a level turn than in straight and level flight?
5. For what two reasons must pitch be controlled?
6. List six uses for torque in every day life.
7. When in a high power, high pitch flight condition, rudder pressure must be applied to which rudder pedal?
8. When in a cruise power, high speed flight condition, to which rudder pedal must pressure be applied?
9. Why does the aircraft manufacturer attach the vertical stabilizer at a point slightly left of the longitudinal axis of the aircraft?
10. Why is the angle of incidence greater on the left wing than on the right wing?
11. If the aircraft is flying "hands-off" with the left wing low, in which direction would you displace the aileron trim tab on the right wing?
12. Why should the pilot keep the aircraft in trim?

# 4

# FLAPS AND OTHER HIGH-LIFT DEVICES

Flight conditions during takeoff, approach, and landing are made safer through the use of flaps or other high-lift devices. With the proper knowledge of these devices we can operate the aircraft at maximum efficiency.

## Flaps

*Flaps* are moveable control panels attached to the trailing edge of each wing, similar in appearance to the ailerons. (See Fig. 26.)

However, there are two very important differences between flaps and ailerons. Both flaps move in the same direction at the same time, and the flaps do not move above the trailing edge of the wing.

Although flaps are commonly called high-lift devices, they are not

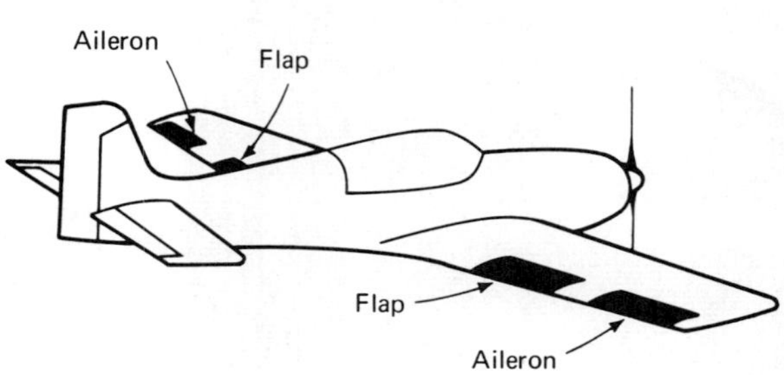


Fig. 26. Flaps.

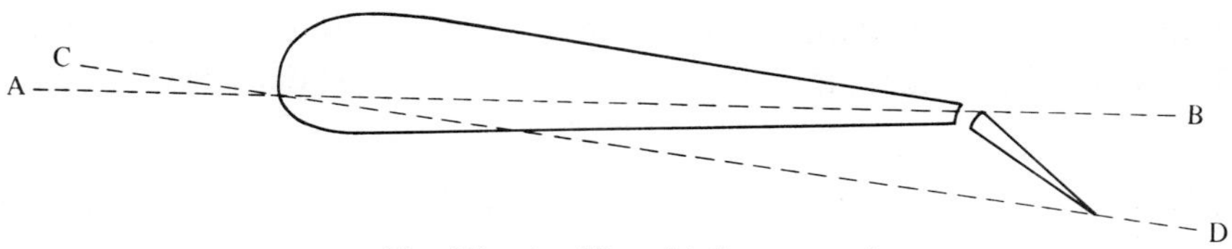

Line AB = chord line with flap retracted.
Line CD = chord line with flap extended.

The extended flap changes the camber and the chord line of the wing.

Fig. 27. The chord line.

normally used to increase the total amount of lift produced by the wing, but to create the same amount of lift at a slower airspeed.

If we look at Fig. 27 we will see that flaps change two important things to produce lift at slow speeds. It is obvious that the extended flap changes the camber of the section of the wing in which it is located. This increase in camber naturally deepens the low pressure area above the wing, creating lift. Another interesting thing happens when the flaps are lowered. The chord line of the wing changes, automatically increasing the angle of attack, which increases lift.

The benefit we receive from extended flaps is counteracted to some extent by the penalty of drag, for every time lift is produced so is drag produced. (This kind of drag, caused by lift production, is called *induced drag*.) For this reason, we must exercise extreme caution and use careful judgement in applying the most beneficial amount of flap for each specific flight condition.

### *Flap Use for Takeoff*

Normally, less than full flap deflection is used for takeoff. The exact percentage of available flap travel to be used in each specific aircraft is defined in the Aircraft Operations Manual. We must familiarize ourselves with this information prior to flying a strange aircraft, to avoid the penalty of too much drag for the amount of lift we can expect to gain at slow speed. Most aircraft takeoff performance charts will reflect a significant savings in takeoff distance required with the use of recommended flap settings, and a great increase in takeoff distance if too much flap is used. (See Chapter 17.)

### *Flap Use for Approach*

During the approach to a landing, we can compensate for changing wind conditions and minor errors in judgement with the judicious use of flaps. Flaps increase the angle of descent without increasing the airspeed. Therefore, we can plan steeper approaches to clear obstacles on the final course to the runway without accelerating to high speeds.

Because of the increased angle of attack and decreased aircraft stability produced by flaps, they should not be used except with extreme caution during windy or gusty conditions. Inexperienced pilots have found themselves in a stalled flight condition on final approach due to careless flap use and pilot technique during gusty wind conditions.

*Flap Use for Landing*

Flaps normally lower the minimum stalling speed of the aircraft. Landing, therefore, can be accomplished at slower speeds, with less energy being absorbed by the landing gear and less runway being used for deceleration.

*Cautions to be Observed*

Flaps may add danger to a flight if certain precautions are not observed.

*Flap Actuation.* We must not lower the flaps at a high airspeed, for the added lift created may cause structural damage to the flap itself or worse yet, to the entire wing. Prior to actuating the flap switch we must be sure that the aircraft is slowed to the required flap operating speed. This speed range is shown on the airspeed indicator as a white arc.

We learned that flaps differ from ailerons in that they both move in the same direction at the same time. Imagine what would happen if only one flap moved downward. The added lift produced by that wing would cause the aircraft to begin an immediate and rapid roll, probably ending in disaster. It is foolhardy to select a desired flap setting in flight without first having tested the flap deflection on the ground. (When in flight, I prefer to just "tickle" the flap actuator switch prior to selecting a specific flap setting. If no rolling tendency shows, I go ahead and lower the flaps.)

When flaps are lowered or raised, the center of the lift forces on the wing changes either forward or aft. Depending on the aircraft design this may cause a rapid pitch change. We must know the characteristics of our aircraft and remain alert in order to counteract possible violent pitch changes when flaps are used.

Power changes with flaps extended may also cause rapid pitch changes. Many aircraft have the nasty tendency to pitch upward violently when power is applied for a full flap go-around. If this is not counteracted with forward control column pressure, an immediate stall will result. Keep alert! Be prepared to counteract pitch changes when power is added or removed while flaps are extended.

Certain aircraft designs are such that full flap deflection blankets out the smooth flow of air passing the tail surfaces. Again, thorough knowledge of the particular aircraft we intend to fly is a must. Be prepared for reduced rudder and elevator efficiency when flaps are lowered.

The increased drag created by flaps causes the aircraft to accelerate slowly and climb with less than normal performance. Hence, flaps are usually retracted if a balked landing and go-around is necessary. Flap retraction when near the surface must be done carefully and exactly as prescribed in the Aircraft Operations Manual. Remember, when flaps are retracted you can expect a serious loss of lift. The aircraft will not accelerate rapidly enough to produce the added lift to make up for that loss. Most pilots "milk" or slowly raise the flaps when near the surface and then only after determining *positively* that the aircraft is under control and in a climbing attitude.

As we can see, the use of flaps demands thorough knowledge of each particular aircraft and the Aircraft Operations Manual. It is imperative that every peculiar characteristic of the aircraft is well learned prior to attempting flight.

## Other High-Lift Devices

As aircraft design becomes more sophisticated so does the design of lift producing devices. Because of the variety of uses and designs, we will take only a short look at the more common types.

*Fowler Flaps*

The Fowler flap increases camber and changes the angle of attack in the same manner as the plain flap. But, as shown in Fig. 28 the Fowler flap also increases the area of the wing. Since one of the factors in determining lift production is wing area, this flap design is quite efficient. It is too sophisticated and expensive for installation on most light aircraft.

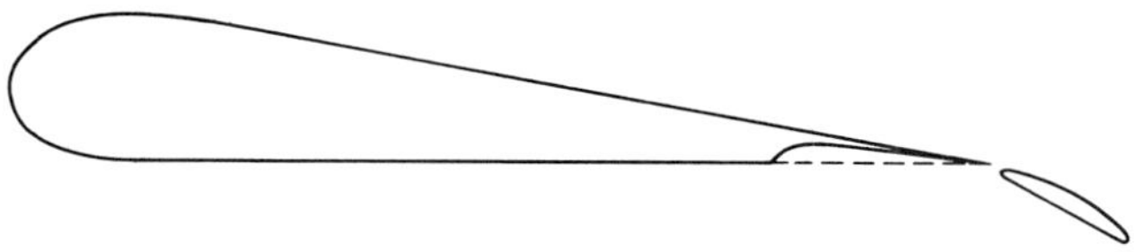

Fig. 28. The Fowler flap.

*Slots*

When the wing reaches a high angle of attack, the smooth flow of air passing the wing begins to tear away and burble. Slots in the leading edge of the wing, as shown in Fig. 29, channel some of the relative wind into a smooth flow across the camber, keeping the burble from occuring until a greater angle of attack is reached and thereby delaying the stall.

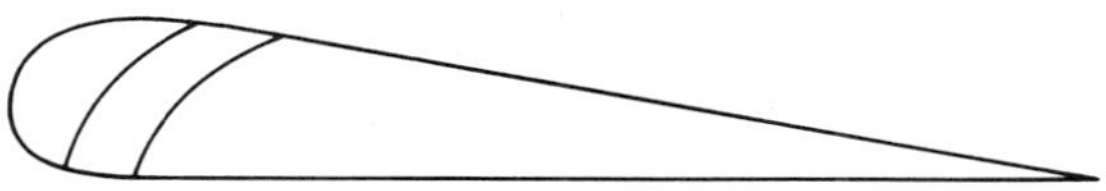

Fig. 29. The leading edge slot.

*Slats*

A *slat* is a moveable portion of the leading edge of the wing. When the slat is extended forward it opens a slot. The slat can be retracted to close the slot and decrease drag. (See Fig. 30.)

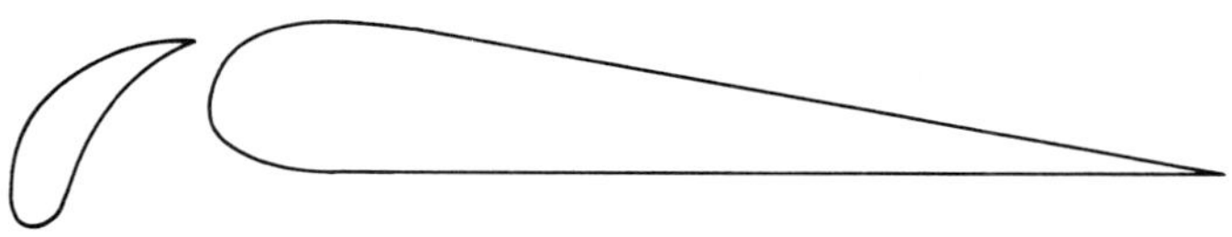

Fig. 30. The leading edge slat.

### CHECK YOUR READING

1. What two things are changed when the flaps are lowered?
2. When lift is produced, drag is also produced. This kind of drag is called what?
3. Where can we find exactly how much flap is recommended for takeoff?
4. Why is less than full flap normally used for takeoff?

5. On approach for a landing, what is the primary use for flaps?
6. What effect do flaps have on the stalling speed of the aircraft?
7. List the cautions which must be exercised when using flaps.
8. Describe the normal procedure for raising flaps on a balked landing or go-around.
9. What is the primary difference between the plain flap and the Fowler flap?

# 5

# LOAD FACTORS

The definition of *load factor* is the actual weight the wings support, divided by the total weight of the aircraft. Think about that for a minute. We learned that lift (the weight that the wings are supporting) equals gravity (the total weight of the aircraft). One question which may come to mind is "How can the total weight being supported by the wings *ever* be any different than the total weight of the aircraft? After all, if the aircraft weighs 4,000 pounds, the wings must be supporting 4,000 pounds in flight."

That is true, but only if the aircraft is in straight and level, unaccelerated flight.

Sometimes the wings are expected to support more than the total weight of the aircraft. For example, as we have seen in a turn, lift not only counteracts gravity but also pulls the aircraft around the turn.

## Load Factor in a Turn

Look at Fig. 31, which shows an aircraft in straight and level, unaccelerated flight in view 1, and shows the same aircraft in a 60 degree banked turn in view 2. If we apply the load factor formula for the aircraft in view 1, we will find that the load factor is 1 $g$.

$$\text{LOAD FACTOR} = \frac{\text{Actual weight wings are supporting}}{\text{Total weight of the aircraft}}$$

$$\text{LOAD FACTOR} = \frac{\text{4,000 pounds}}{\text{4,000 pounds}} = 1\,g$$

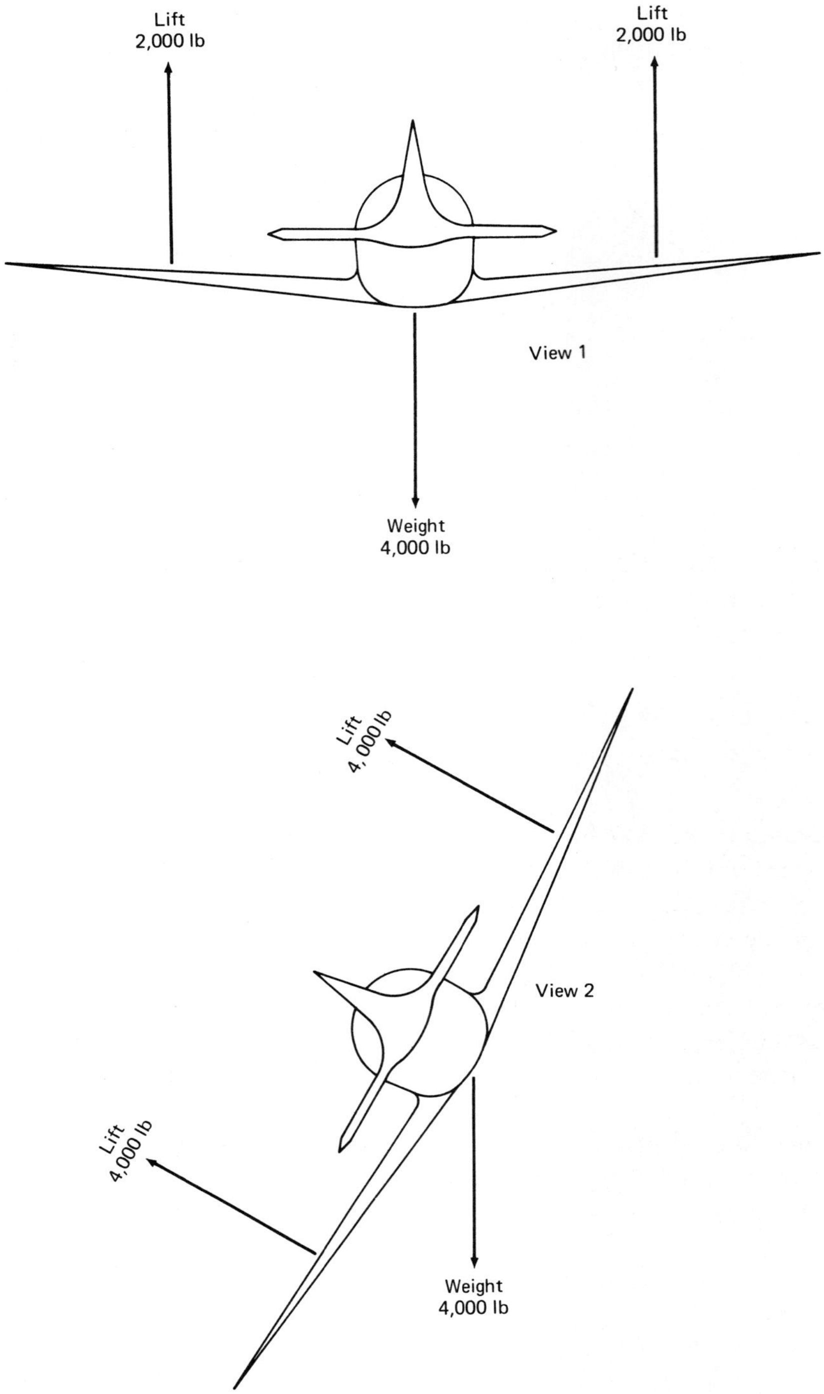


Fig. 31. Load factor. In a 60° banked turn the load factor is 2 *g*.

The load factor on the aircraft in view 1 is 1 *g* just as our load factor when we sit reading this text is 1 *g*.

Looking at the aircraft in view 2, we see that the aircraft still weighs 4,000 pounds, so 4,000 pounds of lift will be required to hold the aircraft up in the air. But, an *additional* amount of lift must be developed to pull the aircraft around the turn. Mathematically, we could prove how much extra lift is required for turning in a 60 degree banked turn. The result of our computations would show that an additional 4,000 pounds of lift must be developed to keep the aircraft at a constant altitude in a 60 degree banked turn. The weight the wings are supporting in view 2 is 8,000 pounds. Let's apply the load factor formula.

$$\text{LOAD FACTOR} = \frac{\text{Actual weight the wings are supporting}}{\text{Total weight of the aircraft}}$$

$$\text{LOAD FACTOR} = \frac{8{,}000 \text{ pounds}}{4{,}000 \text{ pounds}} = 2\,g$$

The load factor of the aircraft in view 2 is 2 *g*. That is, it is twice the force of natural gravity.

### Load Factor Chart

Notice that no mention has been made of the speed of the aircraft during our discussion of load factor. Every aircraft, no matter what size, speed, or shape, will be subjected to the same load factor as long as they are all banked to the

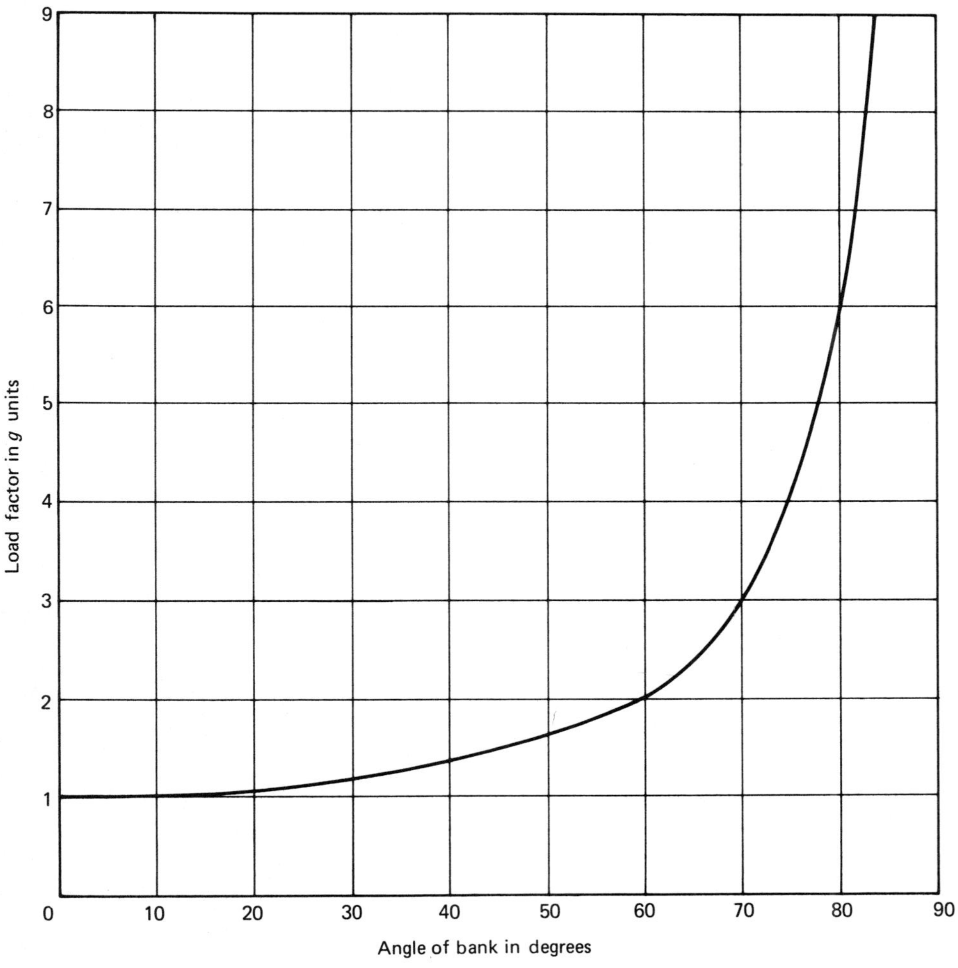


Fig. 32. The load factor chart.

same degree and are maintaining altitude. Mathematical computation for every degree of bank is complicated, so we will use the load factor chart to determine quickly the load factor for any specific amount of bank.

Figure 32 shows the load factor chart. Notice how the load factor increases gradually until a 60 degree bank is reached, then shoots up rapidly as the bank increases, becoming an infinite force at 90 degrees of bank. (A 90 degree banked turn is impossible, if altitude is maintained, because the aircraft would have to be infinitely strong to sustain the infinite load factor.)

## Positive and Negative *g* Forces

Whenever the *g* force is acting in the same direction as natural gravity, it is called "positive *g*." If it counteracts natural gravity, it is called "negative *g*." Most light aircraft are stressed to withstand a load factor of at least 2.5 to 3.8 positive *g*, and at least 1.0 to 1.52 negative *g*. In effect this means that the aircraft structure can support at least 2.5 times its weight in the positive direction, and at least 1.0 times its maximum weight in the negative direction. (Every mass within the gravitational pull of the earth has a load factor of 1 *g*. Hence, one negative *g* will just cancel the effect of natural gravity, leaving the mass *weightless*.)

Some aircraft are stressed for more (or less) than the load factors given here, or they may be stressed for specific load factors when flown in certain conditions. The Aircraft Operations Manual will note specific load factors for which the aircraft is stressed, and any special conditions which must be met to avoid exceeding special load factors.

We have seen that, at times, the wings must support much more than the actual weight of the aircraft. The wings support the extra weight by producing extra lift, either with an increased angle of attack or increased speed. When we practice steep turns in flight we will see how added power (for speed) and back pressure on the control column (for a higher angle of attack) will produce the extra lift.

## Pullout

In many respects a pullout from a dive is the same as a turn, except that it occurs in a different plane of action.

Figure 33 shows an aircraft about to enter a pullout, or start a turn upward away from the surface. Lift must be increased to stop the descent, and change the path of flight. Back pressure on the control column will increase the angle of attack, creating more lift, changing the path of flight. The load factor will become greater than 1 *g* and will remain so until the path of flight becomes a straight line again. We must use caution however, and not increase the angle of attack to the point of stall, or cause the load factor to exceed the maximum allowable, seriously damaging the aircraft.

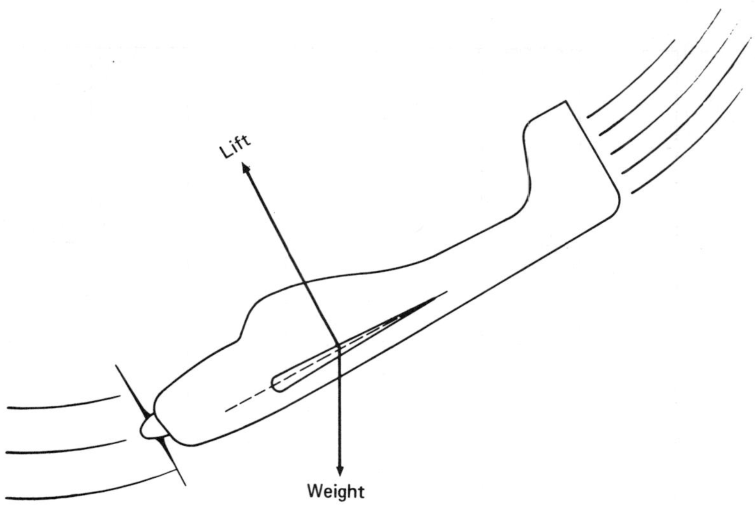


A pullout from a dive increases the load factor.

Fig. 33. The pullout.

## Level-Off

Now let us picture an aircraft in a rapid level-off. (See Fig. 34.)

The path of flight will change from forward and upward to forward and level. When the path of flight changes this time, centrifugal force will pull the aircraft in a direction *opposite to* natural gravity. We will feel light in our chairs as we experience negative *g*.

Whenever there is a change in the direction of flight, there must be a change in the load factor. The more rapid the rate of change in direction, the more severe will be the load factor.

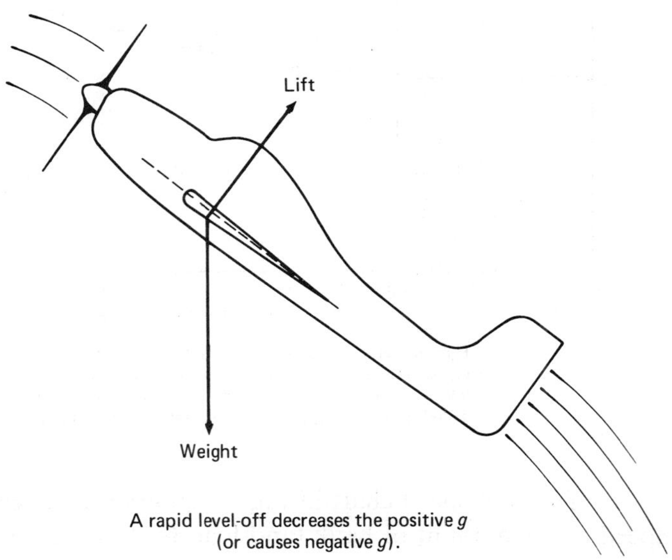


A rapid level-off decreases the positive *g* (or causes negative *g*).

Fig. 34. The level-off.

## Stall Speed

Minimum stall speed also increases with the load factor. (The fact that a load factor greater than 1 *g* exists means that more than the normal amount of lift is being produced. When lift increases, stall speed increases.)

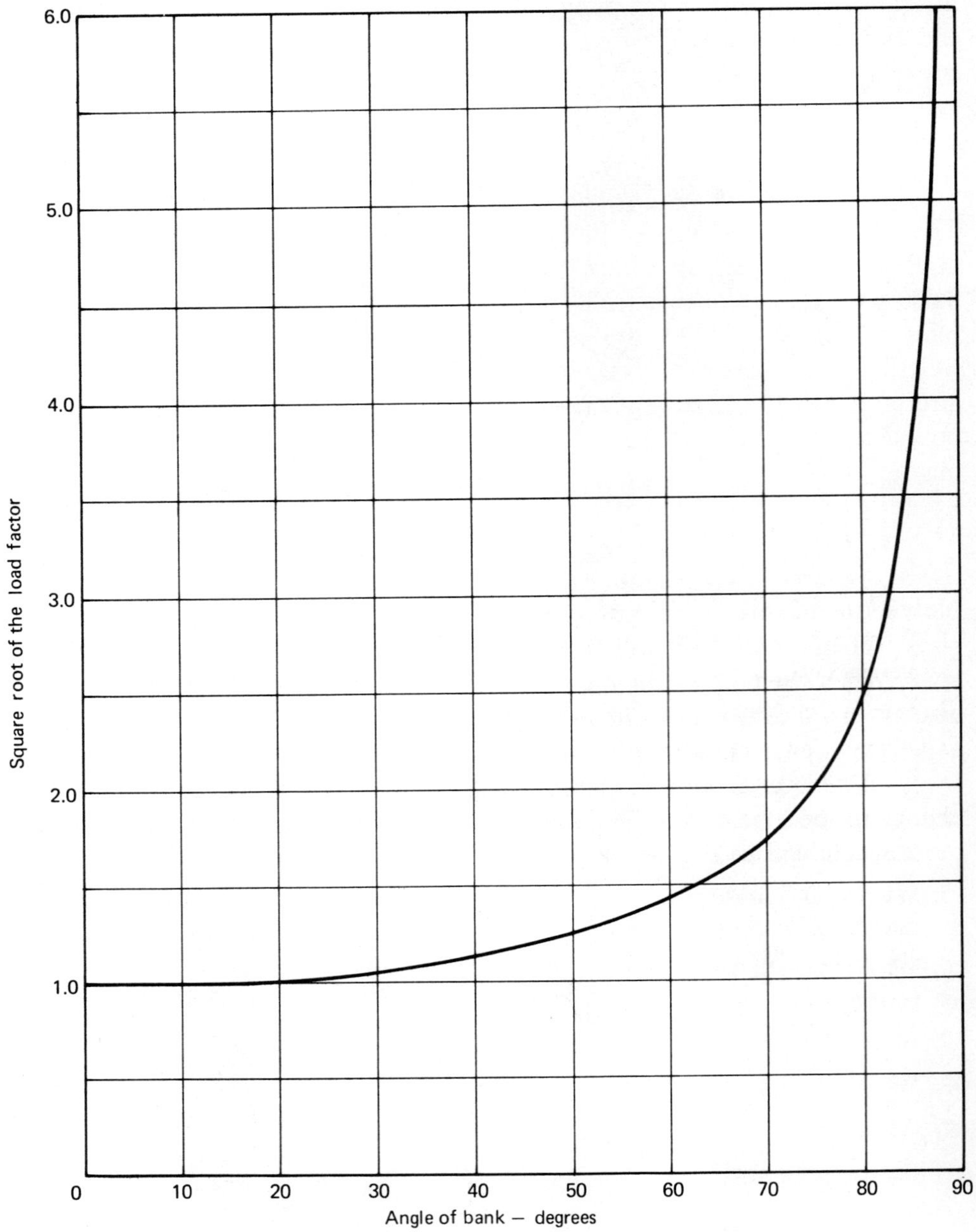


Find the angle of bank on the bottom of the chart, move vertically upward until intercepting curve, and then horizontally left to the edge of the chart. Multiply the wings level (normal) stall speed by that factor to find the stall speed in the bank.

Fig. 35. The stall speed chart.

The stall speed chart in Fig. 35 shows the increase in the minimum stall speed as the amount of bank (and load factor) increases.

*Rate of Change in Direction*

Two important factors limit the rate at which we can change the direction of flight of our aircraft. Load factors in excess of the designed maximum may be imposed, causing structural damage. The increased load factor may cause the aircraft to stall. Of the two possibilities, the stall is better by far, because once the aircraft is stalled, lift is destroyed and the load factor decreases sharply, preventing structural damage.

### Effective Weight

The load factor determines the *effective weight* of the aircraft. A 4,000 pound aircraft in a 60 degree banked turn, maintaining altitude effectively weighs 8,000 pounds (weight multiplied by load factor). We will be aware of the doubling of our own weight, forcing us into the seat. The knowledge of effective weight and increased stalling speeds will prevent us from overloading our aircraft when the temptation arises. One extra passenger weighing in at 200 pounds may equal four heavyweights during a steep turn or when sharply pulling out of a dive.

## CHECK YOUR READING

1. When will the weight being supported by the wings be greater than the total weight of the aircraft?
2. What is the load factor on an aircraft in straight and level, unaccelerated flight?
3. What is the load factor of an aircraft in a 60 degree banked turn? (Assume that it is maintaining altitude.)
4. Use the load factor chart to determine the load factor in a 70 degree banked turn, maintaining altitude.
5. What is the effective weight of a 6,500 pound aircraft in a 70 degree banked turn, maintaining altitude?
6. What is the effective weight of 150 pounds of extra baggage in a 60 degree banked turn, maintaining altitude?
7. What effect does speed have on load factor in a turn if altitude is maintained?
8. What is the definition of *negative g*?
9. What are the maximum allowable load factors for most light aircraft?
10. List the instances in which a load factor of more than 1 *g* may be encountered.

# 6

# WEIGHT AND BALANCE

Every aircraft is designed as a compromise. For instance, we might like an aircraft that could carry ten people, five hours of fuel, and twenty suitcases. Such an aircraft can be manufactured. Its cost will naturally be greater than a two-place trainer because it will be larger. Suppose that the manufacturer finds that very few people wish to buy an aircraft similar to the one we want. He must decide to fill the requirements of a large number of people, and design an aircraft which will satisfy them. While doing so, he may compromise space for efficiency, or baggage storage space for fuel storage tanks. When we fly the aircraft, we must know how much weight it will safely carry, and where the weight should be placed so that the structural limits of the machine will not be exceeded.

## Maximum Allowable Gross Weight

The engineer designs an aircraft which will carry a specific amount of weight, and no more than that amount. This figure is called the *maximum allowable gross weight*. If our aircraft was designed to carry 2,000 pounds, for instance, the *aircraft and everything in it* must not weigh more than 2,000 pounds at the moment of lift-off.

In the previous chapter we learned that a load factor in excess of 1 *g* increases the effective weight of the aircraft. If we cheat just a little and leave the ground with 100 pounds of excess weight we can still expect the aircraft to fly, although accurate measurement would show that more runway length was required for acceleration to flying speed. However, during a steep turn, turbulent flight, or any event producing a load factor, we may impose a load factor of say, 3 *g*'s on the aircraft. Now our 2,100 pound aircraft effectively will weigh 6,300 pounds. If the manufacturer built the machine to withstand a 3 *g* load factor he in fact guaranteed us that the wings would support an effective weight of 6,000 pounds. (Three times the maximum allowable gross weight of 2,000 pounds.) If we subject the wings to a 6,300 pound effective weight we may be entering a hazardous flight condition which will end in disaster.

The point is, every extra pound of weight is multiplied by the load factor encountered in flight. One hundred pounds seems like a small amount of extra weight, but when we multiply it by possible high load factors, it becomes an impressive overload.

### *Hazards of Overweight Flight*

In most light aircraft (and many heavier ones also) it is not possible to fill all the seats, fuel tanks, and baggage space and remain under the maximum allowable gross weight. Remember the compromise of design. It may be possible to use all of the seats, or to fill all of the fuel tanks, but *not* at the same time. Weight limitations cannot be ignored. They are imposed by the aircraft designer to prevent many flight hazards.

*Longer Takeoff Run.* Obviously, the amount of lift required for takeoff will increase with the weight of the aircraft. More lift will require a higher takeoff speed, and acceleration to that speed will require more runway. The takeoff will be followed by a lower than normal rate of climb. The aircraft may not be capable of clearing any obstructions at the end of the runway.

*Slower Cruise Speed.* After the climb and level-off, we will notice that the aircraft cruises at a slower than normal speed, for every pound of weight steals from aircraft efficiency. Since fuel required for a flight is a factor of time aloft, and time aloft is a factor of speed, the total amount of fuel used for an overweight flight will be greater than normal. The initial mistake of carrying too much weight may lead to the complication of fuel starvation prior to reaching the destination.

*Sluggish Control Action.* The engineer who designed the aircraft gave us flight controls which were sufficient to control the aircraft at the maximum allowable gross weight. As the weight of a moving body increases, the force required to change its direction increases. Our flight controls will become less effective as the aircraft weight approaches the maximum allowable gross weight, and may become totally insufficient to control the aircraft when the weight exceeds the maximum allowable.

*Engine Damage.* Engine cooling is dependent upon the air rushing into the engine compartment, washing over the cylinders, and taking heat away. If we overload the aircraft and "mush" along for great distances, insufficient air may pass the cylinders to keep them at a safe temperature. Immediate and severe damage may result, leading to complete engine failure.

*Excessive Area Load.* Aircraft are not only limited as to the total weight they may carry safely, but may also be limited as to the amount of weight which can be placed in certain areas. For instance, the floor loading of our aircraft may be limited to 50 pounds per square foot. If we place a small but heavy article on the floor of the aircraft we may exceed the load limit on that small section of flooring. Then when any severe load factor is imposed on the aircraft, the excess floor load is multiplied by the load factor, making the effective weight of the cargo so great that it will fall through the large hole that it punctured in the floor.

Specific loading restrictions for cargo shelves, storage racks, and the floor area are just as important as the maximum allowable gross weight restriction.

## Weight Computation

To assure that our aircraft is not overloaded, we must know how much the empty aircraft weighs, and the total weight of everything we wish to carry. When we add the two together, the total must not exceed the maximum allowable gross weight.

### *Empty Weight*

The aircraft leaves the factory at a specific weight. The weight of every new item attached to the aircraft after manufacture must be added to the original weight. Hence, we will always know the total weight of the aircraft and all attached equipment. *Undrainable fluids* (fuel, oil, etc.) are also included in this, the aircraft "empty weight." The Aircraft Operations Limitations will list the empty weight.

### *Useful Load*

The *maximum allowable useful load* is the difference between the maximum allowable gross weight and the empty weight. The combined weight of all items not attached to the aircraft (such as crew members, passengers, baggage, *usable* fuel and oil, maps, thermos bottles, etc.) is the *actual useful load.* The actual useful load may not exceed the maximum allowable useful load.

### *Computation of Weight*

A sample weight chart is depicted in Fig. 36. We first deduct empty weight from the maximum allowable gross weight (3,800 pounds minus 2,200 pounds). The remaining figure (1,600 pounds) is the weight available for the useful load. Now we must add the weights of everything we wish to carry.

*Fuel.* Aviation gasoline weighs 6 pounds per gallon. Multiply the number of *usable* gallons of fuel on board by 6 pounds and enter the weight on the chart

| Sample loading form | Sample airplane | | Our airplane | |
| --- | --- | --- | --- | --- |
| | Weight (lb) | Moment | Weight (lb) | Moment |
| Licensed empty weight | 2,200.0 | | | |
| Oil (3 gal) | 22.5 | | | |
| Pilot and front seat passenger | 355.0 | | | |
| Usable fuel (50 gal) | 300.0 | | | |
| Rear seat passengers | 380.0 | | | |
| Baggage | 250.0 | | | |
| Total | 3,507.5 | | | |

Maximum allowable gross weight = 4,000 lb

**Fig. 36. Sample loading form.**

(unusable fuel is already included in the empty weight). Our sample chart shows 50 gallons of usable fuel multiplied by 6 pounds per gallon, or 300 pounds.

*Oil.* Oil weighs $7\frac{1}{2}$ pounds per gallon. Multiply the number of *usable* gallons of oil on board by $7\frac{1}{2}$ pounds and enter the weight on the chart (unusable oil is already included in the empty weight). Most light aircraft oil capacities are given in quarts. We must remember to convert quarts to gallons before multiplying by $7\frac{1}{2}$ pounds. The sample chart shows 3 gallons of usable oil multiplied by $7\frac{1}{2}$ pounds, or 22.5 pounds.

*Pilot and Passengers.* The weight of all crew members and passengers is entered on the chart. If the exact weights are known, we will use them. For long-term flight planning, we may use a figure of 170 pounds as the average weight of each passenger. (We must be prepared to adjust the weight computation if, when the passengers arrive for the flight, they are all heavyweights.) Our sample chart shows a 185-pound pilot and a 170-pound copilot in the front seats, and two passengers weighing 190 pounds each in the rear seats.

*Baggage.* All cargo, food, handbags, cameras, etc. are weighed, and the total weight is entered on the chart. Our sample shows a baggage weight of 250 pounds.

*Final Tally and Adjustments.* Now we add all the weights to find the grand total. This figure is shown on the chart as 3,507.5 pounds. Since the aircraft has a maximum allowable gross weight of 3,800 pounds, we are sure that we are not attempting to carry too much weight. If the total of the desired gross weight exceeds the maximum allowable gross weight, something must be left behind. Circumstances surrounding the flight will dictate the items to be removed. In some cases, all available fuel will be necessary, so some of the baggage will have to be off-loaded. At other times, we may decide to off-load a portion of the usable fuel in order to carry all of the other articles.

We have satisfied the maximum weight limitations, but where shall we place each pound of weight within the aircraft? Can we put all of the baggage in the rear compartment, or must we place some of it in the forward storage area? Can we carry all of the passengers, or must some of them be left behind? In other words, even though we are sure that the aircraft is not overloaded, we must also compute the *balance* of the aircraft.

## Balance

All of us have either seen a teeter-totter in a playground or used one for our own enjoyment. We can use the familiar seesaw to better understand aircraft balancing. (See Fig. 37.)

The balanced seesaw has a balance point or *fulcrum* with an equal weight of plank on either side. If you and I both weigh 200 pounds, we can sit on each end of the plank and keep it balanced. But suppose I weigh 300 pounds and you

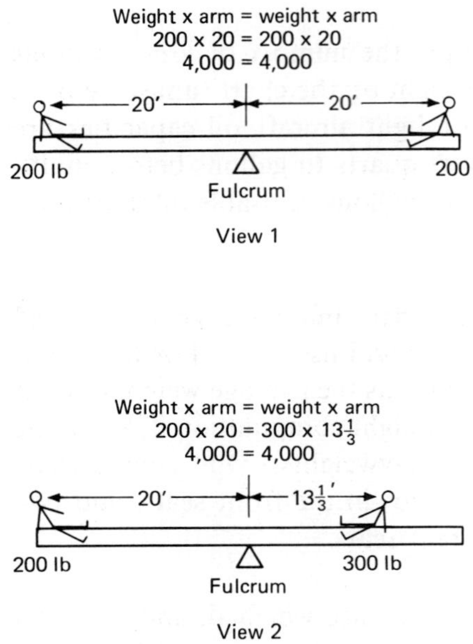


Fig. 37. Weight and balance.

weigh only 200 pounds. I will have to sit closer to the fulcrum if we are to keep the plank balanced (Fig. 37 view 2). In other words, the weight on either side of the fulcrum need not be equal, but the product of the weight and the distance from the fulcrum on one side must equal the same product on the other side. Using the weights in Fig. 37, view 2, we see that the weight (200 pounds) times the distance from the fulcrum (20 feet) on the left side of the fulcrum equals the weight (300 pounds) times the distance from the fulcrum ($13\frac{1}{3}$ feet) on the right side of the fulcrum. As long as the total weight on the plank does not exceed its maximum weight-carrying capability, it is possible to balance unequal weights by simply adjusting the distance between the fulcrum and the weight.

The distance from the fulcrum to the center of gravity of the weighted object is called the *arm*. *Terminology*

The product of the weight and the arm is called the *moment*.

*Computation of Balance*

Let us try to compute a simple balance problem for the seesaw. Suppose two people wish to ride; one weighing 100 pounds, and the other weighing 150 pounds. We will assume that we can place the people anywhere on the plank. Figure 38 shows this sample in picture form.

We will sit the 100-pound person on the extreme left end of the plank, ten feet from the fulcrum. Where must we place the 150-pound person in order to balance the plank?

The weight times the arm on the left side of the fulcrum must equal the weight times the arm on the right side.

$$\text{weight} \times \text{arm} = \text{weight} \times \text{arm}$$
$$100 \text{ pounds} \times 10 \text{ feet} = 150 \text{ pounds} \times x \text{ feet}$$
$$1{,}000 \text{ foot-pounds} = 150\,x \text{ foot-pounds}$$
$$1{,}000 \div 150 = x$$
$$6\tfrac{2}{3} = x$$

The 150-pound person must be seated $6\frac{2}{3}$ feet from the fulcrum in order to balance the plank. Another possibility of course, would be to seat both people on the ends of the plank, but to move the fulcrum toward the heavier person. Or, we could leave the fulcrum in place and require the 100-pound person to carry 50 pounds of sand in his pockets, again causing an equal moment on each side of the fulcrum. Aircraft balancing can be accomplished when the same principles are applied.

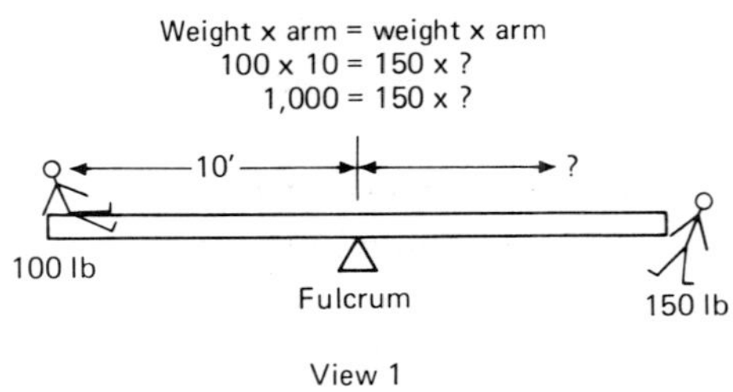


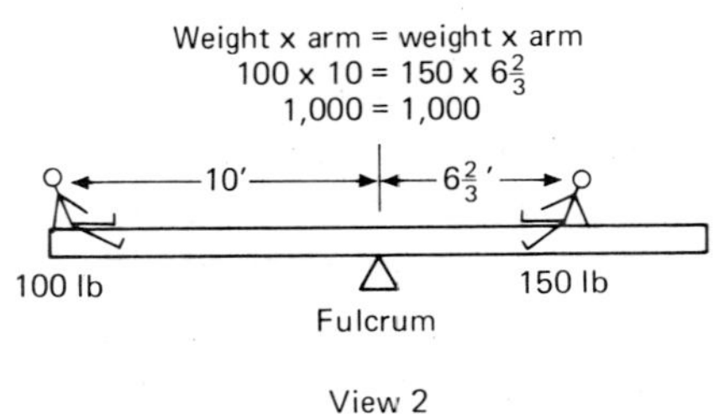


Fig. 38. Weight and balance.

## Balancing the Aircraft

We must balance the aircraft, just as we balanced the seesaw, to prevent too much weight being placed on one side of the fulcrum, which would destroy the stability of the aircraft. The manufacturer tells us the empty weight of the aircraft, a fulcrum or center of gravity, which may be changed within specified limits, and the arm for each seat, shelf, or compartment where weight may be placed. It is then up to us to place the weight in the aircraft in a manner so as to keep the balance within the specified limit.

Aircraft balance differs from the seesaw in that the moments on either side of the fulcrum do not have to be exactly equal because we have flight controls with which to continually correct a slightly unbalanced condition. If we load the aircraft in an unbalanced condition outside of the limits, nothing we can do with the flight controls will keep the aircraft from diving into the ground, or pitching upward into a stalling condition.

Another factor in aircraft balance computation is that the total weight decreases in flight. Fuel of course has weight, so that as we fly, the total weight is decreasing as the fuel remaining decreases. For this reason, we must be sure that the aircraft is in a balanced condition at takeoff and will remain balanced for the duration of the flight.

Just as we began with a known empty weight of the aircraft, and added to it our useful load, so will we begin with the known balance of the empty aircraft and add moments to either side of the center of gravity. The end result must show that we will not exceed the maximum allowable weight *and* that the aircraft is in a balanced condition.

## Weight and Balance Charts

The aircraft manufacturer provides us with a chart which automatically multiplies the weight of each article times its arm, giving us a moment. In order to reduce the size of the numbers used, the chart is normally *indexed* or "decimaled down" by a factor of 1,000. That is, if the moment of a weight is 112,000 inch-pounds, it will be shown on the chart as 112.0 (the decimal point is moved three places to the left). Figure 39 shows the loading graph.

We are also given a chart showing the center of gravity *envelope* which is a pictorial display of the limits in which the center of gravity must fall prior to takeoff, as shown in Fig. 40.

When flying light aircraft, we can be sure that the center of gravity will remain within limits for the entire flight if it is inside the envelope at takeoff. That is how the chart is designed. (Heavy aircraft pilots must compute weight and balance problems for the landing weight also.) We cannot, however, expect to be able to shift high weights in flight without throwing the aircraft out of balance.

## Computing Balance

Let us use the weight and balance charts in Figs. 39, 40 and 41 to compute the balance for a sample airplane.

Notice that the first line of the chart gives the aircraft empty weight (2,200 pounds) and empty moment (35.0). This represents the weight and balance of the empty aircraft. To those figures, we must add the weight of each article we intend to carry and its moment, which we will find on the moment chart.

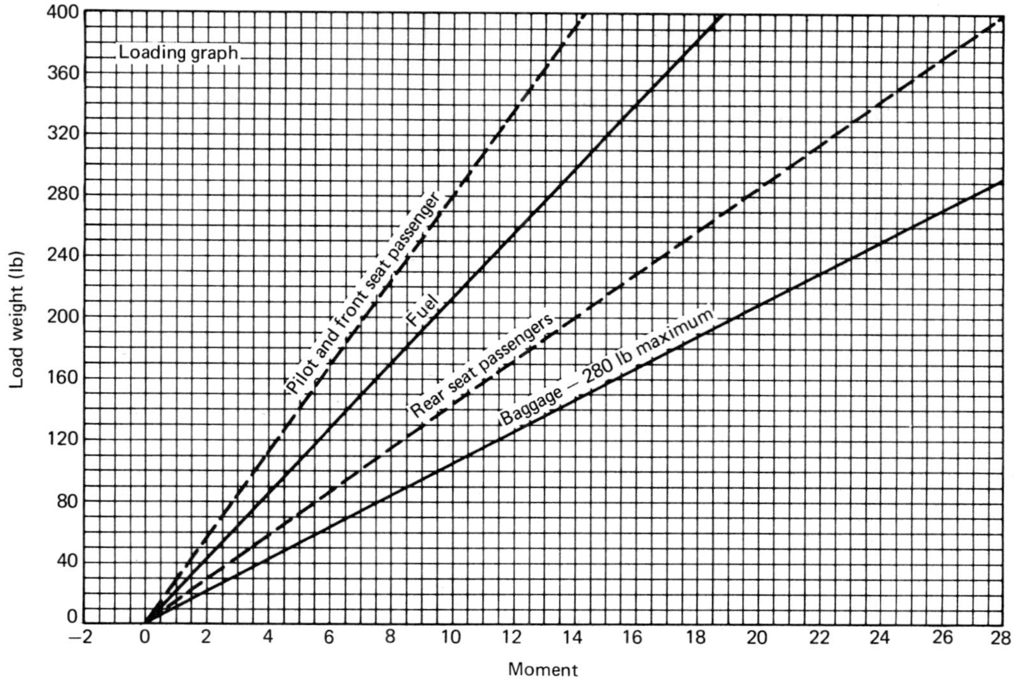


1. Find weight of each object on the left scale
2. Move horizontally to the right until intercepting the appropriate slant line
3. Drop down vertically to find the moment

Fig. 39. Loading graph.

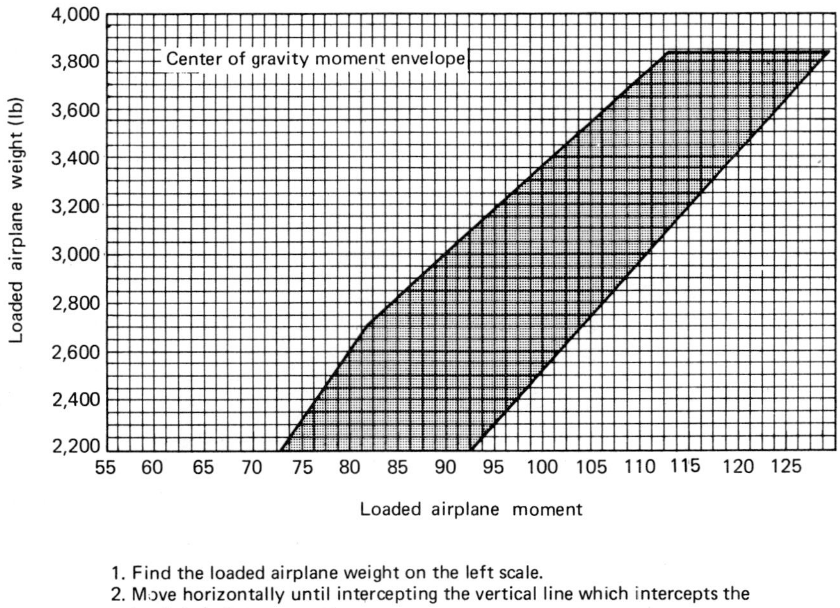


1. Find the loaded airplane weight on the left scale.
2. Move horizontally until intercepting the vertical line which intercepts the loaded airplane moment.
3. The point of interception of the two lines shows the center of gravity.

Fig. 40. Moment envelope.

| Sample loading form | Sample airplane | | | Our airplane | |
|---|---|---|---|---|---|
| | Weight (lb) | Moment | | Weight (lb) | Moment |
| Licensed empty weight | 2,200.0 | 35.0 | | | |
| Oil (3 gal) | 22.5 | −0.7 | | | |
| Pilot and front seat passenger | 355.0 | 12.6 | | | |
| Usable fuel (50 gal) | 300.0 | 14.1 | | | |
| Rear seat passengers | 380.0 | 26.6 | | | |
| Baggage | 250.0 | 24.0 | | | |
| | | | | | |
| Total | 3,507.5 | 111.6 | | | |

Maximum allowable gross weight = 4,000 lb

Fig. 41. Sample loading form.

We will go through this procedure step by step. The weight of the usable fuel is 300 pounds. We will use the moment chart (as directed on the chart) and find that the moment for the fuel is 14.1. The oil weight is 22.5 pounds. This figure will sometimes be printed on the chart, along with the oil moment, which in this case is —0.7. The weight of the pilot and front seat passenger (copilot) totals 355 pounds, and has a moment of 12.6. The two rear seat passengers weigh a total of 380 pounds and have a moment of 26.6. Baggage weight is 250 pounds, with a moment of 24.0.

Notice that some of the moments are minus (—) values and others are plus (+) values. We may use any spot on or near the airplane for the beginning reference line for measuring arms. This reference line is called the *datum line.* All balance problems are computed with the nose of the airplane pointing to the left, for standardization. All moments to the left of the datum line are minus values, and all those to the right are plus values. The total of all minus values will be subtracted from the total of all the plus values, resulting in a net moment which must fall within the center of gravity envelope.

The procedure of using a datum line makes heavy aircraft balance computation more simple than if the center of gravity was used. The datum line can be the nose of the aircraft, in which case all moments will be positive. It makes no difference where the datum line is placed, as long as all arm measurements are taken from the same line.

To complete the sample weight and balance chart, we must add all of the weights and all of the moments (remember to subtract the minus moments from the total of the plus moments). Using the total or gross weight and the net moment, we use the center of gravity moment envelope chart as directed to determine if the center of gravity is within the specified limits. (See Fig. 40.)

When we finish the computations we will see that the gross weight is 3,507.5 pounds, and the total moment is 111.6. When we place these figures on the envelope chart, we will find that the aircraft is properly loaded. If, at this point, the total weight or the net moment was not within limits, an adjustment would be necessary. Final adjustments must be done with care, for if weight is removed from the aircraft, a new moment for the remaining weight must be found. Shifting weight will not affect gross weight but will change the net moment.

*Computation Results*

It is sad but true that some pilots pay little attention to the weight and balance of their aircraft. We may visit the airport any day and watch light aircraft being stuffed with people, baggage, and fuel. Occasionally, the overweight flight will end in disaster. We must learn the capabilities of the aircraft and plan our flight to remain within these capabilities.

A new weight and balance computation need not be done for every flight if the load remains constant. Once the problem is solved for a given load in a specific aircraft, it will remain solved. However, when the load changes a new weight and balance should be computed. As we gain experience in our aircraft, we will know what loads will keep the machine within weight and balance limits. When a new situation arises, a few moments of careful attention to the loading charts will satisfy our desire for safe flight under all conditions.

FEDERAL AVIATION AGENCY
VFR EXAM-O-GRAM *NO. 13

WEIGHT AND BALANCE

Loading the family automobile for a trip requires little serious planning. You can C-R-A-M as much luggage into the trunk as you have space, squeeze as many persons into the seats as you have room, and top off the gas tank with no thought given to Gross Weight or Center of Gravity. A similar approach to loading your "flying machine" could result in a serious accident.

WHAT IS EXCESSIVE WEIGHT? Assume that your airplane is a 4-place airplane with a baggage allowance of 120 pounds, a usable fuel capacity of 39 gallons, and an oil supply of 8 quarts. On a hypothetical flight you take on full fuel and oil servicing, toss the suitcases in the baggage compartment, and you and your three passengers eagerly climb aboard. This seems like a reasonable load, but if you had placed each of them on the scales you might have found that you and the passengers average 180 lbs. each (720 lbs. ), and the four suitcases, 30 lbs. each (120 lbs. ). The usable fuel load weighs 234 lbs. and the oil 15 lbs. Assume, also, that the Weight and Balance Data for the airplane shows an empty weight of 1, 325 lbs. and a maximum allowable gross weight of 2, 200 lbs. NOW, add the weight of the useful load to the empty weight and compare the total to the allowable gross weight. (1, 089 lbs. + 1, 325 lbs. = 2, 414 lbs. ) . . . 214 lbs. excess!

WHAT RESTRICTIONS ARE THERE ON WEIGHT AND BALANCE? In many civilian airplanes it is not possible to fill all seats, baggage compartment, and tanks, and still remain within the approved weight and balance limits. If you do not wish to leave a passenger behind (a normal reaction) you must reduce your fuel load and plan on shorter legs enroute or cut down on the baggage carried, or both. Frequently, restrictions are placed on rear seat occupancy with maximum baggage allowance aboard. By all means follow your airplane's Weight and Balance restrictions. The loading conditions and the empty weight of your particular airplane may differ from those shown in the Owner's Manual, especially if modifications have been made or equipment has been added to the original basic airplane.

IS CRUISE PERFORMANCE AFFECTED BY AN EXCESS LOAD? At normal weight, the airplane requires a certain angle of attack to maintain straight-and-level flight at a given airspeed. To sustain a heavier load at that same airspeed, the angle of attack must be greater to provide the increased lift that is necessary. More power must be added to overcome the increased drag which results from the increased angle of attack. Additional power, in turn, burns more fuel, thereby reducing the range of the aircraft.

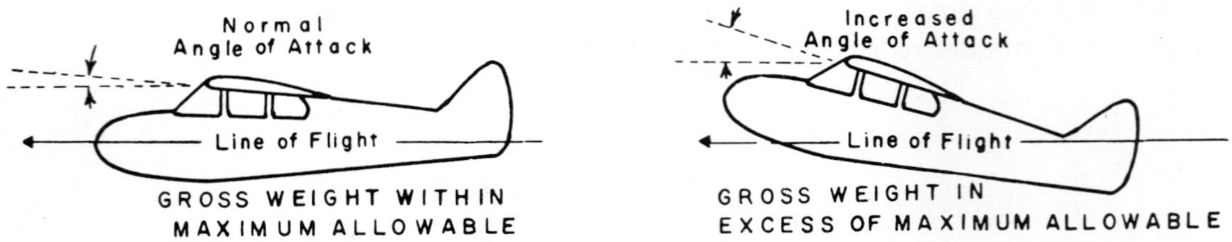


IS CLIMB PERFORMANCE AFFECTED BY AN EXCESS LOAD? Time to climb to a given altitude is lengthened, because the angle of attack is greater and the extra thrust required to carry the additional weight limits the rate of climb and may limit the climbing speed, since this depends on the surplus power available. The additional time in climbing at the higher power setting also increases the fuel consumption.

IS "G" FORCE TOLERANCE AFFECTED? Assume that your airplane has a limit-load factor of 3.8 "G's". If the allowable gross weight is not exceeded, this means the wings can safely support 3.8 times the weight of the airplane and its contents. In accelerated flight (pull-ups, turns, turbulent air) the actual load on the wings would be much greater than the normal load, which of course results in much greater stresses in the wing structure. Overloading, therefore, has the effect of decreasing the "G" load capability of the aircraft and thus could result in the wing being stressed to the point of popped rivets, permanent distortion, or structural failure.

HOW IS AN AIRPLANE BALANCED? An airplane, like a steelyard scale, is in perfect balance when the weight is distributed in such a manner that it remains level when freely suspended. In an airplane, however, as long as the Center of Gravity lies anywhere within specified limits, balance can be maintained in flight. Flight with the CG outside of this range results in unsatisfactory or dangerous flight characteristics. Loading an airplane then, is simply a matter of distributing the load so that the CG falls within the allowable range. This can be accomplished by arranging the load in accordance with the Center of Gravity Envelope provided for each airplane.

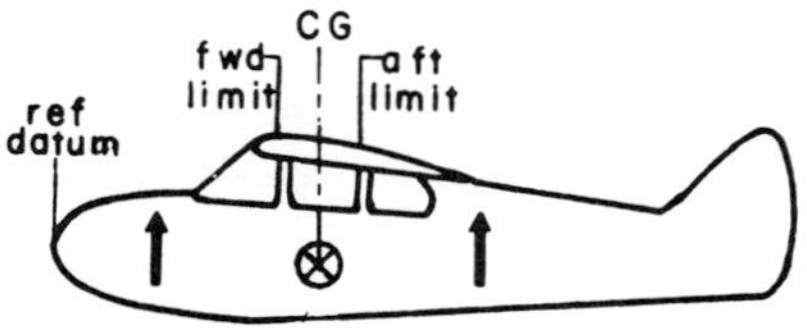

WEIGHT PROPERLY DISTRIBUTED

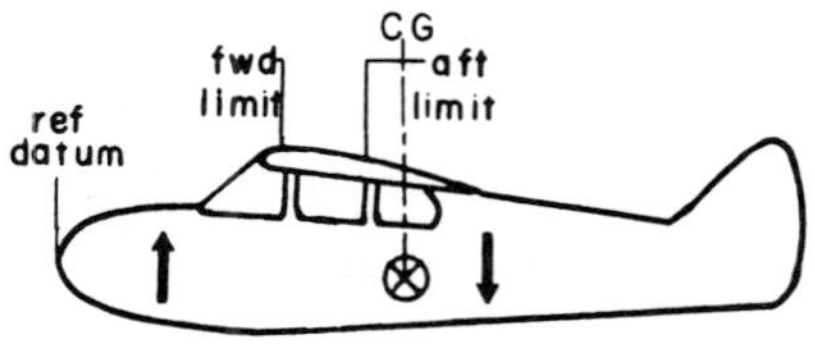

EXCESS WEIGHT PLACED AFT

DOES IMPROPER LOAD DISTRIBUTION AFFECT SAFETY? YES! When loading conditions cause the Center of Gravity to fall outside allowable limits, stability is adversely affected and erratic control forces may develop. Stalling speed, takeoff distance, and landing speed may be increased to the point of actual danger.

Due to the size of many baggage compartments there might be a tendency to fill them to capacity, ignoring the placarded baggage weight limitations. This could produce a Center of Gravity aft of allowable limits creating a highly dangerous flight condition. The result would be a nose high attitude which could lead to a stall from which recovery might not be effected due to inadequate elevator control.

AN AIRPLANE'S BEHAVIOR IN THE AIR
IS DEPENDENT ON WEIGHT AND BALANCE!

FAA Aeronautical Center
Flight Standards Technical Division, AC-700
Operations Branch, AC-740
P. O. Box 25082
Oklahoma City, Oklahoma 73125

Exam-O-Grams available free of charge in limited quantities from this office.



## CHECK YOUR READING

1. What weights are included in the maximum allowable gross weight of an aircraft?
2. What is *empty weight* of an aircraft?
3. What is the *useful load* of an aircraft?
4. List the hazards to flight in an overweight aircraft.
5. How much does aviation gasoline weigh per gallon?
6. How much does oil weigh per gallon?
7. When computing the weight of the fuel for a weight and balance chart, is the unusable fuel included?
8. What is the definition of *arm*?
9. What is the definition of *moment*?
10. What is the definition of the *datum line*?
11. In a light aircraft, if the final center of gravity is within the envelope for takeoff, will it remain within the specified limits for the duration of the flight, if no weights are shifted?
12. When should a weight and balance computation be done?

# 7

# AIRCRAFT SPEEDS

When we discussed load factors in Chapter 5, we learned only of *pilot-induced* loads; that is, load factors in excess of 1 *g* which were produced by a control action by the pilot, either in a turn, pullout, or level-off. The load factor may also be increased by conditions not concerning control use.

## Turbulence

Certain atmospheric conditions, which we will learn about when discussing meteorology, cause the phenomenon called *turbulence* or rough air. As we fly through turbulent areas, the angle of attack on the lifting surfaces may be rapidly changed. (See Chapter 2.) We know that the angle of attack is one factor affecting lift production. Rapid increases in the angle of attack will produce rapid increases in lift production.

A moving aircraft, because it has weight, has momentum. (Momentum is the product of the speed and the mass of a body.) When a body has momentum it is reluctant to change direction or speed, and must be given time to react to any applied force. When the amount of lift produced is rapidly increased in turbulent air, the aircraft does not immediately begin to climb or descend. Momentum tends to keep it on the original path of flight. If enough lift is produced very rapidly, the aircraft wings may be forced toward a new path of flight, but the main mass (the fuselage) may continue on or near the original path of flight. Turbulence then, may induce load factors severe enough to cause structural damage to the aircraft.

## *Maneuvering Speed*

The aircraft engineer designs an aircraft which will withstand a specific load factor (plus a safety factor). He also determines the speed at or below which the aircraft will stall prior to developing a quantity of lift sufficiently high to produce the maximum allowable load factor. It is far better to suffer an unintentional stall as the result of a high load factor than to sustain structural damage or failure of aircraft components.

The speed at or below which the aircraft will not sustain structural damage due to excessive or abrupt control action, or from turbulence, is called the *maneuvering speed.* At or below maneuvering speed, momentum wins the battle over lift, causing the aircraft to stall.

*Finding Maneuvering Speed.* Maneuvering speed for the aircraft will be found in the Aircraft Operations Manual, and may also be found on a placard in the cockpit. It is ***not*** marked on the airspeed indicator. A good rule of thumb for computing maneuvering speed is to increase the minimum power-off stall speed by 70 per cent. For instance, if the minimum power-off stalling speed for our aircraft is 50 miles per hour, we multiply 50 miles per hour by 70 per cent, and add the result to 50 miles per hour (50 mph times 70 per cent = 35 mph. Add 50 mph and 35 mph = 85 mph).

*Use of Maneuvering Speed.* Whenever we suspect a flight condition which may impose high load factors, we should reduce the airspeed to maneuvering speed. Before attempting any flight maneuver requiring abrupt control action, reduce the airspeed to maneuvering speed.

You may ask, "Why not reduce the airspeed even lower than maneuvering speed to increase further the safety factor of the flight?" We can do that, but we will suffer from reduced control efficiency and premature stalling. Particularly in turbulent air, we will need all the control efficiency available to keep the aircraft flying safely. Hence, maneuvering speed is a minimum compromise on control efficiency, but guarantees a stall before structural damage occurs.

## Cruising Speed

The cruising speed is the speed at which the aircraft passes through the air in straight and level flight, with the power output of the engine in a recommended power range. Cruising speed is the result of a compromise between maximum allowable continuous power and much reduced power which causes the aircraft to "mush." High power settings cause rapid engine wear and high fuel consumption rates. Low power settings produce speeds at which control action is sluggish and cause added drag resulting from a constant nose-high pitch attitude.

The important things for us to remember about cruising speed are that we can expect an efficient cruising speed only if we use recommended power settings, and that fuel consumption varies greatly with changing power settings. (See Chapter 11.) The cruising speed range will be shown on the airspeed indicator as a green arc.

## Never Exceed Speed

The *never exceed speed* is the speed above which structural failure may occur *with no induced load factor*. Drag at this airspeed is great enough with only the normal 1 *g* load factor to damage the aircraft. Momentum is so high that any applied force to change the path of flight may structurally damage the aircraft. A red line on the airspeed indicator will show the never exceed speed. (A safety margin is also taken into consideration when specifying the never exceed speed.)

## Caution Speed

The *caution speed* range is marked on the airspeed indicator as a yellow arc. When airspeed is within the caution range structural damage may occur with induced load factors *less than* the designed limits. The higher the airspeed, the less induced load factors will be necessary to cause damage until, at the high end of the caution range, we find the never exceed speed.

## Common Stalling Speeds

Minimum stalling speeds are given in the Aircraft Operations Manual. We must keep in mind that these speeds are *minimums*, for we have learned that the aircraft can stall at any higher airspeed if a load factor is imposed.

### *Power-Off Stall Speed*

The *power-off stall speed* is the minimum stall speed of the aircraft with the engine idled. Normally, different minimum power-off stall speeds will be given in the Aircraft Operations Manual for varying configurations. For instance, the speed will be different with the landing gear and flaps extended than with them retracted. The minimum power-off stall speed will be shown on the airspeed indicator at the low-speed end of the green cruising speed arc. The minimum power-off stall speed with flaps extended will normally be a lower speed, and is shown on the airspeed indicator at the low-speed end of the white flap operating speed arc.

### *Power-On Stall Speed*

The aircraft will stall at a slightly lower minimum speed when the engine is developing power. This is due to two primary factors. First, the spinning propeller hurls air rearward over portions of the lifting surfaces, thereby increasing their true relative wind. Also, power tends to reduce mushing of the aircraft, which allows flight at lower speed before reaching the stall angle of attack.

Power-on stall speed is not marked on the airspeed indicator, but is noted in the Aircraft Operations Manual.

It is important for us to repeat over again that stall speeds are *minimum* speeds, with no load factor being imposed. Load factors in excess of 1 *g* rapidly increase the stall speed of the aircraft.

FEDERAL AVIATION AGENCY
VFR PILOT EXAM-O-GRAM* No. 26

COMMON MISCONCEPTIONS (Series 2)

Each question in FAA Airman Written Examinations offers the examinee a group of four answers from which to select the answer he believes to be correct. Applicants' comments and analyses of the answer sheets indicate that particular incorrect answers are frequently being chosen because of a misconception regarding certain items of required aeronautical knowledge. This Exam-O-Gram, as well as Exam-O-Gram No. 17, attempts to correct a few of these preconceived ideas.

WHAT INDICATED AIRSPEED SHOULD BE USED FOR LANDING APPROACHES TO FIELDS OF HIGHER ELEVATIONS? For all practical purposes, use the SAME indication as you use at fields of lower elevations.

WILL THE SAME INDICATED APPROACH SPEED BE SAFE AT HIGH ELEVATIONS? YES, in relatively smooth air. We all know that as altitude increases, the air becomes less dense, and consequently with decreased drag the airplane travels faster through the air. However, this faster speed creates no increase in impact pressure on the airspeed pitot system because of the lesser air density. In other words, we get a higher True Airspeed with the same Indicated Airspeed. Although the True Airspeed (TAS) at which an airplane stalls in thinner air is higher, the margin of safety is unaffected since the airplane is actually flying at a higher True Airspeed. Nevertheless, for the purpose of maintaining positive control in unstable air, the use of a higher than normal indicated speed is recommended for approaches during the turbulent or gusty conditions prevalent in mountainous areas, just as is used at fields of lower elevations in these conditions.

WHAT EFFECT DOES THINNER AIR HAVE ON APPROACH AND LANDING? Even though using the same indicated airspeed that is appropriate for sea level operations, the True Airspeed is faster, resulting in a faster groundspeed (with a given wind condition). This increase in groundspeed naturally makes the landing distance longer and should be carefully considered when landing at high elevation fields, particularly if the field is short.

WHAT INDICATED AIRSPEED SHOULD BE USED ON TAKEOFF AT HIGH ELEVATIONS? Just as in landing, the groundspeed as well as the takeoff distance, will be greater at high elevation fields. However, don't let this mislead you into P-U-L-L-I-N-G the airplane off the ground. If you do, the airplane will mush and settle back to the ground in a stalled condition. Use the SAME indicated airspeed as you use for takeoff at fields with lower elevations.

* Exam-O-Grams are non-directive in nature and are issued solely as an information service to individuals interested in Airman Written Examinations.

WHAT WOULD YOU THINK IF YOU OVERHEARD THIS AIRPLANE "DISCREPANCY" REPORT? "Hey, Chief, - fix this goofed-up airspeed indicator! I was practicing power-off stalls with the gear and flaps down, but the airplane didn't stall until the pointer was 10 mph less than the white arc painted on the dial."

IS THE AIRSPEED INDICATOR FAULTY OR IS THE WHITE ARC MISPLACED? Not necessarily either one! Remember, the colored arcs on the airspeed dial (see Exam-O-Gram No. 8) mark the Calibrated Airspeed (CAS) and not merely the observed Indicated Airspeed (IAS) limitations.

WHAT IS CALIBRATED AIRSPEED? Calibrated Airspeed, frequently called True Indicated Airspeed (TIAS), is Indicated Airspeed corrected for installation and instrument error. A wide difference between these speeds may exist, particularly at low airspeeds or under landing conditions. Installation error is caused when static atmosphere in certain flight attitudes enters the static system with a different pressure than it does in normal cruise conditions, creating a variance in pitot-static differential. Check the airspeed correction data for each airplane. You may find (as in the typical table below) that an IAS of 60 mph is actually a CAS (or TIAS) of 69 mph. (TRUE INDICATED AIRSPEED is not to be confused with TRUE AIRSPEED which is the actual speed of the airplane through the air.)

**AIRSPEED CORRECTION TABLE**

| | | | | | | | | |
|---|---|---|---|---|---|---|---|---|
| **FLAPS 0°** | | | | | | | | |
| IAS - MPH | 60 | 80 | 100 | 120 | 140 | 160 | 180 | 200 |
| TIAS - MPH | 69 | 82 | 100 | 119 | 139 | 160 | 181 | 202 |
| ***FLAPS 20°** | | | | | | | | |
| IAS - MPH | 40 | 50 | 60 | 70 | 80 | 90 | 100 | 110 |
| TIAS - MPH | 57 | 62 | 68 | 75 | 84 | 93 | 102 | 112 |
| ***FLAPS 40°** | | | | | | | | |
| IAS - MPH | 40 | 50 | 60 | 70 | 80 | 90 | 100 | 110 |
| TIAS - MPH | 57 | 62 | 68 | 75 | 83 | 92 | 102 | 111 |

*Maximum flap speed 110 MPH-TIAS

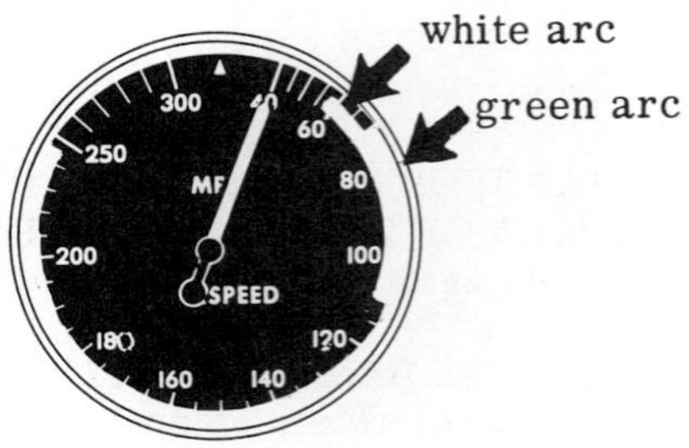


WHAT IS THE RELATIONSHIP BETWEEN AIRSPEED INDICATOR COLORED ARCS AND STALLING SPEEDS? In the above illustrations, the white arc shows a stalling speed of 57 mph (TIAS or CAS), but because of installation error (reflected in the table), this airplane may not stall with power-off and gear and flaps down until the pointer is on 40 mph (IAS). Similarly a variation is noted for the green arc and stalling speed with gear and flaps UP. Since an airplane in flight is operated most of the time within the upper speed range, installation error is normally adjusted so as to be at a minimum in that range. This results in the greatest error at the lower speed range, but provides a corresponding increase in the margin of safety at the critical lower airspeeds.

* * * * * * * * * * * * * * * * * * * * * * * * * * * * * * * * * * * * * * * * * * * * * * * * * * *

CAN NORMAL IN-FLIGHT ASSISTANCE BE RECEIVED FROM ALL VOR STATIONS? NO, many VOR stations can be used only for navigation purposes. These stations with no capability to transmit voice signals are so marked on the chart with the notation "NO VOICE." Stations of this type cannot be used for weather information, position reporting, flight plans, or emergency assistance.

FEDERAL AVIATION AGENCY
VFR EXAM-O-GRAM NO. *28
FACTORS AFFECTING STALL SPEED

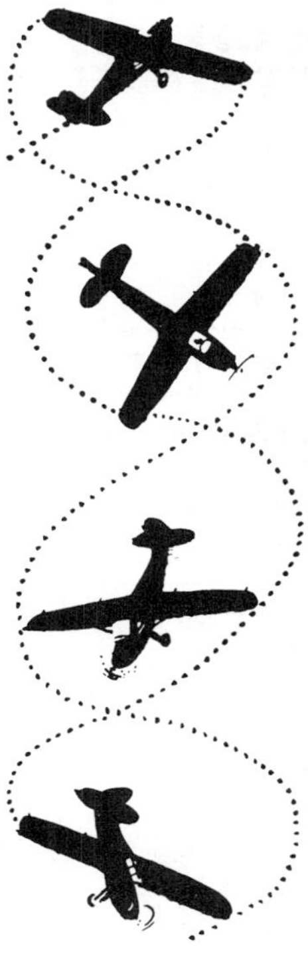
FIG. 1

A recent report indicates that approximately 80% of all accidents are pilot caused. The major cause of fatal accidents is listed as "failed to maintain airspeed (or flying speed) resulting in a stall." Although many of these stalls may have occurred under the stress and duress of other problems such as disorientation during limited visibility or at night, improper division of attention, etc., a review of statistical analyses of written examinations indicates a lack of knowledge and understanding of the various factors that can cause or contribute to a stall. This Exam-O-Gram discusses some of the more important, ever-present factors of which the pilot must have an understanding so that he will instinctively avoid or compensate for situations, conditions, and attitudes which may lead to a stall--even under the stress and duress of additional problems he may encounter in flight.

WHAT CAUSES AN AIRPLANE TO STALL? All stalls are caused by exceeding the critical angle of attack. Knowing this particular fact does not necessarily help the pilot. What is more important to the pilot is to know what factors are likely to contribute to or cause this angle of attack to be exceeded.

IS IT NECESSARY FOR THE AIRPLANE TO HAVE A RELATIVELY LOW AIRSPEED IN ORDER FOR IT TO STALL? No! An airplane can be stalled at any airspeed. All that is necessary is to exceed the critical angle of attack. This can be done at any airspeed if the pilot applies abrupt or excessive back pressure on the elevator control. A stall that occurs at a relatively high speed is referred to as an accelerated or high speed stall.

IS IT NECESSARY FOR THE AIRPLANE TO HAVE A RELATIVELY HIGH PITCH ATTITUDE IN ORDER FOR IT TO STALL? No! An airplane can be stalled in any attitude. Repeating again the statement made above - all that is necessary is to exceed the critical angle of attack. This can occur in any attitude by application of abrupt or excessive back pressure on the elevator control.

DOES WEIGHT AFFECT THE STALLING SPEED? Yes! As the weight of the airplane is increased, the stall speed increases. Due to the greater weight, a higher angle of attack must be maintained to produce the additional lift to support the additional weight in flight. Therefore, the critical angle of attack will be reached at a higher airspeed when loaded to maximum gross weight than when flying solo with no baggage. (See Exam-O-Gram No. 13.)

DOES THE CENTER-OF-GRAVITY LOCATION (WEIGHT DISTRIBUTION) AFFECT STALL SPEED? Yes! The farther forward the center of gravity, the higher the stalling speed. The farther aft the center of gravity, the lower the stalling speed. (See Exam-O-Gram No. 13.)

DOES THIS MEAN THAT THE WEIGHT SHOULD BE DISTRIBUTED IN THE AIRPLANE SO THAT THE CG IS AS FAR TO THE REAR AS POSSIBLE? No! This may present problems with stability that will far outweigh any advantages obtained by the decrease in stall speed. (See Exam-O-Gram No. 13.)

* Exam-O-Grams are non-directive in nature and are issued solely as an information service to individuals interested in Airman Written Examinations.

DO FLAPS AFFECT STALLING SPEED? Yes! The use of flaps reduces stalling speed. The Stall Speed Chart (Figure 2) excerpted from an airplane flight manual illustrates this fact. This also can be readily verified by checking the color coding on any airspeed indicator. The lower airspeed limit of the white arc (power-off stalling speed with gear and flaps in the landing configuration) is less than the lower airspeed limit of the green arc (power-off stalling speed in the clean configuration). (See EXAM-O-GRAM No. 8.) This fact is important to the pilot in that when making no-flap landings, a higher indicated airspeed should be maintained than when landing with flaps. The manufacturers' recommendations should be adhered to as to approach speeds with various configurations.

**STALL SPEED, POWER OFF**

| Gross Weight 3000 lbs. | ANGLE OF BANK | | | |
|---|---|---|---|---|
| CONFIGURATION | 0° | 20° | 40° | 60° |
| GEAR & FLAPS UP | 65 | 67 | 74 | 92 |
| GEAR DOWN, FLAPS 20° | 61 | 63 | 70 | 86 |
| GEAR DOWN, FLAPS 40° | 60 | 62 | 69 | 85 |

SPEEDS ARE MPH, TIAS

FIG. 2. (Note: TIAS identical with CAS)

DOES AN ACCUMULATION OF FROST, SNOW, OR ICE ON THE WINGS AFFECT STALLING SPEED? Yes! Even a light accumulation of frost, snow, or ice on the wings can cause a significant increase in stalling speed. It can increase it so much that the airplane is unable to take off. The accumulation disrupts the smooth flow of air over the wing thus decreasing the lift it produces. To make up for the lost lift, a higher angle of attack must be used or a higher speed must be attained on the takeoff roll. The runway may not be long enough to attain the necessary speed and even though the airplane may become airborne, it could be so close to the stall speed that it would not be possible to maintain flight once the airplane climbs above the comparatively shallow zone where ground effect prevails. DO NOT TAKE OFF UNTIL ALL FROST, SNOW, OR ICE HAS MELTED OR BEEN REMOVED FROM THE AIRPLANE.

FIG. 3

DOES AN INCREASE IN ALTITUDE AFFECT THE INDICATED AIRSPEED AT WHICH AN AIRPLANE STALLS? An increase in altitude has no effect on the indicated airspeed at which an airplane stalls at altitudes normally used by general aviation aircraft. That is, for all practical purposes, the indicated stalling speed remains the same regardless of altitude in this range. This fact is important to the pilot in that the same indicated airspeed should be maintained during the landing approach regardless of the elevation or the density altitude at the airport of landing. (Follow the manufacturer's recommendations in this regard.) If higher than normal approach airspeed is used, a longer landing distance will be required.

DOES AN INCREASE IN ALTITUDE AFFECT THE TRUE AIRSPEED AT WHICH AN AIRPLANE STALLS? Since true airspeed normally increases as altitude increases (for a given indicated airspeed), then true airspeed at which an airplane stalls generally increases with an increase in altitude. Under non-standard conditions (temperature warmer than standard) there is an additional increase in true airspeed above the indicated airspeed.

OF WHAT SIGNIFICANCE IS THIS TO THE PILOT? It is significant in that when landing at higher elevations or under higher density altitudes, he is operating at higher true airspeeds (and therefore higher groundspeeds) throughout the approach, touchdown, and landing roll. This results in a greater distance to clear obstacles during the approach, a longer ground roll, and consequently, the need for a longer runway. If, in addition, the pilot is operating under the misconception that a higher than normal indicated airspeed should be used under these conditions, the situation is further compounded due to the additional increase in groundspeed. (See EXAM-O-GRAM No. 26.)

DOES TURBULENCE AFFECT STALLING SPEED? Yes! Turbulence can cause a large increase in stalling speed. Encountering an upward vertical gust causes an abrupt change in relative wind. This results in an equally abrupt increase in angle of attack which could result in a stall. This fact is important to the pilot in that when making an approach under turbulent conditions, a higher than normal approach speed should be maintained. Also, in moderate or greater turbulence, an airplane should not be flown above maneuvering speed. (See EXAM-O-GRAM No. 8.) At the same time, it should not be flown too far below maneuvering speed since a sudden severe vertical gust may cause an inadvertent stall due to the higher angle of attack at which it will already be flying.

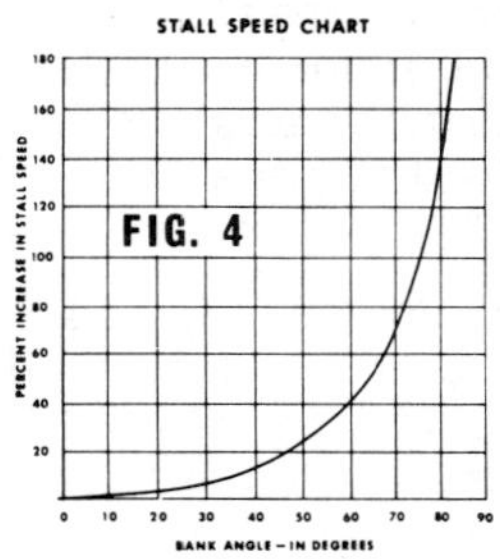

FIG. 4

DOES ANGLE OF BANK AFFECT STALLING SPEED? Yes! As the angle of bank increases in a constant altitude turn, the stalling speed increases. This is easily seen from the STALL SPEED CHARTS (Figs. 2 and 4) which show the increase in stall speed as the angle of bank increases--Fig. 4 in terms of percent, Fig. 2 the actual values for one airplane. At a 60° bank stalling speed is 40% greater than in straight-and-level flight (25-27 mph for the specific example.) At angles of bank above 60°, stall speed increases very rapidly, and at approximately 75° it is doubled with respect to straight-and-level stall speed (Fig. 4).

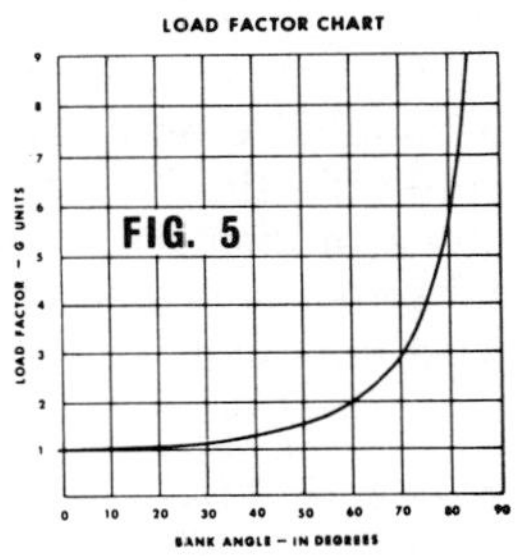

FIG. 5

DOES LOAD FACTOR AFFECT STALLING SPEED? Yes! As the load factor increases, stalling speed increases. When the load factor is high, stalling speed is high. A comparison of the two charts (Figs. 4 and 5) should easily show this relationship. Load factor is the ratio of the load supported by the wings to the actual weight of the airplane and its contents. At a load factor of 2, the wings support twice the weight of the airplane; at a load factor of 4, they support four times the weight of the airplane. Normal category airplanes with a maximum gross weight of less than 4,000 pounds are required to have a minimum limit load factor of 3.8. (The limit load factor is that load factor an airplane can sustain without taking a permanent set in the structure.) Note from the load factor chart (Fig. 5) that this minimum limit load factor is attained in a constant altitude turn at a bank of approximately 75°. Also note from the stall speed chart (Fig. 4) that at this angle of bank, the stall speed is twice as great as in straight-and-level flight. There are two reasons then why excessively steep banks should be avoided--an airplane will stall at a much higher airspeed and the limit load factor can be exceeded. The danger can be compounded when the nose gets down in a steep turn if the pilot attempts to raise it to the level flight attitude without shallowing the bank since the load factor may be increased even more. This is the situation as it generally exists when, due to disorientation, the pilot enters a diving spiral (often referred to as the "graveyard spiral") and attempts to recover with elevator pressure alone.

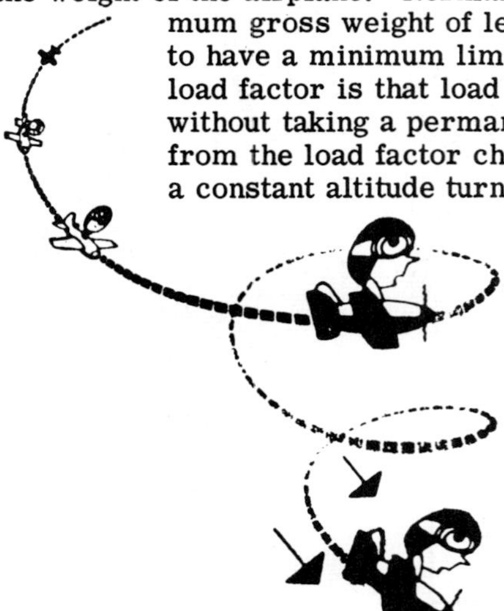

WHAT FACTORS CAUSE AN INCREASE IN LOAD FACTOR? Any maneuvering of the airplane that produces an increase in centrifugal force will cause an increase in load factor. Turning the airplane or pulling out of a dive are examples of maneuvering that will increase the centrifugal force and thus produce an increase in load factor. When you have a combination of turning and pulling out of a dive, such as recovering from a diving spiral, you are, in effect, placing yourself in double jeopardy. This is why you must avoid highspeed diving spirals or if you accidentally get into one--be careful how you recover. Turbulence can also produce large load factors. This is why an airplane should be slowed to maneuvering speed or below when encountering moderate or greater turbulence.

CAN THE PILOT RECOGNIZE WHEN THERE IS AN INCREASE IN LOAD FACTOR? Yes! He can recognize it by the feeling of increased body weight or the feeling that he is being forced down into the seat--the greater the load factor the greater this feeling of increased weight or of being forced down in the seat (Figs. 6 and 7). It is the same feeling one has when riding the roller coaster at the bottom of a dip or going around a banked curve. This feeling of increased body weight is important to the pilot because it should, if it becomes excessive, have the immediate effect of a red flag being waved in his face to warn him that the airplane will now stall at a higher airspeed or that the limit load factor can be exceeded, resulting in structural failure.

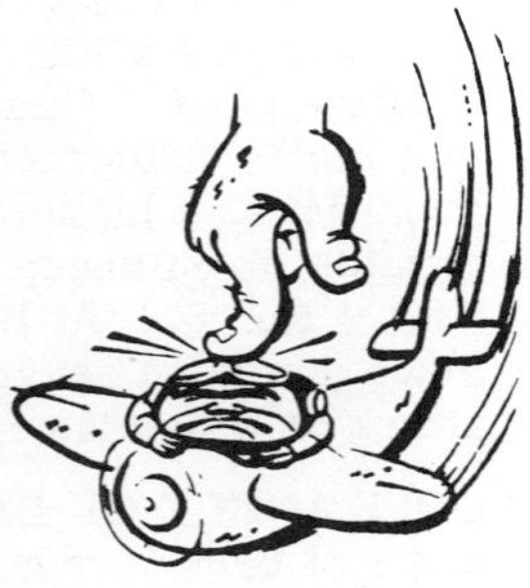

FIG. 6

DOES SPEED AFFECT LOAD FACTOR? Speed does not, in itself, affect load factor. However, it has a pronounced effect on how much of an increase in load factor can be produced by strong vertical gusts, or by the pilot through abrupt or excessive application of back pressure on the elevator control. This is why airspeed should be reduced to maneuvering speed or below if moderate or greater turbulence is encountered. At maneuvering speed or below, the airplane is stressed to handle any vertical gust that normally will be encountered. Also, below this speed, the pilot can make abrupt full deflection of the elevator control and not exceed the maximum load factor for which the airplane is stressed. However, it should be noted that the reason this is possible is because the airplane will stall, thus relieving the load factor. At airspeeds above maneuvering speed, abrupt full deflection of the elevator control or strong vertical gusts can cause the limit load factor to be exceeded. As airspeed continues to increase above maneuvering speed, the limit load factor can be exceeded with less and less turbulence or abrupt use or deflection of the controls.

WHAT IS THE RELATIONSHIP BETWEEN A HIGH SPEED (ACCELERATED) STALL AND LOAD FACTOR? The higher the airspeed when an airplane is stalled, the greater the load factor. When an airplane stalls at a slow airspeed, the load factor will be very little more than one. When stalled at an airspeed twice as great as the normal stall speed, the limit load factor for normal category airplanes probably will be exceeded. This fact can be determined from the stall speed (Fig. 4) and load factor (Fig. 5) charts. See also discussion of "Does Load Factor Affect Stalling Speed" (page 3).

FIG. 7

Prepared by FAA Flight Standards Service
Operations Airman Examination Section
5300 South Portland Avenue
Oklahoma City, Oklahoma 73119

Exam-O-Grams available free of charge in limited quantities from this office.

POOR GUY...

FEDERAL AVIATION AGENCY
VFR EXAM-O-GRAM NO. 17

COMMON MISCONCEPTIONS (Series #1)

The following remarks are actual excerpts from a pilot's written report of an accident in which he was involved.

> "I was climbing at an airspeed of 60 mph. I started a climbing turn to the right. The wind now became a cross wind instead of a head wind. This (lack of head wind) caused the airplane to stall -- to recover from the stall I turned the airplane back into the wind . . . (Later) I was in a quartering tailwind from the right. . . Went into a second stall. . . This is all I remember."

This pilot had over 100 hours, yet stalled and crashed due to an apparent misuse of controls at a slow airspeed (high angle of attack). The inspector who took this pilot's statement decided to pursue this theory with a group of student pilots. He posed this question to them.

> "If the aircraft's stalling speed was 60 mph and you were flying at an airspeed of 70 mph into a 30 mph wind, what would happen if you maintained this airspeed of 70 mph but turned downwind?" Five of the six students said the airplane would stall.

IS THIS ANSWER CORRECT? No.

IS THE STALLING SPEED OF AN AIRPLANE A FUNCTION OF THE AIRSPEED OR THE GROUND SPEED? The airspeed.

DOES THE DIRECTION OF THE WIND HAVE ANY EFFECT ON THE AIRSPEED OF AN AIRCRAFT IN FLIGHT? No.

Now to summarize our point, airspeed is the only speed which holds any significance for an airplane. Once it is off the ground, an airplane feels nothing but its own speed through the air. It makes absolutely no difference what its speed happens to be in relation to the ground. The aircraft in flight feels no wind. It simply proceeds, operating with the same mechanical efficiency, upwind, downwind, crosswind, or in no wind at all. (NOTE: We are referring here to a steady wind. Turbulence, gusts, or wind shears can lead to stalls even though airspeed is being maintained above the normal stalling speed. In such conditions it is wise to add a safe margin to normal climbout or approach speeds.)

*Exam-O-Grams are non-directive in nature and are issued solely as an information service to individuals interested in Airman Written Examinations.

## CHECK YOUR READING

1. Draw a diagram of vertical gusts striking the airfoil, thereby changing the angle of attack of the airfoil.
2. What is *momentum*?
3. What is the maneuvering speed of an aircraft if the minimum power-off stall speed is 62 miles per hour?
4. Why should we not reduce speed lower than maneuvering speed when encountering turbulence?
5. Where is maneuvering speed found in the aircraft?
6. Normal cruising speed is a compromise between at least two factors. Name them.
7. How is the cruising speed range denoted on the airspeed indicator?
8. How is the never exceed speed marked on the airspeed indicator?
9. Should we ever plan on flying our aircraft at speeds within the caution range?
10. Why will the aircraft fly at a slower speed with power on than it will with power idled before stalling?

# 8
# WAKE TURBULENCE

Turbulence produced by aircraft may be more severe than natural turbulence. When aircraft engines produce thrust, they create engine wash or wake turbulence. When an aircraft penetrates the air in flight, it creates *wing-tip vortices* which stretch out behind the aircraft. Both of these man-made forms of turbulence may be hazardous.

## Engine Wash

The spinning propeller of a reciprocating engine or the exhaust nozzle of a turbine engine hurtles a tremendous quantity of air rearward in order to produce thrust. These swirling air currents produce many problems.

Pilots normally check the area in front of the aircraft carefully before starting the engine, for they know that the rotating propeller can be an invisible death-dealing weapon. But too often they tend to ignore the area behind the aircraft.

When the engine starts, a whirlwind of dust-laden air surges rearward. Persons in the path of this man-made dust devil may be seriously injured by flying debris.

While taxiing, we must be careful not to create unnecessary engine wash, particularly when the tail of the aircraft is pointing toward other aircraft, buildings, or people. We must also avoid taxiing behind other aircraft in close proximity when they are producing engine wash, for our aircraft may be overturned by it. It is our responsibility, as command pilots, to conduct a safe operation. Therefore, no matter how congested are the airport ramps, taxiways, or run-up areas, there is no justification for our taxiing behind another aircraft if the slightest doubt exists as to positive safety.

Needless to say, the engine wash from a turbine engine can be much more hazardous because the exhaust gases are very hot (warm enough to melt the asphalt taxiway in some instances).

## Wing-Tip Vortices

Wing-tip vortices are powerful tornadoes of air swirling rearward for many miles from each wing tip, as shown in Fig. 42. They are fast-spinning, compact, and very violent. Unfortunately, they are also invisible. Tests have shown, and accident investigations have proved, that light aircraft may be destroyed if penetration of a vortex occurs under certain conditions.

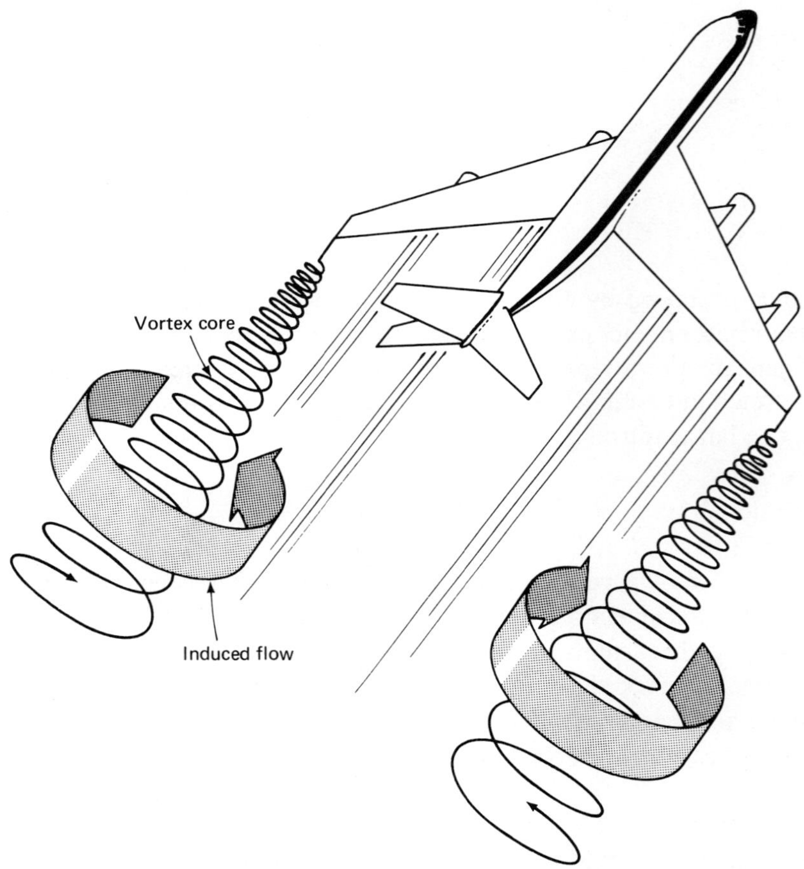


Fig. 42. Wing-tip vortices.

### *Vortex Production*

The wing-tip vortex is caused by the high pressure air under the wing "spilling up" over the wing tip in an effort to fill the lower pressure area above the wing. The greater the differential pressure between the top and the bottom of the wing, the more violent will be the vortex. Heavier aircraft produce stronger, longer lasting vortices than do light aircraft. The hidden danger from vortices may exist for a long period of time, particularly in calm air, and may exist far behind the producing aircraft. (Two or three miles is not uncommon.)

Vortex core intensity varies directly with the speed of the aircraft. The combination of high lift production required for heavy aircraft, and slow speed used in the traffic pattern produces the most severe vortices. Modern aircraft, which are constantly growing in size, will produce very severe vortices, particularly during the takeoff and landing segments of their flights.

### *Vortex Characteristics*

If our aircraft intercepts a vortex at a sharp angle, we will suffer a severe jolt as the aircraft passes through the area of rapidly changing air currents. If we are near the producing aircraft, we may suffer structural damage. When entering the vortex at a shallow angle or on a parallel course, we will experience a rapid roll, and will probably lose control of the aircraft. Normally, the rolling tendency will throw the aircraft out of the narrow core of severe turbulence. If our aircraft is following the producing aircraft but closing on it, the roll will begin slowly and increase in intensity. We should not "fight" the aircraft in an attempt to keep it upright, for this will only increase the stress on its structure. If time permits, we should immediately slow the aircraft to maneuvering speed, and then initiate a climb or a turn in the direction of the roll, to escape the vortex.

Encountering a vortex near the surface is very hazardous for the aircraft may be tossed about and not respond to our control action. We must pay special attention to avoiding vortices at all times, but particularly when at low altitudes.

### *Avoidance of Vortices*

Obviously, we can positively avoid vortices if we never leave the ground. As pilots, we must realize that these invisible hazards exist, and do everything possible to avoid entrapment.

We should avoid passing behind heavy aircraft whenever possible. This is easier said than done, for the producing aircraft may be well out of sight or may even have landed, leaving behind a lingering vortex. However, we must remain alert to spot other aircraft in the vicinity, and plan our flight path to avoid passing behind them, particularly at close range.

When a choice of flight track and altitude exists, we should select flight routes which are not frequented by large aircraft. Constant monitoring of common radio frequencies will give us an indication of the whereabouts of heavy aircraft.

When it becomes necessary to operate behind a large aircraft (in the airport traffic pattern, for instance) we should maintain a flight altitude *higher than* the vortex producing aircraft because wing-tip vortices tend to settle downward towards the surface.

Vortices also drift with the air mass. We should therefore plan our flight so as to remain *upwind* of the path of the heavier aircraft.

Since vortex production is dependent upon lift production, and is most severe when the heavy aircraft is at slow speed, the most severe vortices will be found at the point of lift-off or just before the point of landing. When taking off behind a heavy aircraft therefore, we should plan our takeoff run on the upwind side of the runway, and if possible we should leave the ground at a spot prior to the point where the heavier aircraft left the ground. When approaching for a landing we should plan a descent path *above* that of the heavier aircraft, and if possible land *beyond* the touch-down point of the heavier aircraft.

*Pilot Responsibility*

It is the responsibility of the pilot to conduct the flight with maximum safety. When operating at an airport with air traffic control facilities, we must not allow ourselves to be subjected to unsafe conditions by following the directions of a busy air traffic controller.

For example, when air traffic is at or near the saturation point, it is natural for the controller to suggest close spacing between aircraft. The controller has the responsibility to keep aircraft separated and flowing smoothly, but it is up to the pilot to make the final decision as to the total safety of the flight. If, in our opinion, an air traffic control clearance might subject our aircraft to a potential hazard, it is our duty to advise the controller of our inability to comply with the clearance.

*Cleared for an immediate takeoff* is a common expression used by tower controllers, for instance. The clearance for take off assumes that the pilot agrees with the controller that it is safe to takeoff. If we suspect engine wash or turbulence is lingering over the runway, we need only to advise the controller that we cannot comply due to wake turbulence. It is much more prudent to delay our takeoff for a moment than to lose control of our aircraft due to wake turbulence, subjecting the flight to possible disaster.

In short, we must be alert to the danger of wake turbulence, and use *our* best judgement to insure a safe flight.

## FEDERAL AVIATION ADMINISTRATION
## VFR PILOT EXAM-O-GRAM *NO. 3

## DANGERS OF WINGTIP VORTICES
## (AN INVISIBLE HAZARD TO LIGHT AIRCRAFT)

The Civil Aeronautics Board listed the probable cause of a recent fatal light aircraft accident as structural failure of a light aircraft resulting from excessive airloads created by wing-tip vortices behind a large aircraft. The report also states: "The dangers of wake or vortex turbulence are still unknown to many pilots."

1. WHAT ARE WING-TIP VORTICES? It is unfortunate that vortices are invisible. If you could see them, they would look like a pair of horizontal tornadoes stretching back from each wing-tip. These violent, compact, and fast-spinning air masses extend behind an aircraft for miles. Many pilots refer to this phenomenon as "prop wash" or "jet wash", but engineering studies have revealed this term a misnomer. The main source of this disturbance is not from the power plant; it is from the wing-tips.

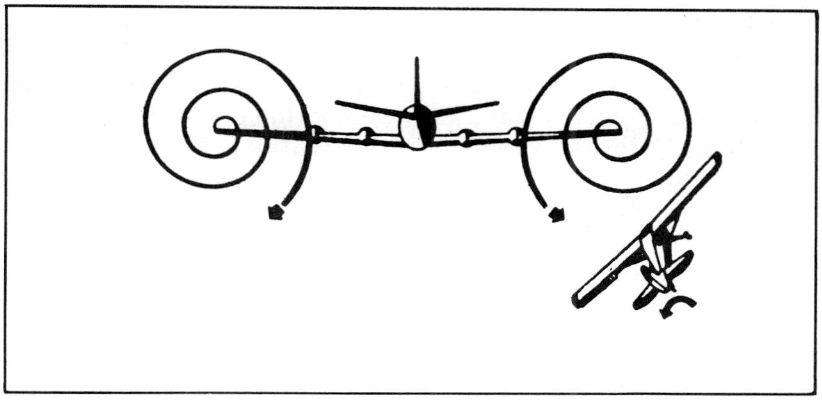

2. WHY ARE THEY DANGEROUS? They are dangerous because all tests to date indicate that structural failure in the air can occur in light aircraft upon penetration of the vortices behind larger transport aircraft. During takeoff or landing, care should be taken to avoid vortex disturbance. Loss of control could be the result at a critical time when control is of prime importance to safety.

3. UNDER WHAT CONDITIONS ARE THEY MOST DANGEROUS? There are many factors affecting the intensity of wing-tip vortices, but it is a safe and practical generalization that the bigger the airplane the more violent and long-lived will be the vortex disturbance. The source of this insidious danger can be out of sight by the time you encounter the wake. For example: A large aircraft could take off and be out of sight, or landed and be on the ramp and the vortices turbulence could still be present near the runway. No practical rule involving a time interval for one aircraft to fly behind another will assure avoidance of the vortices generated by the first. The heavier and slower the aircraft is flying, the greater the intensity of the air circulation in the vortex cores. A modern large transport aircraft will create vortices of maximum turbulence during takeoffs and landings at near maximum gross weights.

4. WHAT ACTION CAN THE PILOT TAKE TO AVOID OR REDUCE THIS HAZARD?

   a. Avoid passing behind any large aircraft.

   b. Avoid, when possible, places and altitudes frequented by large aircraft. Constantly monitor your radio for location of such aircraft.

   c. When it is necessary to operate behind a large heavy aircraft, remain above the flight path of that aircraft. Vortices settle downward toward the surface and also they are affected by the wind and move with the air mass.

   d. When taking off or landing behind large aircraft, be on the alert for the first sign of turbulence and allow adequate spacing.

   e. The best way of avoiding the vortices hazard is to know where they are most likely to be encountered and act accordingly. Since vortices are not formed until lift is produced, they will not be generated by an aircraft taking off until just before lift off -- at the point where rotation is made. Vortices cease to be generated by a landing aircraft when its wings cease to produce lift -- when it has actually landed.

5. RECOMMENDED READING FOR ALL PILOTS. Your attention is invited to the Safety of Flight Section of the Airman's Information Manual which thoroughly and completely explains "Wake Turbulence." This article has been reprinted in FAA Advisory Circular AC 90-23A (which may be obtained free of charge from: Distribution Unit, HQ-438, Federal Aviation Administration, Washington, D. C. 20553). We recommend that all pilots read, study, and think through the contents of this "Wake Turbulence" article.

○○○○○

FAA Aeronautical Center
Flight Standards Technical Division
Operations Branch
P. O. Box 25082
Oklahoma City, Oklahoma 73125

Exam-O-Grams available free of charge in limited quantities from this office.



## CHECK YOUR READING

1. List the areas on the surface where we may expect to find engine wash.
2. What is the cause of wing-tip vortices?
3. Would we expect to find the most severe vortices behind a heavy fast aircraft, or a heavy slow aircraft?
4. When approaching for a landing, behind a heavy aircraft, what will our approach path look like?
5. When taking off behind another aircraft, with a crosswind, which side of the runway should we use?
6. When passing behind another aircraft, should we alter our flight altitude to one higher or lower than the altitude of the other aircraft?

# 9

# AIRCRAFT STABILITY

Contrary to common belief, the aircraft is an extremely stable platform. Aircraft do not "fall" out of the sky when left unattended in flight; in fact, they do a fine job of flying by themselves. There are officially recorded cases of aircraft taking off and flying for as long as four hours with no one at the controls, returning to earth only after running out of fuel!

Aircraft stability is not an accidental product of design. Many hours of engineering are devoted to producing an inherently stable machine. Unfortunately too many pilots spend a great deal of time fighting stability instead of letting it work for them.

Before learning why the aircraft is stable, we must discuss stability in general.

## Static Stability

*Static stability* is the initial tendency of any body to remain in its original position until disturbed. This can be demonstrated by placing a marble inside one half of a rubber ball as shown in Fig. 43. The marble will remain static (at rest) until disturbed. *Positive* static stability causes the marble to return to its original position if displaced. *Negative* static stability can be demonstrated by turning the section of the rubber ball upside down and balancing the marble on top, as shown in Fig. 43. Any displacement of the marble will cause it to move farther from its original position. *Neutral* static stability, also illustrated

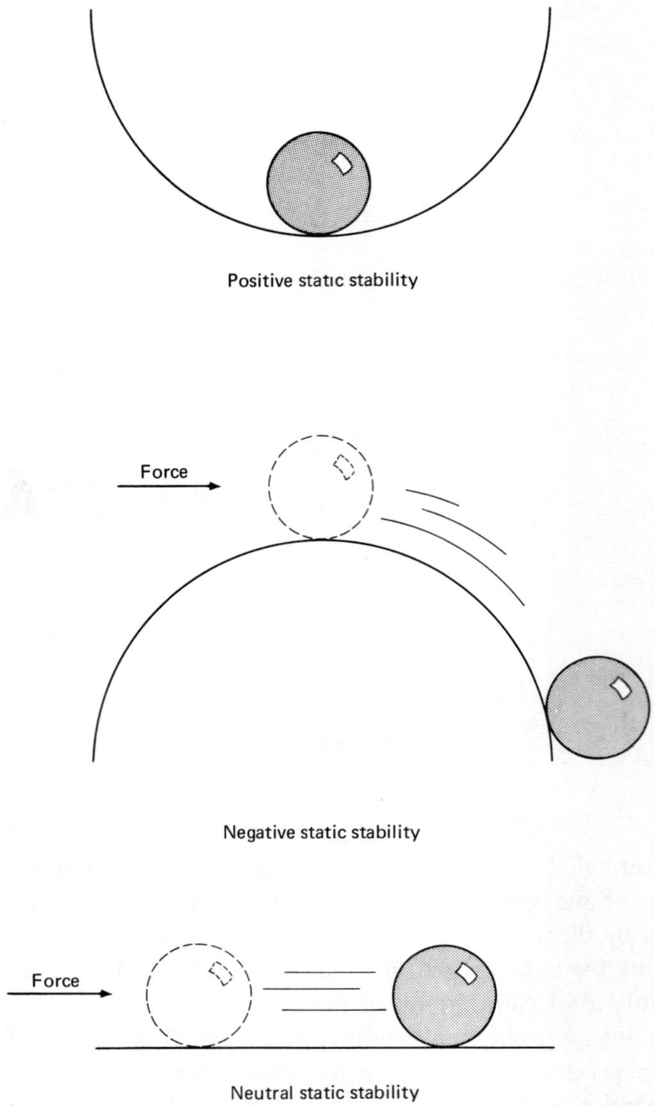


Fig. 43. Stability.

in Fig. 43, is demonstrated by placing a marble on a flat surface. When the marble is displaced, it will move to a new position and stop.

*Dynamic Stability*

The dynamic stability of a body determines its actions following the initial tendency. Positive dynamic stability tends to keep the marble in the center of the curved section of the rubber ball. A force must be applied to the marble to push it up the side of the curved container.

If the marble is forced up the wall of the container and then released, it will roll back toward the bottom, but will overshoot the mark, return, overshoot, etc. in ever-decreasing oscillations until it finally comes to rest in the center of the container. The ability to oscillate less and less on either side of the center before coming to rest is *positive dynamic stability*.

If we rock the curved container with our hand, making the marble oscillate in an equal amount side to side, we can demonstrate *neutral dynamic stability*. If we rock the container even more, we will see the marble oscillations increase in length until the marble is hurled out of the container. This demonstrates *negative dynamic stability*.

*Stability in Summary*

We have seen that a body has positive static stability if its original tendency is to return to its original position, after being displaced. It has positive dynamic stability if it oscillates in decreasing amounts until returning to its original position.

## Stability in Pitch

Now let's see how stability applies to our aircraft as far as pitch is concerned. Longitudinal stability (or pitch stability) is affected more by pilot-induced variables than the stability about either the yaw or roll axes. Aircraft loading is the greatest variable affecting longitudinal stability.

Refer to Fig. 44 which shows an aircraft in straight and level unaccelerated flight.

The wing of this aircraft is designed to produce 5,200 pounds of lift in level flight at cruise speed, which is 200 pounds more lift than is required to counteract gravity. To maintain level flight, 200 pounds of download must be produced by the tail. The horizontal stabilizer is attached to the aircraft at an angle which will produce downward acting lift of 200 pounds at normal cruise

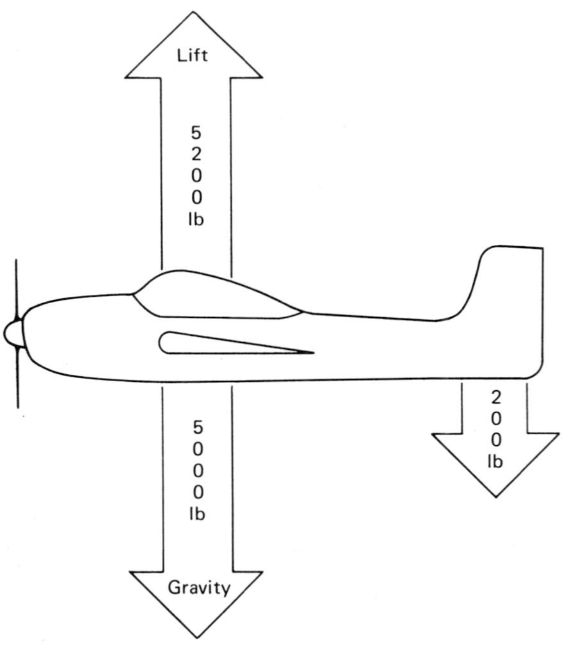


Fig. 44. Tail download.

speed and cruise power. Therefore, the 5,200 pounds of upward lift exactly equals the force of gravity (5,000 pounds) plus the tail download (200 pounds), so the aircraft will maintain level flight.

Suppose we trim the aircraft for straight and level flight. (We trim to adjust for small changes in speed or center of gravity location.) If we reduce the power, the aircraft slows down (because thrust is less than drag) which decreases the amount of lift produced by the wings. The slipstream from the propeller drops to near zero which reduces the tail download. The result of these two factors causes the nose to pitch downward with a resulting increase in airspeed. The tail download will still be less than normal due to the absence of propeller slipstream. The aircraft automatically demonstrates its inherent stability.

The properly loaded aircraft, designed with the center of gravity forward of the center of lift, and a tail download, always returns to a safe flight condition if displaced. Even if we apply back pressure on the control column, forcing the nose upward, it will pitch downward when we release the control pressure. This is because the airspeed (and hence the lift) will decrease (along with tail download) as the nose pitches up and the aircraft's weight will again pull the nose downward to a safe attitude.

Return now to straight and level "hands off" flight. For the next stability demonstration we will simply add power. The increased speed (thrust is greater than drag) and the increased tail download will force the nose to pitch upward, keeping the aircraft from overaccelerating.

One more pitch stability demonstration will certainly convince us of the inherent stability of the aircraft. Again this time we will start in straight and level flight, but we will not change the power setting during the demonstration. Forward pressure on the control column will cause the nose to pitch downward. When we release the pressure, the nose of the aircraft will begin an upward pitch change. Positive static stability will cause this pitch change, which will be to a nose-high attitude. Positive dynamic stability will then cause a series of oscillations, decreasing in length, until the aircraft settles at level flight attitude.

When the nose initially pitched downward the airspeed increased, creating more lift which pulled the nose up beyond the level flight attitude. Then the airspeed decreased, diminishing lift and causing a downward pitch change. Each oscillation became shorter until level flight attitude was maintained.

If we leave a properly trimmed aircraft alone, it will maintain its attitude quite well, requiring only minor guidance.

## Stability in Roll

The primary design factor used to create stability in roll is called *dihedral*. Figure 45 shows dihedral as the angle between the wing spar and the horizontal plane. Remember that lift is always considered to act perpendicular to the lateral axis and to the relative wind. If the aircraft is banked as shown in Fig. 45, the vertical component of the lift produced by the low wing is greater than that of the high wing. Hence, the low wing is lifted until the vertical component of both wings is equal.

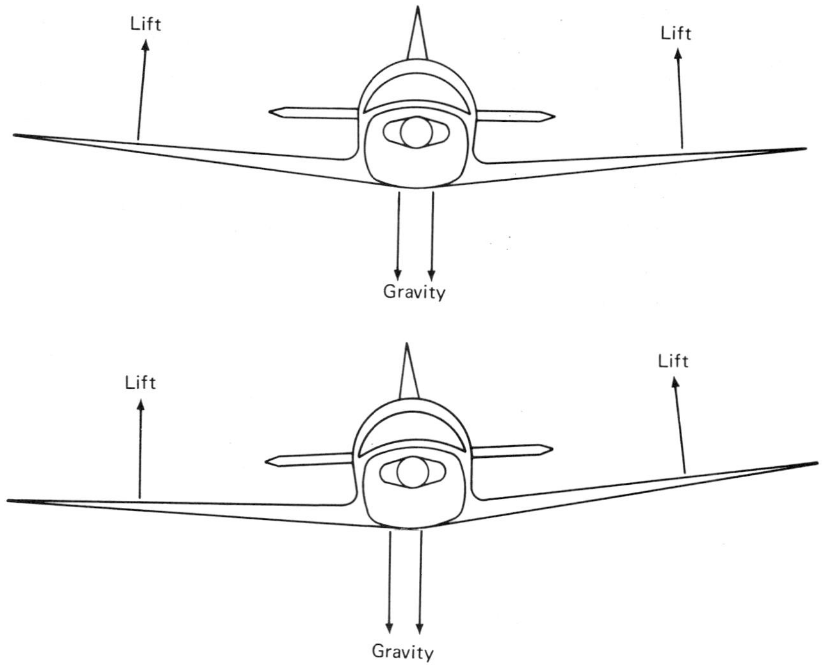


Fig. 45. Dihedral adds stability.

### *Roll Stability in Turns*

When in a shallow banked turn (5° to 20°), dihedral tends to return the aircraft to a wings-level attitude.

In a medium banked turn (20° to 35°), the effect of dihedral is balanced by another factor. That is, the wing on the outside of the turn actually travels farther and faster than the inside wing. Therefore, it creates more lift than the inside wing, which balances the effect of dihedral, and tends to keep the aircraft in a medium bank.

During a steep banked turn (more than 40°), the outside wing develops considerably more lift than the inside wing, overcoming the dihedral effect and causing the aircraft to overbank.

In summary, stability in roll or lateral stability tends to keep the aircraft in a wings-level attitude if a shallow bank is entered, tends to keep the aircraft at a constant medium bank, and is overcome in a steep banked turn.

## Stability in Yaw

The engineer designs the machine to have stability in yaw, or directional stability, by making sure that the center of the side area of the aircraft is aft of the center of gravity. He guarantees this by adding a fin or vertical stabilizer which acts like a ship's rudder in maintaining directional stability.

Some aircraft are built with *sweep back* to the wings. This design also adds to directional stability. Figure 46 will show us in an instant that the wing drag area on the right wing is less than that on the left wing (in a right yaw). The increased drag on the left wing will tend to slow it down, allowing the right wing to "catch up," thus contributing to directional stability.

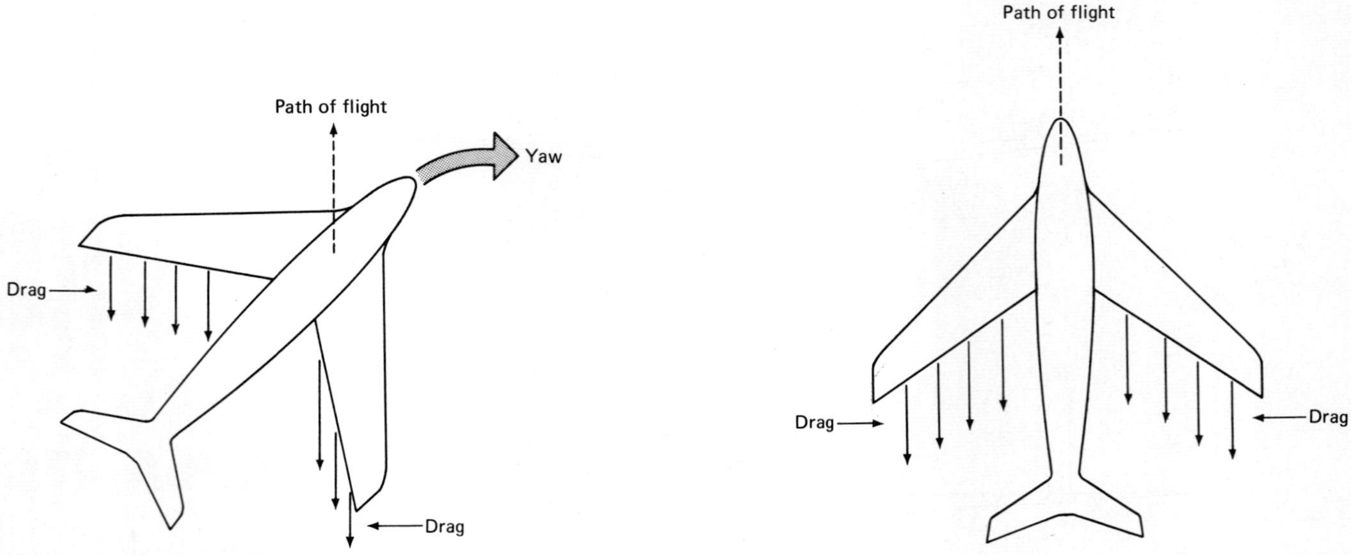


Fig. 46. Sweep back adds stability.

**Summary**

We have seen that the aircraft is designed to be stable. As our flight experience grows, we will be more convinced of this fact. Remember that inherent stability has helped many pilots out of trouble. If we overload our aircraft, or do not pay attention to properly balancing our aircraft, we will destroy its stability. If we keep our aircraft properly loaded and trimmed, we can let it fly "hands off." We do not need to keep the aircraft in the air, for it will do that very well by itself. All we need do is guide it with light pressures on the controls.

## CHECK YOUR READING

1. Define *static stability.*
2. Define *dynamic stability.*
3. Define *positive dynamic stability.*
4. Why is an aircraft longitudinally stable?
5. Why is an aircraft laterally stable?
6. Why is an aircraft directionally stable?
7. Why must an aircraft be properly loaded before flight?

# 10

# POWERPLANTS

## The Reciprocating Engine

The reciprocating engine is the type of powerplant which is used in automobiles, trucks, and trains. It is an internal combustion engine. Combustion means burning of fuel which produces heat and expanding gases. In an internal combustion engine, the burning of fuel is done within the powerplant, as opposed to a steam engine in which combustion occurs outside the engine and heated gas or steam is piped into the engine.

There are many types of internal combustion, reciprocating engines, each filling a specific need. For instance, the powerplant in a railroad locomotive may weigh 40,000 pounds, be 16 feet in length, and develop 1,800 horsepower. An 1,800-horsepower aircraft engine may weigh only 1,500 pounds and be only four feet long. At first glance we might say that the aircraft engine is much more efficient. However, the locomotive engine is designed to last for years, producing 1,800 horsepower at high speed. The aircraft engine on the other hand, may be designed to function for only 1,000 hours before major overhauling, and will operate most of its useful life at less than full speed. The aircraft engine manufacturer must sacrifice engine life in order to meet maximum weight specifications.

It is obviously unnecessary for a pilot to know the engineering and maintenance details of his aircraft engine. We will discuss only those areas in which a sound knowledge of basic operations will aid us in operating the powerplant efficiently.

## The Four-Stroke Engine

Aircraft powerplants are *four-stroke* internal combustion engines. That is, four distinct steps must take place to complete one power cycle. Two revolutions of the engine are required to accomplish all four strokes. Before we can discuss the engine operation, we must become familiar with the major parts. Figure 47 shows these parts. Study it carefully, noting the names of all components.

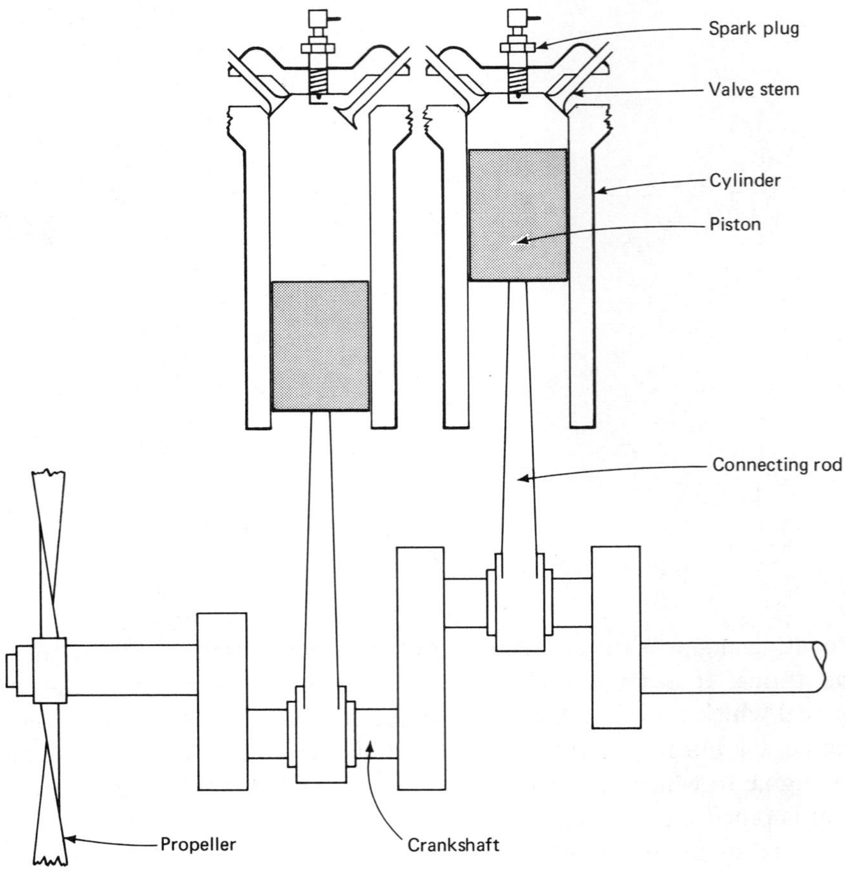


Fig. 47. Principle of the internal combustion engine.

### *The Intake Stroke*

The fuel–air mixture is prepared by the carburetor and is piped to the intake valve of the cylinders. During the intake stroke (Fig. 48, view 1), the piston moves downward, sucking the fuel–air mixture into the cylinder through the intake valve. The exhaust valve is closed, keeping the fuel–air mixture from escaping. When the piston reaches the bottom of the cylinder which is filled with the volatile mixture, the intake valve shuts.

### *The Compression Stroke*

Even though the mixture of fuel and air in the cylinder is burnable, more power will result from burning if it is first compressed. During the compression stroke (Fig. 48, view 2), both the intake and exhaust valves are closed. The piston moves up into the cylinder, compressing the gases to about one-seventh of their original volume. The fuel–air mixture is now ready for ignition.

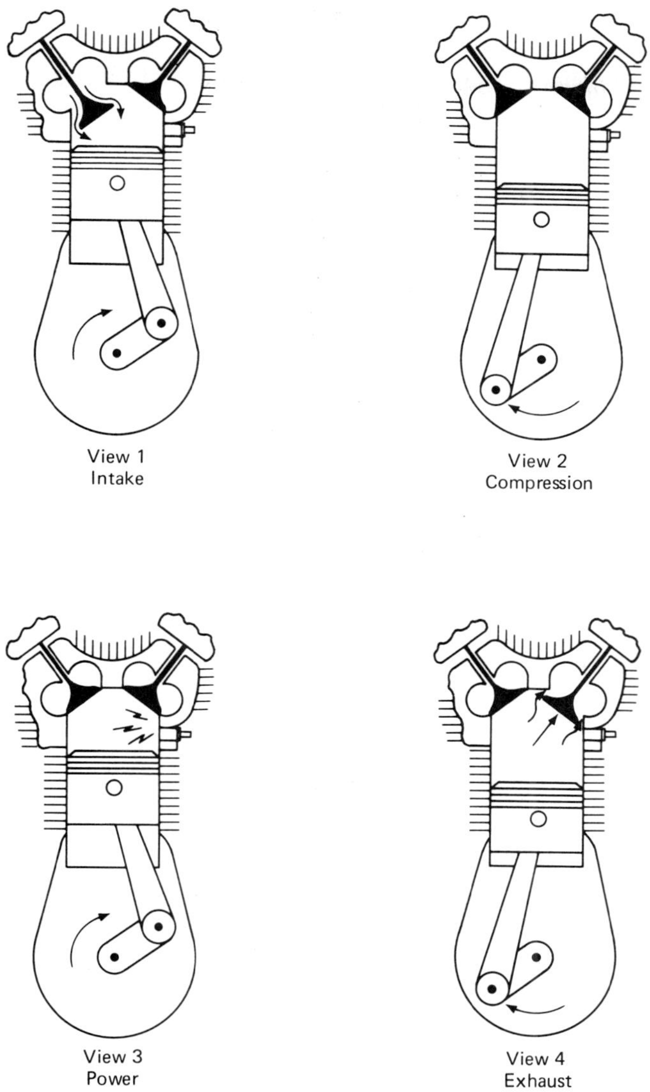


Fig. 48. The four strokes of the internal combustion engine.

### *The Power Stroke*

Just before the piston reaches the upper limit of the compression stroke, the spark plugs receive an electrical charge which causes a spark to jump across the tip. (See Fig. 48, view 3) The spark naturally ignites the highly combustible fuel-air mixture. Rapid burning of the mixture causes heat and tremendous expansion of the burning gases. Meanwhile, the piston has finished its upward stroke and is then forced downward by the expanding gases.

### *The Exhaust Stroke*

When the piston reaches the bottom of the downward stroke, the exhaust valve opens, so that the upward movement of the piston can push the burned gases out of the cylinder (Fig. 48, view 4). The engine has completed two revolutions at the end of this stroke, and the cycle begins again when the exhaust valve closes, the intake valve opens, and the piston begins its downward plunge, pulling another charge of fuel and air into the cylinder.

## The Crankshaft

The pistons are connected to the crankshaft of the engine. The engine is designed so that the power stroke of each piston occurs at the right time to keep a constant flow of power coming to the crankshaft. The crankshaft transmits some of the power to the propeller, while some is used to push the pistons up into the cylinders again in preparation for another power stroke. The propeller may be connected directly to the crankshaft, or a series of gears between the propeller and the crankshaft may be used to allow propeller rotation speed to vary from engine rotation speed.

The total power produced by the engine is affected by the number of revolutions per minute (and consequently the number of power strokes per minute) and the amount of fuel and air taken into the cylinders before ignition.

## The Ignition System

An aircraft ignition system is quite different than that found in an automobile engine. Actually, two complete sets of ignition are used, including wiring, spark plugs, and magnetos. *The ignition system is totally independent of the electrical system.*

### *The Magneto*

In order to keep the ignition system separated from the electrical system, a *magneto* is used to produce electrical energy for ignition. A magneto is simply an arrangement of wiring coils and magnets. We know that electricity is produced if a wire is passed through a magnetic field. Inside the magneto a wire is wrapped around a core many times and is rotated around a magnet. The magneto rotates because it is geared to the engine. When the engine rotates the magneto rotates. The electricity produced is stored up until just the right moment (the end of the power stroke) when it is released to one spark plug.

Although the aircraft magneto is quite reliable, two of them are used to decrease the possibility of ignition failure. The engine will operate on either magneto, but operates more efficiently when both magnetos are producing electrical charges for their respective spark plugs. Prior to flight, we will turn off each magneto in turn to make sure that the engine will operate smoothly, without too much power loss, on either one. If a magneto fails in flight, we will notice a slight power loss, but the other magneto will prevent engine failure.

### *Ignition Wiring*

Each magneto is wired to one set of spark plugs. The wiring must be *shielded* with heavy metal webbing buried inside the rubber insulation to keep the electrical charge from *leaking* into the aircraft radio equipment. (This leakage will cause loud crackles of static on the radio receiver.)

### *Spark Plugs*

The spark plugs used in an aircraft engine are quite similar to automotive plugs. As mentioned, there are two plugs in each cylinder head, for two reasons. First, if one plug fails, the fuel–air mixture in the cylinder will still be ignited by the remaining plug.

Second, the engine will run more efficiently with two plugs in each cylinder firing at the same time. This is because the fuel-air mixture will be ignited by both plugs, one on each side of the cylinder head. Hence, the fuel-air mixture will burn more evenly and quickly from each side of the cylinder toward the center. Rapid, even burning increases the amount of expansion of the burning gases and contributes to smooth power application to the piston.

## Detonation

If the fuel-air mixture explodes upon ignition instead of burning evenly, it is said to be *detonating*. The explosion or detonation of the compressed fuel-air mixture releases all of the energy of the mixture at once, which produces less power and more heat. Cylinder head temperatures will normally climb rapidly, while the power output declines.

## Preignition

Usually preignition will occur along with detonation. When the cylinder head reaches a temperature higher than the maximum desired, it acts as a spark plug; that is, the heated cylinder head is the ignitor of the fuel-air mixture. Preignition refers to the ignition of the mixture prior to regular ignition, before the piston reaches the desired position on its upward compression stroke. Thus a tremendous strain is placed on the piston and cylinder as the piston finishes its upward stroke, working against the force of the expanding gases. We will see a marked decrease in the power output, and should voluntarily reduce power and increase airspeed if possible (to increase cylinder head cooling). An immediate landing may also be necessary.

Preignition can also be caused by carbon deposits in the cylinder which glow with the heat produced from previous power strokes. These carbon particles act as ignitors, causing ignition of the compressing fuel vapors before regular ignition.

## Fuel Octane

The octane rating of fuel is a measure of its ability to be compressed without exploding upon ignition. If the octane rating of the fuel is too low, the heat produced by compression will prepare the fuel-air mixture for detonation instead of the desired even burning. We must be sure to use the recommended octane fuel, or a higher octane, to prevent detonation and preignition.

## Supercharged Engines

It stands to reason that the amount of power developed by an engine varies directly with the volume of fuel-air mixture which is packed into the cylinders.

To increase this volume, some engines are equipped with *superchargers* or air blowers. As the blower rotates, it feeds the cylinders with a high volume of fuel and air. At sea level, many supercharged engines are capable of delivering the combustible mixture to the cylinders at pressures fifty per cent greater than sea level pressure, or more. This system not only enables us to use much higher power settings at sea level, but it also allows us to keep the pressure at much higher altitudes. Without supercharging, the maximum pressure of the fuel-air mixture will be slightly less than the outside atmospheric pressure. As we gain altitude, the outside atmospheric pressure decreases, which means that the engine will produce less power. We can expect full power at much higher altitudes if our aircraft is equipped with a supercharged engine.

## The Propeller

The propeller is simply a rotating airfoil. If we cut the end from one of its blades, the cross section would look quite similar to the airfoil shape of the wing.

### *Propeller Speed*

The crankshaft applies the power developed by the pistons to the propeller. Lift is produced by the rotating airfoil, pulling the aircraft through the air. The faster the propeller turns, the more thrust will be developed. However, the speed at which the propeller can be rotated is limited by its length (and many other factors). If the propeller is turned too fast, the tip of each blade will travel faster than the speed of sound, overstressing the blade.

### *Propeller Pitch*

The *pitch* of the propeller also determines how much thrust it will develop. Propeller pitch is a measure of the distance it will travel through a solid in one full revolution. (Picture a wood screw being turned into a piece of wood. The distance the screw penetrates the wood in one revolution is called the pitch.)

If the propeller has too much pitch, or takes too large a "bite" of air, the angle of attack on the blade may be too high, and the blade will stall. The larger the bite, the more power will be required to turn the propeller.

### *The Fixed-Pitch Propeller*

Just as the aircraft is a compromise of many factors, so is the fixed-pitch propeller. A compromise must be made between large bites (high pitch) which is best for cruising flight, and smaller bites (low pitch) which is best for takeoff and climb performance. The propeller manufacturer must design the propeller to strike a happy medium of pitch. (Climb propellers, or cruise propellers, are also manufactured. These are not compromises, but are most efficient for either climb or cruise conditions.)

If the aircraft is equipped with a fixed pitch propeller, we will notice a decrease in rpm when the flight path is changed from straight and level flight to a climb (assume power output is unchanged) because the engine must work harder in the climbing flight path. Conversely, if we select a specific rpm in level flight, and then change the path of flight to a descent the propeller rpm will increase, because the engine does not need to work as hard in the descent.

## *The Variable Pitch Propeller*

More sophisticated aircraft engines are designed to utilize a propeller on which the pitch can be changed with a control in the cockpit. We can, with a variable pitch propeller, select the best pitch for any given flight condition. Pitch is indicated by the number of revolutions per minute of the propeller. With a constant power setting, a low pitch setting will result in higher rpm, because the propeller can take many "bites" if they are small ones. A high pitch setting will result in lower rpm because the propeller must take fewer big bites.

Propeller speed in revolutions per minute is indicated on the *tachometer* in hundreds of rpm and must be used in conjunction with the manifold pressure gauge to indicate the power output of the engine equipped with a variable pitch propeller.

## *Manifold Pressure*

The propeller pitch control determines the size of the bite the propeller takes, and hence the rpm. The throttle controls the amount of fuel and air entering the cylinders, which determines the amount of power transmitted to the propeller. Therefore, the combination of propeller speed and throttle setting is the true power indication.

Throttle setting is indicated on the manifold pressure gauge. The manifold is the piping system which brings the fuel–air mixture from the carburetor to the cylinders. If we measure the pressure in the manifold, we can tell how much of the combustible mixture is ready to enter the cylinders, and hence how much power will be developed.

## *Brake Mean Effective Pressure*

The *brake mean effective pressure* is a measure of the amount of pressure in the cylinder at the moment of combustion. If we select a high pitch (low rpm) setting on the propeller control, and at the same time select a high-power setting on the throttle, we may cause excessive pressure in the cylinders which may severely damage the engine. The high pressure develops because the propeller is turning too slowly to dissipate the power produced by the engine.

In order to prevent destructive pressure in the cylinders, we must allow the propeller to take many small bites (low pitch) so that the power being produced will be dissipated by the spinning propeller. When increasing power, we must increase rpm *prior to* increasing the throttle setting. When decreasing power, we must decrease the throttle setting *before* reducing propeller rpm.

## *Power Settings*

The Aircraft Operations Manual will indicate power settings (rpm and manifold pressure) for various flight conditions and altitudes. We will enjoy fuel economy and prevent engine damage by using only the recommended power settings for each flight condition.

# Propeller Governing

The rpm of the variable pitch propeller is controlled by the governor, which changes the propeller pitch. Engine oil pressure may be used to counteract the force of a strong spring, twisting the blade to a high pitch setting (low rpm). If the amount of oil pressure within the governor is reduced, the spring forces the propeller blade to a low pitch setting (high rpm).

We can select a desired rpm with the propeller control, and the governor will continually adjust the pitch of the propeller to maintain that rpm, even though we change the throttle setting or change the aircraft pitch attitude. The governor maintains the selected rpm by metering the amount of engine oil pressure which is to work against the spring. When the propeller tries to increase rpm, a set of counterweights in the propeller dome fly outward due to the increased centrifugal force. When the counterweights move, a valve opens, allowing more oil pressure to work against the spring, thereby increasing pitch and decreasing rpm to the desired setting. If the propeller tries to decrease rpm the counterweights move close to the propeller shaft, restricting the oil port, allowing the spring to move the propeller pitch to a higher rpm setting. Because of the continual working of the counterweights and spring, the rpm remains constant.

If the governor fails in flight, the spring forces the propeller into a low pitch (high rpm) condition, for high-power flight.

*Full-Feathering Propeller*

More sophisticated multiengine aircraft are equipped with *full-feathering* propellers. In the event of an inflight engine failure, the propeller blades are twisted until they are approximately 90 degrees to their normal operating pitch setting, thus decreasing the parasite drag on the unused propeller.

The governor for a full-feathering propeller is usually designed so as to automatically twist the propeller blades to a high rpm (low pitch) setting in the event of governor failure, and to the full-feathered position in the event of low engine oil pressure.

*Reversible Pitch*

The propellers of some multiengine aircraft are capable of providing reverse thrust. The engine, of course, always turns in the same direction, but the propeller blades twist past the low pitch limits into a reverse pitch setting. With this pitch, the rotating propeller produces a low pressure area behind the blade and reverse thrust is developed in order to slow the aircraft speed. Microswitches on the landing gear close when the weight of the aircraft settles on the landing gear struts, allowing the propeller control to be moved to the reverse pitch position. This arrangement prevents selection of reverse pitch in flight, which would end in disaster.

## The Turbine Engine

The turbine engine is also an internal combustion engine, but it is not a reciprocating type. It is comprised of four major sections: intake, compression, combustion, and exhaust. The great difference between a turbine and a reciprocating engine is that all four processes are continuous in the turbine, rather than occuring one after the other as in the reciprocating engine.

Thrust is developed because of Newton's third law, which states that for every action there is an equal and opposite reaction. The opposite reaction to high speed escaping exhaust gases is thrust.

Turbine engines develop much power for their size and operate most efficiently at high altitudes. Their useful life (time between major overhauls) may be five or ten times that of a reciprocating engine.

## CHECK YOUR READING

1. Describe each of the four strokes of the internal combustion reciprocating engine.
2. What arrangement in the aircraft ignition system keeps it totally independent of the electrical system?
3. Why are two spark plugs used in each cylinder of an aircraft engine?
4. What is detonation?
5. What is preignition?
6. Should we use a higher or lower octane than recommended if the proper octane fuel is not available?
7. List the results of preignition and detonation.
8. What are the advantages of a supercharged engine?
9. What is the primary indication of power output of an engine equipped with a fixed-pitch propeller?
10. What are the primary indications of power output of an engine equipped with a variable pitch propeller?
11. Describe the procedure to be used when increasing and decreasing power on an engine equipped with a variable pitch propeller.
12. List the primary differences between turbine engines and reciprocating engines.

# 11

# FUEL AND FUEL SYSTEMS

The fuel system in every specific model of aircraft is different from every other in one respect or another. Because of the importance of proper fuel management, we must study the fuel system diagram found in the Aircraft Operations Manual. In this text we will learn only the basic information concerning the most common systems, particularly in light aircraft.

## Aviation Fuel

Fuel used in aircraft reciprocating engines is quite similar to automotive fuel. It is handled and stored more carefully however, to prevent contamination. Light aircraft normally use 80/87 octane or 100 octane fuel. Heavier aircraft with more powerful engines require higher octane fuels. As we learned in Chapter 10, octane rating is a measure of combustibility, or how much compression the fuel-air mixture can withstand and still produce smooth power after ignition. Low octane fuels, if compressed too much, will either detonate upon ignition or will preignite from the heat of compression.

Aviation fuel is color coded to prevent confusion, which might lead to disaster, for it is imperative to use the recommended octane or a higher octane (never lower) for all flight operations.

Aviation 80/87 octane gasoline is tinted red. The next highest octane is 100/130, and is tinted green. More powerful engines will require 115/145 octane, which is purple.

## Fuel Contamination

Aircraft engines are remarkably reliable, but they will not run on water, dirt, or other contaminants. As pilots, we must exercise extreme caution to prevent contamination in the fuel system of our aircraft.

Fuel contamination can be caused by many factors. The most common contaminant is water. Just as dew forms in the early morning hours and covers the outside of the aircraft with tiny water droplets, so does condensation occur within the fuel tanks. Air in the tanks contains water vapor. If the temperature of that air, or of the tank, decreases, some of the water vapor will condense on the sides of the tank and drip into the fuel. Water is heavier than gasoline, so it falls to the bottom of the tank.

The best way to prevent condensation in the fuel tank is to fill it with fuel prior to storing the aircraft overnight. In this way, most of the air will be forced out of the tank, drastically reducing the amount of condensation.

Other sources of water contamination are improper fuel storage facilities (particularly small tanks or drums) and improperly filtered fuel. The chances of being fueled with contaminated fuel are decreased by frequenting standard brand dealers who have underground storage facilities.

Fuel may also be contaminated with dirt, pipe-fitting compound, bits of rubber from the inside of worn fuel hoses, etc. Filtering at the nozzle and within the aircraft fuel system should eliminate most of this debris.

### *Preflight Action to Fight Contamination*

Every aircraft is equipped with fuel sumps situated at the lowest point in each fuel tank. Also a main gasculator drain is located at the lowest point in the fuel system plumbing. Since water does not mix with gasoline and is heavier than it, we can drain any accumulated water in the system by opening each sump drain and the main gasculator drain and allowing a *generous* sample of fuel to collect in a transparent container. We should examine the container for dirt and water. If contamination appears, more fuel must be drained until we are positive that the fuel is clean.

Periodically, the tank strainers, inlet carburetor screens, and carburetor bowl should be carefully cleaned.

Recent tests conducted by the Federal Aviation Administration resulted in some alarming facts coming to light. First, it was proved that most pilots do not drain an adequate amount of fuel from the sumps and gasculator bowl during the preflight inspection. In some of the test cases, as much as a quart of fuel had to be drained before the water which had purposely been introduced to the system appeared.

A second alarming fact came from these tests. Most pilots did not realize that the gasculator petcock drained fuel *only* from the tank which had been selected on the cockpit tank selector. In order to drain all tanks through the gasculator, the tank selector must be switched to each tank in turn.

## The Fuel System

A schematic diagram of a basic light aircraft fuel system is shown in Fig. 49. In high wing aircraft, the fuel flows by gravity through the tank selector and

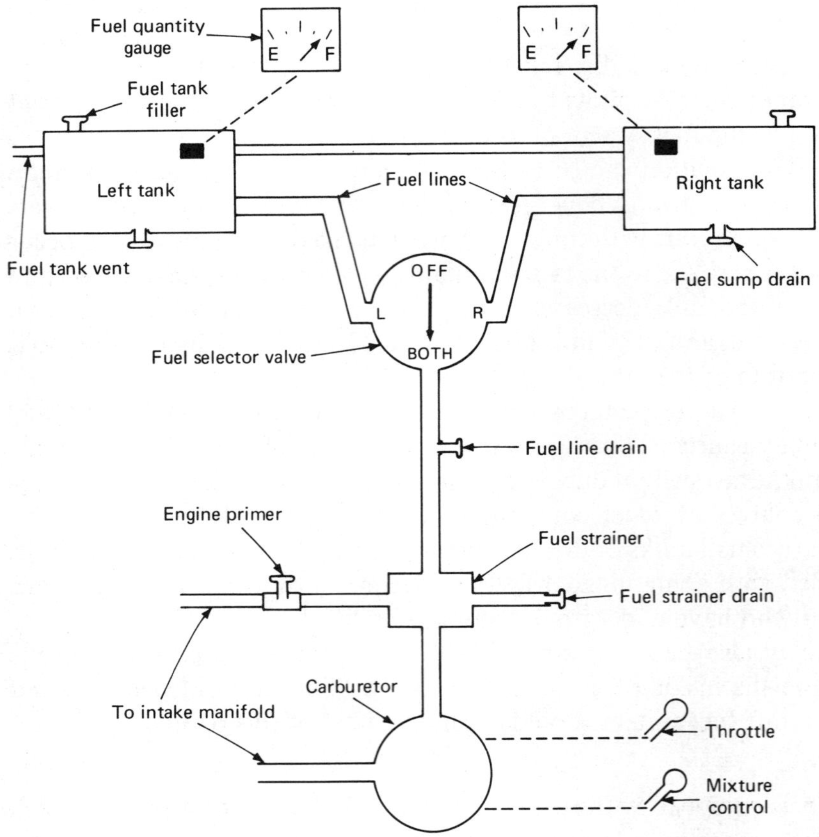


Fig. 49. Typical fuel system schematic.

filtering screens into the carburetor. Low wing aircraft are equipped with electric fuel boost pumps which lift the gasoline from the tank and push it towards the carburetor.

We must become very familiar with the fuel system in every aircraft we intend to fly, checking fuel capacity, on-off selectors, tank selectors, and the location of sump and gasculator drains.

Fuel quantity gauges are now installed in just about all aircraft. However, the wise pilot will check the fuel quantity *visually* prior to the flight, and compute the fuel required, to avoid fuel starvation aloft.

## The Carburetor

The great majority of reciprocating aircraft engines are equipped with carburetors, although fuel injection systems are growing in popularity.

The carburetor is simply a metering and mixing device. Outside air is taken into the throat of the carburetor, where fuel is mixed with it in the form of vapor. The fuel-air mixture is sent through the intake manifold to the cylinders.

Figure 50 shows a cross-section view of a carburetor. If we look closely at the area where the throttle valve is located, we will see that the walls of the carburetor bulge inward. In fact, the cross section view of the walls look very much like the camber of an airfoil. This is called the *venturi* area.

When incoming air passes the venturi section it increases its speed and a

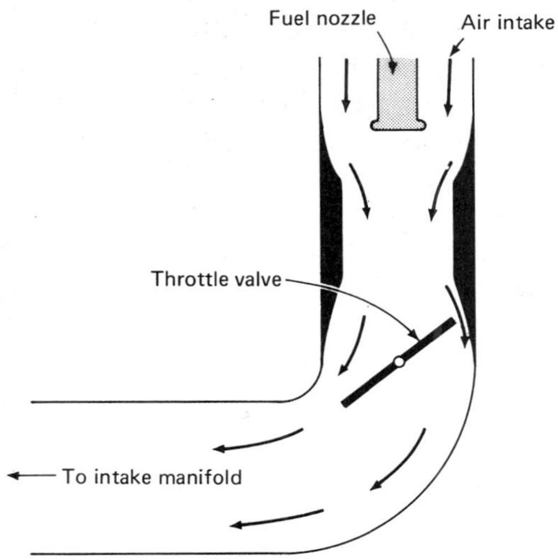


Fig. 50. The carburetor.

low pressure area is produced (similar to the production of a low pressure area above the camber of the wing). The low pressure air allows the gasoline to be more easily vaporized into the airstream.

The throttle valve is a metal plate which can be adjusted from the cockpit to restrict the opening to the venturi, allowing more or less air to pass through the carburetor. The greater the volume of air ingested, the more fuel will be vaporized.

## Carburetor Ice

Although the carburetor is a simple, reliable device, it has one major drawback. During flight in certain atmospheric conditions, ice may form on the throttle plate or on the walls of the venturi. We must learn when to expect carburetor ice formation, so that we can prevent inflight engine failure.

### *Atmospheric Conditions Most Conducive to Carburetor Ice Formation*

Although every carburetor is susceptible to ice in different atmospheric conditions, we can learn the general conditions which most often result in rapid ice formation. The exact temperature required for ice formation in any specific carburetor may not fall within the ranges mentioned here.

Suprisingly, atmospheric temperature must be quite warm; between 35 and 80 degrees Fahrenheit. How can ice form in such warm temperatures? When fuel is vaporized it draws heat from its surroundings. We can prove this by dipping a finger into gasoline, and waving it about. It will immediately feel cool, for as the fuel evaporates it takes heat from the finger. This heat loss in the carburetor is enough to cool the warm air to below-freezing temperatures.

The atmosphere must also be humid. Warm air can contain more water vapór than cold air. As the warm, moist air is cooled, the water vapor is condensed and freezes.

As we have learned, carburetor ice is apt to form rapidly when we least expect it, during warm humid days.

If the outside air temperature is cold, there will be less moisture available to condense, and the temperature drop in the venturi section will bring the temperature far below freezing. Normally, snow will form in these conditions, which will blow harmlessly through the carburetor.

*Effect of Carburetor Ice*

When ice forms in the carburetor it collects on the walls, restricting the airflow, just as the throttle plate restricts the airflow. As a result, the volume of fuel-air mixture delivered to the cylinders is decreased. The resulting power loss is slow but steady until the engine fails from fuel starvation.

Cockpit indications are identical to a slow reduction of throttle. If the aircraft is equipped with a fixed-pitch propeller, we will see a slow decrease in rpm on the tachometer. However, a controllable pitch propeller will maintain rpm, so we must monitor the manifold pressure gauge for indications of a power loss.

If carburetor ice accumulation is severe, the throttle plate may be frozen in position, making any power change impossible.

*Combatting Carburetor Ice*

Fortunately, the carburetor is equipped with a hot air duct which, when we select carburetor heat with the cockpit control, delivers warm air to the carburetor. (Warm air is collected by a heater muff which surrounds the hot exhaust pipes.) When hot air is ingested by the carburetor, a small power loss will result, because warm air is less dense than cooler air and will produce less power. Then, as the ice melts, the engine power output will increase, indicating that the carburetor wall is losing its restriction. When we shut off the heat we will regain the power which was lost due to the ingestion of warmer air.

The carburetor heat will usually be adequate to melt small amounts of ice but it will not always melt away large quantities of already formed ice. Hence, we should apply the heat if carburetor ice is suspected, or at the first sign of ice formation. If the available heat is insufficient to melt the ice, we can create extra heat by increasing the power, and slowing the aircraft by climbing, extending the landing gear, or extending the flaps. (This will diminish engine cooling.)

Preheating the intake air not only reduces engine power but also increases operating temperature. Therefore we should not use the carburetor heat indiscriminately, particularly on hot, dry days, or on takeoff (when full power is required) unless conditions warrant its use. The Aircraft Operations Manual will list any special operating instructions for the carburetor heat.

## Fuel Injection

Some modern aircraft engines incorporate a fuel injection system instead of carburetion. On an engine which is fuel injected, the fuel is piped directly to each cylinder and is vaporized there. Naturally, if the engine is not equipped with a carburetor, we need not worry about carburetor icing. The advantage of ice-free operation is somewhat offset by complicated metering devices which are employed.

We must study the Aircraft Operations Manual carefully to learn the peculiarities of the fuel injection system in each aircraft.

## Mixture Control

One of the most important items in the correct operation of aircraft engines is proper mixture control. Normally the carburetor is calibrated for operation at sea level. That is, the correct mixture of fuel and air will be obtained with the mixture control at the full rich setting when flying at sea level.

As we gain altitude, the atmosphere becomes less dense; that is, it weighs less per given volume. The volume of air passing through the carburetor will remain the same. The quantity of fuel metered by the carburetor depends on the *volume* of air, not the weight. Therefore the amount of fuel used will remain the same even though a lesser weight of air is consumed, causing the fuel–air ratio to become richer as altitude increases.

The fuel–air ratio must remain within certain limits to produce smooth engine operation, most power for a given throttle setting, best range, and long engine life. Gasoline engines require about 15 pounds of air in order to consume 1 pound of fuel. But this perfect mixture may not be desirable for all flight conditions. (For instance, when the engine produces full power, a rich mixture may be required to aid in cooling the cylinders.)

The mixture control compensates for decreased air density by metering the amount of *fuel* passing through the main fuel jet in the carburetor. In the less dense air aloft, the mixture control reduces the fuel flow to maintain the proper fuel–air ratio.

If the fuel–air ratio is too lean (not enough fuel) the cylinders will overheat causing preignition and detonation. Light aircraft engines should never be operated with a mixture which is too lean. If the mixture is too rich (too much fuel) the engine will run rough. It is impossible to design an engine manifold system which will deliver exactly the same fuel–air mixture to every cylinder. Therefore, we may adjust the mixture control to a properly lean setting for all but one of the cylinders, which may overheat until it glows red with heat. For this reason light aircraft engines should always be operated with a slightly rich fuel–air ratio.

## Leaning Procedure

Leaning is normally most effective at altitudes above 5,000 feet. We must always follow the manufacturer's recommendation on leaning for each particular engine. Although the use of the mixture control varies with pilot technique, one generally accepted method is to carefully begin leaning after adjusting the throttle and propeller control to the desired power setting. As the fuel–air ratio becomes leaner, the power output will increase slightly. We may continue leaning until the first indication of power decrease, and then richen the mixture slightly, to return to the "rich side" of best power. Accurate cylinder head gauges may also be used to indicate the proper fuel–air mixture setting. In this instance the mixture is leaned until the manufacturer's specified increase in cylinder head temperature is noted.

We must be sure to return the mixture control to the full rich position before descending below 5,000 feet, because the engine will need extra fuel to develop full power at low altitudes. Failure to do so may cause engine failure at a low altitude.

### Cruise Control

The range of our aircraft will depend to a great extent on the power settings we use and on proper mixture control. The sample cruise control chart in Fig. 51 will show the great difference in fuel consumption between maximum endurance (time aloft) and normal cruise power settings. The cruise control chart should be used for every flight in order to benefit from maximum efficiency.

Cruise performance – lean mixture
(22.0 gal usable fuel)

| Altitude | RPM | % BHP | TAS (mph) | gal/hr | Endurance hours | Range (miles) |
|---|---|---|---|---|---|---|
| 2,500 | 2,700 | 84 | 123 | 6.3 | 3.5 | 430 |
| | 2,650 | 80 | 120 | 6.0 | 3.7 | 444 |
| | 2,550 | 71 | 114 | 5.3 | 4.1 | 467 |
| | 2,450 | 63 | 109 | 4.7 | 4.7 | 512 |
| | 2,350 | 56 | 103 | 4.2 | 5.2 | 536 |
| | 2,250 | 51 | 98 | 3.9 | 5.6 | 549 |
| | 2,150 | 46 | 93 | 3.6 | 6.1 | 567 |
| 5,000 | 2,700 | 79 | 122 | 5.9 | 3.7 | 451 |
| | 2,650 | 72 | 119 | 5.4 | 4.1 | 488 |
| | 2,550 | 66 | 113 | 4.9 | 4.5 | 508 |
| | 2,450 | 59 | 108 | 4.5 | 4.9 | 529 |
| | 2,350 | 53 | 102 | 4.0 | 5.5 | 561 |
| | 2,250 | 48 | 97 | 3.7 | 5.9 | 572 |
| | 2,150 | 44 | 91 | 3.4 | 6.5 | 591 |
| 7,500 | 2,700 | 73 | 121 | 5.4 | 4.1 | 496 |
| | 2,650 | 68 | 118 | 5.1 | 4.3 | 507 |
| | 2,550 | 62 | 112 | 4.6 | 4.8 | 538 |
| | 2,450 | 56 | 107 | 4.2 | 5.2 | 556 |
| | 2,350 | 50 | 101 | 3.8 | 5.8 | 586 |
| | 2,250 | 46 | 97 | 3.6 | 6.1 | 592 |
| | 2,150 | 43 | 92 | 3.4 | 6.5 | 598 |
| 10,000 | 2,730 | 67 | 119 | 5.0 | 4.4 | 524 |
| | 2,650 | 64 | 117 | 4.8 | 4.6 | 538 |
| | 2,550 | 58 | 111 | 4.4 | 5.0 | 555 |
| | 2,450 | 53 | 106 | 4.0 | 5.5 | 583 |
| | 2,350 | 48 | 101 | 3.7 | 5.9 | 596 |
| | 2,250 | 45 | 97 | 3.5 | 6.2 | 601 |
| | 2,150 | 42 | 94 | 3.3 | 6.7 | 630 |

Notice the fuel economy realized when the power is reduced. Although the airspeed decreases, the nonstop range becomes greater with reduced power.

Fig. 51. Cruise control chart.

FEDERAL AVIATION AGENCY
VFR EXAM-O-GRAM *NO. 10

FUEL CONTAMINATION

EXCERPTS FROM A RECENT AIRCRAFT ACCIDENT REPORT: ". . . Subsequent examination of the engine and its components revealed large deposits of foreign material, sediment, and water in the fuel strainer, carburetor bowl, and fuel pump in sufficient quantities to cause stoppage. . . Probable cause of accident: Inadequate preflight action by the pilot; subsequent engine failure due to fuel contamination. . . ."

DO YOU KNOW -- AND PRACTICE -- THE PRECAUTIONS YOU SHOULD TAKE TO AVOID FUEL CONTAMINATION? Perhaps you do, but there are many pilots who obviously do not -- as evidenced by the alarming increase in the number of fuel contamination-caused accidents. The modern aircraft engine is a remarkably reliable and dependable mechanism, but it will not run on water, dirt particles, and other non-combustibles. Let's review this insidious problem by asking -- and answering -- a couple of rather pointed questions about this subject.

1. WHAT CAUSES FUEL CONTAMINATION?

   A. Storing the aircraft with partially filled fuel tanks may cause condensation and water contamination. You have, no doubt, often noticed moisture (or dew) on the outside of your aircraft early in the morning. When you noticed this, did it occur to you that this same moisture could form on the inside walls of your fuel tanks? Water is the worst offender in these contamination cases, and condensation inside the tank is one of the methods by which it finds its way into your fuel system.

   B. Servicing the aircraft from improperly filtered tanks, particularly small tanks or drums, is another principal source of fuel contamination. This practice frequently introduces both dirt and water into the aircraft fuel system.

2. WHAT PRECAUTIONS SHOULD THE PILOT TAKE TO AVOID FUEL CONTAMINATION?

   A. PREFLIGHT ACTION: Drain a generous sample of fuel (several ounces -- not just a trickle or two) into a transparent container from each of the fuel sumps. (Notice that we specified each of the fuel sumps. This includes not only the main gascolator, but also the wing tank sumps.) Examine the sample of fuel from each sump for water and dirt contamination. Water will not mix with gasoline. If present, it will collect at the bottom of the transparent container and will be easily detected. If water or dirt appears, continue to drain from that sump until you are sure the system is clear of all water and dirt. (NOTE: If your aircraft is not equipped with wing tank quick-drain petcocks, it is recommended they be installed. This can make the preflight check of the wing tank sumps much more convenient, as the frequent removal and replacement of wing tank sump drain plugs can be a time-consuming operation.)

   B. POST-FLIGHT ACTION: (1) Top off your tanks at the end of the day to avoid condensation and water contamination inside your fuel tanks. Although this is a desirable procedure to follow at the end of each flying day (assuming your loading schedule for the next day will permit a full load of fuel), it is particularly important that this is done if the aircraft is to stand idle for several days -- whether it is tied down out-of-doors or stored in a hanger. (2) Avoid, if possible, servicing your aircraft from small tanks or drums. Should this become necessary, the fuel should always be strained through a chamois skin that is in good condition.

   C. PRECAUTIONARY MAINTENANCE ACTION: In addition to the previously discussed precautions, the following maintenance precautions should be performed on your aircraft at periodic intervals: (1) Inspect and clean the tank fuel outlet finger strainer. (2) Inspect and clean the inlet carburetor screen. (3) Flush the carburetor bowl.

**BY FOLLOWING ALL OF THESE PRECAUTIONS, YOU CAN GREATLY REDUCE THE HAZARD OF ENGINE FAILURE DUE TO FUEL CONTAMINATION.**

* * * * * * *

To better understand the reasons for the PREFLIGHT ACTION we have recommended, let's take a brief look at an actual water contamination test recently conducted by FAA -- it's a real eye opener!

After all water was removed from the fuel system of a popular make high-wing monoplane, three gallons of water were added to the half-full fuel tank. After a few minutes, the fuel strainer (gascolator) was checked for water. It was necessary to drain ten liquid ounces of fuel before any water appeared. This is considerably more than most pilots drain when checking for water.

In a second test with the same aircraft in flying attitude (to simulate a later tricycle geared model) the fuel system was again cleared of all water; then one gallon of water was added to the half-full tank. Upon checking the fuel strainer (gascolator) quick drain, more than a quart of fuel was drained before any water appeared.

In both of the above-described tests, about nine ounces of water remained in the fuel tank after the belly drain and the fuel strainer (gascolator) had ceased to show any trace of water. This residual water could be removed only by draining the tank sumps.

Two significant findings emerged from the above tests and from tests made on a plastic mockup of a similar fuel system:

1- When water was introduced into the fuel tank, it immediately settled to the bottom, but did not flow down the fuel lines to the fuel strainer until all fuel was drained from the lines. Remember, each fuel tank must be turned ON to drain the tank lines through the gascolator.

2- Since it was found impossible to drain all water from the tank through the fuel lines, it was necessary to drain the fuel tank sumps in order to remove all water from the system.

* * * * * * *

Department of Transportation

FEDERAL AVIATION ADMINISTRATION

## VFR PILOT EXAM-O-GRAM° NO. 38

MIXTURE CONTROL--FUEL/AIR RATIO

It is generally conceded by most leading aircraft engine manufacturers that correct use of the mixture control in flight for adjusting the fuel/air ratio (F/A) is one of the most important items in the operation of aircraft engines. This Exam-O-Gram explains some of the related factors which should be considered when leaning the fuel/air mixture. It is hoped that this brief discussion will serve as a stimulus for pilots to study and search for more information on this subject--in aircraft power plant manuals--and especially the engine manual pertinent to the aircraft they are operating. Certain General Aviation Written Examinations contain test items which are concerned with the results of improper use of mixture control.

The mixture control knob in an aircraft cockpit is usually RED--an indication that it should be used with "caution." Proper leaning of the mixture provides smoother engine operation, more power for a given power setting, best range and endurance; on the other hand, misuse of the mixture control can soon ruin an aircraft engine.

WHAT DOES FUEL/AIR (F/A) RATIO MEAN? It is the ratio between the weight of the fuel and the weight of the air that goes into the cylinders. In general, gasoline engines require approximately 15 pounds of air in order to completely burn 1 pound of gasoline. However, a theoretically perfect mixture ratio is not essential or desired in all cases. Certain conditions may require the use of mixture either richer or leaner than this average ratio. Usually, the useful mixture ratios are between 1 to 11 and 1 to 16. Fuel and air proportions are expressed on the basis of weight rather than volume. Fuel/air ratios may be given as a direct ratio, such as 1 to 12, but in more common usage, they are designated as decimal fractions such as 0.083. For example: $1 \div 12 = 0.083:1$ (0.083 lb. of fuel to 1 lb. of air).

WHAT IS THE FUNCTION OF THE CARBURETOR? It measures the correct quantity of fuel to be supplied to the engine, ATOMIZING and MIXING the fuel with air in the correct proportion (F/A ratio) before the mixture enters the cylinders. This proportioning must be done correctly regardless of the speed, load, and altitude at which the engine is operating. Gasoline cannot ignite or burn when in the liquid state, it first must be vaporized and mixed with the correct amount of air before it can be ignited and combustion takes place. When compared to other gasoline engines, aircraft engines operate at a greater altitude range and therefore are equipped with manual and/or automatic mixture controls.

HOW ARE CARBURETORS NORMALLY CALIBRATED? They are calibrated for sea-level operation, which means that the correct mixture of fuel and air will be obtained at sea level with the mixture control in the "full rich" position. As we climb to higher altitudes, the air density decreases--that is, a cubic foot of air will not weigh as much as it would at a lower altitude. Therefore, the weight of air entering the carburetor will decrease, although the volume remains the same. The amount of fuel entering the carburetor depends on the volume of air and not the weight of air. As the altitude increases, the amount of fuel entering the carburetor will remain approximately the same for any given throttle setting. Since the same amount (weight) of fuel is entering the carburetor, but a lesser amount (weight) of air, the fuel-air mixture becomes richer as altitude increases.

WHAT DOES THE MIXTURE CONTROL DO? It compensates for the decreased air density by metering the amount of fuel that passes through the main jet in the carburetor. In the less dense air at higher altitudes, a leaner mixture reduces fuel consumption and provides smoother engine operation. The mixture control is used to reduce the amount of fuel flow and maintain the proper F/A ratio--this is also true of engines with fuel injection.

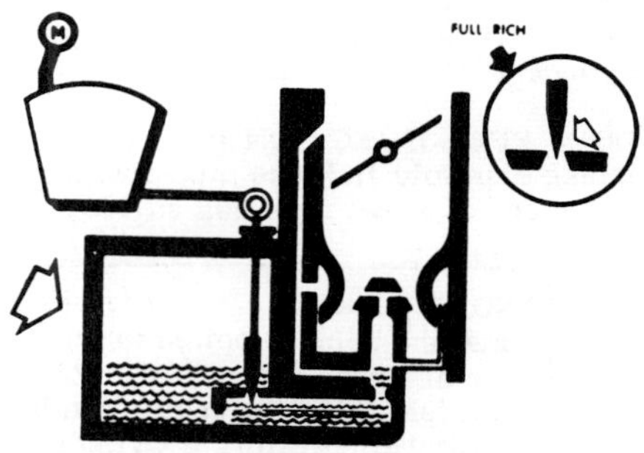

Needle Type Mixture Control System



DISTRIBUTION: ZC-307

AT WHAT F/A RATIO DO THE HIGHEST CYLINDER TEMPERATURES OCCUR? The greatest heat occurs at a fuel to air ratio of 1 to 15, or .067, and this is known as stoichiometric mixture (pronounced like stoy-key-o-metric), which is the chemically correct mixture where all the air and all the fuel is burned. Even though an F/A ratio of .067 is considered a chemically correct mixture for combustion, it produces peak temperatures and this is generally the mixture which will cause all gasoline engines to run the hottest (see Illustration below).

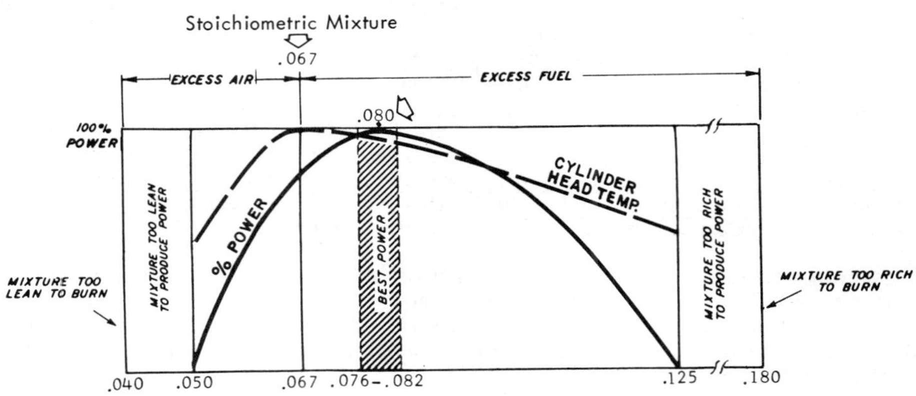


Fuel-Air Ratios vs Power & Temperature

The mixture of .067 is a theoretical point that can be demonstrated only on a single cylinder engine in a laboratory. In engines with more than one cylinder, the variations in fuel distribution between the cylinders makes it difficult to evaluate the F/A ratio in each cylinder. This matter of distributing equal amounts of fuel and air to the various cylinders is one of the greatest problems facing the aircraft engine manufacturers--or the designer of any gasoline engine for that matter. Because of the unequal Fuel/Air ratio in the various cylinders, the pilot who practices using extremely lean mixture settings without reference to proper instrumentation can experience a situation where all cylinders on his engine are operating at normal temperatures--except for one hot cylinder, where the exhaust valve and seat are red hot.

WHAT IS THE "BEST POWER" MIXTURE? The .080 fuel/air ratio is known as the "Best Power" mixture and it is that ratio at which the most power can be obtained for any given throttle setting. "Best Power" mixture is the fuel/air ratio where we can get a given power with the lowest manifold pressure or throttle setting. (See Illustration above.)

WHAT DOES EXCESS AIR AND EXCESS FUEL MEAN IN THE ILLUSTRATION? The illustration shows "Excess Air" on the left side of the .067 mixture, which means there is more air in the cylinders than is needed for normal combustion, and this excess of air absorbs heat and helps to cool the engine. On the right side of .067 mixture, we have "Excess Fuel" which means that there is more fuel in the cylinders than is needed for normal combustion, and this "Excess Fuel" also absorbs heat and provides additional cooling. Large supercharged engines can operate in the "Excess Air" lean mixture side of the .067 mixture, whereas the carburetor equipped, unsupercharged small aircraft engine should never be leaned to this extent. For example: If the manual mixture control of a supercharged engine is moved toward the lean position, cylinder head temperatures will be greatest when the F/A ratio is .067 and as the mixture is leaned still further, cylinder head temperatures will return to cooler normal values. When cylinder head temperatures climb too high while leaning carburetor equipped, unsupercharged engines, the mixture must be richened in order to return to cooler head temperatures.

WHAT IS MEANT BY UNEVEN MIXTURE DISTRIBUTION? In a carburetor equipped engine, the intake manifolds and induction pipes are used to distribute the fuel and air charge to the various cylinders. Those cylinders which are the farthest from the carburetor often receive a slightly leaner mixture than those cylinders close to the carburetor. When the mixture control is used to lean the mixture, the cylinders which are already receiving a leaner mixture will be the first ones to run hot or misfire.

DOES FUEL INJECTION PROVIDE BETTER FUEL DISTRIBUTION? Yes, the fuel is injected into the intake manifold and it is mixed with air just before entering the cylinders. Theoretically, all of the cylinders of a fuel injection engine are receiving an equal amount of fuel.

> NOTE: Fuel injection engines are equipped with Fuel Flow Gauges to indicate the F/A mixture being supplied to the engine. Some of these instruments also show the percentage of power being used. Proper mixture control and better economy in the operation of a fuel injection engine can be achieved best through the use of an Exhaust Gas Temperature Indicator, a Cylinder Head Temperature Gauge and an Oil Temperature Gauge. The two latter instruments have slow response times but the trend of these basic heat references are very meaningful.

WHAT ARE THE RESULTS OF HAVING THE MIXTURE TOO LEAN? When the mixture is too lean there is too little fuel for the amount of air--in terms of weight. Rough engine operation, sudden "cutting out" or "back firing," detonation, overheating, or an appreciable loss of engine power may occur. Lean mixtures must be avoided when an engine is operating near its maximum output.

AT WHAT ALTITUDE IS LEANING THE MIXTURE NORMALLY EFFECTIVE? Leaning is normally effective above 5,000 feet; however, some aircraft engines may be leaned below 5,000 feet. Always follow the manufacturer's recommendations on leaning the fuel mixture for the particular airplane. By leaning the mixture at too low an altitude or leaning the mixture excessively, you could damage the engine at a high power setting. For example: Suppose that a pilot had been cruising at 8,000 feet with a lean mixture and forgot to move the mixture to full rich before entering the traffic pattern of a low elevation airport. The pilot may experience a rough engine or the engine might "cut out" or even worse if he were to exceed approximately 70% power in the pattern or on a go-around he would be in serious trouble with detonation and engine overheating. In general, lean mixtures must be employed with caution when operating aircraft engines at high power settings.

WHY IS 5,000 FEET CONSIDERED A SAFE ALTITUDE FOR LEANING? Certain aircraft engine manuals state that their engines should not be leaned below 5,000 feet. At 5,000 feet the unsupercharged engine is capable of developing only about 75% of its rated power, and at less than 75% power it is much harder to get into trouble using improper leaning techniques, since the cylinders and other engine parts are operating at lower temperatures.

WHAT IS DETONATION? Detonation is the spontaneous explosion of the unburned charge (in the cylinders) after normal ignition. If the temperature and pressure of the unburned portion of the fuel-air charge reach critical values, combustion will begin spontaneously. The result is a sudden and violent explosion of the charge (detonation) rather than the relatively slow burning of normal combustion.

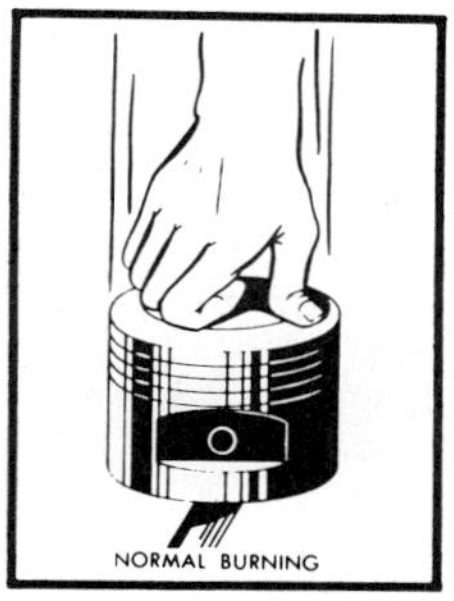


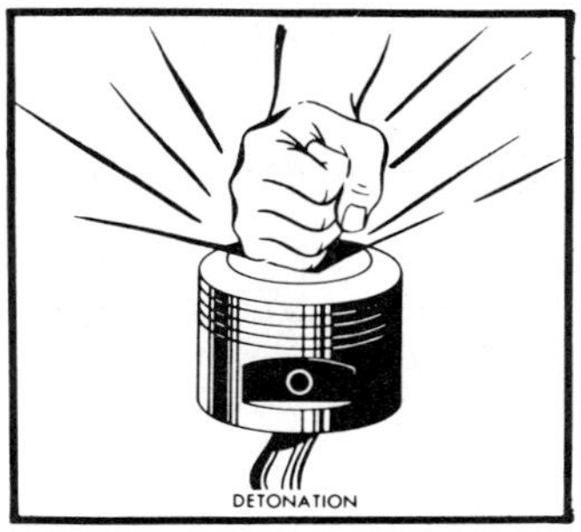


Continued operation when detonation is present can result in dished piston heads, collapsed valve heads, broken rings, or eroded portions of valves, pistons, or cylinder heads, and may terminate in sudden and complete engine failure.

Since it is very important to avoid detonation, it is well to consider the principal factors which contribute to this condition. The anti-knock value of the fuel (octane rating-performance no.), cylinder head temperature, incoming mixture temperature, fuel-air ratio, and intake manifold pressure are the most important factors of greatest significance for the pilot.

Usually detonation cannot be recognized from the cockpit through sound or engine roughness; therefore, protection from its possible occurrence must be provided by the design of the engine and adherence to the engine operating limitations.

HOW IS ADDITIONAL FUEL PROVIDED FOR COOLING THE ENGINE ON TAKEOFF? At full power on takeoff with the mixture "full rich" you are assured of the best combination of power and cooling. The enrichment of the fuel-air mixture at high-power output is accomplished in actual carburetor design by the incorporation of auxiliary fuel-metering devices. Such devices are variously known as economizers, high speed jets, enrichment jets, power compensators, etc. Regardless of the name applied, all such units serve the same general purpose--that is, when full power is used on takeoff, the enrichment jets or valves cut in and provide additional fuel. This additional fuel helps to cool the engine during maximum power operation.

WHAT IS PRE-IGNITION? Pre-ignition is the uncontrolled firing of the fuel-air charge in advance of normal spark ignition. It is caused by the presence within the combustion chamber of an area which is incandescent (red hot, glowing, luminous, with intense heat) and serves as an ignitor in advance of normal ignition. Pre-ignition may result from a glowing spark plug electrode, exhaust valve, or perhaps a carbon or lead particle heated to incandescence. As with detonation, such operating factors as high intake air temperatures, lean mixtures, high manifold pressures, and improper cooling are likely to set the stage for pre-ignition. Pre-ignition may start detonation, and paradoxically, detonation may start pre-ignition because of the high temperatures involved. Moreover, pre-ignition can be fully as destructive as detonation.

WHAT ARE THE RESULTS OF USING AN EXCESSIVELY RICH MIXTURE AT HIGH ALTITUDES? Whenever an unsupercharged engine is operated at a high altitude with an excessively rich mixture, the power will be reduced from that which is available at that altitude with proper mixture. Excessive fuel is not required for combustion chamber cooling at high altitudes. The fouling of spark plugs is one of the greatest "bad effects" of operating with an excessively rich mixture. Spark plugs are designed to operate within certain heat ranges in order to function properly and operate without fouling. The excessively rich mixture will cause a below normal temperature of the spark plug center electrode, which, in turn, results in the formation of carbon and lead deposits. These deposits are electrically conductive and when they reach a sufficient depth, the electric current will flow through the deposit rather than "jumping-the-gap" in the spark plug to ignite the fuel air charge. This is what is known as a "fouled" or "shorted out" plug, since the current flows across the deposits on the ceramic insulator and is grounded instead of jumping the gap. Therefore, it is essential to maintain a fuel-air ratio which will provide sufficient heat in the combustion chamber to vaporize any deposits which may form on the ceramic center of the spark plug.

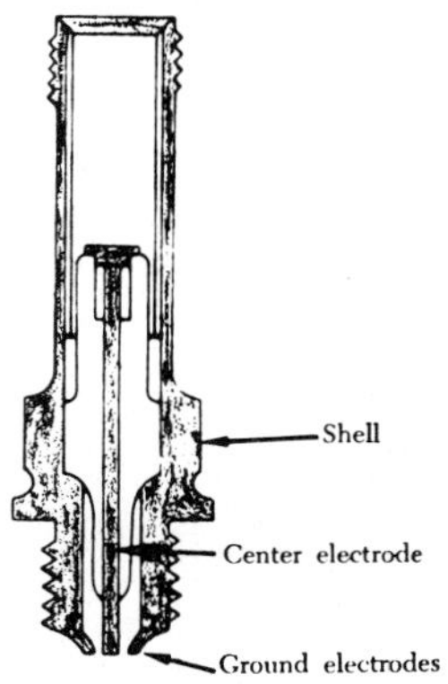


Radio Shielded Spark Plug

●●●●●

## CHECK YOUR READING

1. If the recommended octane fuel is not available, can we use a different octane fuel? Must we use higher or lower octane than recommended?
2. What is the most common fuel contaminant?
3. Why should the aircraft fuel tanks be filled before storing the machine overnight?
4. Why must we drain the fuel tank sumps as well as the main gasculator prior to every flight?
5. State the procedure which should be used when draining the sumps and gasculator.
6. What is the only reliable method to be used to check fuel quantity prior to flight?
7. What is the function of the carburetor?
8. What atmospheric conditions are most conducive to rapid carburetor ice formation?
9. What is the difference between fuel injection and carburetion?
10. How can we increase the amount of carburetor heat available?
11. List the power indications which will occur when ice forms in the carburetor, heat is applied, the ice is melted, and the heat is removed.
12. Why should we not use the carburetor heat at all times?
13. Does the mixture control change the amount of fuel or air passing through the carburetor?
14. At or above what altitude will the mixture control normally be used on a light aircraft engine?
15. Describe the recommended method for adjusting the mixture control.
16. Using the cruise control chart in Fig. 51, list the fuel consumption rate, endurance hours, and range of the aircraft when flying in the following conditions:

    a. 5,000 feet 59% bhp
    b. 2,500 feet 46% bhp
    c. 7,500 feet 2,450 rpm
    d. 10,000 feet 2,730 rpm
    e. 10,000 feet 2,150 rpm

# 12

# THE MAGNETIC COMPASS

It is ironic that with all the sophisticated navigational instruments in today's aircraft, the *magnetic compass* is one of the most dependable and is required by law in every aircraft, and yet the basic design of the magnetic compass has not changed since the first one was placed on board an airplane. It is a simple instrument, and once its peculiarities are understood, it can be used with precision.

## The Design of the Magnetic Compass

The magnetic compass is a self-contained instrument comprised of a pair of magnetized steel needles which are mounted on a float, as shown in Fig. 52.

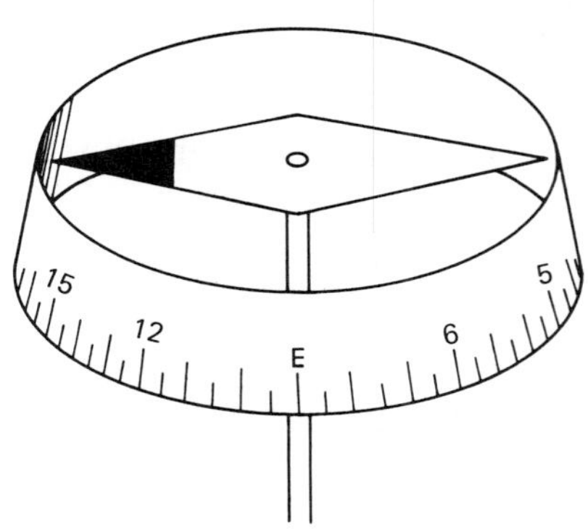


Fig. 52. Magnetic compass float assembly.

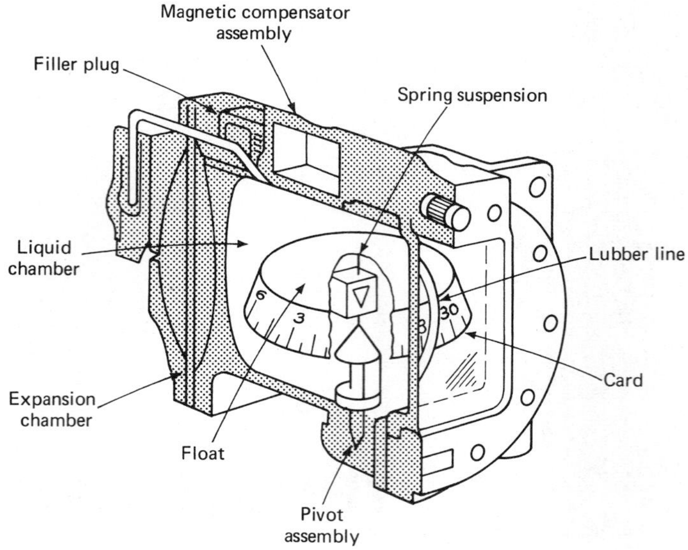


Fig. 53. The magnetic compass.

A compass card is mounted on the edge of the float, on which the cardinal points of the compass (North, East, South, and West) and each thirty-degree interval is engraved. (The last zero of each number is omitted. For instance, 280 degrees appears as 28.) The card is further graduated in five degree increments by lines.

The float assembly is mounted on a pivot pedestal and sealed in a chamber, as shown in Fig. 53.

The sealed chamber is filled with white kerosene which dampens oscillation, reduces the weight of the float assembly on the pivot (due to buoyancy), and lubricates the pivot point.

The float assembly is free to rotate in either direction, but can only be tilted to an angle of eighteen degrees before hitting the sides of the chamber. In the rear of the compass bowl, a diaphragm allows for expansion and contraction of the kerosene.

The front of the instrument case is equipped with a glass window, on the center of which appears a vertical reference line or *lubber line* by which the compass indications are read.

If we study Fig. 53 we will notice one important fact which we must remember when discussing compass errors. That is, from the cockpit, we will be viewing the compass float assembly from the rear. Therefore, although the North seeking end of the float may be pointing toward the nose of the airplane, the compass card engraving "N" faces the cockpit, or the tail of the airplane. In other words, we see an engraved letter or number under the lubber line which indicates where the other side of the float is pointing.

## Compass Errors

The compass suffers from errors, but they are easily understood, and always act in the same manner.

*Variation*

Although we will review variation here, it is not really a compass error. We will learn in Chapter 38 that variation is the angle between True North and Magnetic North, as viewed from any specific geographic location. Magnetic North drifts around the pole area, so that the chart makers designated a permanent geographic location as True North. The perfect magnetic compass would point to the magnetic north pole. Variation is noted on aeronautical charts as easterly or westerly.

*Deviation*

The magnetized needles on the compass float assembly are not only affected by the earth's magnetic field, but also by the magnetic fields generated by electrical equipment in the aircraft and by the aircraft itself. Deviation is the compass error which is induced by the aircraft and its components.

The compass case is equipped with a compensating device composed of tiny magnets. The aircraft is lined up with the cardinal points of the compass using a compass rose painted on the airport taxiway, or a surveyor's transom. The compensating magnets are then adjusted to decrease any error in the compass to the lowest possible amount. The remaining errors are listed on the Compass Correction Card which is posted near the compass.

Deviation may change drastically when the aircraft radio or other electrical components are in use. Therefore, we must know the specific conditions for which the compass was adjusted, and when using the compass for accurate heading information we must duplicate these conditions.

The compass deviation should be checked periodically. We must avoid placing metal computers, flashlights, or any other metal paraphernalia near the compass, for large amounts of deviation may be induced which will seriously affect compass accuracy.

*Oscillation Error*

Oscillation is the result of the swinging compass due to rough pilot technique or turbulence. Therefore, we cannot rely on the compass for heading information while flying in turbulence.

*Acceleration Error*

The acceleration error in the compass is caused by changes in airspeed. The magnetic float assembly is balanced on a jeweled pivot point and is free to rotate and tip in any direction. Its center of gravity is *below* the point of suspension, as shown in Fig. 54.

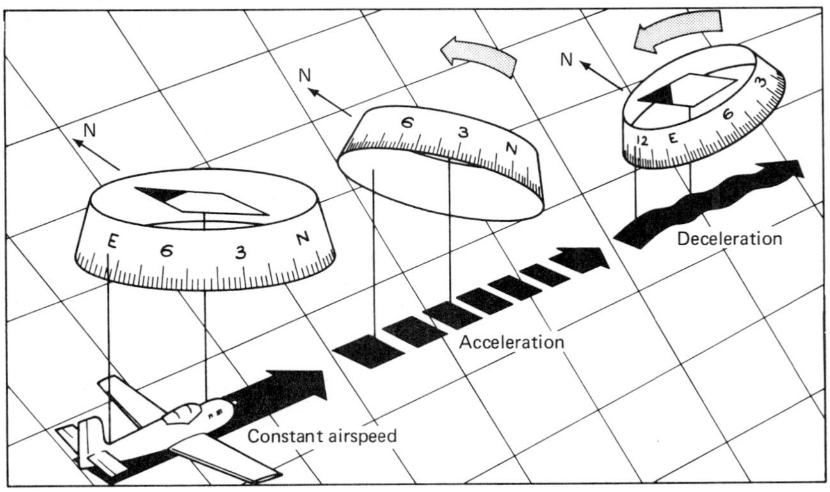


Fig. 54. Acceleration/deceleration error.

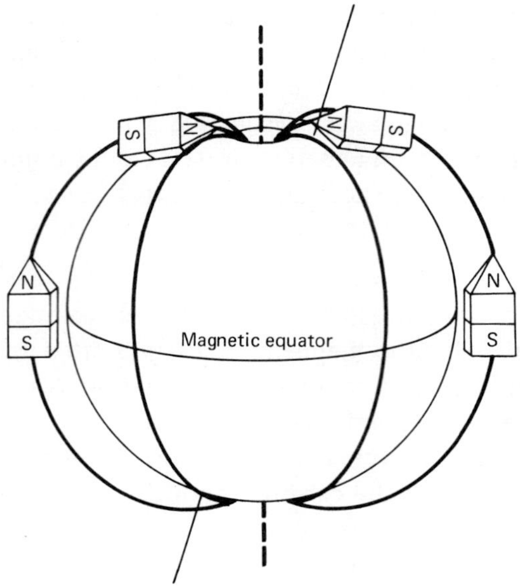

Fig. 55. Magnetic lines of force.

The earth's magnetic lines of force are not parallel to the earth's surface (except near the equator) but in reality slope downward towards the earth's surface as they near the poles, eventually becoming perpendicular to the surface over the poles, as shown in Fig. 55.

The North-seeking needles on the float assembly align themselves with the earth's magnetic lines of force, dipping slightly when near the equator, and severely when near the poles.

The combination of magnetic dip and the fact that the center of gravity of the float assembly is below the pivot point causes acceleration error, which is most apparent on headings of East or West.

When the aircraft accelerates, the magnetic float assembly tips forward away from the pilot. The North-seeking end of the needles dip toward the magnetic pole, causing the compass to swing toward a more northerly heading, as shown in Fig. 54.

When the aircraft decelerates, the float assembly tips rearward toward the pilot and magnetic dip causes the compass to swing toward a more southerly heading. Acceleration-deceleration error is most noticeable on headings of East and West, and diminishes as the aircraft heading approaches North or South.

We must remember that acceleration causes an indication of a more northerly heading, and deceleration causes an indication of a more southerly heading. Some pilots use the word "A-N-D-S" to remember the effect of acceleration-deceleration error (Accelerate-North; Decelerate-South). When referring to the compass for accurate heading information we must be sure that the airspeed is constant.

*Northerly Turning Error*

The same characteristics of the magnetic compass which cause acceleration error cause "northerly turning error" which is most pronounced when turning to or from headings near North or South. This is depicted by Fig. 56.

*Turning From North.* When we initiate a turn in ***either direction*** from North,

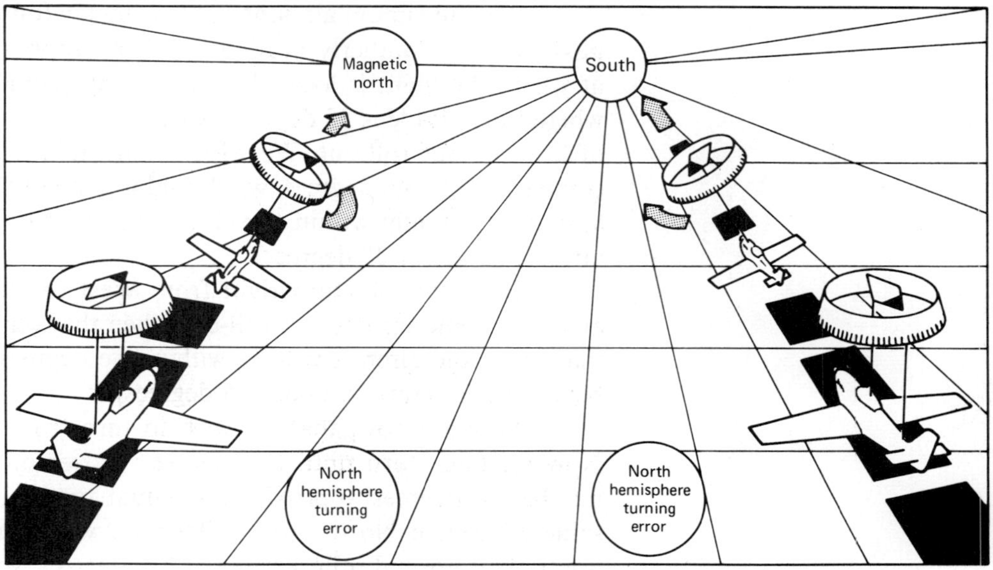


Fig. 56. Northerly turning error.

we will notice that the compass indicates a turn in the *opposite direction.* For example, if we begin on North and turn right toward East, the compass will indicate a left turn. After our turn is established, the compass will begin to indicate a turn in the correct direction but will lag behind the actual turn. The amount of lag will slowly diminish to zero as we approach within two or three degrees of East.

Another example of northerly turning error is called "slow turn error." If, while on a heading of North, we initiate a very shallow banked turn (three or four degrees of bank) the magnetic compass will remain on North, even though the aircraft heading eventually changes twenty degrees or more. As the turn progresses, the compass will begin to indicate the turn, again lagging behind the aircraft.

*Turning From South.* When initiating a turn in ***either direction*** from South, we will see that the compass indicates a turn in the proper direction, but at a rate which is much faster than the aircraft is turning. It *leads* the aircraft. As the turn progresses the amount of lead will diminish, becoming zero lead as we approach two or three degrees of East or West.

*Turning to North.* When we turn toward North, ***from either direction,*** the compass will lag behind the aircraft.

*Turning to South.* When turning toward South ***from either direction,*** the compass will lead the aircraft.

*Turns to Predetermined Headings*

Because of northerly turning error, we must compute the compass reading at which to begin our roll-out of a turn in order to complete the turn at a predetermined heading.

The maximum amount of compass lag (on northerly headings) and compass lead (on southerly headings) will be approximately equal to the latitude at which the flight is located. For example, suppose our flight is in a locality where the latitude is 35 degrees North, and that we normally allow five degrees of lead for our roll-out. We wish to turn left, from East to North. When the compass reads 040 degrees, we should initiate our roll-out, for at that point the aircraft is actually within five degrees of North (the compass lags behind the aircraft by about 35 degrees).

If we wish to turn right from West to North, in the same geographic area, we would initiate our roll-out when the compass reads 320 degrees, for at that point the aircraft will be within five degrees of North (the compass lags behind the aircraft by about 35 degrees).

Suppose now, that we wish to turn from East or West toward South. Now we must turn until the compass *passes* our normal roll-out point by a number of degrees approximately equal to the latitude. For instance, in the same 35 degree North latitude locality, we wish to turn right from East to South. Our normal roll-out lead of five degrees would dictate that we begin our roll-out at 175 degrees. But, due to turning error, we must initiate our roll-out when the compass reads 35 degrees *beyond* our roll-out heading of 175 degrees, or 210 degrees.

If we desire to turn left from West to South, we must initiate our roll-out at 185 degrees, allowing five degrees for the actual roll-out. Since the compass will lead our turn by 35 degrees, it will read 35 degrees beyond 185 degrees, or 150 degrees when we begin the roll-out.

When planning a turn to an intermediate heading (between East or West and North or South), we must make an adjustment as to the time for roll-out, using information as to our proximity to North or South, the direction of the turn, and our own experience.

## Maximum Allowable Bank

The magnetic float assembly is designed to tip to an angle as great as eighteen degrees from the horizontal before hitting the side of the instrument case. Therefore, we must limit our bank angle to about fifteen degrees so that the compass float will not become "hung up" on the instrument case.

## Points to Remember

1. Acceleration–deceleration errors are most pronounced on headings of East or West.
2. The word "A-N-D-S" will remind us that *A*cceleration produces *N*ortherly turn indications, and *D*eceleration produces *S*outherly turn indications.
3. Northerly turning error causes the compass to lag behind the aircraft

when turning to or through northerly headings, and to lead the aircraft when turning to or through southerly headings.

4. Latitude determines the approximate amount of lag or lead which the compass will indicate.
5. When we desire to use the magnetic compass for accurate heading information (to establish a course or to set a gyro heading indicator) while on or near a northerly or southerly heading, we must be sure that the aircraft is not turning (the wings are level) for several seconds prior to taking a compass reading.
6. While on an easterly or westerly heading, the compass will read accurately only if the aircraft is maintaining a constant airspeed.
7. In order to receive accurate compass indications on intermediate headings between North or South and East or West, both the airspeed and the aircraft heading must be constant.

FEDERAL AVIATION AGENCY
VFR EXAM-O-GRAM* NO. 12

THE MAGNETIC COMPASS

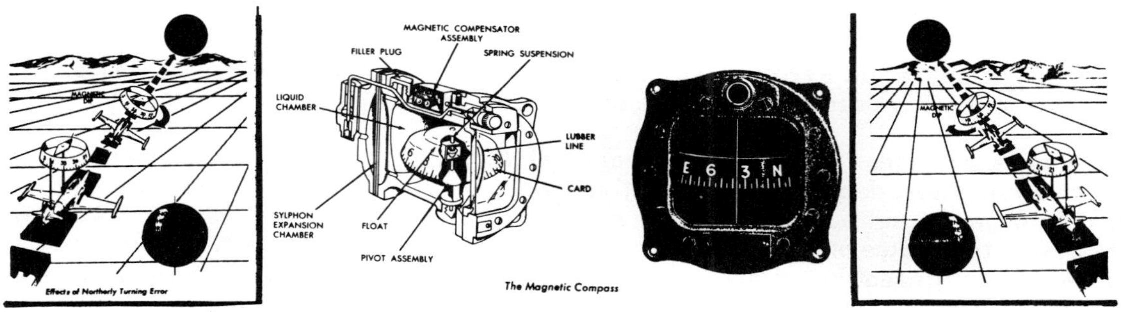

*The Magnetic Compass*

**The magnetic compass, in terms of its errors, limitations, and in-flight characteristics, is one of those aeronautical subjects in which consistently large numbers of pilots fare poorly on FAA written examinations. There is evidence that this veteran instrument -- it was one of the first to be installed in an aircraft -- is one of the least understood instruments in the cockpit of today's modern general aviation aircraft. Many pilots seem to operate on the premise that it is easier to ignore this instrument's errors rather than learn them. However, it should be remembered that (1) this is the only directional seeking instrument in the cockpit of most general aviation aircraft, and (2) it is mechanically a simple, self-contained unit (independent of external suction or electrical power for its operation) that is likely to remain reliable at all times -- reliable, that is, if the pilot understands its inherent errors.**

**WHAT ARE SOME OF THE COMPASS ERRORS THAT THE PILOT SHOULD UNDERSTAND?**

**The pilot should understand:**

**I. VARIATION -- This is the angular difference between true north and magnetic north which is plotted on charts in degrees east or west. The pilot should understand perfectly which to add and which to subtract when converting from true headings or courses to magnetic headings or courses and vice versa. (Many pilots find such**

memory aids as "east is least and west is best" helpful in remembering that east is subtracted and west is added when going from true to magnetic. )

II. DEVIATION -- This is the deflection of the compass needle from a position of magnetic north as a result of local magnetic disturbances in the aircraft. To reduce this deviation, the compass has a compensating device consisting of small adjustable magnets. The compass should be checked and compensated periodically. The errors remaining after "swinging" the compass should be recorded on a compass correction card which should be installed in the cockpit within the view of the pilot. (NOTE: Avoid placing metallic objects such as metal computers, flashlights, etc., on top of the instrument panel near the magnetic compass as this practice may induce large amounts of deviation and seriously affect the instrument's accuracy. )

In addition to these errors, the pilot should have a working knowledge of the following In-flight errors:

III. OSCILLATION ERROR -- The erratic swinging of the compass card which may be the result of turbulence or rough pilot technique.

IV. MAGNETIC DIP -- The tendency of the Magnetic Compass to point down as well as north in certain latitudes. This tendency is responsible for:

A. Northerly Turn Error -- This error is the most pronounced of the in-flight errors. It is most apparent when turning to or from headings of north and south.

B. Acceleration Error -- An error that can occur during airspeed changes. It is most apparent on headings of east and west.

As a quick refresher on this instrument's in-flight dip error, we invite you to accompany us on a simulated demonstration flight around the compass rose. Unless otherwise noted, we will limit our bank during turns to a gentle bank. Also, we will assume that we are in the northern hemisphere because the characteristics which we will observe would not be present at the magnetic equator, and would be reversed in the southern hemisphere.

DEMONSTRATION NO. 1 (Heading -- North; Error -- Northerly Turn Error)

As we start a turn in either direction from this heading, we notice that momentarily the compass gives an indication of a turn opposite the direction of the actual turn. (While the compass card is in a banked attitude, the vertical component of the earth's magnetic field causes the north-seeking end of the compass to dip to the low side of the turn, giving the pilot an erroneous turn indication. ) If we continue the turn toward east or west, the compass card will begin to indicate a turn in the correct direction, but will lag behind the actual turn -- at a diminishing rate -- until we are within a few degrees of east or west. One additional demonstration which we will cover before leaving north is the Slow Turn Error. If, while holding a compass indication of north, we sneak into a very gradual and shallow banked turn -- say 3 or 4 degrees of bank -- it is possible to change the actual heading of the aircraft by 20° or more while still maintaining an indication of north by the compass.

DEMONSTRATION NO. 2 (Heading -- East; Error -- Acceleration Error)

The Northerly Turn Error that we previously experienced is not apparent on this heading (or on a west heading). However, let's see what happens when we accelerate and decelerate by changing the airspeed. With the wings level, we will increase the airspeed by increasing the power setting or by lowering the nose -- or both. Result --

although we are holding the nose of the aircraft straight ahead, our compass card erroneously indicates a turn toward north. On the other hand, if we decrease the airspeed by reducing the power setting or raising the nose of the aircraft -- or both, the compass will give an erroneous indication of a turn toward south. (Because of the pendulous-type mounting, the end of the compass card which the pilot sees is tilted

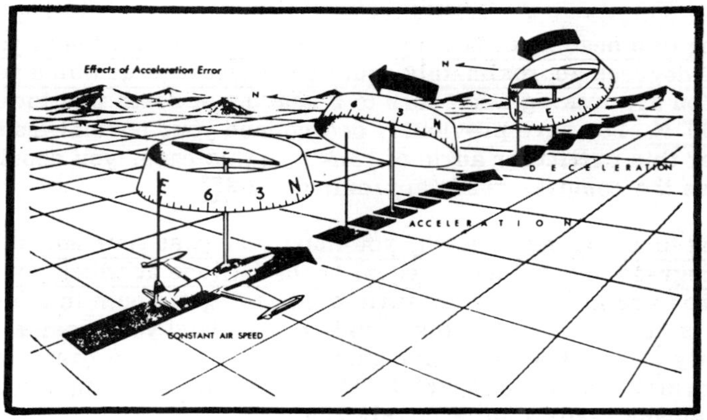


upward while accelerating, and downward while decelerating during changes of airspeed. This momentary deflection of the compass card from the horizontal, results in an error that is most apparent on headings of east and west.)

DEMONSTRATION NO. 3 (Heading -- South; Error -- Northerly Turn Error)

Again we are presented with the Northerly Turn Error problem that we encountered in DEM. NO. 1. Although the same set of forces that caused the erroneous indication when we banked the aircraft while on a north heading will likewise be working against us on this heading, the compass indications will appear quite different. For example, as we roll into a turn in either direction, the compass gives us an indication of a turn in the correct direction but at a much faster rate than is actually being turned. As we continue our turn toward west or east, the compass indications will continue to precede the actual turn -- but at a diminishing rate -- until we are within a few degrees of west or east. (It might be noted that the Acceleration Error is not apparent on this heading or on a north heading.)

DEMONSTRATION NO. 4 (Heading -- West; Error -- Acceleration Error)

On this heading we encounter the exact same errors that we have previously covered on a heading of east in DEM. NO. 2. If we increase the airspeed, we will get an erroneous indication of a turn toward north. If we decrease the airspeed, we will get an erroneous indication of a turn toward south. (A memory aid that might assist you in recalling this relationship between airspeed change and direction of the error is the word "ANDS" -- Accelerate-North, Decelerate-South.)

WHAT ARE THE MAIN POINTS THAT SHOULD BE REMEMBERED CONCERNING THESE FOUR DEMONSTRATIONS?

The points we are trying to get across are these: (1) WHEN TAKING READINGS FROM THE MAGNETIC COMPASS WHILE ON A NORTHERLY OR SOUTHERLY HEADING (for establishing a course, setting the gyro-driven heading indicator, etc.), REMEMBER THAT IT IS ESSENTIAL TO HAVE THE WINGS PERFECTLY LEVEL FOR SEVERAL SECONDS PRIOR TO TAKING THE READING. (2) IF YOU ARE ON AN EASTERLY OR WESTERLY HEADING, IT IS IMPORTANT THAT THE AIRSPEED IS CONSTANT IN ORDER TO GET AN ACCURATE READING. (3) ON AN INTERMEDIATE HEADING, BOTH OF THE ABOVE CONDITIONS SHOULD BE MET. (Note: If your aircraft is equipped with a gyro-driven heading indicator, check it frequently with your magnetic compass.)

* * * * * * *

## TURNS TO HEADINGS BY REFERENCE TO THE MAGNETIC COMPASS

**For the pilot who would like a general set of rules for determining his lead point for making turns by reference to the Magnetic Compass, the following is submitted:**

**(Note: The angle of bank should not exceed 15° in order to minimize dip error.)**

1. **When turning to a heading of north you must allow, in addition to your normal lead, a number of degrees approximately equal to the latitude at which you are flying. Example: You are making a left turn to a heading of north in a locality where the latitude is 30° N. You have previously determined your normal lead to be approximately 5° for this particular angle of bank. In this case, you should start your roll-out when the compass reads approximately 35°.**

2. **When turning to a heading of south, you must turn past your normal lead point by a number of degrees approximately equal to the latitude at which you are flying. Example: You are making a right turn to a heading of south in a locality where the latitude is 30° N. You have previously determined your normal lead to be approximately 5° for this particular angle of bank. In this case, you should turn past your normal lead point of 175° (180° - 5°) by 30°, and start your roll-out when the compass reads approximately 205°.**

3. **The error is negligible when turning to east or west, therefore, use the normal amount of lead when turning to an east or west heading.**

4. **For intermediate headings that lie between the cardinal headings, use an approximation based on the heading's proximity to north or south, the direction of the turn, and your knowledge of the compass's lead and lag characteristics in these areas. In other words, use an "educated guesstimate."**

**We won't guarantee you that the above method will roll you out on the exact heading every time -- at best, it is an approximate method. But it will get you reasonably close to your desired heading, and this beats having no method at all.**

### CHECK YOUR READING

1. Why is the magnetic float assembly in the compass surrounded by kerosene?
2. Explain *variation.*
3. What causes *deviation*? How can it be minimized?
4. What causes *oscillation error*?
5. We depart from runway 01 and initiate a left turn when reaching an altitude of 400 feet above the surface. What compass indication would we expect?
6. We wish to turn right from East to South, in a locality where the latitude is 030 degrees North. At what magnetic compass reading should we begin the roll-out, if ten degrees will be required to complete the roll-out?
7. We wish to turn left from North to West. Assuming that the airspeed is constant, and that five degrees will be required for actual roll-out, at what magnetic compass reading must we initiate the roll-out?
8. We wish to establish a course in a southerly direction, using the magnetic compass. What precautions must be taken to insure an accurate compass reading?
9. We wish to roll out on a heading of North after turning right from West. We are flying in an area where the latitude is 45 degrees North. We will require ten degrees for the roll-out. At what magnetic compass reading must we initiate the roll-out?
10. When accelerating on a heading of East or West, what can we expect of the magnetic compass indications?

# 13

# THE PITOT-STATIC SYSTEM AND ASSOCIATED INSTRUMENTS

The pitot–static system, although simple in design, is very important, for it is the system which drives three of our primary flight instruments. The airspeed indicator, vertical velocity indicator, and altimeter all depend on the pitot-static system. Because of the importance of these instruments, we must have a working knowledge of the system which makes them operate. Each of the instruments depends on and responds to a difference in air pressure, either between static and ram air pressure, or in static and trapped air pressure.

## The Pitot System

The pitot system is comprised of the pitot head, a heating element, and plumbing which connects the head to the airspeed indicator, as shown in Fig. 57.

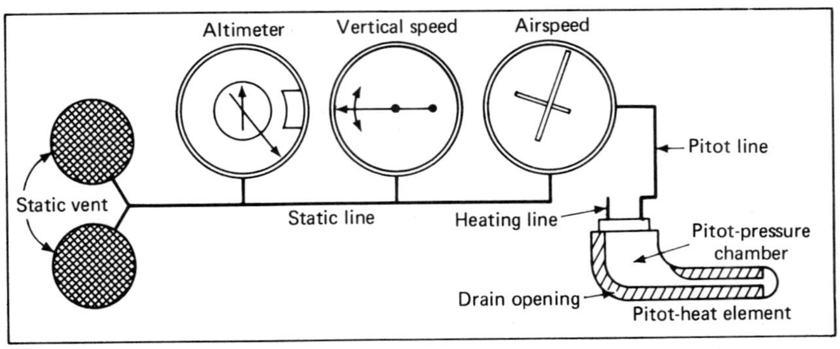


Fig. 57. The pitot-static system.

Ram air creates impact pressure in the pitot head which is normally located near the wing leading edge, nose section, or vertical stabilizer. In order to keep the pitot head free of structural ice, a heating element is installed which is controlled from the cockpit.

## The Static System

The airspeed indicator, altimeter, and vertical velocity indicator are all connected to the static system which is comprised of plumbing and static vents as depicted in Fig. 57. The static vent is normally flush mounted on the fuselage or nose section, and must be kept clean at all times. An alternate source of static pressure may be provided for emergency use.

### *Alternate Source of Static Pressure*

If the static system vents become clogged, the system is vented through the alternate source. In unpressurized aircraft this vent will normally be inside the aircraft. Because atmospheric pressure is usually slightly lower inside the aircraft than it is outside, the altimeter will read higher than normal, and the vertical velocity indicator will momentarily show a climb, when the alternate static pressure source is in use.

## The Airspeed Indicator

The airspeed indicator measures the difference between ram air pressure from the pitot head and atmospheric pressure from the static vents. A diaphragm within the instrument case is sensitive to pressure changes. As the ram air pressure increases, the diaphragm expands, and a series of levers and gears transmits this movement to the needle on the instrument face. A cutaway view of the airspeed indicator is shown in Fig. 58.

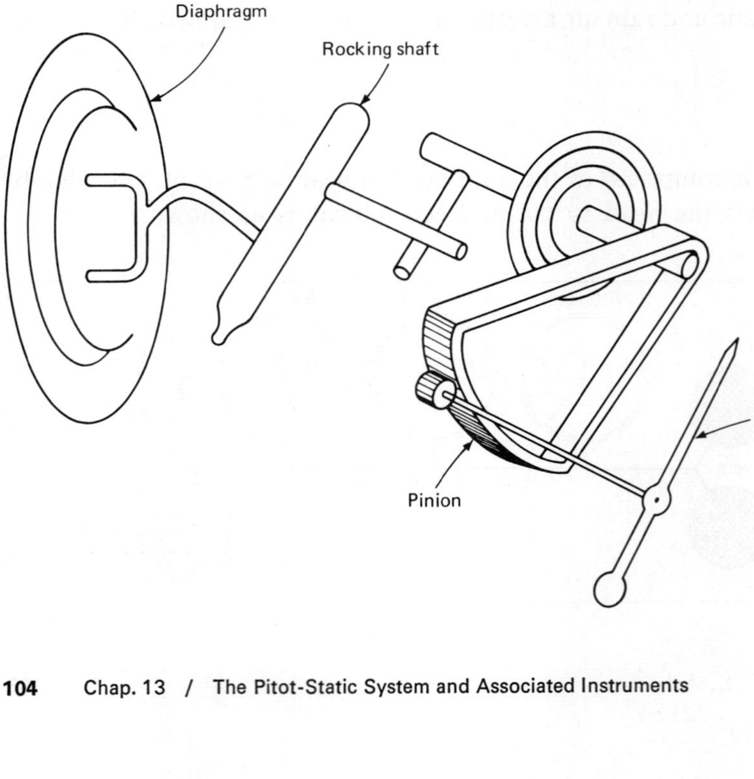


Fig. 58. The airspeed indicator (exploded view).

The dial on the face of the airspeed indicator may be calibrated for statute miles per hour, or for knots (nautical miles per hour).

*Airspeed Errors*

Whenever conditions vary from those for which the instrument was designed, errors will result in the readings.

*Installation Error*. Variations of the angle with which the airstream strikes the pitot head (due to changing airspeed) causes a change in the differential pressure established by the pitot head and the pressure which should be developed at a specific airspeed. Pipe fittings and bends in the plumbing cause friction which may change the airspeed readings.

*Density Error*. The standard airspeed indicator cannot compensate for changes in air density. Therefore, as altitude and/or air temperature differs from standard, an error will occur in the airspeed indicator.

*Compressibility Error*. At high airspeed, the ram air packs into the pitot head resulting in higher than normal airspeed readings. This problem is most common at airspeeds in excess of 180 knots. At low altitudes, the error is almost nonexistent.

*Types of Airspeed*

When discussing airspeed, it is important to denote the specific type of airspeed to which we refer.

*Indicated Airspeed.* The value read from the face of the instrument is the *indicated airspeed.* It is our immediately available reference for speed in flight.

*Calibrated Airspeed. Calibrated airspeed* is the airspeed to which aircraft operating data is referred in the Aircraft Operations Manual. It is indicated airspeed corrected for installation error, and is sometimes called *true indicated airspeed.*

*Equivalent Airspeed.* The calibrated airspeed corrected for the compressibility factor is called *equivalent airspeed.* Since the compressibility factor in light aircraft is negligible, equivalent airspeed is assumed to equal calibrated airspeed in light aircraft.

*True Airspeed.* The equivalent airspeed (calibrated airspeed in light aircraft) corrected for air density error gives us the *true airspeed.*

*Airspeed vs. Altitude*

Suppose our aircraft has a minimum power-off stall speed of 60 knots indicated airspeed at sea level, standard air density. What indicated airspeed will be the minimum power-off stall speed at 10,000 feet?

The *indicated* minimum stall airspeed will be exactly the same, 60 knots.

The aircraft will stall at a higher *true* airspeed because the air is much less dense at 10,000 feet than it is at sea level, but the same factors which allow the aircraft to stall at a higher true airspeed also affect the airspeed indicator, making it read a lower value when the aircraft stalls. As a result of these opposing factors, the aircraft will stall at the same *indicated* airspeed at any altitude, assuming all other factors to be equal.

Therefore, if all other factors are equal, we will maintain the same *indicated* airspeed for final approach or any other maneuver no matter what our altitude, remembering that the *true* airspeed will be higher than indicated as the altitude increases.

*Apparent–Airspeed Lag*

The airspeed indicator reports airspeed instantly. The *apparent lag* in the instrument readings is due to the fact that the aircraft requires time to accelerate or decelerate. Therefore, as we change pitch attitude or thrust the airspeed does not instantly change.

## The Vertical Velocity Indicator

The vertical velocity indicator (commonly called the *rate of climb* or *vertical speed* indicator) is an instrument which measures the *rate of change* in atmospheric pressure. The instrument case is sealed and connected to the static port by a *calibrated leak*. As the outside air pressure changes, it affects a diaphragm which is connected to the indicating needle on the instrument face by a series of levers and gears. The rate of pressure change between the trapped static pressure in the instrument case and the instantaneous static pressure in the diaphragm is thus measured and transmitted to the cockpit as a rate of climb or descent in hundreds or thousands of feet per minute.

When differential pressure changes occur, the indicating needle will immediately show a climb or descent. However, until the rate of differential pressure change stabilizes, the instrument readings will be in error, because of a six to nine second delay caused by the calibrated leak.

*Limits of Operation*

Although the vertical velocity indicator provides reliable information to establish and maintain level flight and rate climbs or descents, sudden or abrupt changes in the aircraft attitude cause erroneous readings as the airflow fluctuates over the static ports. Turbulence and rough control action, therefore, cause unreliable indications.

## The Instantaneous Vertical Speed Indicator

Acceleration pumps have been installed in newer designs of instantaneous vertical speed indicators which eliminate the six to nine second delay which is associated with the vertical velocity indicator. During level flight and steady rates of climb or descent, the instrument operates on identical principles as the vertical velocity indicator.

## The Altimeter

The pressure altimeter is one of the most important instruments in the aircraft, and yet, it is far from being an accurate height-measuring instrument. It provides information which is essential to us for maintaining aircraft separation and terrain clearance, but it is designed to operate in atmospheric conditions which rarely occur. However, years of experimentation have failed to produce a more accurate or more reliable pressure-measuring device for reporting altitude. We must devote special attention to the peculiarities of the instrument and operating procedures so that we can live with it until, hopefully, a better one is invented.

### *Principles of Operation*

The altimeter is an aneroid barometer. That is, it is constructed with a stack of aneroid wafers, which are paper thin capsules that flex when the trapped atmosphere inside expands and contracts. If we transport the altimeter to a lower pressure area, the atmospheric pressure within the instrument case bleeds out via the static port. With reduced pressure surrounding the aneroid wafers, the atmosphere within them is free to cause expansion of the wafers. Conversely, if we increase the atmospheric pressure inside the instrument case, by transporting the altimeter to an area of higher pressure, outside atmosphere bleeds into the instrument case and tends to collapse the wafers. The movement of the aneroid wafers is magnified by levers and gears and is transmitted to the pointers on the face of the instrument.

The altimeter is constructed to read the amount of atmospheric pressure *above* it, pushing down. In reality it does not measure height, but rather it measures the amount of atmosphere above it. The conversion of atmospheric pressure to feet of altitude is based on the United States Standard Atmosphere. Atmospheric conditions are standard when sea level pressure is 29.92 inches of mercury, and the sea level temperature is 15 degrees Celsius. The temperature lapse rate must be 2 degrees Celsius per one thousand feet of altitude.

Standard atmospheric conditions rarely exist however. The altimeter presents standard information, even though it senses nonstandard conditions. These two factors are what cause difficulty in understanding and properly using the altimeter.

### *Reading the Altimeter*

Tests have proven that even the most experienced pilot occasionally misreads the altimeter, particularly when rushed or confused. As we see on the face of the altimeter shown in Fig. 59, the instrument has three hands or needles.

The shortest pointer indicates tens of thousands of feet, and should always be read first. Because light aircraft pilots normally fly at less than 10,000 feet, they tend to ignore the short pointer entirely. We must form the habit

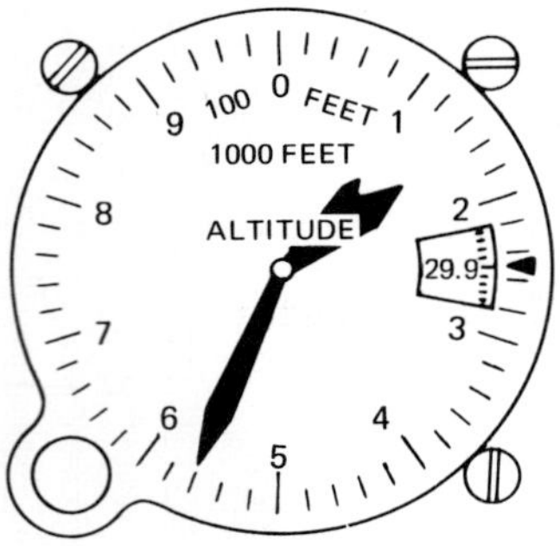


Fig. 59. The altimeter.

from the very beginning, however, of reading the 10,000-feet pointer first. Then, as we progress to high altitude aircraft, we will have no difficulty in maintaining proper altimeter reading habits.

If the short pointer lies between numbers one and two on the instrument face, it indicates that the altitude is between 10,000 feet and 20,000 feet.

The medium-length pointer indicates thousands of feet, and should be read second. If the pointer is midway between one and two on the instrument face, it indicates that the altitude is between 1,000 and 2,000 feet above the lowest altitude depicted by the short pointer.

The longest pointer indicates hundreds of feet, and should be read last. If the pointer is between five and six on the instrument face, it indicates that the altitude is between five hundred and six hundred feet above the lowest altitude depicted by the medium-length pointer.

Therefore, when reading the altimeter we should begin with the 10,000 foot level and build on it, the 1,000-foot and 100-foot levels. Five spaces between each one-hundred-foot number allow a further breakdown to twenty-foot increments.

Referring to Fig. 59, we see that the short pointer lies between the 10,000 and 20,000 foot digits. The medium-length pointer indicates between 1,000 and 2,000 feet. The longest pointer indicates three spaces more than 500 feet. Thus the altimeter shows an altitude of 11,560 feet.

### *The Altimeter Setting*

Notice that our discussion thus far has not been concerned with altitude *above* anything, but simply "altitude." Does the altimeter indicate altitude above the surface, above sea level, or above a standard pressure altitude?

Remember that we learned that the altimeter is designed to indicate an altitude based on a conversion from standard atmospheric pressure. If standard atmospheric pressure always existed at sea level (and the lapse rate was always standard also), the altimeter would report height above sea level. However, this situation rarely occurs. Due to changing weather systems and pressure patterns, both the atmospheric pressure at sea level and the lapse rate constantly change.

Therefore, an altimeter-setting dial is provided in which we can set the *actual* sea level barometric pressure for our geographic location. This will compensate for one error in the altimeter; that which is caused by nonstandard atmospheric pressure.

At weather observatories all over the world, the atmospheric pressure is recorded at least hourly. (If the pressure is changing rapidly, the record may be taken more often.) Since every weather observer is not located at sea level, each pressure reading must be adjusted to what it *would be* if the station were at sea level. Then the station atmospheric pressure *reduced to sea level* is broadcast for aeronautical use. Hence, we always have a current *altimeter setting* available to us. As soon as we adjust the altimeter to the atmospheric pressure reduced to sea level, the altimeter will report height above sea level (assuming a standard lapse rate).

It is important that we maintain a current altimeter setting in our instrument. If our aircraft is equipped with a two-way radio, we should receive a current altimeter setting when departing from an airport and every 75 to 100 miles enroute. Then, prior to landing, a new altimeter setting should be requested from the weather observer at the destination. Without two-way radio, we can set our altimeter only at the point of departure. If no altimeter setting is available, we can set the altimeter to read the field elevation prior to departure.

Because the altimeter will record height above sea level, it will read airport elevation (above sea level) when the aircraft is on the airport surface. All obstructions to flight, such as mountains and television or radio broadcast towers, are shown on the aeronautical charts with their height above sea level, so that we can determine whether we will clear the obstacles while flying.

Suppose we fly to an area in which the atmospheric pressure is *higher than* that set in the altimeter setting window and we fail to reset the altimeter. The following sample problem will illustrate the results.

Assume that we depart an airport which is 1,000 feet above sea level, and the current altimeter setting is 29.80 inches of mercury. We wish to land at an airport which is 1,500 feet above sea level and where the current altimeter setting is 30.10 inches of mercury. We will *not* reset the altimeter. When we land at the destination, what will the altimeter indicate?

Atmospheric pressure decreases about one inch of mercury for each 1,000 feet of altitude gain, and conversely increases one inch of mercury for every 1,000 feet of altitude decrease. This is what the altimeter measures. In the sample problem above, we have transported the altimeter to an area of higher sea level barometric pressure. The altimeter senses this increase in sea level pressure, but does not "know" if the increased pressure is caused by flying to a lower altitude (which will result in higher pressure) or by flying into a geographic area in which a higher sea level pressure weather pattern exists. Hence, when we land in the higher sea level pressure area, the altimeter will indicate an altitude which is *lower* than the actual altitude. We transported the altimeter into an area in which the atmospheric pressure at sea level was 0.30 inches of mercury higher. If one inch of mercury equals 1,000 feet of altitude, then 0.30 inches of mercury equals 300 feet of altitude. Therefore, the altimeter will read 300 feet *lower* than our actual height above sea level, or 1,200 feet, when we land at the destination airport which is 1,500 feet above sea level. In other words, the altimeter reported an altitude of 1,500 feet when the aircraft was actually at 1,800 feet, and will indicate 1,200 feet when we land.

Now let us see the results of a flight into an area in which the sea level pressure is lower. Suppose we depart an airport which is 500 feet above sea level, and where the current altimeter setting is 30.00 inches of mercury. We wish to fly over a mountain which is 3,500 feet above sea level, and which is in an area of barometric pressure reduced to sea level of 29.50 inches of mercury. We plan, for this example, to maintain an indicated altitude of 3,800 feet, thus supposedly having a 300-foot clearance over the mountain top. (We will see why this amount of clearance above the mountain is *not* sufficient.) We will assume that the altimeter setting is not changed on the altimeter, remaining at 30.00 inches of mercury for the entire flight.

We are transporting the altimeter to an area of lower barometric pressure (reduced to sea level), from 30.00 to 29.50 inches of mercury. When we move the altimeter into lower barometric pressure it will show an increase in altitude. But we have decided to *maintain* an indicated altitude of 3,800 feet to clear the mountain top. As the flight progresses into the lower barometric pressure area the altimeter will slowly indicate a higher altitude. We will continually ease back down to an indicated altitude of 3,800 feet.

The total pressure (reduced to sea level) change between the departure point and the mountain is 0.50 inches of mercury, which equals 500 feet. Therefore, the altimeter will indicate an altitude which is 500 feet *higher* than actual, or to put it the other way around, the aircraft will be 500 feet lower than indicated on the altimeter. Figure 60 depicts this situation.

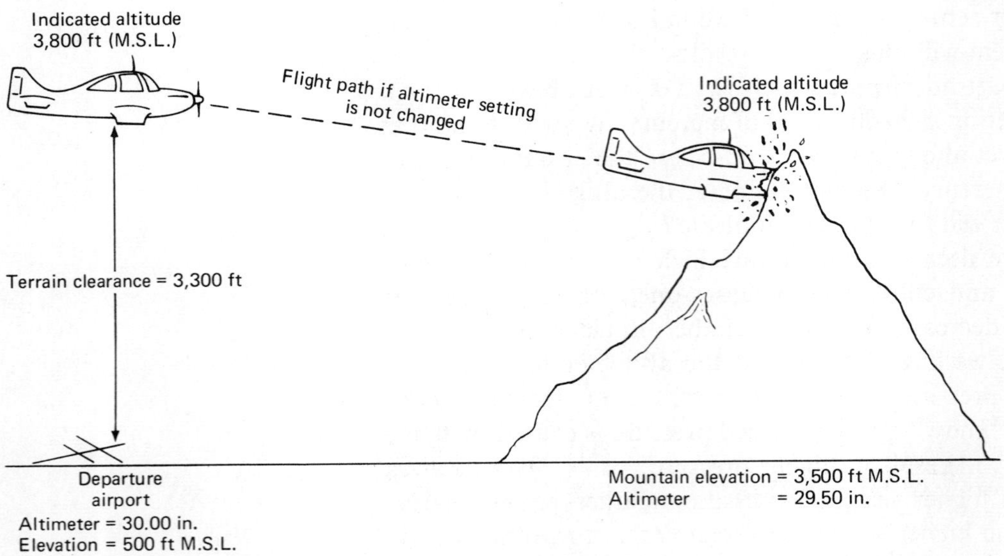


Fig. 60. The importance of using a current altimeter setting.

It is obvious that the flight would *not* clear the top of the mountain even though we maintained an indicated altitude of 3,800 feet.

Some pilots find this simple rhyme easy to remember. "When flying from *HIGH* to *LOW* (pressure) LOOK OUT BELOW." In other words, when flying from an area of higher barometric pressure (reduced to sea level) into an area of lower barometric pressure (reduced to sea level) the altimeter will read a higher altitude than is fact, so we must "look out below."

We have seen that a current altimeter setting is important to safe flight. An altimeter setting which indicates an airport elevation which is in error by 75 feet or more when the current altimeter setting is used, should not be depended upon for flight. We should make every effort to receive and use the latest altimeter setting while enroute.

*Temperature Effect*

The altimeter is also affected by nonstandard temperature. On a warmer than standard day the atmosphere expands. The pressure level at which the altimeter will read 5,000 feet for example is higher than it would be under standard atmospheric temperature conditions. On a colder than standard day, the pressure level is lower than normal due to the contraction of the atmosphere. Therefore, the same handy rhyme holds true when discussing temperature variation from standard, as it applies to atmospheric pressure changes. "When flying from HIGH (temperature) TO LOW (temperature) LOOK OUT BELOW." Unfortunately, we have no adjustment on the altimeter for nonstandard temperatures.

### *Altimeter Errors*

Mechanical, installation, and elastic errors also plague the altimeter. Manufacturing and installation specifications should reduce these errors to negligible amounts.

### *Types of Altitude*

We have learned about indicated altitude above sea level. There are four other kinds of altitude with which we must be familiar.

*Pressure Altitude.* If we adjust the altimeter to the Standard Atmospheric Pressure of 29.92 inches of mercury, we will read *pressure altitude* directly from the altimeter. Pressure altitude is used to determine aircraft performance and (for high altitude flights) for assigning flight levels enroute.

*Density Altitude.* If we adjust pressure altitude for nonstandard temperature, we will have *density altitude.* If the air is standard as far as temperature is concerned, there is no difference between pressure altitude and density altitude. Aircraft performance data is based on standard air density. We must compute the density altitude to find true aircraft performance. For instance, suppose the Aircraft Operations Manual states that the aircraft requires 780 feet for ground run prior to taking off, at a field which is 5,000 feet above sea level in standard atmospheric conditions. However, we note that our temperature at 5,000 feet is 20 degrees Celsius above standard for that altitude. If we use a table or compute the density altitude we will find that it is above 7,000 feet, and the ground run required will be closer to 1,000 feet.

*Absolute Altitude.* The exact height *above the surface* is called the *absolute altitude.* A radio/radar altimeter indicates this altitude by measuring the time interval of a signal which is bounced vertically from the aircraft to the surface and back. In light aircraft, we can approximate the absolute altitude by referring to the altimeter and the surface elevation as noted on the aeronautical chart. Absolute altitude is essential when flying near obstructions.

*True Altitude.* The true height *above sea level* is called *true altitude.* It can be determined by computer, but the solution is not apt to be accurate because computations must be based upon standard atmospheric pressure conditions which rarely occur.

## FEDERAL AVIATION AGENCY
## VFR EXAM-O-GRAM* NO. 8

### AIRSPEED INDICATOR MARKINGS

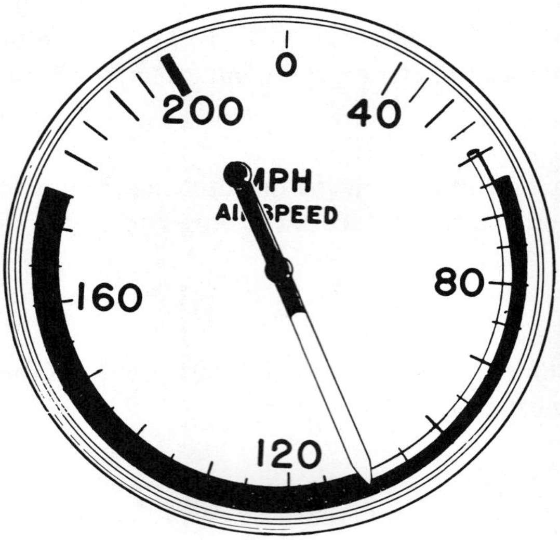


**The above airspeed indicator depicts the airspeed limitation markings of a late model civilian airplane. How many of the airspeed questions below can you answer by studying the airspeed indicator pictured above?**

1. WHAT IS THE FLAP OPERATING RANGE?
2. WHAT IS THE POWER-OFF STALLING SPEED WITH THE WING FLAPS AND LANDING GEAR IN THE LANDING POSITION?
3. WHAT IS THE MAXIMUM FLAPS EXTENDED SPEED?
4. WHAT IS THE NORMAL OPERATING RANGE?
5. WHAT IS THE POWER-OFF STALLING SPEED "CLEAN" - (GEAR AND FLAPS RETRACTED)?
6. WHAT IS THE MAXIMUM STRUCTURAL CRUISING SPEED?
7. WHAT IS THE CAUTION RANGE?
8. WHAT IS THE NEVER EXCEED SPEED?

NOTE: ANSWERS TO THE ABOVE QUESTIONS AND AN EXPLANATION OF THE STANDARD AIRSPEED INDICATOR MARKING SYSTEM APPEAR ON THE BACK OF THIS PAGE.

Airplanes manufactured after 1945 and certificated under the provisions of FAR 23 (12, 500 lbs. or less) are required to have the standard system of airspeed indicator markings described in this Exam-O-Gram. In the interest of safety, it is important for you as a pilot to recognize and understand these airspeed limitation markings. And too, this information will come in handy if you are planning to take a written examination for a pilot's certificate; current FAA written examinations contain questions on this subject. A short explanation of the airspeeds and airspeed ranges you need to know follows. The descriptions, through choice, are limited to simple layman language. (For the more technical engineering nomenclature, refer to Federal Aviation Regulations, Part 23.)

| Starting with the lower speeds and working up we have: | Airspeeds (See Illustration) |
|---|---|
| 1. FLAP OPERATING RANGE (the white arc) . . . . . . . . . | 59 to 110 mph |
| 2. POWER-OFF STALLING SPEED WITH THE WING FLAPS AND LANDING GEAR IN THE LANDING POSITION (the lower limit of the white arc) . . . . . . . . . . . . . . . . . | 59 mph |
| 3. MAXIMUM FLAPS EXTENDED SPEED (the upper limit of the white arc). This is the highest airspeed at which you can put down full flaps. If flaps are operated at higher speeds, severe strain or structural failure may result . . . . . . . . . . . . | 110 mph |
| 4. THE NORMAL OPERATING RANGE (the green arc) . . . . | 65 to 175 mph |
| 5. POWER-OFF STALLING SPEED "CLEAN" - WING FLAPS AND LANDING GEAR RETRACTED (the lower limit of the green arc) . . . . . . . . . . . . . . . . . . . . . . . . | 65 mph |
| 6. MAXIMUM STRUCTURAL CRUISING SPEED (the upper limit of the green arc). This is the maximum speed for normal operation . . . . . . . . . . . . . . . . . . . . . . . | 175 mph |
| 7. CAUTION RANGE (the yellow arc). You should avoid this area unless you are in smooth air. . . . . . . . . . . . . | 175 to 200 mph |
| 8. NEVER EXCEED SPEED (the radial red line). This is the maximum speed at which the airplane can be operated in smooth air. No pilot should ever exceed this speed intentionally. . . . . . . . . . . . . . . . . . . . . . . . . . | 200 mph |

There are other airspeed limitations not marked on the airspeed indicator which you should know. They are generally found on placards in view of the pilot or in the Airplane Flight Manual. One of these speeds, a very important one, is the MANEUVERING SPEED. This is your "rough air" speed and the maximum speed for abrupt maneuvers. If during flight you should encounter severe turbulence, you should reduce your airspeed to maneuvering speed or less in order to reduce the stress upon the airplane structure.

KNOW YOUR AIRSPEED LIMITATIONS
THIS KNOWLEDGE MAY SAVE YOUR LIFE

FEDERAL AVIATION AGENCY
VFR EXAM-O-GRAM* NO. 9

## ALTIMETRY

Your altimeter is a vitally important instrument. You will agree that flight without this instrument would indeed be a haphazard undertaking -- yet, HOW WELL DO YOU KNOW YOUR ALTIMETER? Take this short quiz on altimetry; grade yourself by checking the answers and explanations on the opposite side of this sheet.

1. Check your ability to quickly interpret your altitude by jotting down the readings of the following 6 altimeters. Allow yourself 1 minute.

(1)

(2)

(3)

(4)

(5)

(6)

2. FAR requires that you maintain your cruising altitudes (VFR as well as IFR) by reference to your altimeter. What do regulations require concerning the setting (or adjustment) of your altimeter?

3. If you are flying in very cold air (colder than standard temperatures), you should expect your altimeter to read
   a) higher than your actual altitude above sea level.
   b) lower than your actual altitude above sea level.
   c) the same as your actual altitude above sea level.

4. Here are 4 altitudes with which you should be familiar. Briefly give the meaning of each. (1) Indicated altitude. (2) Pressure altitude. (3) Density altitude. (4) True altitude.

5. Assume that your proposed route crosses mountains with peaks extending to 10,900 feet above sea level. Prior to crossing this range, you adjust the altimeter setting window of your altimeter to the current altimeter setting reported by a Flight Service Station located in a valley near the base of this mountain range. If you maintain an indicated altitude of 11,500 feet by your altimeter, can you be assured of at least 500 feet vertical clearance of these mountain peaks?

*Exam-O-Grams are non-directive in nature and are issued solely as an information service to individuals interested in Airman Written Examinations.

## ANSWERS TO QUESTIONS ON ALTIMETRY

1. 1) 7, 500 ft. 2) 7, 880 ft. 3) 1, 380 4) 8, 800 ft. 5) 12, 420 ft. 6) 880 ft.

   If your altimeter is the three-pointer-type sensitive altimeter such as those pictured on the reverse side of this sheet, an orderly approach to reading your altitude is to first glance at the smallest hand (10, 000-ft. hand); next read the middle hand (1, 000-ft. hand); and last, read the large hand (100-ft. hand). For the two-pointer altimeter, simply read the small hand first and the large hand next.

2. Your altimeter should be set to the current reported altimeter setting of a station along the route of flight (Flight Service Stations, Control Towers, etc.). If your aircraft is not equipped with a radio, you should obtain an altimeter setting prior to departure if one is available, or you should adjust your altimeter to the elevation of the airport of departure.

3. If you are flying in cold air, you should expect your altimeter to indicate HIGHER than you actually are. There is an old saying -- one well worth remembering -- that goes something like this: "WHEN FLYING FROM A HIGH TO A LOW OR HOT TO COLD, LOOK OUT BELOW!" In other words, if you are flying from a high pressure area to a low pressure area or into colder air, you had better be careful because you probably aren't as high as you think -- assuming, of course, that no compensations are made for these atmospheric conditions.

4. (1) Indicated altitude--That altitude read directly from the altimeter (uncorrected).
   (2) Pressure altitude--The altitude read from the altimeter when the altimeter setting window is adjusted to 29. 92. (This altitude is used for computer solutions for density altitude, true altitude, true airspeed, etc.).
   (3) Density altitude--This altitude is pressure altitude corrected for nonstandard temperature variations. (It is an important altitude as this altitude is directly related to the aircraft's take-off and climb performance.)
   (4) True altitude--The true height of the aircraft above sea level - the actual altitude. (Often you will see a true altitude expressed in this manner: "10, 900 ft. MSL"--the MSL standing for MEAN SEA LEVEL. Remember that airport, terrain, and obstacle elevations found on charts and maps are true altitudes.)

5. NO, you are not assured of 500 feet vertical clearance with these mountains. As a matter of fact, with certain atmospheric conditions, you might very well be 500 feet BELOW the peaks with this indicated altitude. (To begin with, 500 feet is hardly an adequate separation margin to allow on flights over mountainous terrain -- 1, 500 to 2, 000 feet is recommended in order to allow for possible altitude errors and downdrafts.)

A majority of pilots confidently expect that the current altimeter setting will compensate for irregularities in atmospheric pressure. Unfortunately, this is not always true. Remember that the altimeter setting broadcast by ground stations is the station pressure corrected to Mean Sea Level. It does not reflect distortion at higher levels, particularly the effect of nonstandard temperature.

When flying over mountainous country, allow yourself a generous margin for terrain and obstacle clearances.

## KNOW YOUR ALTIMETER

## CHECK YOUR READING

1. List the instruments which are dependent upon the pitot–static system for operation.
2. What instruments depend on the static system only?
3. If an alternate source of static pressure is in operation, what variable errors will result concerning instrument readings?
4. What errors are there in the airspeed indicator and associated plumbing?
5. What is the difference between indicated airspeed and true airspeed?
6. What is calibrated airspeed?
7. If an aircraft has a recommended indicated airspeed for approach of 75 knots at sea level, what will the recommended indicated airspeed be for an approach at 5,000 feet? Why?
8. Why does the vertical velocity indicator *lag* behind the aircraft about 6–9 seconds?
9. Explain the correct procedure to be used when reading the altimeter.
10. What is *barometric pressure reduced to sea level*?
11. Upon what kind of altitude is aircraft performance based?
12. How can we determine pressure altitude in flight?
13. What computations are required to convert indicated altitude to density altitude?
14. In the following problems, assume that the altimeter setting at the destination airport is *not* set in the altimeter window.
    A. Altimeter setting at departure airport = 30.12 inches
    Field elevation at departure airport = 195 feet
    Altimeter setting at destination airport = 30.00 inches
    Field elevation at destination airport = 220 feet
    What will the altimeter read when landing?
    B. Altimeter setting at departure airport = 29.95 inches
    Altimeter setting at destination airport = 30.05 inches
    Field elevation at departure airport = 2,420 feet
    Field elevation at destination airport = 3,700 feet
    What will the altimeter read when landing?
    C. Altimeter setting at departure airport = 29.90 inches
    Field elevation at departure airport = 2,002 feet
    Altimeter setting at destination airport = 29.75 inches
    Field elevation at destination airport = 1,750 feet
    We fly a downwind leg at 2,550 feet indicated altitude, which should be 800 feet above the field. What will be the actual height above the airport?

# 14

# GYROSCOPES AND GYROSCOPIC INSTRUMENTS

Three of our flight instruments are of gyroscopic design. The *turn needle*, *directional gyro*, and *attitude indicator* (artificial horizon) all use the gyroscope in their performance as flight instruments. Before we discuss each of these instruments, we should learn how the gyroscope works and its limiting factors.

## The Gyroscope

Any spinning mass has gyroscopic properties, be it the earth, a propeller, or a wheel on an axle. The gyroscope in the flight instrument is a small wheel which is mounted on three rings (gimbals) so that only one point, the center of gravity,

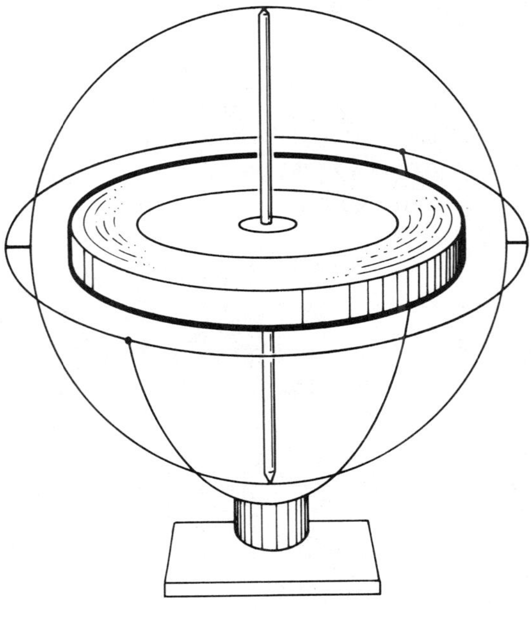

Fig. 61. The gyroscope.

is in a fixed position while the wheel is free to turn in any direction about the point, as shown in Fig. 61.

### *Gyroscopic Properties*

Because the gyroscope is a rotating body, it exhibits certain properties dictated by Newton's laws of motion.

The first law says that a body at rest will remain at rest, or if in motion in a straight line it will continue so unless acted upon by an outside force.

The second law says that a deflection of a moving body is proportional to the deflective force applied. The amount of deflection is inversely proportional to the weight and speed of the mass. That is, the heavier the body and the faster its speed, the less it will be deflected by a given force.

### *Gyroscopic Inertia*

Gyroscopic inertia (the ability to resist deflection) depends upon the following design factors.

*Weight.* A heavier mass will be deflected to a lesser degree than a lighter mass (by an equal force).

*Angular Velocity.* Rigidity, or resistance to deflection, increases as the speed of rotation of the mass increases.

*Radius at Which the Weight Is Concentrated.* Maximum gyroscopic inertia will be obtained when the principal weight of the mass is concentrated near the outer rim.

*Bearing Friction.* Friction on the gyroscopic bearings causes deflection. Minimum bearing friction keeps deflection to a minimum.

### *Gyroscopic Precession*

Gyroscopic precession causes the gyro to react to a disruptive force or pressure in a way which is different from a nongyroscopic mass. If a force is applied to the gyro about the horizontal axis, the force is met with resistance and the gyro, instead of turning about its horizontal axis, turns or *precesses* about its vertical axis.

In other words, when a disruptive force is applied to a gyroscope, it will precess in a direction ninety degrees from the applied force.

### *Gyroscopic Mounting*

The gyroscope in the flight instrument may be mounted so as to use specific properties. A freely mounted gyro is mounted on three gimbals and the gyro is free to rotate in any plane. The semirigid mounting uses two gimbals limiting the gyro to two planes of rotation.

## Sources of Power

The gyroscopic rotor must be made to spin at very high speed (18,000 rpm is not uncommon). Two power sources are available in aircraft to drive the rotor.

In most light aircraft, a vacuum system will normally be used to drive the attitude indicator and the directional gyro. An electrically operated gyro will be used in the turn and bank indicator so that even with vacuum system failure, we will have a reliable standby instrument for emergency operation.

*The Vacuum System*

The vacuum system sucks air through the gyro instrument case where it moves over rotor vanes and, just as a river current drives a water wheel, causes the rotor to spin. Either a venturi or vacuum pump can be used to provide the suction. Depending on the instrument, suction must be between 1.8 and 5.0 inches of mercury.

*The Venturi.* A venturi is simply a restricted tube, as shown in Fig. 62, which faces into the airstream. As the air rushes through the tube, its pressure is reduced in the area of the restriction.

A pipe is connected to the venturi at the point of the restriction, and air is sucked through it which tries to "fill" the low pressure area. Hence, suction is produced on the other end of the pipe.

The venturi is a simple and inexpensive device, but it suffers from definite limitations. As airspeed varies, so does the amount of suction produced. Structural ice may clog the venturi. Full efficiency is not realized until the aircraft approaches cruising speed. Therefore, the gyroscopes do not rotate at full speed until well after takeoff.

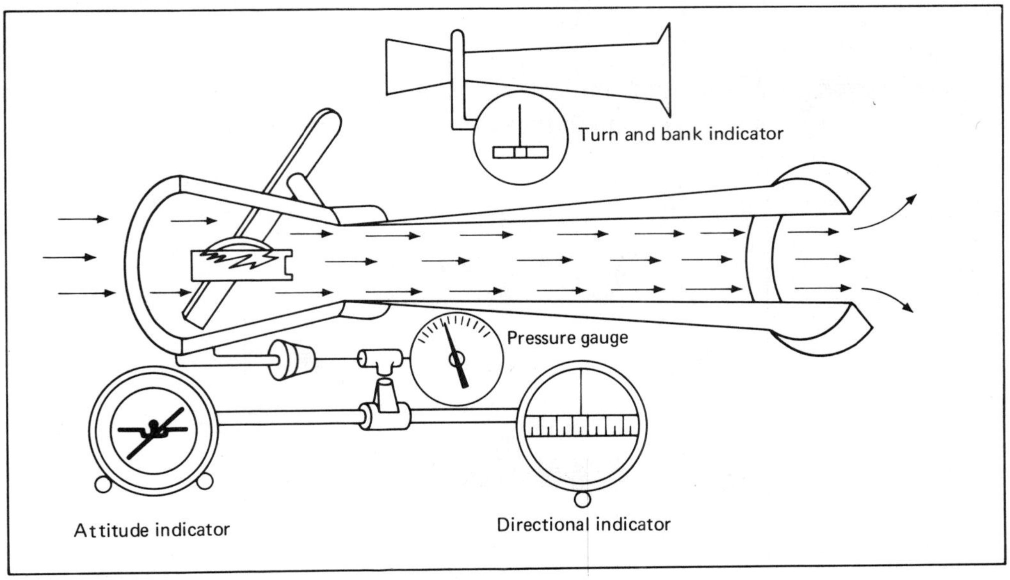


Fig. 62. The venturi.

*The Engine–Driven Vacuum Pump*

The engine-driven vacuum pump is another common source of vacuum. A vane type pump is connected to the accessory drive shaft of the engine. Pump size and capacity naturally vary with different aircraft, but most engine-driven vacuum pumps are capable of producing at least ten inches of mercury suction at engine speeds above 1,000 rpm.

The engine-driven vacuum pump is more complicated than a venturi, with an air–oil separator, relief valve, restrictor valve, etc., but it allows gyro instruments to function prior to takeoff, is not affected by engine or aircraft speed, and is not susceptible to structural ice.

A suction gauge in the cockpit indicates the amount of vacuum pressure in the system. The attitude indicator and the directional gyro require 3.5 to 5.0 inches of mercury suction. The turn and bank indicator is equipped with a restrictor valve which reduces suction to the desired operating range of 1.8 to 2.1 inches of mercury.

### *The Electrical System*

Electrical energy may also be used to drive the rotor within the gyroscopic instruments. The gyro is simply a small electric motor and flywheel. Normally, if the turn and bank indicator is electrically powered, it will be connected directly to the battery so that it will operate regardless of generator failure, as long as current is avàilable at the main bus bar.

## The Attitude Indicator

The *attitude indicator* or *artificial horizon* is designed to show the attitude of the aircraft in relation to the natural horizon. There is some variety in the design of modern attitude indicators, but all relate a gyro controlled horizon line to a pitch and roll reference. A typical attitude indicator is depicted in Fig. 63.

### *Description*

The attitude indicator is equipped with a horizon bar (representing the natural horizon), a miniature aircraft, a bank indicator, and a bank index. The bank

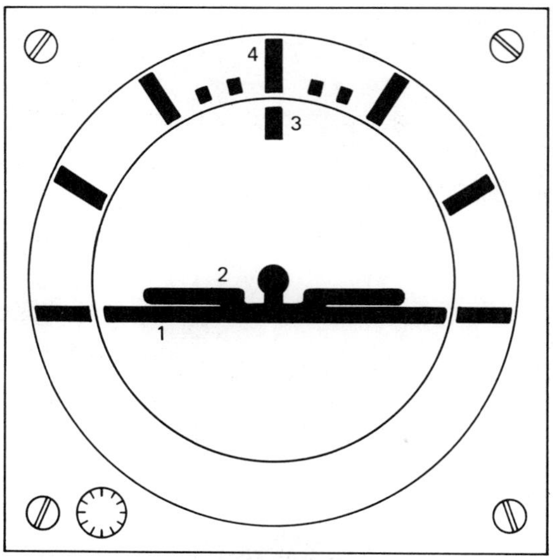


Fig. 63. The attitude indicator.

1. Horizon bar
2. Miniature aircraft
3. Bank indicator
4. Bank index

index is engraved to show ten degree increments of bank up to thirty degrees, and thirty degree increments up to ninety degrees. Some models are equipped with a *cage knob* which is used to lock all of the gimbals in a rigid position.

The miniature airplane is adjustable vertically so that each pilot may adjust its height for his own eye level.

The attitude indicator presents an instantaneous picture of our pitch and bank attitudes. The direction of the bank is shown by reading the relationship between the wings of the miniature airplane and the horizon bar. The bank indicator and bank index show the amount of bank, but in the opposite direction. Therefore, we must use caution in referring to the **miniature airplane wings** in order to determine the direction of bank and to use the bank index to determine *only* the amount of bank. Figure 64 shows an attitude indicator which is depicting a *left* turn.

Fig. 64. The attitude indicator–a left turn.

*Limits of Operation*

If we rotate the gimbals of the gyro beyond their limits, the gyro will tumble. The bank limits of the vacuum driven attitude indicator are about 100 to 110 degrees, and the pitch limits vary from 60 to 70 degrees. As the rotor gimbal hits the stops, the rotor will precess abruptly causing excessive wear on the bearings. The rotor will normally precess back to the horizontal plane at a rate of about eight degrees per minute.

Since the limits of light plane gyros are well beyond the attitude restrictions of *normal category* operations, gyro tumbling does not normally occur.

*Precession of the Attitude Indicator*

The attitude indicator precesses during normal flight but is designed to return to the normal erected position after the precession force stops.

In a normal turn the gyro will precess towards the inside of the turn. The maximum amount of precession will occur in 180 degrees of turn. If the turn is continued, the precession will automatically be cancelled out during the second 180 degrees of turn. If we roll-out after completing 180 degrees of turn, the bank index will depict 3.5 degrees of bank, even though the wings are level. The erecting mechanism within the gyro will correct this precession in about thirty seconds.

Acceleration and deceleration also produce precession errors. During acceleration, such as for takeoff, the attitude indicator will erroneously indicate a nose-high attitude. This is particularly unnerving on the climbout after takeoff because the attitude indicator depicts a steeper climb attitude than is occuring in reality, causing a tendency to lower the nose. During deceleration, the attitude indicator will depict a nose-low attitude. Again, this can be a dangerous indication because, as we slow the aircraft speed we receive an erroneous indication from the attitude indicator which tells us to raise the nose, producing a potentially dangerous combination of low airspeed and high pitch attitude.

## The Directional Gyro

We learned that the magnetic compass is quite susceptible to oscillation error due to turbulence or rough pilot technique. It is also plagued by turning errors and acceleration errors. Therefore, the *directional gyro* or *heading indicator* is used for directional information. The directional gyro is *not* a magnetic instrument. It simply uses the gyroscopic principle of rigidity in space to present a heading reference which is not as susceptible to oscillation, turning, or acceleration errors. A typical directional gyro is depicted in Fig. 65.

### *Description*

The instrument case is equipped with a window through which the horizontal gimbal is viewed. A compass card is attached to the gimbal which remains rigid in space. As the aircraft, and thus the instrument case, rotates around the rigid gimbal, the compass card provides accurate directional information.

Since the directional gyro is not magnetic we must set it to match the magnetic compass reading. We must use caution when setting the directional

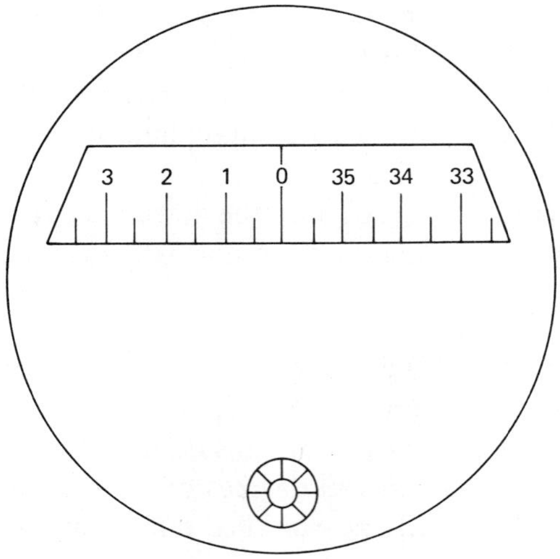


Fig. 65. The directional gyro.

gyro to be sure that the magnetic compass is not under the influence of any error. For instance, we cannot set the directional gyro when the magnetic compass is oscillating. The aircraft must be in straight and level, unaccelerated flight.

After setting the directional gyro, we must uncage it by pulling out or releasing the setting knob. Some pilots make it a habit to turn the setting knob after uncaging the gyro to be sure that the gyro is uncaged.

*Limits of Operation*

The directional gyro is designed to operate normally with up to 55 degrees of aircraft pitch and bank. If more extreme attitudes are encountered, the precessional forces will cause the compass card to spin rapidly. We may correct this action by caging, resetting, and uncaging the gyro.

Bearing friction causes the compass card to drift slowly. An unbalanced gyro will show increased precession. We should check the directional gyro every 15 minutes during flight and reset it to match the magnetic compass if required.

## The Turn and Bank Indicator

The *turn and bank indicator* (sometimes called the *turn and slip indicator*, or *needle and ball*) is actually two instruments combined in one case. The turn needle is gyro operated, while the ball reacts to gravity and/or centrifugal force. A typical turn and bank indicator is shown in Fig. 66.

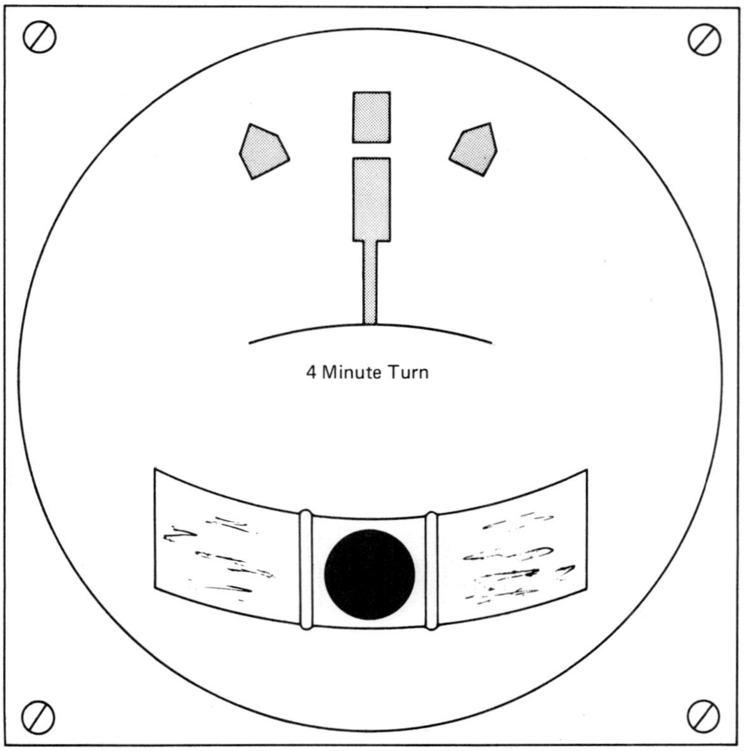


Fig. 66. The turn and bank indicator.

*The Turn Needle*

The gyroscope in the turn and bank indicator is mounted semirigidly, so that the gyro rotates freely about the lateral and longitudinal axes.

Gyroscopic precession causes the rotor to tilt when the aircraft turns. The rotor is mounted so as to spin up and away from the pilot. Precession is presented to the face of the instrument with linkages which activate the turn needle.

The turn needle indicates the *rate* at which the aircraft is turning (by the amount of its displacement from center) and the direction of the turn (by the direction of its displacement).

*The Ball (Slip Indicator)*

The slip indicator consists of a curved glass tube which is sealed at both ends and contains white kerosene and a black ball, which is free to move. The fluid provides a dampening action to the ball. Because the tube is bent upwards on each end, the ball tends to seek the middle of the tube (the lowest portion), when the instrument is perfectly horizontal. Two reference lines mark the center of the tube.

When the aircraft is straight and level, gravity acts to keep the ball in the center (lowest point) of the tube. During a coordinated turn, both gravity and centrifugal force keep the ball centered. An uncoordinated turn, however, will cause the ball to drift from center.

*Using The Turn and Bank Indicator*

When in flight, the turn needle may be controlled by either the aileron or rudder, for either control will cause horizontal movement of the aircraft nose. So may the ball be controlled by either the aileron or rudder, because either control

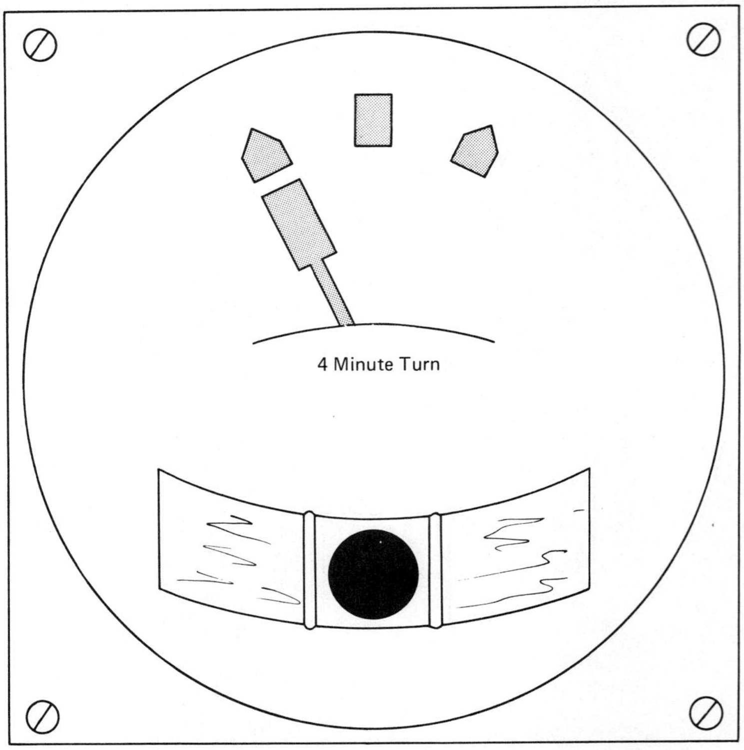


Fig. 67. The turn and bank indicator—a coordinated left turn.

affects the amount of centrifugal force produced, which affects the ball. However, most pilots prefer to think of the aileron as the primary control for the turn needle, and the rudder as the primary control for the ball.

If the aircraft is properly trimmed in straight and level flight, the needle and the ball will be perfectly centered.

During a coordinated turn, the needle will be displaced from the center, in the direction of the turn. The ball will remain centered due to the combination of gravity and centrifugal forces. This is pictured in Fig. 67.

If the ball moves toward the inside of the turn, it indicates that the force of gravity is predominant over centrifugal force. In other words, the rate of turn is insufficient to produce enough centrifugal force to hold the ball centered. The ball and the aircraft are falling toward the inside of the turn, or the aircraft is said to be *slipping*. This condition is depicted in Fig. 68.

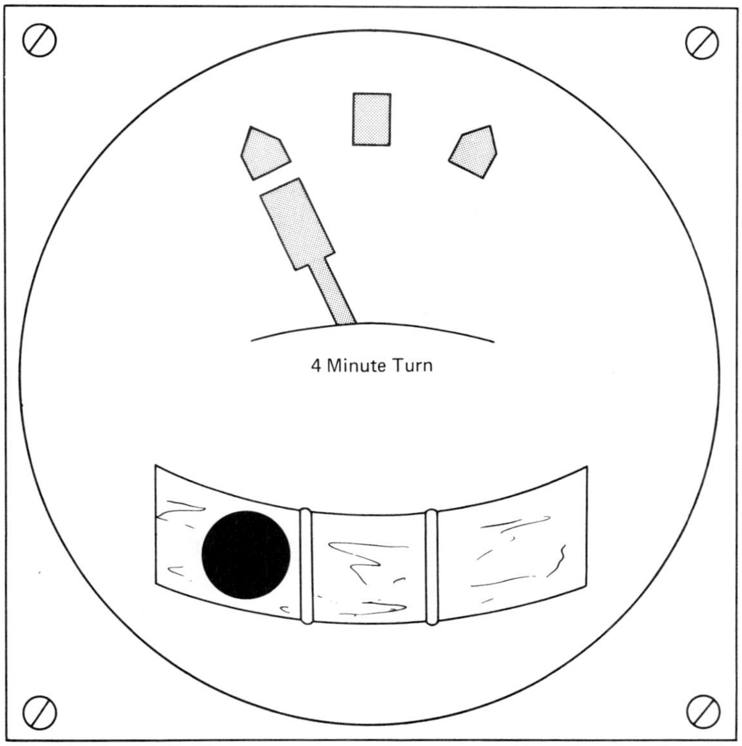


Fig. 68. The turn and bank indicator–a slipping left turn.

The ball may be centered by returning to coordinated flight in either of two ways. The ball may be centered by increasing the rate of turn (by increasing rudder pressure). The increased rate of turn will produce more centrifugal force which will move the ball to the center. The ball may also be centered by shallowing the bank, thus, in effect, allowing gravity to pull the ball toward the center of the tube.

If, conversely, the ball moves towards the outside of the turn, it indicates that too much centrifugal force is present. In other words, the rate of turn is too rapid considering the amount of bank present. In this instance, the aircraft is said to be *skidding*. This condition is depicted in Fig. 69.

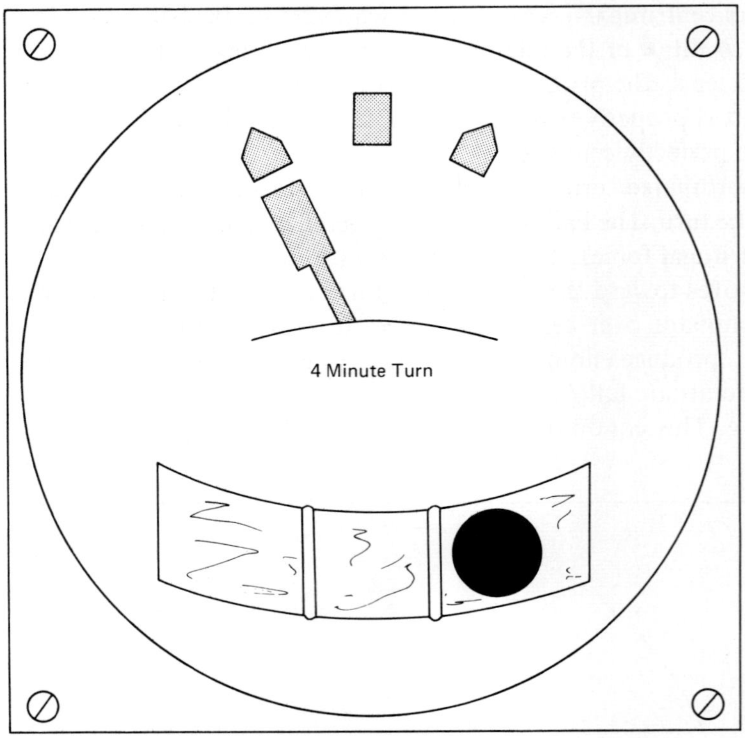


Fig. 69. The turn and bank indicator—a skidding left turn.

We can center the ball by reducing the amount of rudder pressure or by increasing the bank by applying aileron pressure.

In short, the ball can be compared to an automobile which is negotiating a steep curve. If the road is banked to the proper degree, the auto will remain in the center of the road. If the road is not banked sufficiently, the auto will tend to skid toward the outside of the turn. If the road is overbanked, the auto will slip toward the inside of the turn.

Many pilots make the mistake of trying to "fly the instrument" instead of flying the aircraft and using the instrument as a reference to determine the quality and quantity of the turn. We should fly the aircraft, and use the instrument to determine how well we are doing.

*Limits of Operation*

The gyro assembly is designed with stops which prevent it from tilting more than 45 degrees either side of the vertical. The ball is limited only by the end of the glass tube in which it is contained. Normally flight in a light aircraft will not approach the limits of the turn and bank indicator.

## CHECK YOUR READING

1. What are the three gyro operated flight instruments?
2. List and describe the four factors which determine gyroscopic inertia.
3. Why is the turn and slip indicator in most light aircraft operated electrically, while the attitude indicator and directional gyro are vacuum operated?

4. Describe how a venturi works.
5. What does the attitude indicator tell us?
6. What part of the attitude indicator properly depicts the direction of bank?
7. What instrument depicts the quality of a turn?
8. Why must the directional gyro be reset every 15 minutes?
9. In reference to the magnetic compass, what precautions must be taken prior to resetting the directional gyro?
10. What does the needle portion of the turn and bank indicator depict?
11. What is the common primary control for correction of the position of the ball on the turn and slip indicator?
12. What are the operating limits of the attitude indicator?
13. What potentially dangerous condition may be caused by acceleration error within the attitude indicator?

# 15

# THE PREFLIGHT INSPECTION

One of the most important procedures we can accomplish to guarantee safe flight is the preflight inspection of our aircraft. Unfortunately, it seems that new, "inexperienced" pilots are more apt to perform an adequate preflight inspection than the old timers. Maybe many of them are discouraged by the fact that so few discrepancies are ever found on the aircraft, day in and day out. However, this thinking is wrong, because just one substandard item on the aircraft may lead to serious difficulties in the air. We must get into the habit of performing a thorough preflight inspection, and we must complete one prior to every flight. Naturally, our preflight inspection will be more detailed prior to the first flight of the day than the one we might make after stopping for lunch. Perhaps, prior to the second and subsequent flights of the day, a quick *walk around* inspection will suffice.

Because of the many different aircraft models in use today, it is impossible to discuss here every item on every aircraft which should be checked on the preflight inspection. This duty will fall to the Flight Instructor, who is familiar with the idiosyncrasies of each particular aircraft he flies, and who will show us specific check points to which we should pay particular attention during the preflight. Neither is any attempt made here to define the specific order in which the inspection should be accomplished. Most of the items described here are common to most light aircraft, however, and will be included in the preflight inspection of the aircraft we fly.

## The Cockpit

The preflight inspection should begin in the cockpit, where we must determine that all switches are in the *off* position. We will be inspecting the engine compartment and propeller area shortly, and will not want the engine to start inadvertently.

*Paperwork*

While in the cockpit, we should check the paperwork for the aircraft to see that those documents which are required by Federal Air Regulations 91.27 and 91.31 are in the aircraft and properly filled out.

The following documents are required by *federal* law:

1. A registration certificate issued to the aircraft owner.
2. An appropriate and current airworthiness certificate. (The airworthiness certificate is renewed annually by maintenance personnel and certifies that the aircraft is airworthy.)
3. A current FAA approved Aircraft Flight Manual for the aircraft, placards, listings, instrument markings, or any combination thereof, containing each operating limitation prescribed for that aircraft. (Notice that Aircraft Operations Limitations may be comprised of *placards* and *instrument markings*. On many light aircraft, the colored speed range arcs on the airspeed indicator, for instance, are part of the operations limitations. The weight and balance documents and data are also part of the Aircraft Flight Manual.)

In addition to these documents, state and city laws may require registration certificates. If the aircraft is equipped with an operational two-way radio, it will need to be licensed by the Federal Communications Commission.

*Control Locks*

We should unlock the control locks while in the cockpit so that we can check the aileron, elevator, and rudder movement. The flight controls should be moved in both directions, to their full deflection, to insure that the control surfaces are free and moving in the right direction.

## Outside The Aircraft

Most pilots prefer to begin their preflight inspection at the same point on the aircraft every time they perform it, so as to develop a habit pattern. Each of us can determine for himself the most convenient starting point on the aircraft, but, for the purpose of this discussion, we will begin at the nose of the aircraft.

*The Propeller*

The propeller can be a lethal weapon. Even though we checked to be sure that the magneto switches were off, a frayed or disconnected wire might have left the magneto "hot." Therefore, we must use ***extreme caution*** when near the propeller, and never place ourselves within the turning radius of the blades.

The leading edge of the propeller should be inspected for nicks, dents, or cracks. The propeller dome must be securely fastened and free of cracks.

*Cowling*

The engine cowling serves two purposes. It directs ram air around the cylinders to cool them and it streamlines the engine, reducing drag. It must be securely fastened at all attach points. Baffles, openings, and slots in the cowl must be free of all debris to insure proper engine cooling. It is surprising how fast birds can build a home of straw and twigs in the openings of the cowl. Naturally, when the engine begins to develop heat, these nests constitute a fire hazard.

*Oil Cooler Screen*

Most light aircraft engines are equipped with a ram air oil cooler situated in the front of the engine cowl. The cooler screen must be clean if proper oil cooling is to result. Bugs, dirt, and grass clippings must be removed from the oil cooler screen.

*Nose Wheel Assembly*

The nose wheel strut should be checked for proper inflation, as well as the security of nose wheel linkages, cables, cotter pins, and bushings. Most nose wheel assemblies include a *shimmy dampener* which is designed to keep the nose wheel from oscillating. We should attempt to wiggle the nose wheel to test the shimmy dampener.

Proper tire inflation is important, as is even wear. If the nose wheel is equipped with a wheel pant, we should roll the aircraft a foot or two so that the entire wheel becomes visible.

*Engine Compartment*

Most new light aircraft are very closely cowled. Therefore, while the engine compartment hatch is open, we should carefully inspect the entire engine area for loose wires, oil spills, and exhaust leaks. Many of these flaws will go undetected if we do not spend an extra minute or two in a careful search for them.

The oil quantity must be checked to be sure that it is within the engine manufacturer's specifications. When we check the oil, it is a good idea to *never let the oil filler cap out of our hands*. If we lay it aside, we may forget to replace it, and when in flight, the oil will siphon out of the engine. Replace the filler cap tightly.

*Fuel Strainer Drains*

As we learned in Chapter 11, which concerns fuel and fuel systems, every aircraft is equipped with fuel strainer drains. During the preflight inspection we should drain fuel from each strainer to determine its cleanliness and water content. We should use a small glass jar to collect a fuel sample from the strainer. Since water is heavier than fuel, it will settle to the bottom of the jar and be quite visible. Generous amounts of fuel must be drained, particularly if the aircraft has not been flown recently, to guarantee that no water remains in the fuel system.

The windshield, side, and rear window panels must be clean, and free of cracks. *Windows*

When inspecting the main landing gear, we should pay particular attention to the hydraulic brake lines and connections, making sure that there is no evidence of leaking fluid. The condition of the tire and wheel assembly is important enough to roll the aircraft in order to inspect the portion which may be hidden by wheel pants. *Main Landing Gear*

As we move along the leading edge of the wing, we should be critical of any dents or tears in the skin. Wrinkles on the top or bottom of the wing indicate possible structural damage and should be inspected by a licensed airframe mechanic. *Wing*

The navigation light on the wing tip should be inspected for cleanliness. Prior to any night flight, we should check the light with the power on.

The object of our aileron inspection will be to determine its structural integrity, including hinges, cotter pins, and control rods. We should also test the movement of the aileron to see that it moves freely and to the full extent of its travel. Use caution when moving the aileron by hand, to keep it from slamming against the stops. *Aileron*

The flaps should be inspected for structural integrity as well as clean and crack-free slide tubes or slots. The trailing edge of the flaps often shows damage from vibration in the form of loose or missing rivets. *Flaps*

We should inspect the entire fuselage for damage or wrinkled skin. The belly of the aircraft will normally show signs of oil staining due to oil vapors from the engine oil breather, but no wet or dripping oil should appear. *Fuselage*

When we inspect the tail of the aircraft, we should carefully check the cables, turnbuckles, and lock nuts of all moveable surfaces. Again, the controls should have free movement to their limits. The condition of the tail light and anti-collision beacon should also be noted. *Tail*

The landing/taxi light lens should be carefully inspected for cracks and cleanliness, particularly prior to a night flight. *Landing/Taxi Light*

Many modern aircraft are equipped with a stall warning indicator, normally located on the leading edge of the wing. This moveable vane detects the approaching stall by measuring the angle of attack of the relative wind, and *Stall Warning Indicator* activates a signal (buzzer or light) in the cockpit. During the preflight inspection, we should check to determine the cleanliness of the stall warning device.

*Pitot-Static System*

The pitot-static system is essential for the operation of the airspeed indicator, vertical speed indicator, and altimeter. The pitot head is an open tube which faces into the relative wind. Because it is a favorite nesting place for flying insects, it deserves our special attention during the preflight inspection. However, *under no circumstances should we attempt to clean the pitot head by blowing into it*. This action may seriously damage the sensitive instruments to which the pitot tube is attached.

The static port vents outside atmospheric pressure to the instrument cases. It, too, must be free of all dirt. The location of the static port varies on different aircraft, but it is normally found on the side of the fuselage, near the cabin. It is a small hole, about the size of a pin, and therefore can easily be clogged by dirt, soap, or wax.

*Fuel Quantity*

The fuel quantity must be checked *visually* during the preflight inspection. On high-wing aircraft, we may need a stepladder in order to get a good look into the fuel tank filler neck. As we check the fuel quantity, we should also check to see if the fuel is the proper color (which denotes fuel grade) for the aircraft.

It is imperative that the fuel tank caps be tightened properly, for fuel will siphon out of the tank in flight if the covers are not properly secured.

## Summary

A thorough, conscientious preflight inspection is worth every minute of the time it takes to complete. After we form an habitual pattern, the inspection should be completed in about 5 minutes on most light aircraft.

Each pilot will eventually develop a list of items for inspection which are not listed here. Experience will add to the list of favorite items to be checked.

### CHECK YOUR READING

1. What should be the first item on the preflight inspection of every light aircraft?
2. What papers are required by *federal law* to be aboard an aircraft prior to flight?
3. Describe the engine cowling and propeller "critical" check points.
4. What precautionary action is required when inspecting an aircraft tire and wheel assembly which is equipped with wheel pants?
5. What precaution must be observed when checking the engine oil quantity?
6. What items are of concern when inspecting any moveable control surface?
7. Why must we refrain from blowing into the pitot head in order to clean it?
8. Modern aircraft are equipped with fuel quantity gauges. Why then must we visually check the fuel quantity?

# 16

# ENGINE STARTING AND TAXIING

Many aircraft are seriously damaged on the surface because of poor engine starting techniques or improper taxiing habits. Our flight begins well in advance of the takeoff. From the moment we arrive at the airplane, we must exercise all of the knowledge and experience at our command.

## Engine Starting

Because the exact procedure used to start our aircraft engine will depend on the particular engine, it is impractical to specify in this text the details of that procedure. However, a few practical tips can be discussed here, which are applicable to every aircraft.

### *The Checklist*

No pilot can commit to memory every item which should be checked prior to starting the engine and guarantee that he will never delete one or more of the items in repeated starting attempts. Therefore, a detailed checklist should be used every time the engine is started. Many pilots prefer to say aloud each item on the checklist in an effort to conscientiously complete the prestart procedure.

### *The Propeller*

The propeller can be a deadly weapon. Before we activate the magneto switch or the starter, we must be absolutely positive that no person is within range of the propeller. We should shout "clear" loudly, and wait for a possible reply before cranking the engine.

If possible, the engine should be started when the propeller is over a hard surface, which is cleaned of all pebbles and dirt, for the propeller will suck up foreign objects and hurl them rearward. The leading edge of the propeller blades may be damaged by flying debris.

The area directly behind the aircraft may also be critical. We should be sure that persons or property to our rear are not close enough to be damaged by the debris which is hurled rearward or by the propeller blast.

Most light aircraft are now equipped with electric starters. It is a good idea to check a spring-loaded starter handle to determine that it returns to the disengaged position immediately after the engine starts. An engaged starter may cause permanent damage to the engine in a very short period.

*The Throttle*

We will soon discover that more power will be required to begin our taxi roll than to continue motion. For that reason, power should be decreased slowly and smoothly once the aircraft begins to move. Abrupt throttle changes abuse the engine.

Many light aircraft are very closely cowled. High power settings for an extended period of time while on the ground may seriously overheat and damage the engine. We should use only the amount of throttle necessary to accomplish our goal. The habit of using brakes and throttle simultaneously is a bad one. Throttle causes the aircraft to move, and brakes cause the aircraft to slow down or stop; they should not be used concurrently.

If our engine is not equipped with a cylinder head temperature gauge, we must be especially careful to avoid overheating the cylinders. Severe damage may be inflicted by cylinder heat before the oil temperature gauge reflects the excess heat. A good indication of sufficient engine heat for takeoff is that the engine may be accelerated smoothly but rapidly without skipping or overloading.

## Taxiing

The throttle, brakes, and flight controls may be used quite effectively to provide positive control of the aircraft while taxiing. They may also be abused, which will aggravate a potentially dangerous situation until disaster strikes.

*Taxiing into a Headwind*

When we taxi into a direct headwind, we will notice that the flight controls are quite effective. More power will be required to move the aircraft into the wind than will be required when moving downwind.

The ailerons should be held in neutral, and the elevator should be held in neutral or slightly forward.

*Taxiing into a Quartering Headwind*

If we taxi into a quartering headwind, the upwind aileron should be up; that is, aileron should be applied into the wind. This will help in holding the upwind wing down. The aircraft will tend to turn into the wind, just as a weathervane.

*Taxiing Crosswind*

When taxiing with a direct crosswind, the tendency of the aircraft to weathervane will be most pronounced. The upwind aileron should be held up, and the elevator should be neutral or slightly forward.

*Taxiing with Direct Tailwind*

Taxiing downwind will require very little power. The ailerons should be held in neutral, and the elevator should be held slightly forward of neutral.

*Taxiing with a Quartering Tailwind*

The most dangerous of all wind conditions for taxiing, particularly with a tricycle geared airplane, is the quartering tailwind. The aircraft has a tendency to weathervane and to tip over, towards the direction in which the wind is blowing. The upwind aileron should be held down; that is, aileron should be applied with the wind. The elevator should be held slightly forward of neutral. Abrupt throttle movement must be avoided.

*Turning*

All turns must be accomplished at a low speed because as the aircraft turns, centrifugal force tends to tip it towards the outside of the turn.

Turning from a downwind condition is particularly dangerous and requires extreme caution.

Experience in each specific type of airplane will increase our ability to judge the turning radius required. If we are not absolutely sure that sufficient room exists for a safe turn, we must not attempt it.

*Taxiing Speed*

It is difficult to specify an exact speed for safe taxiing. Wind, surface, and congestion conditions vary from day to day, and from location to location. Taxi speed must not be so great as to leave any doubt as to loss of control or directional stability.

*Taxiing Near Obstructions*

Those of us who are accustomed to driving an automobile sometimes forget that our aircraft is equipped with two very large wings, which require much room for maneuvering. Also, the tail of many airplanes swings through a large arc as the aircraft turns and in some cases actually moves in a direction opposite to the aircraft nose during a tight turn.

When nearing a congested area, we must double our watchfulness. There is no excuse for taxiing into an area which is so congested as to cause any shadow of doubt concerning a collision.

## CHECK YOUR READING

1. Why should a checklist be used during the prestart procedure?
2. Why should the aircraft be started while the propeller is over a hard, clean surface?
3. Why must we shout "clear" prior to engaging the electric starter?

4. What hazard exists if the aircraft engine is run at high power for long periods of time while on the ground?
5. What is the most dangerous wind condition while taxiing?
6. Describe the position of the flight controls while taxiing in a quartering tailwind from the right.

# 17

# TAKEOFF CONDITIONS AND PROCEDURES

The takeoff segment of each flight is one of the more critical periods. The aircraft is at the highest weight expected for the particular flight; the aircraft is on or near the surface; and the airspeed and power control are of critical importance.

Therefore, to insure safe flight, we must devote very special study to the problems concerned with takeoff.

## The Effect of Wind on Runway Length Required

The natural wind greatly affects the length of the runway which will be required for takeoff. Let us assume that the airspeed of our aircraft at lift-off must be 60 knots. In effect, this means that 60 knots of wind must pass over the lifting surfaces in order to produce sufficient lift for flight. Assume also that our aircraft will accelerate to 10 knots in 5 seconds, or 60 knots in 30 seconds.

If there is no natural wind blowing down the runway, the aircraft will accelerate to lift-off airspeed (60 knots) in thirty seconds. Our *average* speed (from zero to 60 knots) will be 30 knots, or about 50 feet per second. After 30 seconds, we will have travelled 1,500 feet down the runway.

Now let's see what effect a 20-knot wind blowing directly down the runway will have on our takeoff, if we take off *downwind.*

As we point our aircraft down the runway with the wind on our back, we actually have a negative 20-knot wind passing our wings. In other words, our aircraft will have to travel down the runway at 20-knots just to cancel out the tailwind, which would result in no wind passing the wings. Then the aircraft will need to accelerate an *additional* 60-knots, to a total groundspeed of 80 knots, in order to create 60 knots of airspeed for lift-off. Let's see how much runway will be consumed in this takeoff using the identical aircraft performance figures.

With a rate of acceleration of 10 knots every 5 seconds, we will require 40 seconds to accelerate to 80 knots groundspeed (60 knots airspeed). Our average ground speed (from zero to 80 knots) will be 40 knots, or about 66 feet per second. After 40 seconds, we will have travelled 2,640 feet down the runway. This represents an increase of 1,140 feet required to accelerate the identical aircraft to an identical airspeed of 60 knots, with a 20-knot tailwind.

Finally, let's look at the figures for a takeoff *into* the 20-knot wind, which is the correct procedure.

When facing down the runway, with a 20-knot headwind, we have in effect a 20-knot wind passing the lifting surfaces, even though our aircraft is standing still. We need only to accelerate to forty knots *groundspeed* to produce a 60-knot relative wind over the lifting surfaces. A simple computation, again using identical aircraft performance figures, will show us the amount of runway required for acceleration to liftoff speed.

We will accelerate to 40 knots groundspeed at 10 knots per 5 second interval, which will take 20 seconds. Our average groundspeed (from zero to 40 knots) will be 20 knots, or about 34 feet per second. After 20 seconds, we will have travelled only 680 feet down the runway.

The difference in runway required to accelerate to 60 knots lift-off speed with a 20-knot tailwind as compared to a 20-knot headwind is 1,960 feet. That is a severe penalty to pay for taking off downwind (especially if the runway is only 1,500 feet long).

### *Headwind Component*

In the previous discussion, we have been speaking of a 20-knot wind blowing directly down the runway. More often than not, the wind will be blowing across the runway to varying degrees. The amount of the crosswind which acts as a headwind can be computed by using a Headwind Component Chart, as shown in Fig. 70.

## The Effect of Wind on Landing

For instance, as shown on the chart, a 30-knot wind which is blowing across the runway at a 30-degree angle has the same effect on the takeoff distance required as a 26-knot wind which blows directly down the runway.

Similar computations concerning aircraft deceleration would show that the wind greatly affects the runway length required for landing, shortening the landing roll considerably. This is reflected in note number 2 on the sample Landing Distance Table shown in Fig. 71.

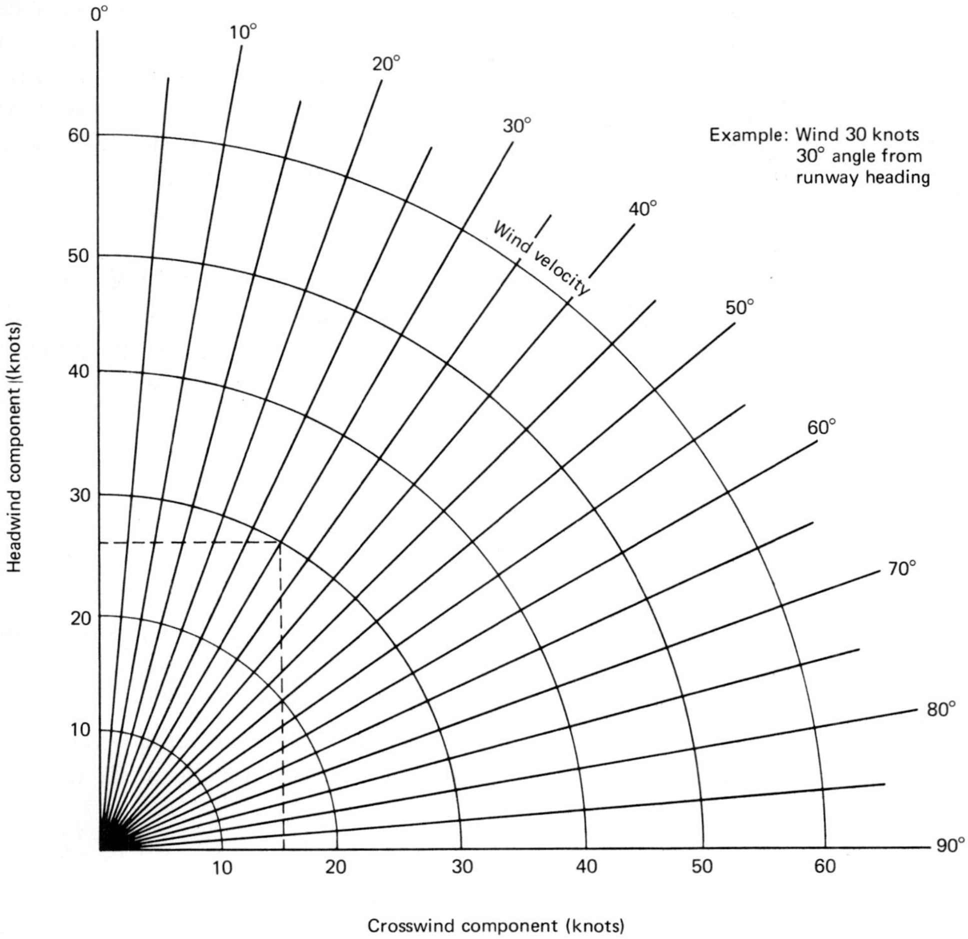


**Fig. 70. Headwind component chart.**

| Landing distance table<br>Landing distance — 45° flaps — hard surfaced runway | | | | | | | |
|---|---|---|---|---|---|---|---|
| Gross weight (lb) | Approach I.A.S. (knots) | Sea level at 59° F | | 2,500 ft at 50° F | | 5,000 ft at 41° F | |
| | | Ground roll | Total to clear 50-ft obst. | Ground roll | Total to clear 50-ft obst. | Ground roll | Total to clear 50-ft obst. |
| 2,800 | 72 | 490 | 1,360 | 515 | 1,440 | 585 | 1,620 |

Notes: 1. Distances shown based on zero wind, power off, max. braking.
2. For each 4 knots headwind, reduce landing distance by 10%.
3. For operation on dry grass runway, increase distances by 20%.
4. Increase distances 10% for each 15° F above standard.

**Fig. 71. Landing distance table.**

Even more interesting, however, is a simple formula which will show us the amount of *energy* which must be absorbed by an aircraft if it strikes an unmoveable object such as the ground or a stone wall on the end of the runway. The formula is $E = MV^2$, where $E$ equals energy, $M$ equals mass (or weight) and $V$ equals velocity. An important thing to remember is that the velocity in the formula is the speed of the object in relation to the unmoveable object it is about to strike, i.e., the ground. (Therefore, groundspeed will be used in the formula, *not* airspeed.)

For this example, we will assume that our aircraft weighs 2,000 pounds, and that we are on takeoff into a 20-knot headwind. Just at the time of lift-off, at an airspeed of 60 knots, we strike a solid object. Because of the 20-knot headwind, our *groundspeed* will be 40 knots or about 66 feet per second.

$$E = MV^2$$

$$\text{Energy} = 2{,}000 \text{ pounds} \times 66^2 \text{ feet per sec}$$

$$\text{Energy} = 2{,}000 \text{ pounds} \times 4{,}356 \text{ feet per sec}$$

$$\text{Energy} = 8{,}712{,}000 \text{ foot/pounds}$$

When the aircraft strikes the solid object, 8,712,000 foot/pounds of energy must be absorbed. How much energy would have to be absorbed under the same circumstances except for a downwind takeoff?

The only figure in the formula which will change is the velocity (groundspeed) which, with the 20-knot tailwind will be 80 knots, or about 132 feet per second.

$$E = MV^2$$

$$\text{Energy} = 2{,}000 \text{ pounds} \times 132^2 \text{ feet per sec}$$

$$\text{Energy} = 2{,}000 \text{ pounds} \times 17{,}424 \text{ feet per sec}$$

$$\text{Energy} = 34{,}848{,}000 \text{ foot/pounds}$$

This time, with the 20-knot tailwind, the energy which must be absorbed when the aircraft strikes the solid object is 34,848,000 foot/pounds—an increase of 26,136,000 foot/pounds.

Hence, in the event of an unexpected power loss on takeoff, resulting in contact with the surface, we would be far better off if we performed the takeoff into the wind.

## Atmospheric Conditions Affecting Takeoff Performance

Mother Nature controls three additional factors which affect aircraft performance on takeoff, all having to do with the "thickness" or density of the atmosphere. Naturally, the more dense the air, the more lift will be developed by the lifting surfaces at a specific airspeed, because there will be more molecules of air present in each cubic foot of atmosphere.

*Temperature Effect*

As the air temperature increases, the molecules of air spread farther apart. Therefore, we will need to accelerate our aircraft to a higher airspeed in order to cause enough air molecules to pass the lifting surfaces to produce lift-off.

*Humidity Effect*

Humidity is a measure of the number of water vapor molecules in a given volume of air. Every water vapor molecule displaces an air molecule. Since water vapor molecules do not produce lift, their presence in the atmosphere decreases the amount of lift produced by a given volume of air. Again, we must make our aircraft penetrate the atmosphere at a higher speed in order to develop sufficient lift for takeoff.

*Altitude Effect*

The density of the air is affected by altitude. As we increase our height above sea level, the air molecules spread out. Hence, the height of the runway has an important effect on takeoff performance. If all other factors remain constant while we compare takeoff distances required at sea level and at 5,000 feet, we will find that approximately *twice* the runway will be consumed for acceleration to lift-off speed at the 5,000 foot elevation.

If the effects of temperature, humidity, and elevation are combined, aircraft performance may suffer to an extent which will prevent takeoff from many airports. Therefore, we must pay particular attention to the takeoff performance chart or computer contained in the Aircraft Operations Handbook and the density altitude computer which combines all three atmospheric variables in computing density altitude. Remember, aircraft performance depends on the density altitude; that is, the density of the air. (See Chapter 41 for density altitude computations with the navigation computer.) A typical takeoff performance chart is shown in Fig. 72.

Notice the large differences in runway length required for varying weight, headwind, altitude, and temperature conditions.

Takeoff Performance

Obstacle takeoff data 15 degree flaps

| | Sea level | | | 2,000 ft | | | 4,000 ft | | |
|---|---|---|---|---|---|---|---|---|---|
| Wind velocity (mph) | Temperature (F) | Ground run (ft) | To clear 50-ft obstacle (ft) | Temperature (F) | Ground run (ft) | To clear 50-ft obstacle (ft) | Temperature (F) | Ground run (ft) | To clear 50-ft obstacle (ft) |
| 0 | 30 | 785 | 1,175 | 20 | 900 | 1,340 | 15 | 1,060 | 1,580 |
| | 59 | 890 | 1,320 | 52 | 1,035 | 1,535 | 45 | 1,215 | 1,810 |
| | 90 | 1,005 | 1,490 | 80 | 1,160 | 1,720 | 75 | 1,380 | 2,065 |
| 10 | 30 | 620 | 955 | 20 | 715 | 1,095 | 15 | 850 | 1,300 |
| | 59 | 705 | 1,080 | 52 | 830 | 1,260 | 45 | 975 | 1,495 |
| | 90 | 805 | 1,220 | 80 | 935 | 1,425 | 75 | 1,110 | 1,715 |

Note: Decrease distance approximately 15 per cent for each 100 pounds under maximum allowable gross weight.

Fig. 72. Takeoff performance chart.

## Mechanical Effects on Takeoff Performance

We have control over the mechanical effects on takeoff performance. It is for our benefit that we control them carefully.

*Throttle*

The engine manufacturer specifies power settings for use during takeoff in the Aircraft Operations Handbook. We must follow these specifications to avoid engine damage while developing the most power allowable.

On most aircraft, the throttle friction may be adjusted in the cockpit. It is important that we keep the throttle from slipping or "creeping" to a decreased power setting during the takeoff run. Many pilots prefer to keep one hand on the throttle until well above the surface to prevent throttle creeping.

*Engine Operation*

The power output of the engine must be monitored carefully during the takeoff roll and until a safe altitude is attained. It is during takeoff that the engine must develop high power and is most apt to fail. Rarely do engines fail without giving advance warning.

*Flap Operation*

The aircraft manufacturer will specify flap settings to be used for takeoff. As we learned in our discussion of flaps, we may lengthen the takeoff run with erroneous flap use, and, under some circumstances, shorten the takeoff run with proper flap use.

## The Climb Out

After lift-off, we must execute a safe and efficient climb. When flying in single-engine aircraft, we will be concerned with two factors during the climb out. We must clear all obstacles on the surface, and we must gain as much altitude as possible in as short a time as possible.

*Best Angle of Climb*

If we are primarily interested in clearing an obstacle at the end of the runway, we must climb at the *best angle of climb* speed. This speed will be published in the Aircraft Operations Handbook and will result in the most altitude gain for the shortest distance travelled. Since this is normally a slower than normal climb speed, engine cooling may be affected. Therefore, we must note and abide by any restrictions on the amount of time spent at best angle of climb speed.

*Best Rate of Climb*

When we are primarily concerned with the greatest amount of altitude gain in a given period of time, we must climb at the *best rate of climb* speed, which will be published in the Aircraft Operations Handbook.

During the climb out, we may use the best angle of climb speed until

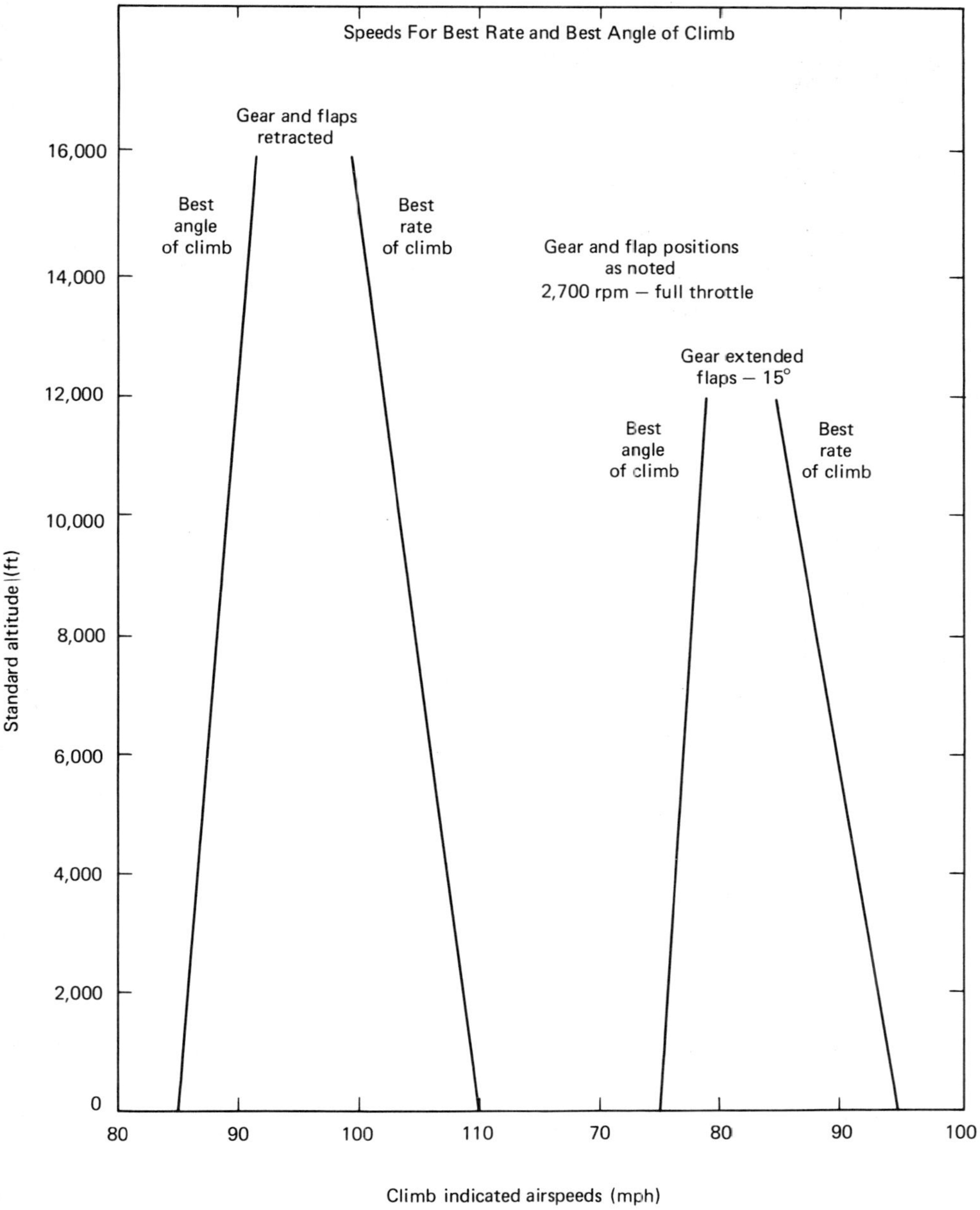


Fig. 73. Speeds for best rate and best angle of climb.

clear of all obstacles, and then change to the best rate of climb speed to reach a specific altitude in as short a time as possible.

Figure 73 shows a sample chart for computing the best rate of climb and the best angle of climb.

Figure 74 depicts a *Time to Climb Chart* which is sometimes provided the pilot by the aircraft manufacturer.

Figure 75 shows a commonly used *Climb Rate vs. Standard Altitude and Weight Chart* with which we can determine the rate of climb to expect under various conditions.

Figure 76 depicts a more sophisticated ground run distance chart which is sometimes provided by aircraft manufacturers.

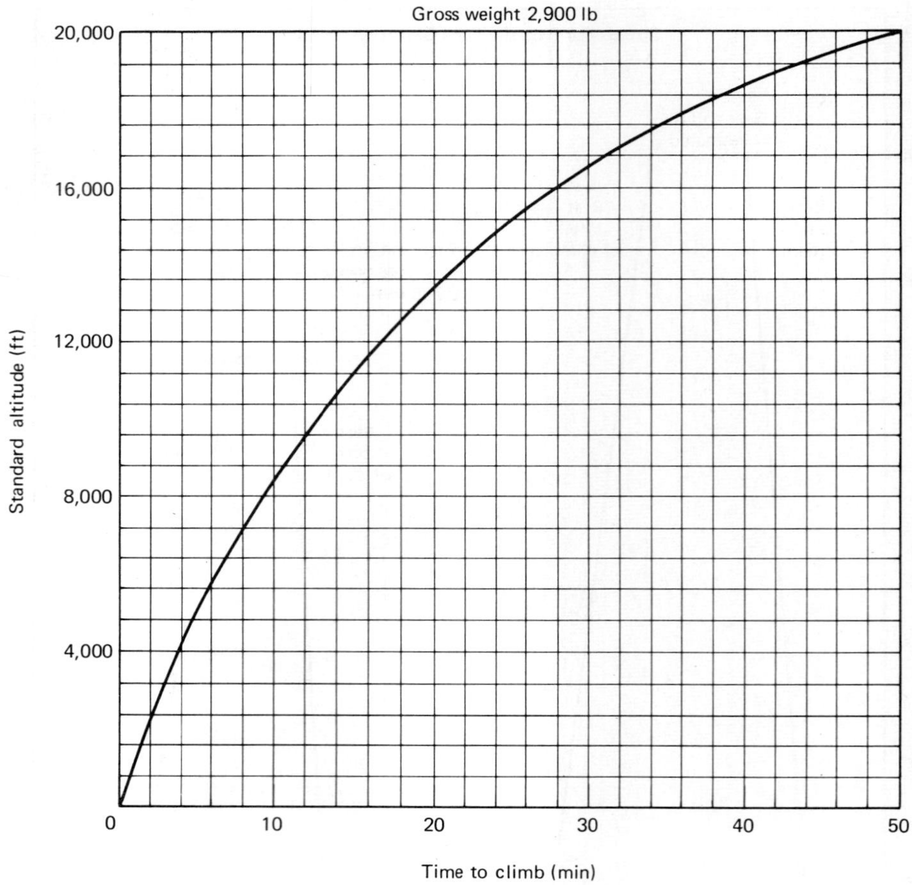


Fig. 74. Time to climb.

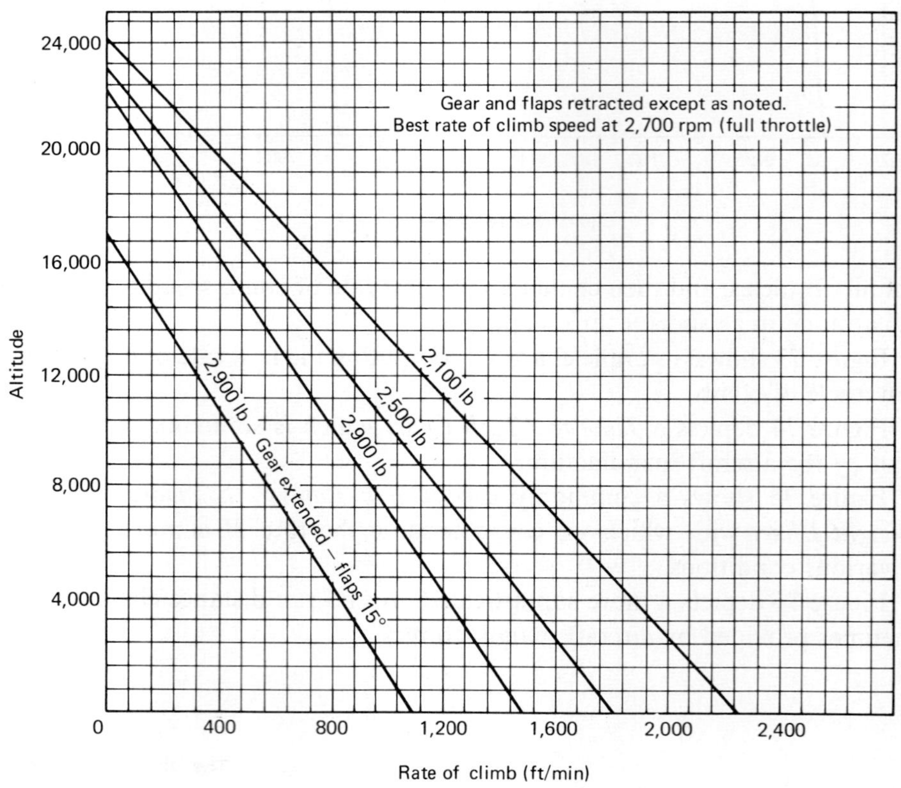


Fig. 75. Climb rate versus standard altitude and weight.

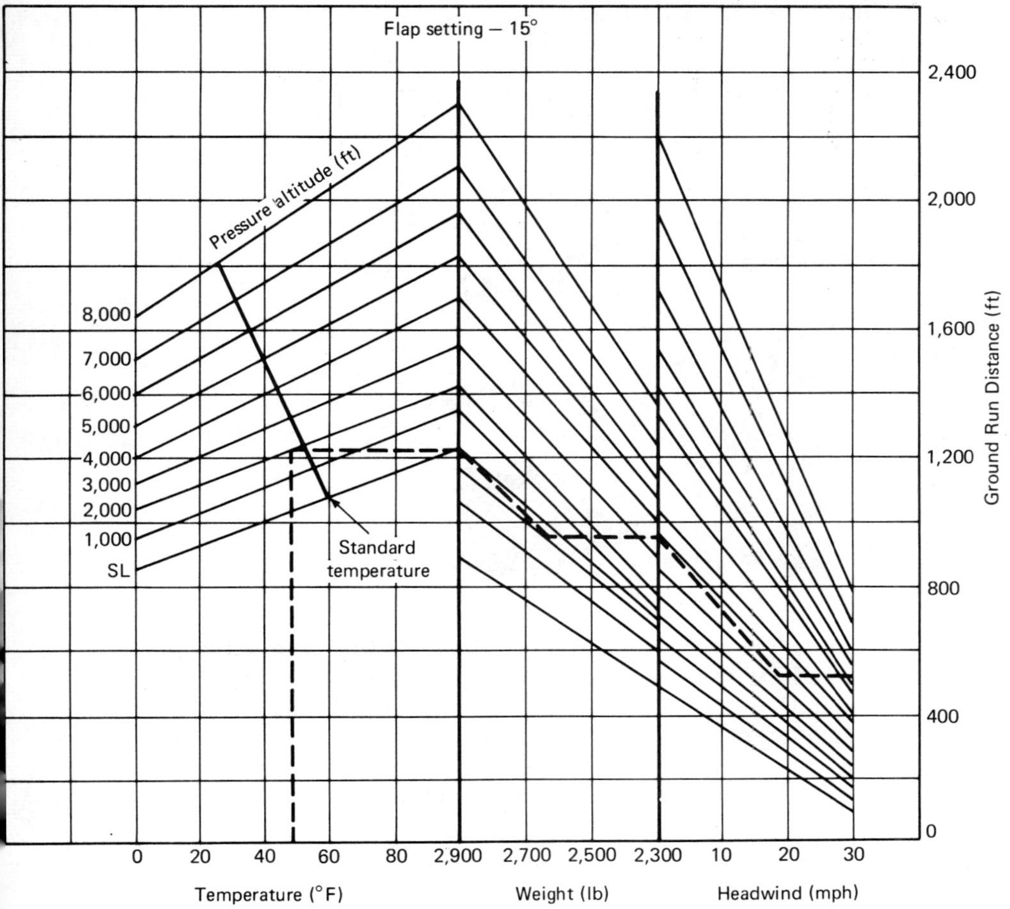


Fig. 76. Takeoff ground run distance.

To compute the ground run distance required, we take into consideration the temperature, pressure altitude, aircraft weight, and headwind component, by completing the following steps.

1. Find the outside air temperature on the bottom of the chart.
2. Move vertically up until intercepting the current pressure altitude line.
3. Move horizontally to the right until intercepting the weight segment of the chart.
4. Move right, paralleling the slant lines until intercepting a vertical line which drops downward to the proper aircraft weight (on the bottom of the chart).
5. Move horizontally to the right until intercepting the headwind component segment of the chart.
6. Move parallel to the slant lines, to the right, until intercepting a vertical line which drops downward to the proper headwind component.
7. Move horizontally to the right to find the ground run distance.

FEDERAL AVIATION AGENCY
VFR EXAM-O-GRAM *NO. 11

## DENSITY ALTITUDE AND ITS EFFECT ON AIRCRAFT PERFORMANCE

A report of a recent accident was stated in the following words: "Takeoff was attempted on a 1, 600-foot strip; the airplane cleared the fences but sank back and struck a ditch." The pilot states that he failed to consider the effect of the grassy, rough field, the 90° temperature, heavy load of fuel and passengers, and the calm wind.

This EXAM-O-GRAM discusses the effect that high temperature and other factors had on this takeoff.

1. WHAT IS DENSITY ALTITUDE? It is a measure of air density. Under non-standard conditions, density altitude will differ from the elevation. As the air density decreases (i. e., air becomes thinner), density altitude increases, and vice versa. Low atmospheric pressure, high temperature, and high humidity all result in a decrease in air density and an increase in density altitude. (Contrary to prevailing opinion, moist air is less dense than dry air. Water vapor actually weighs less than dry air - approximately 5/8 as much.)

2. WHAT EFFECT DOES AN INCREASE IN DENSITY ALTITUDE HAVE ON AIRCRAFT TAKEOFF PERFORMANCE?

   A. Engine horsepower decreases (unless it is a supercharged engine).
   B. The propeller loses some of its efficiency as it will not take as much of a bite out of the thinner air.
   C. Takeoff distance is increased and rate of climb is decreased because of the loss of engine power and propeller efficiency, and the higher true airspeed necessary to obtain the required lift in the thinner air. (In other words, if the density altitude is 8, 000 feet at an elevation of 5, 000 feet, the aircraft flies as though it were at 8, 000 feet.)

*Exam-O-Grams are non-directive in nature and are issued solely as an information service to individuals interested in Airman Written Examinations.

FEDERAL AVIATION AGENCY
VFR EXAM-O-GRAM *NO. 33

# USE OF PERFORMANCE CHARTS

A report of an accident was stated in the following words: "Takeoff was attempted on a 1,600-foot strip; the airplane cleared the fences but sank back and struck a ditch." The pilot stated that he failed to consider the effects of the grassy, rough field, the 90° temperature, heavy load of fuel and passengers, and the calm wind. COULD THE USE OF THE TAKEOFF PERFORMANCE CHART FOR HIS AIRCRAFT HAVE PREDICTED THE SAD ENDING TO THIS FLIGHT?

WHAT ARE PERFORMANCE CHARTS? They are charts that describe or predict the performance of an aircraft under a given set of conditions or ground rules. They may be in tabular or graph form. (Because of their importance to safety, all applicants are being tested, and will continue to be tested, on use of performance charts in the written examinations.)

WHERE DO YOU FIND PERFORMANCE CHARTS? You can find them in the FAA-approved Airplane Flight Manual and the Owner's Manual or Handbook prepared by the manufacturer. In many cases, the FAA-approved Flight Manual must be carried in the aircraft at all times.

ARE THE CONDITIONS OR GROUND RULES UNDER WHICH YOU USE A PARTICULAR TYPE PERFORMANCE CHART ALWAYS THE SAME? No. The particular set of conditions or ground rules, as well as format, will vary with the manufacturer. Although ground rules for their use may be different, the information obtainable is essentially the same--takeoff and landing distance (ground run or roll and to clear a 50-foot obstacle), fuel consumption, rate of climb, true airspeed, etc.

HOW ACCURATE SHOULD YOU CONSIDER THE PREDICTIONS OF PERFORMANCE CHARTS? You will be headed in the safe direction if you always consider the performance of the airplane you fly to be less than predicted by the performance charts. The following statement is contained in one airplane flight manual: "Flight tests from which the performance data was obtained were flown with a new, clean airplane, correctly rigged and loaded, and with an engine capable of delivering its full rated power." You can expect to do as well only if your airplane, too, is kept in the peak of condition.

IS IT NECESSARY THAT YOU ALWAYS CONSULT PERFORMANCE CHARTS PRIOR TO TAKEOFF OR LANDING? No. Obviously, if you are taking off or landing on a 10,000-foot runway in a light airplane, you need not check the takeoff or landing data charts. But where is the dividing line--6,000? 4,000? 2,000? This depends on a lot of factors which include the equipment you are flying; pilot skill, proficiency, and familiarity with equipment; and the relative values of the 3 major factors affecting aircraft performance (density altitude, gross weight, and wind) plus the type and condition of the runway.

WHEN SHOULD YOU CHECK YOUR PERFORMANCE CHARTS? Any time there is doubt in your own mind, whether it be due to the length and/or condition of the runway, the high density altitude, a recognition of your own limitations or a lack of familiarity with the equipment you are flying--which will be alleviated through the use of performance charts. You should begin an operation with complete confidence in its success. Use everything at your disposal to establish this confidence. Charts do not cover all conditions that might have an effect on performance; but by making adequate allowances to the information obtained, you can ensure a greater margin of safety.

WHAT CAN YOU OBTAIN FROM TAKEOFF PERFORMANCE CHARTS? You can find the predicted length of the takeoff ground run and/or the predicted distance necessary to clear a 50-foot obstacle (which includes the ground roll). For example:

Chart 1: At an elevation of 4,000 feet, zero mph wind, 75° F, 15° of flaps, and maximum gross weight (2,300 lbs. for this airplane) the predicted ground run is 1,380 feet and the predicted distance necessary to clear a 50-foot obstacle is 2,065 feet. If the airplane weighed 200 lbs. less than maximum gross weight, these distances would be reduced by 30% and become 966 feet and 1,445 feet, respectively. (See NOTE at bottom of chart.)

OBSTACLE TAKE-OFF DATA — 15° FLAPS

| | Sea Level | | | 2000 Ft. | | | 4000 Ft. | | | 6000 Ft. | | | 8000 Ft. | | |
|---|---|---|---|---|---|---|---|---|---|---|---|---|---|---|---|
| Wind Vel. mph | Temp. °F | Ground Run Ft. | To Clear 50' Obst. Ft. | Temp. °F | Ground Run Ft. | To Clear 50' Obst. Ft. | Temp. °F | Ground Run Ft. | To Clear 50' Obst. Ft. | Temp. °F | Ground Run Ft. | To Clear 50' Obst. Ft. | Temp. °F | Ground Run Ft. | To Clear 50' Obst. Ft. |
| | 30 | 785 | 1175 | 20 | 900 | 1340 | 15 | 1060 | 1580 | 10 | 1260 | 1895 | 0 | 1475 | 2305 |
| 0 | 59 | 890 | 1320 | 52 | 1035 | 1535 | 45 | 1215 | 1810 | 38 | 1430 | 2170 | 30 | 1695 | 2735 |
| | 90 | 1005 | 1490 | 80 | 1160 | 1720 | 75 | 1380 | 2065 | 70 | 1640 | 2560 | 60 | 1890 | 3275 |
| | 30 | 620 | 955 | 20 | 715 | 1095 | 15 | 850 | 1300 | 10 | 1015 | 1570 | 0 | 1195 | 1920 |
| 10 | 59 | 705 | 1080 | 52 | 830 | 1260 | 45 | 975 | 1495 | 38 | 1160 | 1810 | 30 | 1380 | 2290 |
| | 90 | 805 | 1220 | 80 | 935 | 1425 | 75 | 1110 | 1715 | 70 | 1335 | 2135 | 60 | 1575 | 2780 |

CHART 1

NOTE: Decrease distance approximately 15% for 100 pounds decrease in gross weight.

Chart 2: At an elevation of 4,000 ft., 75° F, flaps up, and gross weight of 2,800 lbs., the takeoff distance is 1,600 ft. If you get 1,275 ft., it is because you used the same set of ground rules that you used in Chart 1. Since Chart 2 is based on standard altitude (standard temperature and pressure), you must first convert the elevation (to be completely accurate, the pressure altitude at that elevation) and temperature to a density altitude. A temperature of 75° F at an elevation (pressure altitude) of 4,000 ft. results in a density altitude of approximately 6,000 ft. (see Density Altitude Chart, page 4). Using an altitude of 6,000 ft. in Chart 2, you obtain the predicted takeoff distance of 1,600 ft. (75° F = 24° C)

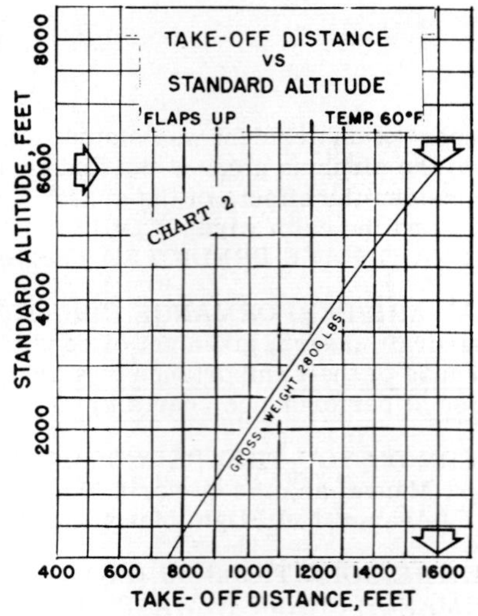


WHAT CAN YOU OBTAIN FROM CLIMB PERFORMANCE CHARTS? Primarily, the rate of climb under various conditions. The information from these charts becomes exceedingly important when you have to cross high mountain ranges relatively soon after takeoff. Some charts also give the best climb airspeed and fuel consumed during the climb. For example:

**CLIMB DATA**

| GROSS WEIGHT LBS. | AT SEA LEVEL & 59°F. | | | AT 5000 FT. & 41°F. | | | AT 10000 FT. & 23°F. | | |
|---|---|---|---|---|---|---|---|---|---|
| | BEST CLIMB IAS MPH | RATE OF CLIMB FT/MIN | GAL. OF FUEL USED | BEST CLIMB IAS MPH | RATE OF CLIMB FT/MIN | From SL FUEL USED | BEST CLIMB IAS MPH | RATE OF CLIMB FT/MIN | From SL FUEL USED |
| 2100 | 87 | 1470 | 1.5 | 82 | 1200 | 2.8 | 78 | 925 | 4.3 |
| 2400 | 88 | 1210 | 1.5 | 84 | 960 | 3.1 | 80 | 710 | 5.0 |
| 2650 | 90 | 1030 | 1.5 | 86 | 795 | 3.5 | 83 | 560 | 5.9 |

Note: Flaps up, full throttle and 2600 RPM. Mixture leaned for smooth operation above 5000 ft. Fuel used includes warm-up and take-off allowance.

Chart 3: At 5,000 ft., 41° F, and 2,100 lbs. gross weight, the rate of climb is 1,200 ft./min.; best climb speed is 82 mph; and fuel used to climb from sea level to 5,000 ft. is 2.8 gal. At a gross weight of 2,650 lbs. under the same conditions, the rate of climb is 795 ft./min.

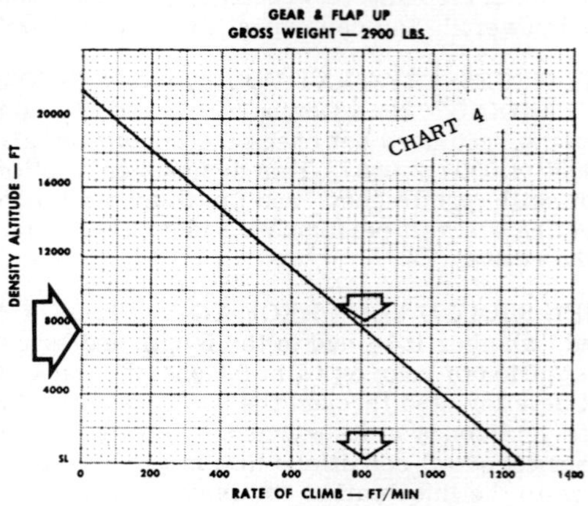


Chart 4: At 5,000 ft., 86° F, and 2,900 lbs. gross weight, the rate of climb is approximately 810 ft. -- not 970 ft. Note that you must first convert the altitude and temperature to a density altitude using the Density Altitude Chart, page 4. The density altitude at this altitude and temperature is approximately 7,750 ft. (86° F = 30° C)

WHAT CAN YOU OBTAIN FROM CRUISE PERFORMANCE CHARTS? Some of the items you can obtain include recommended power settings at various altitudes, along with percent of brake horsepower at these settings, rate of fuel consumption (gal/hr), true airspeed, hours of endurance with full tanks, and range in miles under standard conditions and zero wind. Not all of these values are obtainable from all charts. For example:

**CRUISE AND RANGE PERFORMANCE**

| Altitude | RPM | M. P. | BHP | %BHP | TAS MPH | Gal/Hr. | End. Hours | Mi/Gal. | Range Miles |
|---|---|---|---|---|---|---|---|---|---|
| 5000 | 2450 | 23 | 179 | 78 | 163 | 14.5 | 3.8 | 11.2 | 615 |
| | | 22 | 169 | 73 | 159 | 13.6 | 4.0 | 11.7 | 640 |
| | | 21 | 161 | 70 | 156 | 13.0 | 4.2 | 12.0 | 660 |
| | | 20 | 150 | 65 | 151 | 12.2 | 4.5 | 12.5 | 685 |
| | 2300 | 23 | 167 | 73 | 158 | 13.4 | 4.1 | 11.8 | 650 |
| | | 22 | 158 | 69 | 155 | 12.6 | 4.4 | 12.2 | 675 |
| | | 21 | 148 | 64 | 151 | 11.9 | 4.6 | 12.7 | 700 |
| | | 20 | 139 | 60 | 146 | 11.2 | 4.9 | 13.1 | 720 |
| | 2200 | 23 | 157 | 68 | 155 | 12.4 | 4.4 | 12.5 | 685 |
| | | 22 | 148 | 64 | 151 | 11.7 | 4.7 | 12.9 | 710 |
| | | 21 | 138 | 60 | 146 | 11.0 | 5.0 | 13.3 | 730 |
| | | 20 | 131 | 57 | 143 | 10.5 | 5.2 | 13.6 | 750 |

Cruise performance shown is based on standard conditions, zero wind, lean mixture, 55 gallons of fuel, no fuel reserve, and 2650 pounds gross weight.

Chart 5: At 5,000 ft., 2,300 RPM, and 21 inches of manifold pressure, you should get 64% rated power, approximately 151 mph true airspeed, and consume approximately 11.9 gal./hr. of fuel which will give you an endurance of 4.6 hrs. and a range of 700 miles under standard conditions, zero wind, and full fuel tanks.

Chart 6: At 8,000 ft. you can obtain 55% rated power and 10.3 gal./hr. fuel consumption with 2,200 RPM and 19 inches of manifold pressure.

**Power Setting Table –**

| Press. Alt. 1000 Feet | Std. Alt. Temp. °F | 138 HP — 55% Rated Approx. Fuel 10.3 Gal./Hr. RPM AND MAN. PRESS. | | | | 163 HP — 65% Rated Approx. Fuel 12.3 Gal./Hr. RPM AND MAN. PRESS. | | | |
|---|---|---|---|---|---|---|---|---|---|
| | | 2100 | 2200 | 2300 | 2400 | 2100 | 2200 | 2300 | 2400 |
| SL | 59 | 21.6 | 20.8 | 20.2 | 19.6 | 24.2 | 23.3 | 22.6 | 22.0 |
| 1 | 55 | 21.4 | 20.6 | 20.0 | 19.3 | 23.9 | 23.0 | 22.4 | 21.8 |
| 2 | 52 | 21.1 | 20.4 | 19.7 | 19.1 | 23.7 | 22.8 | 22.2 | 21.5 |
| 3 | 48 | 20.9 | 20.1 | 19.5 | 18.9 | 23.4 | 22.5 | 21.9 | 21.3 |
| 4 | 45 | 20.6 | 19.9 | 19.3 | 18.7 | 23.1 | 22.3 | 21.7 | 21.0 |
| 5 | 41 | 20.4 | 19.7 | 19.1 | 18.5 | 22.9 | 22.0 | 21.4 | 20.8 |
| 6 | 38 | 20.1 | 19.5 | 18.9 | 18.3 | 22.6 | 21.8 | 21.2 | 20.6 |
| 7 | 34 | 19.9 | 19.2 | 18.6 | 18.0 | 22.3 | 21.5 | 21.0 | 20.4 |
| 9 | 27 | 19.4 | 18.8 | 18.2 | 17.6 | — | 21.3 | 20.7 | 20.1 |
| 8 | 31 | 19.6 | 19.0 | 18.4 | 17.8 | — | — | 20.5 | 19.9 |
| 10 | 23 | 19.1 | 18.6 | 18.0 | 17.4 | — | — | — | 19.6 |

CHART 6

CRUISE PERFORMANCE

| ALT. | RPM | % BHP | TAS MPH | 58.8 Gal Endurance Hours | 58.8 Gal Range Miles |
|---|---|---|---|---|---|
| | 2500 | 75 | 130 | 6.0 | 773 |
| 2500 | 2350 | 63 | 118 | 7.1 | 832 |
| | 2200 | 53 | 107 | 8.4 | 894 |
| | 2525 | 75 | 131 | 6.0 | 775 |
| 3500 | 2400 | 65 | 121 | 6.9 | 827 |
| | 2250 | 55 | 110 | 8.0 | 874 |
| | 2550 | 75 | 132 | 6.0 | 780 |
| 4500 | 2400 | 63 | 120 | 7.0 | 841 |
| | 2250 | 53 | 109 | 8.3 | 905 |
| | 2600 | 77 | 135 | 5.8 | 775 |
| 5500 | 2450 | 65 | 123 | 6.8 | 837 |
| | 2300 | 55 | 112 | 8.0 | 887 |

CHART 7

Chart 7: At 5,500 ft. and 2,450 RPM, you have 65% rated power, should obtain approximately 123 mph true airspeed, have an endurance of 6.8 hrs., and a range of 837 miles.

Use cruise performance charts to plan refueling stops. If you learn that your airplane performs differently than predicted by the chart, use this information; especially when performance is worse than predicted by the chart.

WHAT CAN YOU LEARN FROM STALL SPEED CHARTS? Chart 8 is a typical example of a Stall Speed Chart taken from an airplane flight manual. Note and continually be aware of the wide variation in stall speed between straight-and-level flight and various angles of bank. Note that the stall speed in a 60° bank with flaps up and power off (102 mph) is almost double the stall speed in straight-and-level flight with flaps down and power on (55 mph). Even with power on in the 60° bank, the stall speed is reduced only 4 mph to 98 mph. Study this chart and be aware of its significance, especially during traffic patterns and landings. You will find similar charts in any airplane flight manual.

**STALL SPEEDS IAS**

| CONFIGURATION | ANGLE OF BANK 0° | 20° | 40° | 60° |
|---|---|---|---|---|
| Flaps Up — Power Off | 72 mph | 74 mph | 82 mph | 102 mph |
| Flaps Up — Power On | 69 mph | 71 mph | 79 mph | 98 mph |
| Flaps Down (30°) — Power Off | 64 mph | 66 mph | 73 mph | 91 mph |
| Flaps Down (30°) — Power On | 55 mph | 57 mph | 63 mph | 78 mph |

CHART 8

WHAT CAN YOU OBTAIN FROM LANDING PERFORMANCE CHARTS? The same type of information that you get from Takeoff Performance Charts--distance required to clear a 50-foot obstacle, length of the ground run, and in some cases, the recommended approach speed on which these figures are based. Landing Performance Charts will generally be used in the same way as Takeoff Charts for any given airplane, since each manufacturer usually follows the same format in these two charts. If you can read Takeoff Charts, you should have no difficulty reading Landing Charts.

HOW CAN YOU OBTAIN VALUES FROM PERFORMANCE CHARTS FOR CONDITIONS INTERMEDIATE TO THOSE GIVEN? By interpolation. For example, in Chart 1 (page 1) find the ground run required at an elevation of 5,000 ft., 72.5°F, zero wind, and maximum gross weight:

| | |
|---|---|
| Ground run at 4,000 ft., 75°F, zero wind = 1,380 ft. | 1,640 - 1,380 = 260 |
| Ground run at 5,000 ft., 72.5°F, zero wind = ? | 1/2 x 260 = 130 |
| Ground run at 6,000 ft., 70°F, zero wind = 1,640 ft. | 1,380 + 130 = 1,510 |

Since 5,000 ft. is halfway between 4,000 and 6,000 and the temperature is halfway between 75° and 70°, the ground run should be halfway between 1,380 and 1,640, which is 1,510.

Find the distance to clear a 50-foot obstacle at 4,000 ft., 65°F, zero wind, and maximum gross weight:

| | |
|---|---|
| Distance at 4,000 ft., 45°F, zero wind = 1,810 ft. | 2,065 - 1,810 = 255 |
| Distance at 4,000 ft., 65°F, zero wind = ? ft. | 2/3 x 255 = 170 |
| Distance at 4,000 ft., 75°F, zero wind = 2,065 ft. | 1,810 + 170 = 1,980 |

Since 65° is two-thirds of the way between 45° and 75°, the distance should be two-thirds of the way between 1,810 and 2,065 which is 1,980 ft.

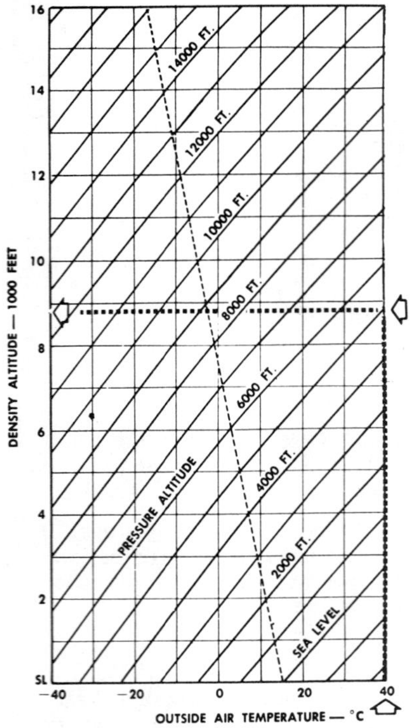


IF INTERPOLATION IS DIFFICULT OR YOU ARE IN DOUBT ABOUT YOUR COMPUTATION, HOW CAN YOU ENSURE BEING ON THE SAFE SIDE? Use a condition more adverse than the one that actually exists--one that you can read directly from the chart without interpolating. Suppose, for example, you were taking off from an airport at an elevation of 5,200 ft. with a 5 mph headwind, a temperature of 65° F, and maximum gross weight. By using an elevation of 6,000 ft., a zero mph wind, and 70° F, you can read the takeoff distance directly from Chart 1. The conditions you are using are more adverse than the actual conditions. If the results indicate that takeoff is feasible, then you should have no difficulty taking off under the actual conditions.

(NOTE: Charts 1, 3, 5, 6, and 7 are excerpts from charts. Charts 2, 4, 8, and 9 are complete. Charts 1, 2, and 4 have been reduced.)

Chart 9: Density Altitude Chart. At an elevation of 5,000 ft. (assuming pressure altitude and elevation are identical) and a temperature of 40° C (104° F) the density altitude is approximately 8,750 ft.


FAA Aeronautical Center
Flight Standards Technical Division, AC-700
Operations Branch, AC-740
P. O. Box 25082
Oklahoma City, Oklahoma 73125

Exam-O-Grams available free of charge in limited quantities from this office.




## CHECK YOUR READING

1. Briefly describe why the takeoff should always be planned into the wind.
2. Briefly explain why landings, either normal or emergency, should be planned into the wind.
3. What are the atmospheric conditions which affect takeoff performance?
4. What is the worst combination of atmospheric conditions concerning takeoff performance?
5. Of the three atmospheric conditions affecting takeoff, which has the greatest effect?
6. Using the takeoff performance chart on page 141, compute the takeoff distance required to clear a 50-foot obstacle with the following conditions:
   wind: 10 mph
   altitude: 2,000 feet
   temperature: 52°F
   maximum allowable gross weight: 3,000 pounds
   takeoff gross weight: 2,800 pounds
7. What are the mechanical controls which affect takeoff performance?
8. If the pilot wishes to climb to 10,000 feet as quickly as possible, which climb speed should he use?
9. If the pilot is faced with an obstacle at the end of the runway, which climb speed should he use?
10. Where can the pilot find the best rate and best angle of climb speeds?

# 18

# TRAFFIC PATTERN AND AIRPORT INDICATORS

The traffic pattern at an airport is a predetermined path which all aircraft follow when taking off or landing so that an orderly flow of traffic is maintained. It is obvious that pandemonium would exist at most uncontrolled airports if every pilot did not conscientiously follow the traffic pattern. The tremendous increase in air traffic over the past few years has increased the need for diligence in the air, particularly when near an airport.

Less than ten per cent of the airports within the United States are equipped with Air Traffic Control towers. Without a control tower, we have no one to assist us in maintaining an orderly traffic flow. For that reason, we must become familiar with traffic patterns and abide by them. Although most airports publish traffic patterns which are quite similar if not identical to the one presented here, local geographical features near some airports demand the use of nonstandard patterns. Local authorities normally design the traffic pattern, which will be described in detail and pictured on the airport bulletin board. If, however, we land at an unfamiliar airport, we will be sure that our traffic pattern is not to be criticized if we follow the instructions learned here.

## Determination of Runway In Use

Prior to planning a particular traffic pattern, we must determine the runway in use. We may have a choice of methods to use.

### *Two-Way Radio*

Many airports are equipped with a *Unicom* transceiver. Although the Unicom *cannot* be used to issue traffic clearances, it can be effectively used for advisories. If the aircraft is equipped with a communications radio, we may contact Unicom on 122.8 MHz and request the runway in use. The airport operator or manager will tell us the runway in use and may include an estimate of the amount of traffic near the airport. We do not have to use the runway which the Unicom operator suggests. The final choice of the runway is the responsibility of the pilot in command. (For example, we may require a longer runway than that suggested.) Normally, however, we will use the same runway that other aircraft are using.

### *Surface Wind Indicators*

Whether or not we receive a Unicom advisory concerning the runway in use, we may elect to fly near enough to the airport to observe the surface wind indicator and make our own determination of the most suitable runway. We will normally wish to land on the runway which is most nearly aligned with the surface wind and will plan the landing into the wind.

If we elect to fly over the airport prior to entering the pattern, we must

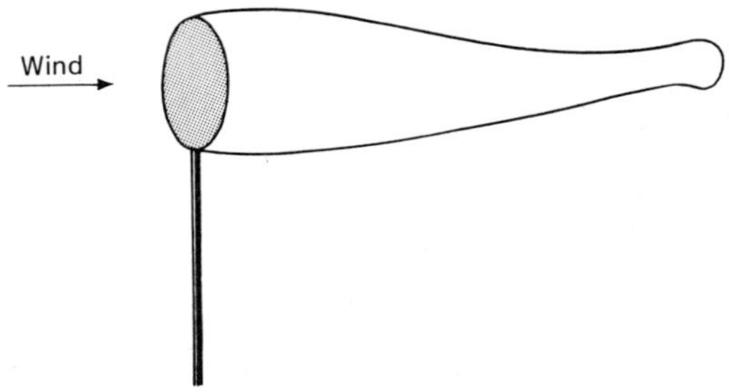


View 1 — The wind sock

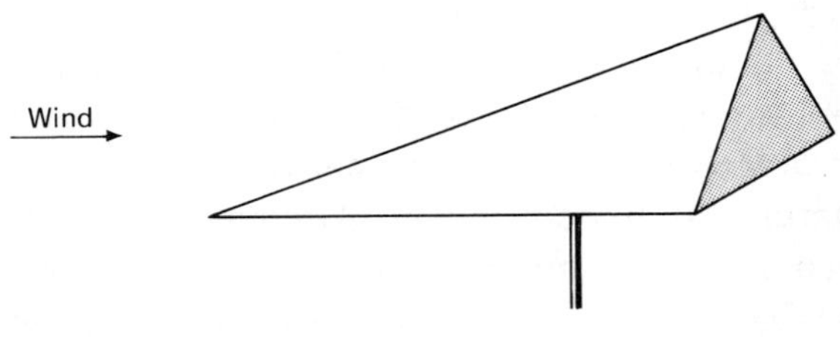


View 2 — The tetrahedron

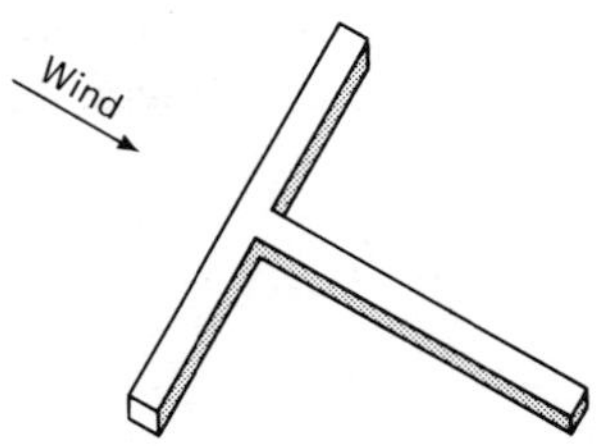


Fig. 77. Wind indicators.

View 3 — The wind tee

be sure to remain well clear of the pattern so as to avoid disrupting traffic. It is normally a good idea to remain at least 1,500 feet above the airport. We may refer to the Sectional Chart or the Airman's Information Manual to find the airport elevation. If we add at least 1,500 feet to the field elevation, we will know the safe minimum circling altitude.

There are three surface wind indicators commonly used. Although there is no standard location on the airport for these indicators, they will usually be found near the center of the airport. Figure 77 depicts the three indicators.

*The Wind Sock.* The wind sock is shown in Fig. 77, view 1. The wind blows in the large end of the sock and out the small end. The wind sock also gives an indication of the strength and gustiness of the wind. A strong wind will cause the sock to stand straight out, parallel to the surface. A gusty wind will cause the sock to flutter and whip about.

*The Tetrahedron.* The tetrahedron in Fig. 77 view 2, is similar to a large arrowhead, which aligns itself with the wind like a wind vane. It points into the wind, or to the direction of landing or takeoff. Depending on the particular installation, the tetrahedron may be left to swing free with the wind or may be secured in a position parallel to the runway in use. The tetrahedron may be outlined with white lights so as to be visible at night.

*The Wind Tee.* Figure 77 view 3, shows the wind tee, which resembles an aircraft. It points to the direction for landings and takeoffs and, like the tetrahedron, may be outlined with white lights.

### *Other Aircraft*

If other aircraft are in the process of landing, we may *tentatively* assume that they are using the proper runway. *However, it is the responsibility of the pilot in command to determine for himself which runway will best suit his particular needs.* After a landing accident, due to a poor runway choice, the excuse that other aircraft were using the same runway will *not* be accepted.

### *Runway Marking*

Most paved runways will have large numbers painted on each end which indicate the *magnetic direction* (in reference to Magnetic North) of the runway. The magnetic direction of the runway is rounded off to the nearest ten degrees, and the last digit (zero) is omitted. For instance, 03 indicates that the magnetic direction of the runway is between 026 and 034 degrees. The number 27 would indicate that the magnetic direction of the runway is between 266 and 274 degrees.

One or more "×" marks on the runway indicates that the runway is closed and cannot be used for flight operations.

### *Airport Lighting*

Many airports are equipped with lighting for night operations. We may check the Airman's Information Manual or Navigation Chart to determine the availability of lighting.

Runways are outlined in white light. The threshold on the approach end may be lighted green, and the far end may be lighted red.

Taxiways, if lighted, will be outlined in blue.

Red lights will indicate obstructions on or near the airport.

## Left and Right Hand Patterns

A standard traffic pattern will be comprised of all left hand turns, with all traffic circling to the left. However, geographical conditions (cities, mountains, towns, etc.) may create a hazardous condition if the standard pattern is used. Local authorities may require the traffic pattern for one or more of the runways to be right hand, with all traffic circling to the right. One of the following two indicators will be used to inform us that a nonstandard right hand pattern is in use. An amber-colored flashing light may be located in a prominent position on the airport. When the light is flashing, the runway in use is equipped with a right hand pattern.

The segmented circle, shown in Fig. 78, will be equipped with an L-shaped spoke aligned with each runway.

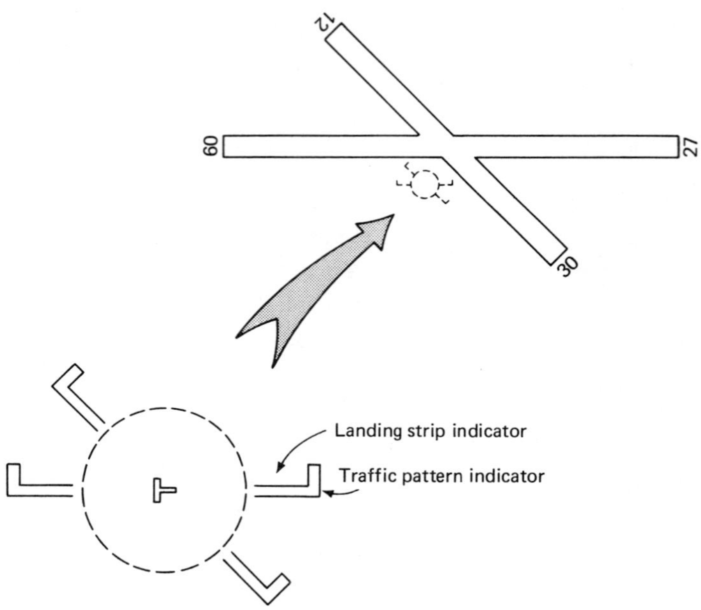


Fig. 78. The segmented circle.

The "L" indicates the direction of the turn to the final approach. For instance, the L-shaped spoke aligned with runway 09 in Fig. 78 shows a left turn to the final approach. The L-shaped spoke which is aligned with runway 30 shows a right turn to the final approach. If the turn to the final approach is left hand, *all* turns in the pattern will be left hand. If the turn to final approach is right hand, *all* turns in the pattern will be right hand.

## The Traffic Pattern

The traffic pattern is comprised of seven segments or "legs." It is in the shape of a rectangle as shown in Fig. 79.

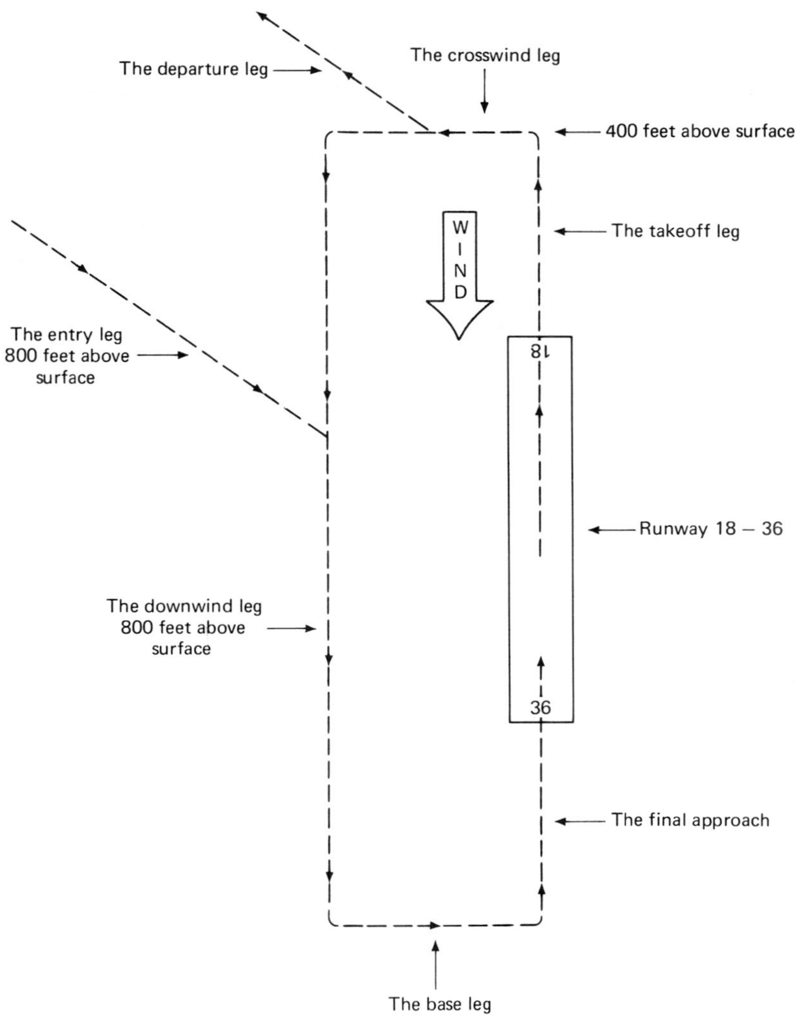


Fig. 79. The traffic pattern.

*The Takeoff Leg*

After takeoff, we should attempt to follow the extended centerline of the runway until a safe altitude has been reached. Normally, the first turn is made at 400 feet *above the surface*. In order to maintain a ground track over the extended runway centerline, we must be prepared to establish a drift correction angle after lift-off. The traffic pattern must be a perfect rectangle over the ground.

*The Crosswind Leg*

After reaching 400 feet above the surface, a climbing left turn (if the pattern is standard) is made. If we take off into the wind, the 90-degree left turn will place the aircraft in a position which is approximately 90 degrees to the wind. The crosswind will tend to drift the aircraft towards the airport, so our turn will have to be somewhat less than 90 degrees in order to compensate for drift.

### The Downwind Leg

The downwind leg parallels the runway in use. When remaining in the traffic pattern, we will normally execute a climbing left turn from the crosswind leg to the downwind leg. Again, we must be conscious of drift and establish a downwind heading which will keep the aircraft on a course which is parallel to the runway. The distance between the downwind leg and the runway varies with the speed of the aircraft and individual technique, but a good rule of thumb is to plan a downwind leg which is about one-half of the runway length away from the runway. Our altitude on the downwind leg will normally be 800 feet *above the surface.*

### The Base Leg

The base leg is 90 degrees to the runway, and is entered by a turn of approximately 90 degrees from the downwind leg. The wind will tend to drift the aircraft away from the airport, so a drift correction angle must be established. The distance between the base leg and the runway threshold varies with aircraft speed, surface wind and pilot technique.

### The Final Approach

The final approach is aligned with the centerline of the runway. For most pilots, it is the most critical portion of the traffic pattern, for the aircraft is nearing the surface and power, airspeed, and attitude control are all critical.

A number of important points should be mentioned concerning flight on the final approach.

*Power Control.* We must remember that to rely on the aircraft engine for safe flight is to invite disaster, particularly in single-engine aircraft. The final approach should be planned so as to allow enough altitude to establish a glide angle to a safe landing area (preferably the runway) in case of complete power failure. Too many pilots plan a long, low final approach, "hanging on the prop," dragging the aircraft to the runway. In the event of an engine failure, they have no alternative to a forced landing in what may be very rough or densely populated terrain.

*Airspeed Control.* Every aircraft manufacturer specifies a final approach airspeed, which allows for sufficient control response without excessive airspeed. If we maintain an airspeed lower than specified, we increase the hazard of control inefficiency at a critical period. If we maintain an airspeed higher than specified, we will use many unnecessary feet of runway during our "flare out" and landing.

Experience will aid us in learning the atmospheric conditions which will dictate a higher than normal approach speed, such as gusty surface winds, strong crosswinds, or structural ice on the aircraft.

*Altitude Control.* The final approach to every runway is unique. That is what makes landing interesting and fun. Not only will we find great variety in topographic features (hills, gullies, mountains, oceans, cities, farms, etc.) but we will experience variety on the same final approach from day to day, or even minute to minute.

The type of terrain over which we fly determines the amount and intensity of thermal activity we will experience on the final approach. Freshly plowed soil, for instance, absorbs and radiates more heat than a lush green forest. As we fly at low altitude from one type of terrain to the next, we will experience varying thermal activity. Flight over a cool surface (foliage, water, snow, etc.) may be primarily in stable or subsiding air. When our aircraft passes over a warmer surface (roads, buildings, parking lots, etc.) it may be subjected to unstable convective currents. Needless to say, all this adds up to a situation on final approach which requires practice and constant vigilance.

## The Visual Approach Slope Indicator

Many airports are now or will soon be equipped with a Visual Approach Slope Indicator (VASI) which is a system of lights that are installed near the ap-

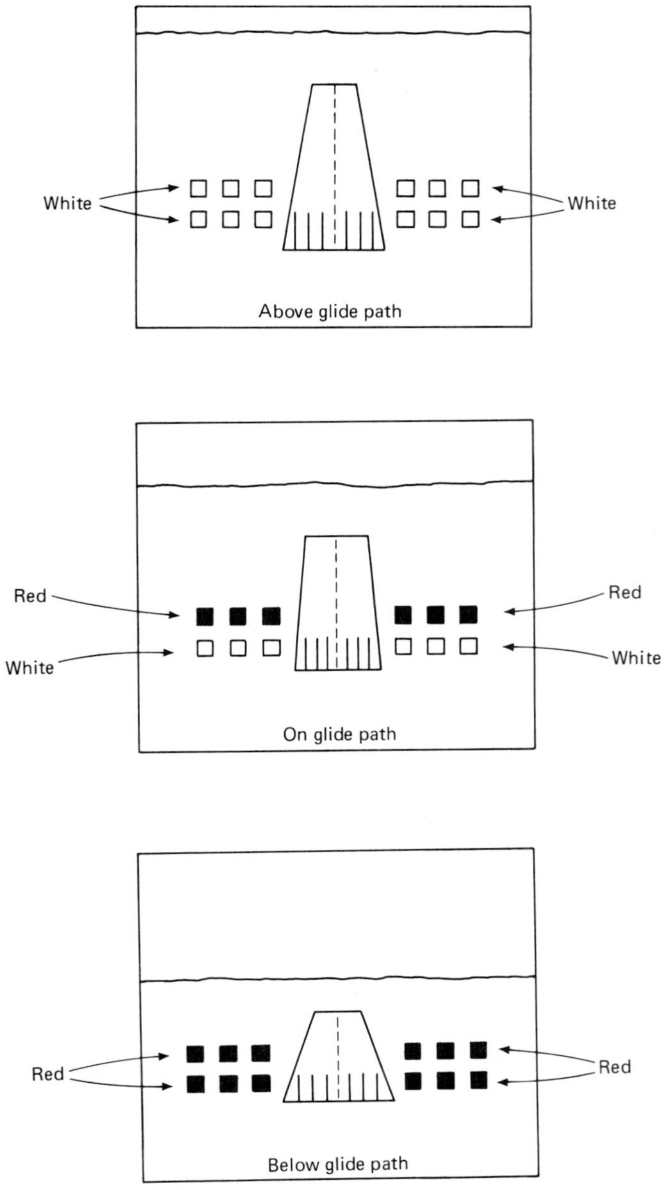


Fig. 80. The visual approach slope indicator.

proach end of a runway, to provide a *visual glide path* during the final approach. The VASI is particularly helpful during approaches over water or snow surfaces where depth perception is most difficult.

A standard VASI installation will incorporate twelve lights which are arranged in two split bars stradling the runway, as shown in Fig. 80.

Three lights are placed on each side of the runway about 600 feet from the threshold. Another set of three lights is placed on each side of the runway about 1,300 feet from the threshold. When viewed from the final approach, the near lights (nearest to the runway threshold) are called the *downwind bars*, and the far lights (farthest from the runway threshold) are called the *upwind bars*. The visual approach reference point is located midway between the upwind and the downwind bars.

*The Principle of VASI*

The light units are arranged with filters and apertures so as to show pure white light when viewed from a high angle, pure red when viewed from a low angle, and pink when viewed from the horizontal center of the aperture.

When we are on a proper approach path, we will be overshooting (flying too high) in reference to the downwind (nearest) bars and undershooting (flying too low) in reference to the upwind (farthest) light bars. Therefore, when on a proper approach glide path, we will see the downwind bars as white, and the upwind bars as red. If we fly below the VASI approach glide path both bars will appear red. If we fly above the VASI glide path, both bars will appear white.

An indication of our departure from the glide path will be a changing of the color of either bar through a shade of pink. When we begin to go too high, the red upwind bar will change from red to pink and eventually to white. When we begin to fly below the glide path, the white downwind bar will change from white to pink and eventually to red.

*Use of VASI*

Airports equipped with VASI are listed in the Airport/Facility Directory in the Airman's Information Manual. There is no VASI information presented on the Sectional Chart.

VASI is normally operated day and night whenever the runway on which it is installed is the landing runway or when requested by any pilot. Federal Air Regulation 91.87 states in part: "When operating to an airport with an operating control tower, each pilot of. . . . (3) an airplane approaching to land on a runway served by a Visual Approach Slope Indicator shall maintain an altitude at or above the glide slope until a lower altitude is necessary for a safe landing."

The intensity of the VASI lights may be adjusted from the control tower, so we may request the control tower operator to turn down the brightness or increase their intensity. Naturally, the distance from which the VASI lights can be seen varies with the amount of daylight and the atmospheric conditions, but they can normally be seen for a distance of four to five nautical miles.

When landing on a VASI equipped runway, we should proceed inbound on the final approach at traffic pattern altitude. When the downwind (near) bars change in color from red to pink to white, we should begin our descent. Normal bracketing maneuvers will then keep us on the glide slope.

As we approach the runway threshold we may experience some deterioration of guidance due to the spreading of light when close to the source. However, since VASI is an *approach* aid (*not* a landing aid) we can be sure that we will pass safely through the *gate* at the threshold from which we may execute a normal landing.

### *Entering The Traffic Pattern*

When we enter the traffic pattern we join the normal flow of traffic around the active runway. We must use extreme caution to avoid entering that flow in a manner which might subject our aircraft or others to a dangerous situation.

The entry leg should be long enough to allow us to get a good long look at other traffic before nearing the pattern. The entry leg extends from a point three or four miles from the downwind leg to the midpoint of the downwind leg, joining it at a 45-degree angle. We must establish traffic pattern altitude and maintain that altitude while on the entry leg, searching constantly for other aircraft.

### *Leaving The Traffic Pattern*

As shown in Fig. 79, the exit leg begins at about the midway point on the crosswind leg and extends outward a mile or two at a 45-degree angle. After turning crosswind, if we desire to leave the traffic pattern, we should make a climbing right turn of about 45 degrees. This maneuver will give the following aircraft definite indication that we are departing from the pattern.

## Caution

Whenever we fly we must watch for other aircraft. When flying in the vicinity of an airport where many aircraft converge, we must double our watchfulness.

The traffic pattern will result in a smooth flow of traffic only if every pilot makes a conscientious effort to follow the pattern and is willing to give the other fellow the right of way, even when not required by the law.

Final approach must be at an altitude commensurate with safe flight. Long, low final approaches, particularly in single-engine aircraft are dangerous.

The traffic pattern is a rectangular course in relation to the surface. We must not attempt to memorize traffic pattern headings, for varying wind conditions will demand varying aircraft headings.

Before entering any traffic pattern *THINK* and *LOOK*.

Department of Transportation

FEDERAL AVIATION ADMINISTRATION

# VFR PILOT EXAM-O-GRAM* NO. 40

## VISUAL APPROACH SLOPE INDICATOR (VASI)

Within the National Airspace System, currently there are 67 airports equipped with the standard Visual Approach Slope Indicators, and 7 are equipped with Abbreviated Visual Approach Slopes. The abbreviated systems contain fewer light units which may be installed on one side of the runway only. It is apparent that misconceptions and a lack of knowledge concerning this aid exist among the general aviation public. To assist those who are not familiar with VASI, particularly those taking FAA written examinations, this Exam-O-Gram briefly explains the system, and answers questions commonly asked regarding the purpose, availability, and use of the device.

SYSTEM DESCRIPTION. The Visual Approach Slope Indicator (VASI) is a ground device which uses lights to define a predetermined visual glide path during the approach to a runway. As soon as the VASI lights are visible on final approach--day or night--a pilot receives the same information by visual reference that the glide slope unit of the Instrument Landing System (ILS) provides electronically. Once the principles and color code of the lighting system are understood, flying the VASI is as simple as looking out through the windshield and establishing and maintaining the proper rate of descent to stay on the glide slope.

This facility emits a visual light path within the final approach zone, at a fixed plane inclined from a minimum of $2\frac{1}{2}°$ to a maximum of $4°$ from horizontal, which gives the pilot visual descent guidance during an approach to landing. The beam width is 15 degrees on each side of the extended runway centerline, but actual runway alignment is not provided by VASI. Course guidance to assure runway alignment should be obtained by reference to the runway lights, the runway itself, or by other approach aids.

Standard installation of the system requires twelve light-source boxes arranged in two split bars of light that straddle the runway. A set of 3 light units (boxes) is placed on each side of the runway approximately 600 feet from the threshold, and a second set of 3 is placed on each side at approximately the 1,300-foot point from the threshold. These light unit arrays are called the "downwind" and "upwind" light bars, respectively. The visual approach slope reference point is located midway between the upwind and downwind bars.

●●●

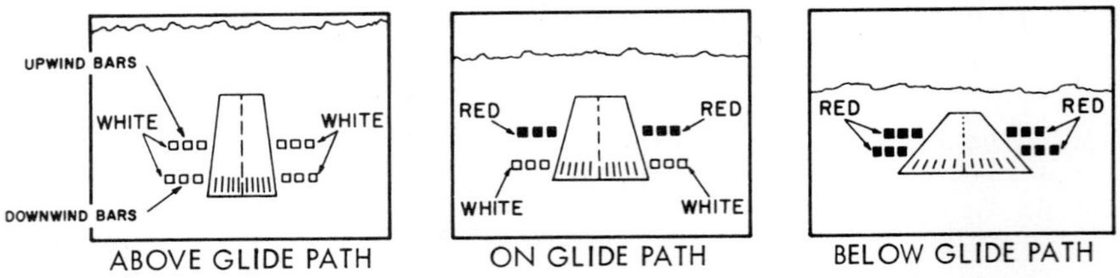


DISTRIBUTION: ZC-307.

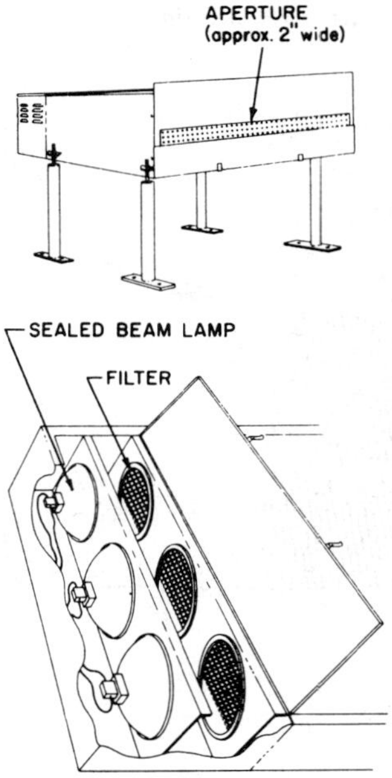


VASI LIGHT UNIT

<u>LIGHT UNITS</u>. Each light unit consists of a metal box 4.5 feet square and about 1 foot thick, housing three high intensity sealed beam lamps. Immediately in front of the lamps is a color filter assembly, the upper two-thirds of which is red and the lower portion is clear. Across the front of the box at the focal point of the lamps is a 2-inch aperture that extends across the width of the box. The "lens" effect, thus achieved, causes the light to appear white when viewed from a high angle, red when viewed from a low angle, and pink when viewed from the horizontal center of the aperture. The light intensity of each light unit is approximately 40,000 candlepower in the white light zone.

<u>PRINCIPLE</u>. The basic principle of VASI is that of color differential between red and white. The VASI provides obstruction clearance in the final approach area only. It is especially effective during approaches over water or featureless terrain where other sources of visual reference are lacking or misleading.

●●●

HOW ARE THE VASI APPROACH LIGHTS USED? For VFR conditions, proceed inbound maintaining normal traffic pattern altitude. When the downwind (near) bars transition from red through pink to white, commence descent. When on the proper approach path, the pilot is, in effect, overshooting the downwind bar light beam, and undershooting the upwind bar light beam. Thus, he sees the downwind bars as white and the upwind bars as red. When below the glide path, both bars are red; when high, both bars are white.

Departure from the glide path is indicated to the pilot by a transition in color of one of the light bars. If the departure is to the high side, the upwind bars will change from red through pink to white, leaving a completely white display. If the departure is to the low side, the downwind bar will change from white through pink to red, warning of a descent below the approach slope--by a completely red array of lights.

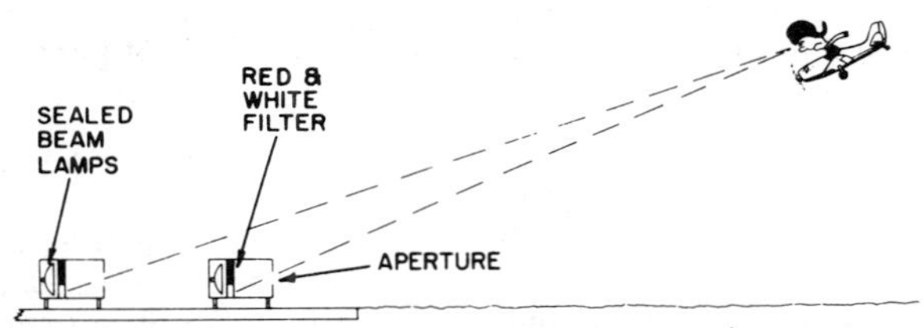


ABOVE GLIDE SLOPE ( PILOT SEES WHITE/WHITE )

NOTE: Some deterioration of system guidance may occur as the pilot approaches the runway threshold due to the spread of light sources and narrowing of individual colors. However, the VASI will bring the pilot safely through a "gate" at the threshold where he may accomplish a normal flareout and landing. Since deterioration of system guidance occurs close in, the VASI is an <u>approach aid</u> rather than a landing aid.

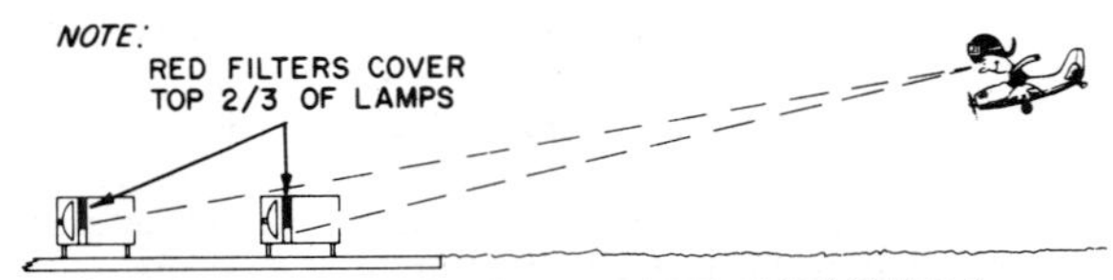


ON GLIDE SLOPE ( PILOT SEES RED/WHITE )

●●●

WHAT ARE THE USABLE DISTANCES OF VASI? During daylight hours the VASI lights normally can be seen at distances of approximately 4 to 5 nautical miles. With bright sunlight or snow conditions the range is decreased. During the hours of darkness the lights may be seen at greater distances.

CAN THE LIGHT INTENSITY OF VASI BE ADJUSTED? Yes, the standard VASI system includes a light intensity control switch in the control tower. This remote control may be a two position "HIGH-LOW" switch or a three position "LOW-MEDIUM-HIGH" selector. If a pilot making a night VASI approach desired to have brightness of the lights turned down, he can request the controller in the tower to do so.

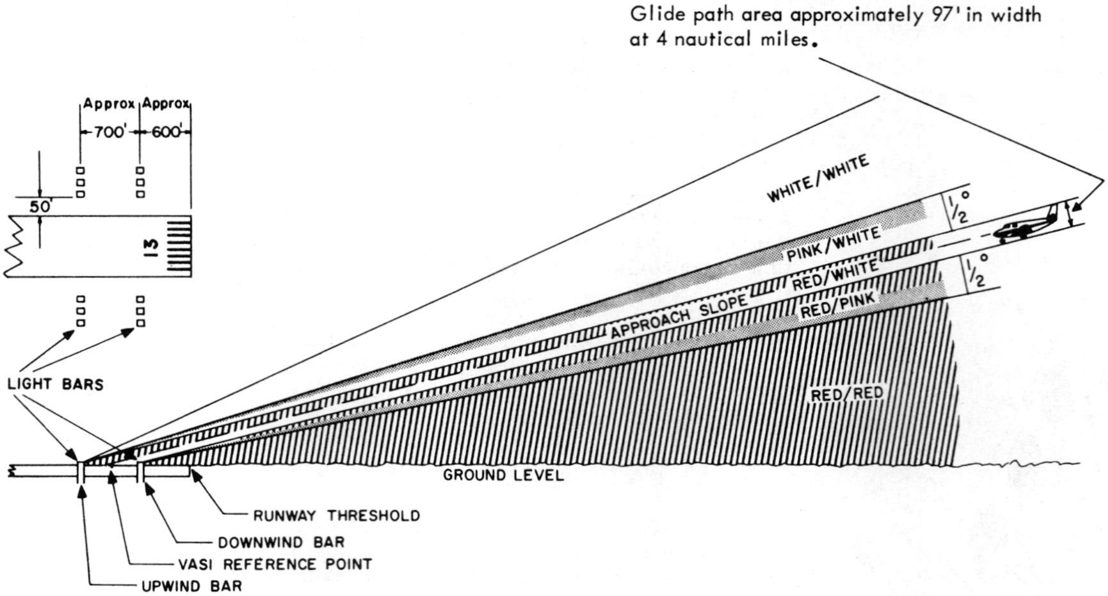


TYPICAL VASI COLOR INDICATION PROFILE

WHAT ARE SOME OF THE FACTORS THAT AFFECT A PILOT'S COLOR INTERPRETATION? Such factors as snow, dust, precipitation, and color of background terrain affect the pilot's color interpretation of the VASI. Atmospheric conditions may distort the color the pilot is actually seeing or preclude the determination of a well-defined glide path or transition area. However, there is no distortion in the Red/Red area. When all lights are solid red, the aircraft is definitely below the glide slope.

HOW DOES A PILOT DETERMINE IF AN AIRPORT IS EQUIPPED WITH VASI? By referring to the Airman's Information Manual (AIM) Airport/Facility Directory listing for that airport. Note in the excerpt to the right that the VASI is on Runway 29 only. There is no VASI information listed on WAC or Sectional Charts, however, it does appear on the Instrument Approach Procedure Charts.

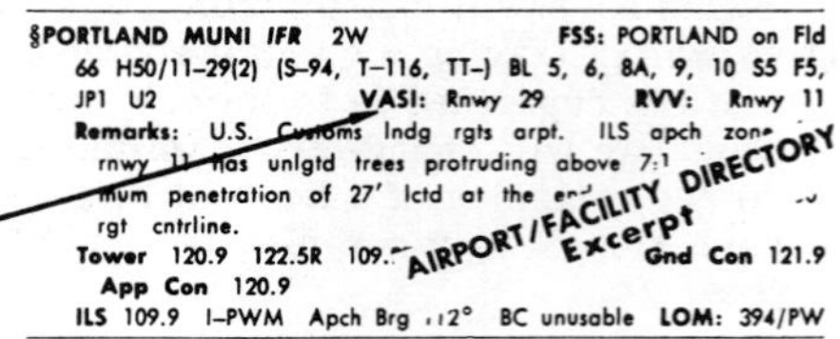
§PORTLAND MUNI IFR 2W FSS: PORTLAND on Fld
66 H50/11-29(2) (S-94, T-116, TT-) BL 5, 6, 8A, 9, 10 S5 F5, JP1 U2 VASI: Rnwy 29 RVV: Rnwy 11
Remarks: U.S. Customs lndg rgts arpt. ILS apch zone [illegible] rnwy 11 has unlgtd trees protruding above 7:1 [illegible] mum penetration of 27' lctd at the [illegible] rgt cntrline.
Tower 120.9 122.5R 109.[illegible] Gnd Con 121.9
App Con 120.9
ILS 109.9 I-PWM Apch Brg [illegible]2° BC unusable LOM: 394/PW



WHEN ARE THE VASI LIGHTS IN OPERATION? The VASI shall normally be operated, day or night, when the runway it serves is the landing runway--or--when requested by the pilot.

ARE THERE ANY REGULATIONS THAT DEAL WITH VASI? Yes, FAR 91.87 states in part: "(d) Minimum altitudes. When operating to an airport with an operating control tower, each pilot of --. . . (3) an airplane approaching to land on a runway served by a visual approach slope indicator, shall maintain an altitude at or above the glide slope until a lower altitude is necessary for a safe landing." (NOTE: Regulations permit normal bracketing maneuvers above or below the glide slope that are conducted for the purpose of remaining on the glide slope.)

●●●

VASI provides unquestionable obstruction clearance in the approach area which is very comforting during night landings at strange airports. It reduces the chance of overshooting or undershooting, and it aids in making a landing in the first portion of the runway. VASI is also an effective noise abatement procedure where large airplanes and turbine-powered airplanes are concerned.

★★★★★

## CHECK YOUR READING

1. What use does the traffic pattern serve?
2. After landing at an unfamiliar airport, where can we normally find a detailed description of the traffic pattern?

3. What are the common surface wind indicators?
4. When landing on runway 33, what will be the magnetic heading on final approach, if there is no crosswind?
5. What will the approximate magnetic heading be on base leg for runway 03 in a standard traffic pattern with a 20-knot wind blowing directly down the runway?
6. What do blue lights indicate on an airport surface?
7. How are runways lighted at night?
8. How will we know that a nonstandard (right hand) pattern exists at an airport?
9. Normally, after takeoff, the first turn is made only after reaching what height above the surface?
10. Why is a long, low final approach dangerous?
11. Why is airspeed control important on final approach?
12. When on final approach to a runway which is equipped with VASI, what indications will warn us of flight below the glide path?
13. When on the glide path to a VASI equipped runway, what will the VASI indications be?
14. When must VASI be used by a pilot?
15. What cautions must be exercised when in or near the traffic pattern?

# 19

# NIGHT FLIGHT

Night flight can be a safe and exhilarating experience. It may also be a dangerous and frightful adventure. In this chapter we will discuss only the preparations for and the more important features of night flight, leaving our discussion of physiological factors to Chapter 47.

## Preparations

Prior to beginning night flight, or a flight which may end after dark, we should be positively sure that our aircraft and equipment are in good operating condition.

Special equipment must include a good flashlight with at least one set of extra batteries. Some models of flashlight are equipped with a magnetized handle for easy attachment to the instrument panel. As good as this idea sounds, a magnetic flashlight may prove to be a hazardous accessory, because the prime instrument we will use for directional information is the magnetic compass which is easily affected by *any magnet*. Too many pilots have flown a perfect compass heading for hours, only to find that they had maintained a heading to their own flashlight!

The fuse box for the aircraft electrical system should be carefully inspected to determine that no fuses are blown. At least two spare fuses of every magnitude in the fuse box should be stowed in an easily accessible place.

We must be sure that we are familiar enough with the cockpit to be able to touch all controls, switches, and fuel selectors without looking at them.

In the event of a complete electrical system failure, we will appreciate the fact that we spent a little extra time on cockpit familiarization.

During our preflight inspection, we should pay particular attention to the condition of the aircraft lights. After turning on the master switch and each individual light switch, we must inspect each light. It is a good idea to tap each light fixture lightly to be sure that the electrical power at the light is not intermittent with vibration. Landing lights, taxi lights, and rotating beacon or anti-collision lighting should also be checked for proper operation and cleanliness.

*Flight Planning*

Night flying must be preceded by a complete weather briefing, because the weather tends to change more rapidly as night falls, and after dark it is very difficult to detect the telltale signs of approaching weather.

A detailed inspection of the navigation chart will alert us as to the obstructions enroute. We should devote more than the usual amount of time on investigation of airports enroute in case of possible power loss.

*Taxiing Operations*

When we taxi at night, we must take special precautions. Naturally, visibility is much reduced compared to daytime. Taxi speed must be slower so as to see and avoid any obstacles on the taxiway.

As a courtesy to other pilots, we must avoid shining our taxi lights directly at another aircraft if it is at all possible. When we discuss the physiological factors of night flying (Chapter 47), we will learn that night vision is immediately destroyed by bright lights. Therefore, as we taxi, we must remain conscious of other aircraft; and if it becomes necessary to turn towards another pilot, we should shut off the taxi lights until the scan of our taxi lights no longer points in his direction.

## Night Flight

Not only does the law demand certain precautions prior to night flight, but so does common sense. Inexperienced pilots do not always realize the hazards involved in night flight and attempt it without prior instruction.

*Disorientation*

After takeoff, particularly in sparsely populated areas, it is quite easy to confuse the lights on the surface with the stars in the sky. This seems difficult to believe until it is actually experienced, but it certainly has been a factor in aircraft accidents in the past. For this reason, night flying instruction should incorporate basic knowledge of instrument flying, so that, in case of disorientation on the climb out, we may refer to our instruments to insure safe flight.

*Obstruction Clearance*

Obstructions on or near the airport will be marked with red warning lights. When flying at night at an unfamiliar airport, we must pay particular attention to the obstructions and ask a pilot who is familiar with that airport if he is aware of any obstructions which are not noticeable but might present a hazard under certain conditions.

Night takeoffs will normally be followed by a *best angle of climb* airspeed in order to gain as much altitude as possible before leaving the immediate airport boundary. Then, the *best rate of climb* airspeed should be established in order to gain as much altitude as possible in the shortest time period. Remember, the choice of a landing area in the case of an emergency is drastically reduced at night. Altitude will provide time to find the best possible landing area.

### *Enroute Flight*

Cross-country night flight is very rewarding. The air is usually quite stable because of the reduced surface heating, and surface lights provide an interesting panorama of scenery. However, certain precautions must be exercised.

The shape of cities may appear quite different at night than during daylight, because only the lighted areas will be visible. Identification of cities and towns will therefore be more difficult. Another interesting fact is that the size of each city and town appears to shrink as the night wears on. This is because more and more lights are extinguished with the passing hours, until, in the small hours before dawn, only the lights at the core of the city and on major outlying streets are visible. Again, this makes positive identification of these areas difficult.

Naturally, most of the identifying features on the surface will be invisible. We must depend upon dead reckoning and radio navigation rather than pilotage.

Clouds, too, are more difficult to see at night. What begins as a flight in clear, smooth air may end in inadvertent instrument flight if we do not keep a constant watch for clouds or precipitation. Frequent radio contact with Flight Service stations along our route will help us to determine sky cover and weather in the vicinity.

Visibility is a factor of atmospheric content (such as dust, soot, haze, etc.) and the amount of visible moisture in the air. It is also largely dependent upon the amount of light available. Therefore, a visibility of five miles prior to sunset may deteriorate to less than three miles after the sun sets. This one factor has contributed to the woes of many pilots who departed an airport with near-marginal visibility at dusk, only to find themselves in visibilities of near zero after night sets in.

We discussed in Chapter 8 the hazard of wingtip vortices and jet-wash in flight. One of the precautions we learned to use to avoid vortices was to watch for other aircraft and to remain well clear of their path of flight. Night flight makes this more difficult. Our watchfulness for other aircraft must be increased so that we will be constantly aware of others who share the sky. It is more difficult at night to determine the altitude of other aircraft; so, we must plan our path of flight so as to remain well clear of any possible vortex from another aircraft. In other words, night flying requires more vigilance than ever to remain safe.

## Aeronautical Beacons

The Federal Aviation Administration and local authorities maintain a system of aeronautical beacons which provide a visual display of flashing white and/or colored lights to point out the location of airports, landmarks, and some points along the federal airways (in mountainous terrain) to point out hazards to flight.

### *Airport Rotating Beacon*

The airport rotating beacon will be comprised of a white lens and a colored lens which rotate at a constant speed, producing 12 to 15 flashes per minute. The rotating beacon at a lighted land airport will display one white beam (a split white beam for a military airport) and one green beam. If a second lighted land airport lies not too distant from the primary airport, its beacon may show two green flashes. A lighted water airport will display a white and a yellow light. All yellow beacons will denote a lighted water airport near another.

Navigational points or landmarks will display a white and red beacon, and red alone indicates a hazard.

*Daylight Beacon Operation.* When the rotating beacon is operated at an airport during the hours of daylight, it shows that the surface visibility is less than three statute miles and/or the ceiling at the airport is less than 1,000 feet, meaning that a clearance from Air Traffic Control is required for any operation within the airport traffic area.

### *Lighted Tetrahedron and Traffic Indicators*

During the hours of darkness, flashing lights outlining the tetrahedron or wind tee indicate that the surface visibility is less than three statute miles and/or the ceiling at the airport is less than 1,000 feet. A clearance from Air Traffic Control is required prior to operating within the airport traffic area.

## Depth Perception

As we will learn when discussing the physiological changes to the body in flight, the eye operates in an entirely different manner at night than it does during daylight. For this reason, depth perception at night is not as accurate. The results of reduced depth perception will be most noticeable near the surface, during the approach and landing phases of flight. Most pilots prefer to maintain a slightly higher than normal approach path at night to avoid any unseen obstructions and to be on the safe side if their depth perception is not accurate.

## Summary

It must be emphasized again, that night flying is a special art, and requires training prior to being attempted. We must not make the mistake of thinking that it is similar enough to daytime flying so as to be routine. The aircraft must be properly equipped and in good working condition. Preflight planning must be extensive. Constant alertness and watchfulness is a must.

## CHECK YOUR READING

1. List the special preflight precautions which are necessary prior to night flight.
2. What courtesy should we show towards other pilots when operating at night?
3. What potentially dangerous condition exists after the takeoff at night, particularly during flight in sparsely populated areas?
4. What do steady red lights on or near the airport indicate?
5. What potential hazards exist during night flight enroute?
6. During a night cross-country flight, we see a rotating beacon which shows one white and one green beam. What does this beacon tell us?
7. What is the difference between the rotating beacon at a lighted civilian airport and that at a lighted military airport?
8. What is indicated by an operating airport beacon during the daylight hours?
9. What is depicted by the flashing lights which outline a tetrahedron or wind tee?
10. Why should we receive night flight instruction prior to attempting night flight?

PART II

# METEOROLOGY

# 20

# THE ATMOSPHERE

We live at the bottom of an ocean of air called the *atmosphere*. The atmosphere is divided into a number of layers, each having peculiar characteristics.

To man, the most important layer is the one which lies on the surface of the earth, called the *troposphere*. This layer extends upward from the surface to about 55,000 feet over the Equator and to about 35,000 feet over the poles. Its depth varies with the seasons, being slightly greater in summer. The temperature within the troposphere normally decreases with altitude.

The top of the troposphere, or the boundary between it and the next layer is called the *tropopause* and is characterized by an abrupt change in the *rate* of temperature decrease. Above the tropopause lie the stratosphere, the mesosphere, and the thermosphere.

We will discuss in detail only the layer of the atmosphere of greatest importance to pilots, the troposphere. This is the unsettled region of the atmosphere in which most weather exists.

## Composition of the Atmosphere

The atmosphere is composed of a mixture of gases which have weight, are elastic, and are compressible. Nitrogen accounts for 78 per cent of a given volume of pure dry air, 21 per cent is oxygen, and ten other gases fill 1 per cent of the total volume. Air also contains water vapor in amounts varying from

0–5 per cent of volume, and a tremendous quantity of impurities (salt, dust, etc.).

## Temperature of the Atmosphere

Heat is a form of energy which causes molecular activity. Just as people in a cold room tend to huddle together, and drift apart when it is warm, so do the molecules of any substance. Every substance has a different molecular structure, so that an equal amount of heat applied to different substances will cause one substance to become warmer than another. The degree of "coldness" or "hotness" of a substance is measured with a thermometer and is called its temperature.

The major force producing our weather is heat energy which radiates from the sun.

### *Measurement of Temperature*

As pilots, we will use two different temperature scales, Fahrenheit and Celsius (or Centigrade). The Fahrenheit scale is used when recording surface temperatures. The freezing point of water on this scale is 32 degrees, and the boiling point of water is 212 degrees. On the Celsius scale, the freezing point of water is 0 degrees, and the boiling point of water is 100 degrees. The ratio of Fahrenheit degrees to Celsius degrees is 9 to 5. We can convert one temperature to the other by using one of these two formulas:

$$F = 9/5(°C) + 32$$
$$C = 5/9(°F - 32)$$

Most navigation computers can be used to convert temperatures also.

### *Temperature Range*

As we know, the surface temperature varies considerably between night and day, and from season to season. At about 4,000 feet above the surface and higher, there is little temperature change between night and day.

The large variation in air temperature at the surface is caused by the rapid heating of the surface substances during the day (which in turn radiates heat back into the air above), and the rapid cooling of the surface at night (which cools the air above).

The temperature aloft normally decreases with altitude. This variation is called the *lapse rate*, and is expressed in degrees of change per 1,000 feet of altitude. The average lapse rate is two degrees Celsius per 1,000 feet, but rarely will we see this average, for the lapse rate varies greatly, depending on the amount of heat radiating from the earth and the vertical motion of the air.

### *The Inversion*

Sometimes the situation will occur when the temperature increases with altitude rather than decreasing. This phenomenon is called an *inversion*. It occurs quite frequently but is confined to a shallow layer.

An inversion may be caused by rapid cooling of the earth's surface, which cools the air next to it. The amount of cooling is greatest near the surface, decreasing with altitude, so the air a short distance above the earth is not affected. Hence, the lower levels of air are colder than the air aloft.

An inversion may also occur due to the insurgence of cold air moving under warmer air, which will be covered in detail in Chapter 27.

Sometimes a wide area of air aloft will settle, leaving a lower level unchanged. The settling air is heated by compression and becomes warmer than the air under it, causing an inversion.

## Pressure of the Atmosphere

The atmosphere has weight which exerts a pressure of almost 15 pounds per square inch on the surface of the earth. As we increase altitude the atmospheric pressure decreases because there is less air above us pushing down. In the lower levels, this pressure decrease is about one-half pound per square inch. Although the atmosphere has no sharply defined upper limit, we know that even at 100,000 feet there is still some air present. However, when we pass through about the 18,000-foot level, one-half of the atmosphere will lie below us. This is because the molecules of air are packed more tightly near the earth's surface due to the weight of the air above.

### *Pressure Measurement*

The most accurate method used to measure atmospheric pressure is to balance its weight against the weight of a column of mercury. The device used for this method is called the *mercurial barometer*.

The mercurial barometer is simply a dish of mercury into which an evacuated tube is placed. The weight of the atmosphere pushing down on the mercury in the dish forces it up into the tube. At sea level, standard atmospheric pressure will cause the mercury to rise 29.92 inches up the tube. This shows that the weight of the atmosphere pressing on the mercury in the dish is equal to the weight of a column of mercury almost 30 inches high. If we changed the fluid in the barometer from mercury to water, the height of the column which would be supported by standard atmospheric pressure would be 32 feet!

If we moved our mercurial barometer up to an altitude of 1,000 feet, the length of the mercury column would decrease by about one inch, because there would be less air above the dish pushing downward. In order to use this device as an altitude indicator, we only need to paint 1,000-foot indications on the tube instead of inch gradients. This barometer would, of course, be difficult to carry in an aircraft, so a different type is used in the aircraft.

The *aneroid barometer* is comprised of a cell made of thin metal which is partially evacuated. On one end of the cell a pointer is attached which moves on a dial marked in hundreds or thousands of feet. As the aneroid barometer is moved aloft, the decreased atmospheric pressure causes the cell to expand, which affects the pointer indications. This type of barometer can be encased in a small container and installed on the instrument panel of an aircraft. It is the heart of the altimeter.

### *Surface Pressure Variations*

Atmospheric pressure at the surface is very important to us as pilots, for it affects aircraft performance and indicates weather trends.

The pressure of the atmosphere varies from day to day due to the movement of differing air masses, changing intensity of low pressure areas, and, in equatorial latitudes (south of 60 degrees latitude), the normal daily variations. All over the world, weather observers record and publish their barometric pressure, *after reducing it to sea level,* for use by forecasters and pilots.

Notice that the *actual* atmospheric pressure is not issued. This is because every station is not at sea level. An observer at a weather station which is 1,000 feet above sea level will always read a barometric pressure about one-inch lower than the sea level station. If station pressure was published, we would not know if it was abnormally low due to the proximity to a low pressure weather pattern or due to station altitude. In order to publish a barometric pressure reading which, when compared to other stations, will show a meaningful atmospheric pressure distribution pattern, all station pressures are *reduced to sea level.* (That is, the observer converts the current station pressure to the pressure that **would be** recorded in that geographic position *if the station were at sea level.*) With this system, we can glance at the published barometric pressure for any station and instantly know what the pressure is in relation to all other stations, without having to know the altitude of the recording station.

### *Standard Sea Level Pressure*

Since aircraft performance depends on atmospheric pressure, a standard pressure had to be determined so that all comparisons of high and low pressure would be from the same base. We should memorize the standard barometric pressure in the most common units of measurement.

| Standard Pressure | Units of Measure |
|---|---|
| 29.92 | inches of mercury |
| 14.7 | pounds per square inch |
| 1013.2 | millibars |
| 760.0 | millimeters |

## Circulation of the Atmosphere

The atmosphere is attracted to the earth by gravity and rotates as the earth rotates. Other forces upset its equilibrium, causing it to circulate over and around the earth.

The primary force causing atmospheric circulation is heat. Different surface substances and geographic areas are heated by the sun to varying temperatures and hence radiate varying amounts of heat. Hot air, being less dense than cold air, is lighter and tends to rise. Cooler air rushes in to fill the void left by the rising air, causing circulation. The amount of heat absorbed by the earth's surface (and thus the amount of heat that is radiated back into the air) is dependent upon the surface material and the angle with which the sun's rays strike the surface. A much greater amount of heat energy will be absorbed if the sun's rays strike the surface at an angle of 90 degrees than if the rays strike the surface at a shallow angle.

## *Ideal Circulation*

The angle at which the sun's rays strike the earth is most nearly perpendicular near the Equator. At the poles, the angle is quite shallow. Therefore the surface of the earth near the equator absorbs and radiates a great deal more heat than at the poles, causing the air above the Equator to rise, as shown in Fig. 81. The cooler air over the poles moves toward the equator to fill the low pressure area created by the rising warm air. Consequently, the primary world circulation pattern shows air moving upward vertically from the Equator. The air cools as it rises and moves horizontally toward the poles, settling towards the earth as it moves.

This is ideal or primary circulation and never exists, due to many modifying factors.

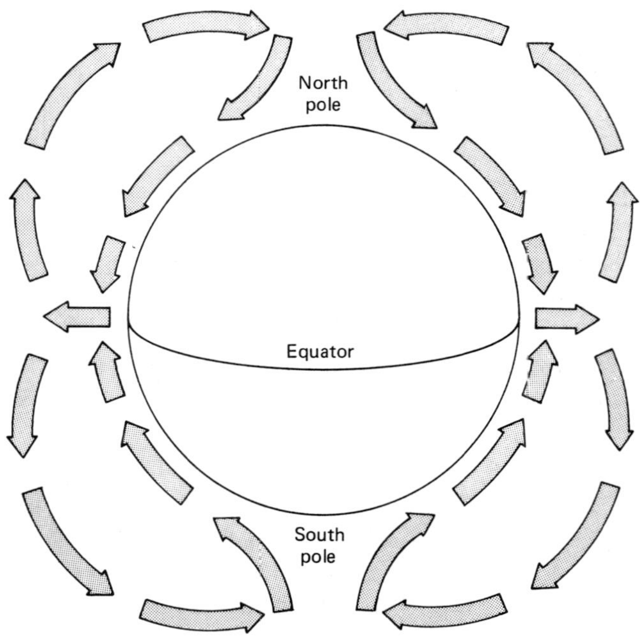


Fig. 81. Theoretical world circulation (if the earth did not rotate).

## *Factors Affecting World Circulation*

The horizontal movement of air is called the wind. There is a close interrelationship between the wind, pressure, and temperature because the wind has an effect on these, the same factors that cause it. Just as the oceans try to maintain a constant level, so does the atmosphere attempt to maintain a state of equilibrium.

*Rotation of The Earth.* The earth's rotation causes an *apparent* force on the atmosphere which deflects the wind toward the right in the Northern Hemisphere and toward the left in the Southern Hemisphere. This apparent force is called *Coriolis force*. We can demonstrate this effect with a record player turntable and a piece of chalk. If we turn on the machine, rotate the turntable, and quickly draw the chalk in a straight line across the record, the result will be a curved line on the record. We drew a straight line, but the surface on which we drew it was moving under the chalk.

In fact, any free moving body is apparently deflected because of Coriolis force. The path of a space capsule in earth orbit is a straight line, but when viewed from earth it appears to be a curved path.

To simplify our discussion, we will limit it to circulation in the Northern Hemisphere. As the air moves northward from the equator, it is deflected toward the east by Coriolis force. By the time about a third of its northward journey is completed, the air is moving eastward, causing a "pile up" of air in the area of 30 degrees north latitude. In the polar regions, southward moving air is also deflected to the right until it finally becomes a west wind.

*Pressure Gradient Force.* Air is a fluid and just as any fluid tends to flow from high to low. Air flows from high pressure to low pressure areas. The speed of this moving air, or wind, increases as the pressure gradient increases between high and low pressure areas. However, as the wind blows towards the low pressure area, it is deflected to the right which starts a counterclockwise circulation around the low pressure center and a clockwise circulation around a high pressure center.

*Friction.* Friction acts to slow the movement of air. Coriolis force varies directly with wind speed, so it will decrease as friction decreases the wind speed. The decrease in Coriolis force allows the air to turn inward toward the low pressure center, or outward from the high pressure center. Naturally, friction is greatest near the surface, and decreases with altitude until, at about 2,000 feet, it is negligible.

*Centrifugal Force.* The curved path of the wind causes it to be subjected to centrifugal force, which hurls the air outward from the center of rotation.

*Other Factors.* Gravity, differing heat transfer properties of surface substances, irregular distribution of oceans, irregular terrain, and seasonal changes all affect atmospheric circulation. We can see, therefore, that the atomsphere around us is in a constant state of movement, in a never-ending search for equilibrium. It is this circulation which keeps the equator from becoming warmer and warmer and the poles from becoming increasingly colder.

## CHECK YOUR READING

1. What is the name of the atmospheric layer in which most weather occurs?
2. What is the atmosphere composed of?

3. What is *heat?*
4. What Celsius temperature equals 78 degrees Fahrenheit?
5. What causes the air near the surface to change temperature from day to night?
6. What is an *inversion?*
7. List the causes of an inversion.
8. List the standard pressure units of measure.
9. What is the amount of pressure change in inches of mercury per 1,000 feet?
10. What is *station barometric pressure?*
11. What is station pressure *reduced to sea level?*
12. Why must the weather observer adjust his atmospheric pressure reading to sea level pressure before publishing it?
13. What causes atmospheric circulation?
14. List the factors which affect ideal circulation.
15. Why does the wind blow counterclockwise around a low pressure area, and clockwise around a high pressure area?

# 21

# WIND SYSTEMS AND TURBULENCE

Our almost constant weather change is caused by the movement of high and low pressure systems. Normally, these pressure systems move from west to east with the prevailing westerly winds aloft. Low pressure systems are usually accompanied by large areas of precipitation. High pressure brings clear weather. Large and rapid pressure changes and shifting winds occur with the passage of either pressure area.

When the pressure area is depicted on a weather chart, points of equal barometric pressure are connected by lines called *isobars*. The isobars will curve around the pressure area, and the wind will generally blow parallel to them, in a clockwise direction around the high and counterclockwise around the low (in the Northern Hemisphere). Therefore, if we stand with our back to the wind, the lowest pressure will be to our left.

The speed of the wind will increase as the pressure gradient increases. That is, the closer the isobars, the stronger the wind. Greatest pressure gradients occur in low pressure areas, so the strongest winds will accompany them.

## Local Winds

Complicating the general wind systems are the local winds, created by bodies of water, mountains, or valleys.

### *Sea and Land Breezes*

If we are fortunate enough to live near the ocean, or other large body of water, we may enjoy a delightfully cool *sea breeze* on a warm summer day. This local wind is caused by the different rate of heat absorption and radiation between the land and the water.

During the day, the land area becomes warmer than the water surface. The air above the land is heated and rises. Cooler air over the water flows toward the shore to fill the void left by the rising warm air, giving us the welcome relief from the heat which makes the shoreline so popular in the summer.

At night, the situation is reversed. The land area cools more rapidly than the water. Air above the land is cooled and settles. The air above the water is heated by it and rises. An off-shore or *land breeze* then develops. Figure 82 shows this phenomenon.

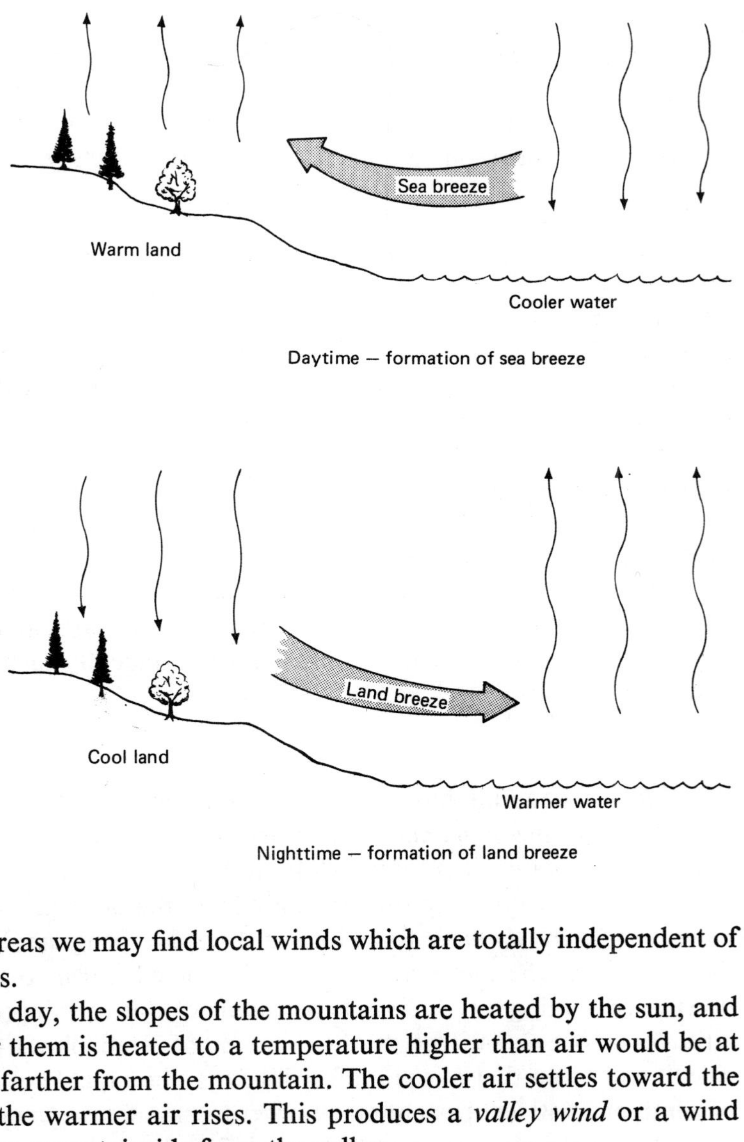


Fig. 82. Local winds.

### *Mountain and Valley Winds*

In mountainous areas we may find local winds which are totally independent of circulation systems.

During the day, the slopes of the mountains are heated by the sun, and the warm air near them is heated to a temperature higher than air would be at the same altitude farther from the mountain. The cooler air settles toward the valley surface as the warmer air rises. This produces a *valley wind* or a wind which blows up the mountainside from the valley.

In the evening, the mountain slopes cool, as does the air next to them. The cooler air slides down the mountainside filling the valley.

*Katabatic Wind*

Any wind which blows down an incline is called a *katabatic wind.* If the downslope wind is warmer than the air in the lowland, it is called a *foehn* (pronounced fawn). If it is colder than the air in the valley, it is called a *gravity wind.*

## The Jet Stream

Although the jet stream wind is normally found only at high altitudes (25,000 to 40,000 feet), we will discuss it briefly here because some meteorologists believe that it is in some cases the powerful *steering force* which controls the direction of movement of low pressure areas.

The jet stream is a narrow core of very high speed wind. Although wind speed may reach 300 miles per hour, it is more often 100 to 200 miles per hour. The main jet stream lives a cycle of formation, intensification, movement, and dissipation. Its position varies both seasonally and daily.

## Turbulence

Turbulence is caused by air in motion, either vertically or horizontally. We have seen how air is warmed by the underlying surface, becomes less dense, and rises. When we fly through these vertical currents we experience a series of bumps or jolts. The reaction of the aircraft varies with the intensity of the vertical current, the speed of the aircraft, wing loading, and aircraft attitude. Convective currents may exist from the surface up to about 10–12,000 feet.

The horizontal wind also causes turbulence. As the wind strikes surface objects or mountainous terrain, it swirls and eddies about, causing extremely variable air currents. This variable wind not only causes turbulence but is very important to pilots when flying near the ground, particularly on takeoff, approach, and landing, because it increases the difficulty of flight path planning and aircraft control.

*The Mountain Wave*

When flying in mountainous terrain, we must exercise extreme caution to avoid entrapment in the *mountain wave.* When the wind blows across a mountain or a line of mountains, it is forced aloft, and then surges down the *lee* side of the mountain. If we approach the mountain from the same general direction as the wind, our aircraft will be forced upward by the rising air. When approaching the lee side of the mountain, we may be forced down to the surface by the downward surging air. A 30 mile per hour wind is certainly not uncommon at altitudes above 2,000 feet. If this amount of wind blows down a mountain side, it will mean that our aircraft will have to be capable of maintaining a rate of climb of 2,640 feet per minute in order to maintain a predetermined altitude. Not many light aircraft have this capability.

Since the mountain wave may extend for a hundred miles or more downwind from the mountain range, we may expect turbulence and vertical currents far from the mountains. If any moisture is present in the air, a cloud may form over the mountaintop as the moisture is condensed due to the cooling of the air as it rises. The crest of each wave downwind of the mountain will also be cloud capped. *Rotor clouds* or small rounded puff-type formations may mark the low spot in each wave of air currents. Figure 83 shows the mountain wave graphically.

Fig. 83. The mountain wave.

## *Wind Shear*

When the wind changes its speed and/or direction, a *shearing* or tearing effect takes place in either the horizontal or vertical plane. This shearing produces turbulence which varies directly with the rate of change in wind speed or direction. The wind shear area is normally a narrow band and is often encountered when we climb or descend through an inversion.

A particularly strong wind shear may be associated with a temperature inversion near the ground. Occasionally at night, rapid cooling of the surface causes a pocket of cold air over the airport. Warmer air aloft traps the cold air, and a strong wind shear may be present in the transition zone between the two. This may present a hazard to flight as the aircraft intercepts this wind shear area, causing turbulence-produced control problems and fluctuating airspeed.

## Turbulence Classification

There is no direct measurement of turbulence intensity. No two pilots will classify turbulence in the same way. Yet, it is important for us to know how severe the expected or actual turbulence is. The effect of it on our aircraft will vary with the speed, wing loading, and altitude of the aircraft. Through the years, the following definitions of the degrees of turbulence have been adopted. However, we must remember that light turbulence encountered in a four-engine jet aircraft may have a moderate or severe effect on a light aircraft.

*Light Turbulence.* Occupants of the aircraft may be required to use seat belts; objects within the aircraft remain at rest.

*Moderate Turbulence.* Occupants are occasionally thrown against the seat belts; unsecured objects in the aircraft move about.

*Severe Turbulence.* The aircraft may be momentarily out of control; occupants are thrown about violently. Objects not secured in the aircraft are tossed about.

*Extreme Turbulence.* The aircraft is violently tossed about and is practically impossible to control; structural damage may result.

## CHECK YOUR READING

1. What are *isobars*?
2. What effect does pressure gradient have on wind speed?
3. Why does a sea breeze form near the coast line of large masses of water?
4. Why does a land breeze form at night on the coast line of large water masses?
5. What is the cause of mountain winds?
6. What is the cause of turbulence?
7. List the variables which effect the reaction of the aircraft to turbulence.
8. When flying near mountainous terrain, on which side of the mountain will dangerous downdrafts be found (windward or lee)?
9. What causes wind shear?
10. List the classifications of turbulence and the description of each.

# 22
# MOISTURE

The most important element in the production of our weather is moisture. Without moisture the earth could not support life. Moisture is found in the atmosphere in various forms including invisible water vapor, liquid, and solid. Most of our weather occurs in the troposphere because that is where most of the atmospheric moisture is found. Only rarely will significant amounts of moisture in any form be found above the tropopause.

The oceans, lakes, rivers, wetlands, and snow or ice fields are the primary sources of moisture in the air. When it is first introduced into the air, it is in the form of invisible water vapor, which may be carried great distances by circulatory systems. Then it is removed from the atmosphere in one form of precipitation or another.

## Changes of State

### *Evaporation*

Molecules of liquid are constantly moving about. Some of the molecules move into the air in the form of water vapor. This is called *evaporation* and increases as the temperature of the liquid increases.

When the molecules do the work of jumping from the surface of the liquid, they use energy in the form of heat, which lowers the temperature of the liquid. The heat is not lost forever for it remains latent in the water vapor, ready to be released when the water vapor changes back to visible moisture. Hence, we can say that evaporation is a cooling process. (Remember how fuel evaporation cools the air in the carburetor?)

We feel uncomfortable on warm humid days because the perspiration from our body does not evaporate rapidly. We will understand why it evaporates slowly on a humid day after reading the coming paragraphs concerning humidity. If we use a fan to circulate more air past our body, we can increase the rate of perspiration evaporation, which has a cooling effect.

*Condensation*

*Condensation* is the opposite of evaporation. That is, when invisible water vapor changes back to visible moisture it has condensed. Condensation returns the heat which remained dormant in the water vapor. So, we can say that condensation is a warming process. It is caused by cooling the air in which water vapor is present, because cold air cannot hold as much water vapor as warmer air. In a given volume of air, the molecules move farther apart as the air is warmed and closer together as the air is cooled. Therefore, in the warm air there is room between the air molecules for the water vapor molecules. As the air is cooled, the air molecules move close together and "squeeze out" the water vapor molecules which band together to form water droplets.

Condensation may also occur if too many water vapor molecules evaporate into the air. We will study more about this in the text concerning humidity.

*Sublimation*

We have all seen snow or ice disappear from the fields without first melting. This process by which solid water evaporates without first changing to the liquid state is called *sublimation*. The reverse process, when water vapor changes directly to frozen water, is also called *sublimation*. (For instance, frost forms directly from water vapor on a cold evening.)

The same amount of heat energy is required for sublimation to occur as would be required if the frozen liquid first melted and then evaporated, even though the liquid state is bypassed.

## Moisture Content

As mentioned, there is a definite limit to the amount of invisible water vapor the air can hold. This limit is determined primarily by the temperature of the air.

*Saturation*

When the air contains the maximum amount of water vapor it is said to be *saturated*. Condensation will occur if either of two things happens to saturated air. If the temperature decreases, some of the water vapor will condense; if additional water vapor is added to the air, it will also condense. The condensation may be in the form of clouds, fog, or precipitation.

Naturally, if unsaturated air is cooled sufficiently, it will become saturated at some temperature.

### Relative Humidity

The ratio of the amount of water vapor actually in the air to the amount of water vapor the air could hold is called the *relative humidity*. If a parcel of air contains all the water vapor it can hold without condensation occuring, its relative humidity is 100 per cent. If the parcel of air contains only one quarter of the water vapor it could hold, its relative humidity would be 25 per cent.

On warm humid days, the air is nearly saturated. We feel uncomfortable because there is little room between the molecules of air around us which is not already filled with water vapor. Our perspiration, therefore, cannot easily evaporate to cool our body. We can increase evaporation by moving more air past our body, or moving drier, less humid air past it.

### The Dew Point

The *dew point* is the temperature to which the air must be cooled before becoming saturated (assume constant pressure). If we know the current air temperature and the current dew point, we can tell how near the air is to being saturated. The difference between the air temperature and the dew point is called the *spread* and is very significant to us as pilots. When the spread is great, the humidity is low, which indicates that fog, cloud, or precipitation is not likely. As the spread diminishes the chance of visibility-reducing moisture formation increases.

Normally the spread is decreased by falling temperatures, although it can be narrowed if more water vapor is added to the air, raising the dew point.

The air temperature and the dew point are published by every weather observer. The current spread and the temperature/dew point *trend* will aid us in determining the weather in the future. If we check the spread for three or four hours prior to the current report and see that the spread is diminishing, we might suspect that it will continue to narrow, causing deteriorating weather conditions. When the spread is within four degrees, the humidity is such that fog, low cloud, or thick haze may form rapidly.

### Condensation Nuclei

When the air becomes saturated, condensing water droplets will collect on the tiny particles of salt, dust, and soot which are in the air. These small contaminants are called *condensation nuclei*. Without these particles in the air, supersaturation may occur. That is, the air may hold more water vapor than is required to saturate it, but, because of the lack of water absorbent particles in the air, the water vapor does not condense out. In this atmospheric condition the introduction of any condensation nuclei will touch off a chain reaction of instant condensation.

### Clouds, Fog, and Precipitation

If condensation occurs, clouds or fog will form. Gentle lifting of the air, due to convective currents, keeps the tiny water droplets suspended in the air. If the water droplets grow in size (by colliding with each other) they may become heavy enough to fall to earth in the form of liquid or solid precipitation. Some of the precipitation may evaporate as it falls, adding to the water vapor content of the air under the clouds. Strong vertical currents, if existent, will keep the precipitation from falling until the drops become quite large and heavy. Gravity then will overcome the updrafts, and the drops or ice will fall to earth.

The form of the precipitation (liquid or solid) depends on the air temperature and the intensity of the vertical currents. Although moisture is sometimes found in the liquid state in air well below freezing temperature, generally it will freeze to form solid precipitation. Also the form of the precipitation may change from liquid to solid or vice versa when it falls through air which is warmer or colder than freezing. For instance, snow may form in cold air, but melt as it falls through a layer of warmer air below. Liquid drops may freeze as they fall through a colder layer of air.

Strong vertical currents may carry the precipitation up and down through layers of air having above- and below-freezing temperature. This is how hail is formed.

## CHECK YOUR READING

1. Why does most of our weather occur in the troposphere?
2. Where does water vapor come from?
3. What is *evaporation*?
4. What is *sublimation*?
5. What is *condensation*?
6. Why does evaporation remove heat from the surrounding area?
7. Where does the heat go which is lost due to evaporation?
8. What is the determining factor concerning the amount of water vapor a parcel of air can hold?
9. List the ways we can make saturated air condense.
10. What is the *dew point*?
11. Why is the spread between the temperature and the dew point important?
12. What are *condensation nuclei*?
13. What factors determine the form of precipitation?
14. How is hail formed?

# 23

# STABILITY

Stability of the atmosphere may be likened to aircraft stability. A stable aircraft, when disturbed, tends to return to its original attitude. An unstable aircraft will move farther from its original attitude when disturbed. The atmosphere acts in the same manner. If the air is stable, it will resist vertical motion. If the air is unstable, it will easily be moved vertically and will increase its vertical motion once it is initiated.

Atmospheric stability depends on the weight distribution (vertically) of the air. That is, if a parcel of air is lighter than the surrounding air, it will tend to rise. If it is equal in weight to the surrounding air, it will remain stationary. Heavier air will sink or subside. Since the weight of the air depends on its temperature, we can say that stability depends on air temperature distribution (vertically).

## Lapse Rate

A comparison of the temperature of a parcel of air to the temperature of the surrounding air will show its stability.

We learned that the average lapse rate is $3\frac{1}{2}$ degrees Fahrenheit (2 degrees Celsius) per 1,000 feet of altitude. When unsaturated air is forced aloft by either convection or by orographic lifting, it will cool at the rate of $5\frac{1}{2}$ degrees Fahrenheit (3 degrees Celsius) per 1,000 feet of altitude. This is the dry adiabatic lapse rate. *Adiabatic* means the temperature change occured without either taking heat away from or adding heat to the surroundings. The adiabatic temperature change occurs simply because of expansion or compression of the air (expansion cools the air; compression warms it).

When saturated air rises, either because of convection or orographic lifting (being forced up a slope), it cools at a much slower rate of from 2 to 5 degrees Fahrenheit (1.1 to 2.8 degrees Celsius) per 1,000 feet of altitude. The slower cooling rate of saturated air is due to the fact that the water vapor in the air condenses as it cools. Condensation releases heat to the air. The net effect is a reduced rate of cooling. The exact rate depends upon the amount of moisture present. The amount of heat released by condensation is greater in warm air because more water vapor may be present.

### *Dew Point Lapse Rate*

The dew point temperature decreases at the rate of 1 degree Fahrenheit (0.56 degree Celsius) per 1,000 feet during the lifting process. Therefore, in unsaturated air which is lifted, the temperature and the dew point approach each other at a rate of about $4\frac{1}{2}$ degrees Fahrenheit ($2\frac{1}{2}$ degrees Celsius) per 1,000 feet. If we know the surface air temperature and dew point, we can compute the level at which cloud-forming condensation will occur when the air rises.

For instance, if the surface temperature is 70 degrees Fahrenheit and the dew point at the surface is 52 degrees Fahrenheit, the height of the cloud base will be about 4,000 feet if the air is lifted ($70 - 52 = 18$ degrees $\div 4\frac{1}{2} = 4$).

## Stability Characteristics

Unsaturated air normally will be stable. Saturated air at low temperature normally is stable. Saturated air at high temperature is normally unstable.

In other words, air at high temperature is either stable or unstable depending upon its moisture content.

## Stability Determination

We can determine the degree of stability of each layer of air by comparing the actual lapse rate with the dry and moist adiabatic lapse rates.

### *The Adiabatic Chart*

Atmospheric measurements may be made by balloons sent aloft, and the information received is plotted on an adiabatic chart.

*Absolute Stability.* When the actual lapse rate is less than the moist adiabatic lapse rate, the air is *absolutely* stable, no matter how much moisture it contains. Absolutely stable air, when lifted, becomes cooler than the air around it and sinks to its original position when the lifting force stops.

*Absolute Instability.* When the actual lapse rate is greater than the dry adiabatic lapse rate, the air is unstable no matter what amount of moisture is present. In this instance, a parcel of air will be warmer than its surroundings even if lifted slightly and will continue to rise rapidly.

### *Conditional Stability*

If the actual lapse rate is somewhere between the dry and the moist adiabatic lapse rate, the air is said to be conditionally unstable. That is, stability will depend upon the amount of moisture in the air.

### Effects of Stability

The type of clouds that form will be determined by the stability of the air. Clouds formed in unstable air will be vertically developed, and the air will be turbulent.

Clouds formed in stable air will be in layers, with no vertical development. The air will be smooth.

Subsiding air is warmed by compression. This frequently causes an inversion which will increase stability. It is this type of inversion which serves to intensify a surface inversion, trapping smoke and haze in the lower atmosphere.

## CHECK YOUR READING

1. On what factors does atmospheric stability depend?
2. Why is atmospheric stability important to the pilot?
3. What is the dry adiabatic lapse rate in Fahrenheit degrees?
4. What is the moist adiabatic lapse rate in Fahrenheit degrees?
5. Why is the moist adiabatic lapse rate less than the dry adiabatic lapse rate?
6. What is the dew point lapse rate?
7. If the surface temperature is 80 degrees Fahrenheit and the dew point is 55 degrees Fahrenheit at what height will cloud bases be found?
8. Warm air is either stable or unstable depending upon what factor?
9. What type of clouds will form in stable air?
10. What type of clouds will form in unstable air?

# 24

# CLOUDS

Clouds are formed when the invisible water vapor in the air is condensed into visible moisture. To most people, it is fascinating to realize that enough moisture is in the air to form a dense cloud, and yet, most of the time the moisture remains invisible. It has been said that the visible moisture in a cloud one-mile square and 10-feet thick could be placed in a water tumbler.

We can picture how invisible water vapor can suddenly become quite visible if we think of a huge football stadium packed with 200,000 people. Suppose 10,000 of the people are wearing red hats and are scattered throughout the crowd. An observer would not particularly notice the red hats. But, if all of the red-hatted people were to come together in one section of the grandstand, the observer would suddenly see a mass of bobbing red hats. They were there all the time, but remained *invisible* until they massed together.

Water vapor molecules act just like the red-hatted people. If they are separated they are invisible. When merged together, they suddenly appear in the form of clouds or fog.

## Basic Cloud Formations

We have seen that atmospheric stability determines the severity of vertical air movement. This in turn determines the type of clouds which will be formed.

Cumulus clouds are clouds of vertical development having the appearance of puffs of cotton, large cauliflower, or towering mountains. They are produced by strong vertical currents in unstable air which can carry moisture-laden air to great heights. (In some cases, the vertical currents will extend to the tropopause.) It is the churning vertical currents which cause the billowing appearance of the cumulus cloud. The air may be extremely turbulent both under and within the cloud. Precipitation, if it occurs, will be localized and in the form of showers.

*Cumulus Clouds*

When stable air is forced aloft, any cloud formation will be blanket-like in appearance. These layered clouds may cover huge geographic areas, but they will have little or no vertical development. Because of the absence of vertical currents, any precipitation falling from stratus clouds will be light (drizzle or light snow) but may cover large areas. The air under and within the cloud will normally be quite smooth.

*Stratus Clouds*

If the air is conditionally unstable, a mixture of cumulus and stratus clouds may form. The portion of the air mass which is lifted violently will create cumulus-type clouds. Other areas, in which little or no lifting occurs, will be covered with stratus-type clouds.

*Cumulostratus Clouds*

## Cloud Classification According to Height

Clouds are further classified and described according to the altitude at which the *base* of the cloud appears.

*High Cloud*

High clouds have bases between 16,500 and 45,000 feet. The prefix denoting high clouds is *cirrus*.

*Cirrus Clouds.* Cirrus clouds are often described as "mare's tails." They form in thin, white cloud patches, composed of ice crystals and have a feather-like appearance.

*Cirrocumulus Clouds.* Thin sheets of clouds, having patches of white puffs, are called *cirrocumulus* clouds. They are normally composed of ice crystals but may contain supercooled water droplets.

*Cirrostratus Clouds.* The cirrostratus cloud is sometimes called the *halo* cloud, for it causes a huge halo around the sun or moon when light penetrates its transparent veil. It is whitish and forms in sheets which cover wide areas of the sky.

*Middle Clouds*

The middle clouds have bases between 6,500 and 16,500 feet. The prefix for middle clouds is *alto*.

*Altocumulus Clouds.* These clouds are white or gray, occur in patches or in solid layers, and often have a wave-like look. Sometimes they resemble ocean waves and rounded masses. They may be composed of small supercooled water droplets, or may be a mixture of water and ice crystals.

*Altostratus Clouds.* The altostratus clouds are usually gray or blueish in color. They are sometimes associated with altocumulus clouds, and are thin enough to allow the sun to be dimly visible. Often very light precipitation will fall from them.

### Low Clouds

Low cloud bases lie between the surface and 6,500 feet. They are given no prefix which denotes low altitude, but may be used with the word *nimbus* to indicate that continuous precipitation is falling.

*Nimbostratus Clouds.* The massive and dark nimbostratus cloud is composed of large water droplets in suspension and falling drops of water or snow.

*Cumulonimbus Clouds.* The cumulonimbus clouds are the ultimate of towering cumulus clouds. They often take the shape of huge towers or mountains, sometimes having plume-like tops. Precipitation will be heavy and may be accompanied by thunder and lightning. (See Chapter 28, Thunderstorms.)

*Fog.* A cloud having a base 50-feet or less above the surface is called *fog*. Fog is important enough to warrant a complete chapter. (See Chapter 25, Fog.)

## Summary

We must become very familiar with the various cloud formations as they will provide us with clues concerning the stability of the air, the type of precipitation to be expected, and the amount of turbulence we may encounter.

FEDERAL AVIATION AGENCY
VFR EXAM-O-GRAM * NO. 32

SIGNPOSTS IN THE SKY

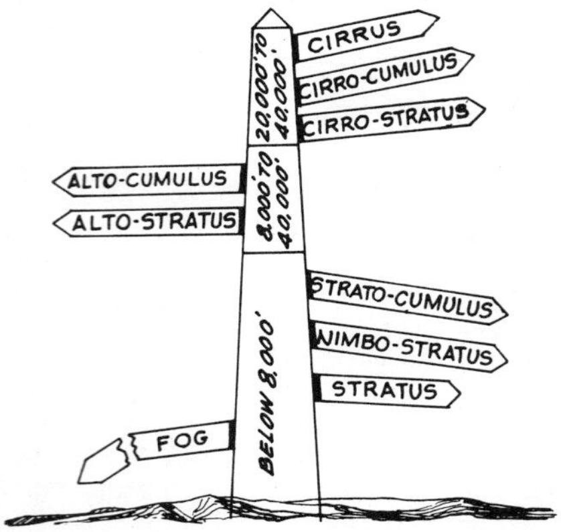


Clouds are nature's signposts in the sky. Long before meteorology became established as a science, man learned to read the coming storm in the gathering thunderhead, and see the promise of rain in a mackerel sky or a halo around the moon. The various forms that clouds take are still among the important indicators of future weather even for modern forecasters. To those who take the trouble to read these signs, a change in the weather seldom comes as a surprise.

WHAT CAUSES WEATHER? Most visible weather, when reduced to its simplest term, is determined by the amount of moisture in the air, and as surely as night follows day, clouds and precipitation are accompanied by high moisture content. The signs pointing to weather change, usually in a high to low cloud sequence, are as obvious as the changing cloud formations themselves.

WHAT CAN A PILOT DETERMINE FROM THE WEATHER SIGNS? Any informed airman can borrow a leaf from the book of the skilled mariner or woodsman and learn to recognize cloud types and understand what these different signs mean in terms of:

a. Probable weather
b. Interpreting weather reports
c. Judging the weather while in flight
d. Turbulence
e. Precipitation
f. Icing hazards

Before we look at the signs, however, we should examine a few things about their make-up.

THE SHAPE OF CLOUDS

STRATUS

CUMULUS

CIRRUS

HOW CAN A PILOT IDENTIFY THE THREE BASIC CLOUD FORMS?

1- STRATUS - Does it sound like straight? Well, it is! Remember the flat streaks or layers of smoke that hung over the room during a "friendly" game of pinocle? Yep--that's your stratus cloud form.
2- CUMULUS - Reminds you of accumulated, doesn't it? Shades of a weather prophet, it is! Remember the wooly, billowy shape of the accumulated puffs of the cigar smokers during the pinocle game? Right you are--they were making cumulus shapes.
3- CIRRUS - Sound kinda icy? It is! Remember the webby, fibrous "smoke" which flows from the open food locker? Yes, sir--that's your cirrus cloud form!

THE VOCABULARY OF CLOUDS

HOW CAN A PILOT EASILY LEARN THE NAMES OF CLOUDS? Just add a descriptive suffix or prefix to the three basic cloud forms. For example:

(a) Alto (high) * (b) Fractus (broken) (c) Nimbo (raincloud)

*Refers actually to middle clouds.

We can have a cloud vocabulary adequate for everyday needs by combining the names of the three basic cloud types with one another, as well as with these qualifying terms; like so:

(a) Cirro-cumulus
(b) Cirro-stratus
(c) Alto-cumulus
(d) Alto-stratus
(e) Strato-cumulus
(f) Cumulus-fractus
(g) Nimbo-stratus
(h) Stratus-fractus
(i) Cumulo-nimbus

THE CLOUD SIGNS

HOW CAN A PILOT READ THE CLOUD SIGNS? Just put the components together and by observing three important steps, go about the task of interpretation.

STEPS: (a) Identify the clouds.
(b) Observe their altitude and direction of movement.
(c) Observe the sequence of cloud formations.
(d) Observe the surface wind direction.

Typical major cloud signs appear below and on the opposite page along with the caption forecasts for each. Each one in its appearance, its altitude, and its direction of movement usually foretells a particular kind of weather when related to a steady surface wind direction. Study them carefully. Understand that exceptions can and will occur, and that there is no substitute for a thorough weather briefing from a trained forecaster or briefer.

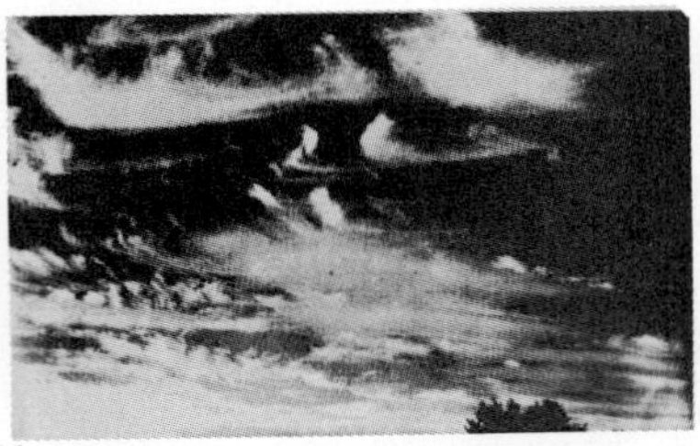

**CIRRUS 20,000 TO 40,000 FEET**

Ice clouds. Good weather if winds from W NW to N. Precipitation likely within 20 to 30 hours if winds steady from NE E to S.

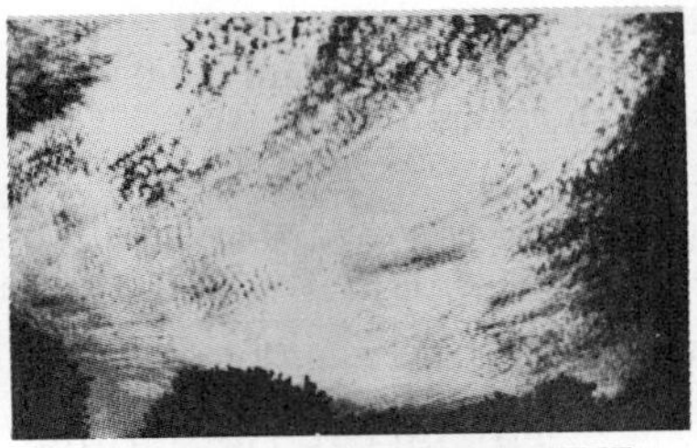

**CIRROCUMULUS 20,000 TO 40,000 FEET**

Ice clouds. Precipitation likely 15-20 hours if wind NE to S. Early summer A.M. often afternoon thunder showers. Other winds overcast.

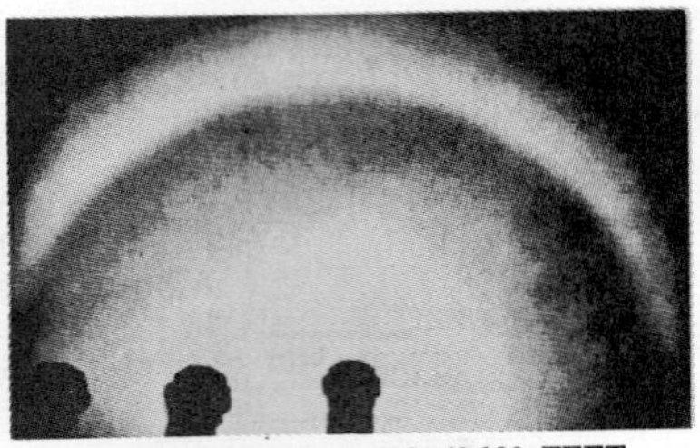

**CIRROSTRATUS 20,000 TO 40,000 FEET**

Precipitation likely within 15 to 24 hours if wind steady from NE to S. Prismatic effect of sun or moon through ice crystals cause halo.

**ALTOCUMULUS 8,000 TO 20,000 FEET**

Cloud tops forming altocumulus cumulogenitus usually bring precipitation in 5-10 hours, gusty winds, thunder showers, wind squalls.

**ALTOSTRATUS 8,000 TO 20,000 FEET**

Precipitation likely within 10 to 20 hours if wind is steady from NE to S. Sun appears behind frosted glass. Other winds overcast sky.

**LENTICULAR 8,000-20,000 FEET**
A sign of danger. This type of cloud is associated with high winds, standing waves, and severe to extreme turbulence in mountainous regions.

**NIMBOSTRATUS BELOW 8,000 FEET**

Rain or snow clouds. Precipitation of long duration if winds NE to S, or shorter duration if winds are from SW to W to N.

**STRATOCUMULUS BELOW 8,000 FEET**

Immediate threatener of bad weather from a Sprinkle to heavy precipitation. If at head of cold front, gusty winds or thunder shower.

**STRATUS BELOW 8,000 FEET**

Winds from NE to S may bring heavy precipitation. Other winds bring only light drizzle or an overcast sky.

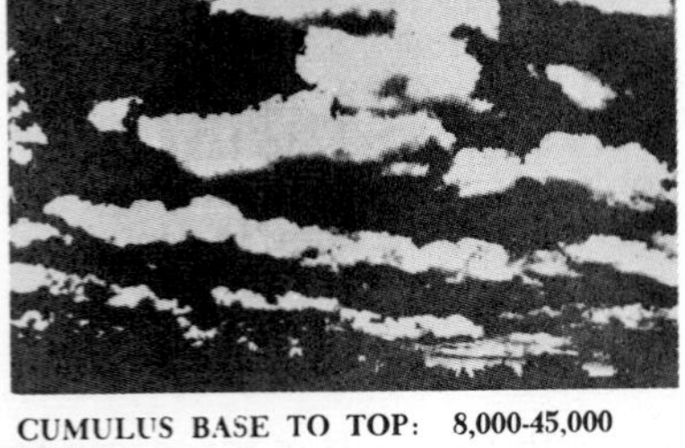

**CUMULUS BASE TO TOP: 8,000-45,000 FEET**

Fair weather clouds if they show no vertical development. Can build up and develop into cumulus congestus or cumulonimbus clouds.

**CUMULONIMBUS MAMMA 8,000-10,000 FEET**

Seldom seen low to middle clouds. Associated with severe wind squalls, hail, heavy precipitation, tornadoes and thunderstorms.

**CUMULONIMBUS BASE TO TOP: 10,000-60,000 FEET**

Precipitation likely and soon coming usually from SW W to N. Distant clouds often show an anvil-shaped cirroform cap.

TYPICAL STABLE AIR WARM FRONT

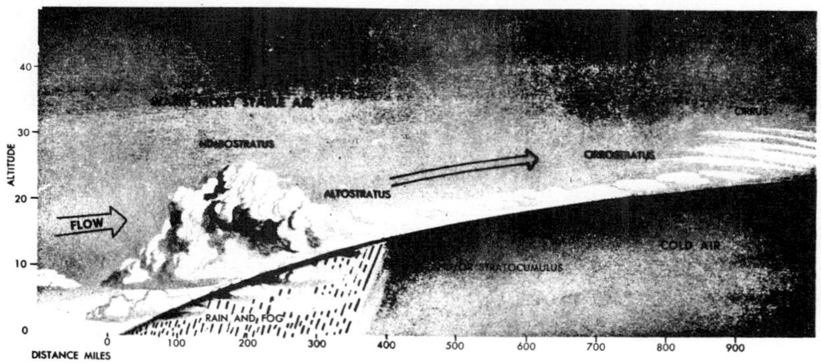


WHAT CLOUD SIGNS GENERALLY PRECEDE A TYPICAL SURFACE WARM FRONT? Imagine yourself at ground point "A" above. You have really been there many times. Observe and remember the following:

(a) Preceding the surface position of the front and approaching from a westerly direction are high cirrus clouds, thickening and lowering to cirro-stratus, alto-cumulus, alto-stratus, and finally nimbo-stratus or cumulo-nimbus.
(b) Precipitation can usually be expected in one to three days after the first sign of cirrus, and
(c) Weather is widespread and of comparatively long duration.

WHAT OTHER SIGNS PRECEDE THE WARM FRONT? Surface warm fronts are generally preceded by (a) a slowly falling barometer, and (b) surface winds blowing steadily from a southeasterly direction.

TYPICAL FAST MOVING COLD FRONT

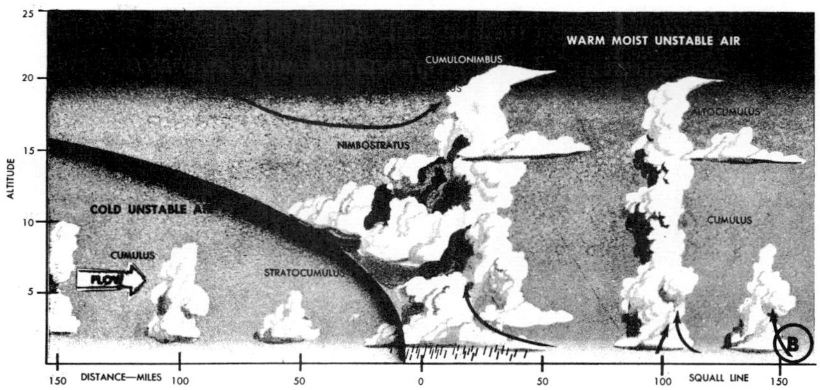


WHAT CLOUD SIGNS GENERALLY PRECEDE A TYPICAL SURFACE COLD FRONT? Imagine yourself at ground point "B" and observe the following: (a) the transition of clouds is likely to take place in a few hours, (b) usually there is little warning from cirrus clouds--only cumuliform clouds appear, (c) heavy precipitation is likely in 5-10 hours after appearance of clouds, and (d) weather is typically more violent and short-lived than warm front weather.

WHAT OTHER SIGNS PRECEDE THE COLD FRONT? The typical surface cold front is preceded by (a) a rapidly falling barometer, and (b) surface winds blowing steadily from a southwesterly direction.

CONCLUSIONS. Two inescapable and often exasperating facts mark a pilot's personal relations with the weather. One is that from his first flight to his last it is always there, in one form or another. The second absolute is the certainty of change without even token reference to the needs or wishes of the pilot. If a pilot is capable of "eye balling" the "signposts in the sky" with discrimination and judgment, he is better equipped to live with the weather as he must--on its terms.

## CHECK YOUR READING

1. What is the name of the broad classification of clouds which form in stable air?
2. Describe the clouds which are formed in unstable air.
3. Which type of cloud will be accompanied by heavy precipitation?
4. Which type of cloud may be accompanied by steady light precipitation?
5. What is the major difference between *fog* and *clouds*?
6. At what height will the bases of high clouds be found?
7. What is the prefix used to denote high clouds?
8. At what height will middle cloud bases be found?
9. What is the prefix used to denote middle clouds?
10. What does the word *nimbo* attached to the basic cloud name denote?

# 25

# FOG

Fog is a cloud formation within 50 feet of the surface, composed of water droplets or tiny ice crystals.

## Fog Formation

Three factors must be present to produce fog:

1. *High Relative Humidity*. The air must be saturated with water vapor before fog-producing condensation can occur. This is why the temperature/dew point spread is of great concern to the pilot. A narrow spread indicates high relative humidity. The air temperature may drop due to radiational cooling, orographic lifting, or by movement over a cool surface. The dew point may rise if more water vapor is added to the air by evaporation. In either case, as the spread narrows, the chance of fog increases, until, when the spread becomes only one or two degrees, the possibility of fog forming becomes quite definite. We must be aware of the *trend* in temperature/dew point spread for two or three hours prior to flight time in order to accurately forecast the spread tendency. For instance, if the spread is four degrees and decreasing one hour before sunset, we will be safe in assuming that it will continue to narrow as the air is cooled radiationally after dark. If the same four-degree spread exists just prior to sunrise, we would be justified in forecasting a larger spread after sunrise due to

convectional heating (assuming all other factors remain constant).

2. *Condensation Nuclei.* There must be present in the atmosphere, water absorbent particles of salt, dust, soot, or other contaminants to which the water vapor molecules can cling. If the air is absolutely pure it will become supersaturated, without condensation occuring.
3. *Wind.* The presence of wind is extremely important when forecasting fog formation. The speed of the wind is critical, for if too much wind exists, the air will be mixed through a deep layer and will be lifted, causing low clouds instead of fog. If too little wind exists, the air will not be mixed at all, and radiational cooling will be limited to a very thin layer of air near the surface. In this case, dew will form as the water vapor condenses into tiny water droplets which fall to earth.

## Types of Fog

There are six common types of fog, each being formed by specific circumstances. The first three types listed here are caused by cooling temperatures. The second group of three are caused by the addition of water vapor to the air.

### *Radiation Fog*

This type of fog (sometimes called *ground fog*) is caused by radiational cooling of the land, usually on clear calm nights, or during the early morning hours. It will usually disappear soon after the land is heated by the rising sun. A light wind (less than 5 miles per hour) will deepen the layer of ground fog by spreading the cooling effect as the air is mixed. With less wind, the ground fog will be quite shallow, sometimes measuring only two-or three-feet thick. Ground fog may be very deceiving because visibility is good when looking straight down into it, but visibility will diminish rapidly to near zero upon entry into it. We must, therefore, exercise extreme caution when approaching an airport for a landing if ground fog is present.

### *Advection Fog*

Advection fog forms when warm moist air moves over a colder surface. This is common along the coastline of large bodies of water. Land areas will radiate heat rapidly while the nearby water retains its heat. If the horizontal movement of the air is toward the land, the warm moisture-laden air will cool radiationally and fog will spread inland. The thickness of the fog bank will depend on the wind speed. A wind of 15 miles per hour or less will thicken the fog bank, but wind in excess of 15 miles per hour will cause lifting and produce low clouds.

### *Upslope Fog*

Upslope fog is common in mountainous areas when warm moist air is gently lifted orographically. The upslope wind must continue to exist if the fog is to exist. Again, a strong upslope breeze will cause rapid lifting and low clouds.

### *Steam Fog*

If very cold air moves over warmer water, it will cause rapid evaporation from the surface. This may cause saturation of the cold air, forming very shallow fog.

Its smoke-like appearance gives it the nickname of "smoke fog." The air around the steam fog tends to be unstable because of heating from below. Therefore, turbulence will be common in the area of steam fog.

*Ice Fog*

If the air is very cold and moist, tiny ice crystals may form by sublimation. The ice particles remain suspended in the air, giving the appearance of shimmering needles. Visibility in ice fog will decrease rapidly when we look toward the sun or other light source because of the intense light reflection by the ice crystals. Sudden formation is a characteristic of ice fog, particularly if the wind is light and the temperature is below −30 degrees Celsius. The exhaust gases of aircraft engines or surface vehicles may provide the contaminants to start a chain reaction, forming ice fog almost instantly.

*Precipitation-Induced Fog*

When precipitation falls through the air, a portion of it evaporates, increasing the relative humidity of the air. Evaporation may also occur after the precipitation has reached the surface. Precipitation-induced fog often forms near a weather front (see Chapter 27, Fronts) so it is commonly called "frontal fog."

## CHECK YOUR READING

1. What is the difference between clouds and fog?
2. List the three requirements for fog formation.
3. What is your forecast for a coastal station at which the temperature is 72 degrees Fahrenheit and the dew point is 69 degrees Fahrenheit, with an onshore wind of seven miles per hour?
4. During what hours of the day is radiation fog most likely?
5. List the steps required for the formation of advection fog.
6. Why is the air turbulent near steam fog?
7. Where is precipitation-induced fog most likely to form?

# 26

# AIR MASSES

An air mass is a large body of air having approximately uniform temperature and moisture characteristics throughout (horizontally). That is, any layer of air has about the same temperature and moisture content throughout the air mass.

Weather at a geographic area depends on the prevailing air mass or the movement of an air mass into the area. Local geographical features such as oceans, mountains, and valleys may cause minor variations in the weather.

## Source Region

The source region of an air mass is where it acquires its identifying features of temperature and moisture content. If an air mass lingers over an area having fairly uniform characteristics, it will take on those characteristics. The cold of the polar region, moisture of the ocean, heat of the tropics, or dryness of the desert will be transmitted to the air mass. The intensity of the air mass characteristics will depend upon the length of time it remains over the source region and the difference between the temperature of the air and the surface. If the air is colder than the surface, it is heated by convection which produces rising currents (warm air rises) which carry the heat and moisture aloft, modifying the air mass to a great height. If the air is warmer than the surface, it is cooled from below. Since cool air settles, convective currents will not develop and only the lower levels of the air mass will be modified.

Naturally, as seasonal changes occur in the source region, so will seasonal changes occur in the air mass characteristics.

There are two general source regions: (1) polar (designated P) and (2) tropical (designated T). The surface of the source region is either maritime (designated m) or continental (designated c). Hence, the four basic air mass designators are cP, cT, mP, and mT.

A third identifying letter follows the two-letter source region identifier and denotes the temperature of the air mass in relation to the surface over which it is passing. An air mass which is colder than the underlying surface is called *cold* and is designated by the letter k. An air mass which is warmer than the surface over which it passes is called *warm* and is designated w. Notice that the temperature of the air mass itself does *not* determine whether it is called cold or warm. It is the temperature of the air mass *relative to* the underlying surface which determines its nomenclature, and its characteristics.

## Characteristics of Air Masses

A cold air mass will be heated from below by heat radiating from the surface. Warm air (being lighter than cold air) rises, causing specific characteristics. The rising vertical currents produce turbulence up to 10–12,000 feet. Good visibility will prevail because the convective currents will carry the dust, smoke, and other contaminants aloft, spreading them through a thick layer of air. Clouds of vertical development may form depending upon the moisture content of the air. If precipitation occurs, it will be localized and in the form of rain or snow showers, or thunderstorms.

The warm air mass is warmer than the surface over which it passes and is cooled by that surface, particularly in the lower levels. When the air near the surface is cooled, it becomes heavier and settles. Due to the absence of vertical currents the air will be smooth. Visibility will be poor because the contaminants will settle to the lower altitudes. Cloud formation will be horizontal, with little vertical development, and precipitation will be widespread but light.

## Air Mass Modification

As an air mass moves to a new geographical area, it is modified according to the surface over which it passes. The amount of modification will depend on the nature of the underlying surface, the speed with which the air mass moves, and the difference in temperature of the air mass and the surface.

### *Warming From Below*

The air mass may be warmed due to movement horizontally over a warmer surface or heating of the underlying surface by the sun. Warming from below results in instability or rising air currents.

*Cooling From Below*

The air mass may be cooled by moving horizontally over a colder surface or because of the cooling of the underlying surface by radiation. Cooling of the air results in stable air and no vertical currents.

*Change in Moisture Content*

The water vapor content of the air may be increased by evaporation from water surfaces or from precipitation. Snow or ice on the surface may also evaporate into the air. (The process of snow and ice turning to water vapor without melting is called *sublimation*.) Water vapor may be reduced by condensation or precipitation. The air becomes more stable as the amount of water vapor decreases.

*Lifting and Sinking*

If air is lifted, its stability decreases. Lifting may occur due to convection, by being forced up a mountain slope (called *orographic lifting*), or by being forced over colder air.

Sinking air is stable and is caused by air either being forced down a mountain slope or being forced down from above a colder air mass. Air may also sink due to its own weight (this is called *subsidence*). When air sinks it is compressed by its own weight. Compression causes heat. Since warm air can contain more water vapor than cold air, the relative humidity of descending air decreases.

Although each of the modifying factors have been treated separately here, we must remember that more than one of these factors may be in progress at one time.

## CHECK YOUR READING

1. List the factors which determine the intensity of air mass characteristics.
2. Can we expect more complete modification of a cold or a warm air mass? Why?
3. List the characteristics of a cold air mass.
4. List the characteristics of a warm air mass.
5. If we were to take a passenger on his first flight, would we plan on making the flight in a warm or a cold air mass? Why?
6. How may the water vapor content of the air be increased?
7. How may the water vapor content of the air be decreased?
8. What is the name of the process by which air is forced up a mountain slope?
9. List the factors which might cause air to sink.

# 27

# FRONTS

A *front* is a line of discontinuity between air masses. The air in differing air masses does not readily mix. Like oil and water in a container, each air mass tries to keep its identity. Between the two is a zone of transition or a front. Because air masses cover wide areas, their fronts will be hundreds of miles long. The width of the front will depend on the amount of difference between clashing air masses and the manner in which they meet. Narrowest frontal zones occur when two air masses meet head-on.

As pilots, we will find that frontal weather presents many serious problems because it may be quite severe and very variable.

## Frontal Structure

Remember that air masses are modified according to their source region and the surfaces over which they move. The greatest modification occurs near the surface, so the contrast between masses will be greatest in the lower levels, diminishing to insignificance above 20,000 feet (except for temperature differences).

Air masses differing in temperature and moisture content will have different densities, cold air masses being more dense or heavier than warm air masses. Therefore, if a cold air mass intercepts a warm air mass, the cold air will force its way under the warmer air, creating a steep frontal zone. If a warm air mass moves toward colder air, it will be forced aloft, creating a shallow frontal zone of great width.

## Frontal Discontinuity

When we traverse a frontal zone, the differences of each air mass will be quite apparent. Cloud types, moisture, wind, temperature, and atmospheric pressure will all usually change during the transition from one air mass to another. If the frontal zone is narrow, these properties will change rapidly. If the frontal zone is wide, the changes will be more subtle.

### *Temperature*

A front will always be accompanied by a temperature change. The rate of change is an indication of the intensity of the front and will be greatest near the earth's surface. Strong fronts having narrow transition zones will be accompanied by abrupt and large temperature fluctuations. The passage of a weak, diffuse front will cause a slower temperature change of less magnitude.

### *Wind*

A wind shift will accompany frontal passage. In the Northern Hemisphere, the wind will change direction clockwise as the front is penetrated. When we fly through a front in any direction, we must correct the aircraft heading to the right to maintain our course.

Wind speed may not change during frontal passage. Quite often, however, when we fly from a warm air mass to a cold one, the wind speed will increase abruptly because wind speeds in cold air masses are generally higher.

### *Pressure*

Normally a front will be located in a low pressure area called a *trough*. Therefore, as the front approaches, the barometric pressure will slowly decrease. After frontal passage, the pressure will begin to rise. Naturally, when we enter a cold air mass the pressure will increase more rapidly than in a warm air mass because cold air is more dense.

Not all fronts are associated with a major low pressure trough. Sometimes the primary trough becomes stationary while the front moves on within a second minor trough.

Troughs are not always associated with fronts. A trough is an area where air is converging from different directions. If there is no marked difference in the converging air, no front will exist. Since convergence causes lifting of the air, troughs sometimes contain quite violent weather even though no front exists. We should not conclude, therefore, that all weather is associated only with frontal activity.

## Factors Influencing Frontal Weather

Many factors determine the severity of frontal weather, which may vary from little or none to violent, hazardous combinations of all weather known to man.

*Moisture.* Obviously, without sufficient moisture in the air, no clouds will form.

*Stability.* Cloudiness will be either stratiform or cumuliform depending upon the stability of the lifting air. Stratiform clouds will produce steady precipitation, but little or no turbulence will be present. Cumuliform clouds will produce showery precipitation (sometimes heavy), and the air will be turbulent.

*Frontal Slope.* Steep frontal slopes will produce narrow weather bands as they move along rapidly. Shallow frontal slopes will be accompanied by wide areas of cloudiness and steady, widespread precipitation.

*Speed.* Faster moving fronts usually produce more violent weather of shorter duration than slow moving fronts.

*Temperature and Moisture Content.* The intensity of frontal weather varies directly with the contrast in temperature and moisture content between the clashing air masses.

Figure 84 shows a typical plan view of the surface positions of both a warm front and a cold front, extending from an area of low pressure.

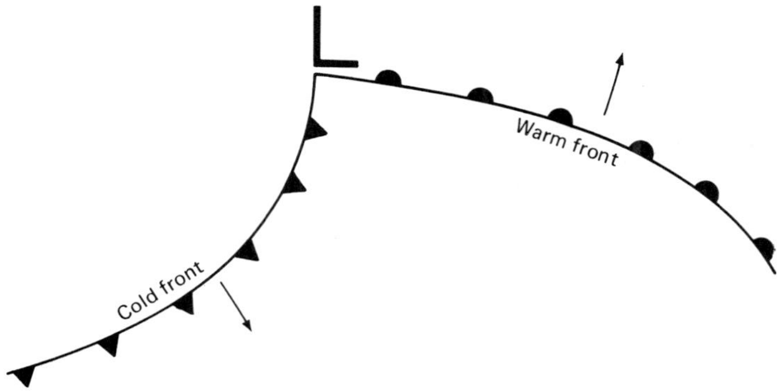


Fig. 84. Surface positions of warm front and cold front.

## The Cold Front

A cold front is the leading edge of a mass of air which is colder than the surface over which it is moving. In the Northern Hemisphere the cold front is normally orientated along a northeast-southwest line and moves across the surface at an average speed of 30 miles per hour. (The speed of cold fronts varies from 5 to 60 miles per hour.)

Winds preceding the front are usually from the southwest, and will shift clockwise to the west or northwest after frontal passage. Wind speed may remain constant but will often increase in the cold air.

The temperature will decrease after frontal passage, particularly in the lower levels.

*Associated Weather*

Friction between the incoming colder air and the surface causes the air at the surface to move slower than the air aloft. (Picture a pencil, with the eraser end on a table top, being pushed along. The friction of the eraser tends to slow the low end of the pencil, while the upper end tries to overrun the eraser.) The cold air, being heavier than the warmer air, forces its way under the warm air.

As shown in Fig. 85, which depicts a cross-sectional view of a cold front, the associated weather will normally be contained in a narrow band. However, very marked changes in the weather will occur.

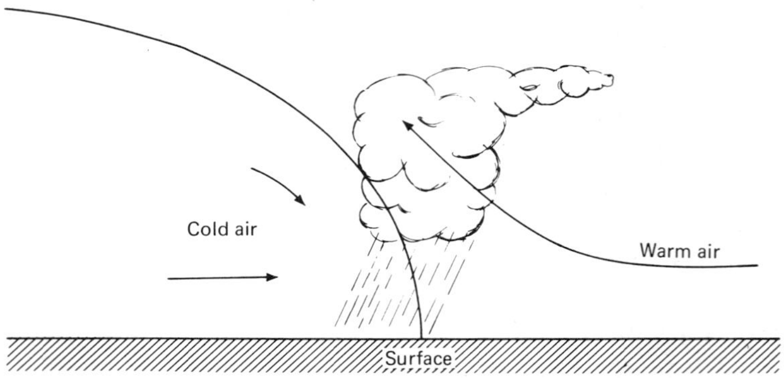


Fig. 85. The cold front.

*Typical Sequence of Weather*

Imagine that we are flying northwestward in a warm air mass approaching a cold front. The warm southwesterly wind will increase as we near the frontal zone, while the barometric pressure decreases. Ahead, altocumulus clouds appear on the horizon. When we get quite close to the front we will see cumulus and cumulonimbus clouds, some of which tower to great heights. Rain or snow will begin as the air increases its turbulent churning. The precipitation beats a wild pattern on our windshield as we are tossed about in violent vertical currents. Suddenly, the temperature decreases, and the wind direction shifts to northwest, indicating frontal passage. Another few miles will bring us into rapidly clearing weather with falling temperature, increasing northwest surface winds, and a decreasing dew point.

This imaginative flight through a cold front is typical but by no means standard, for the weather associated with every front varies drastically.

*Fast Moving Cold Fronts*

As the speed of the cold front increases, so does the severity of the associated weather. The front becomes much steeper, and in many cases a solid line of thunderstorms develops which moves out ahead of the front. This is called a *squall line* and may contain monstrous thunderheads towering to tremendous heights. (Cloud tops have been observed at 70,000 feet.) Squall lines contain some of the most violent weather known as well as turbulent, gusty surface winds.

*Slow Moving Cold Fronts*

If the cold front moves slowly, the warm air preceding it will be lifted gently over the shallow slope, resulting in a wider weather pattern. Depending upon the stability of the warm air, clouds of vertical development may be imbedded in stratiform clouds.

## The Warm Front

A warm front is the leading edge of a mass of air which is warmer than the surface over which it is passing. In the Northern Hemisphere, the warm front is usually orientated along a northwest-southeast line and generally moves at about 20 miles per hour. (The speed of warm fronts may vary between 5 and 40 miles per hour.)

Winds preceding the front will be from an easterly or southeasterly direction and will shift clockwise to southwesterly after frontal passage.

The temperature will usually be somewhat warmer after the front passes.

### *Associated Weather*

Advancing warm air will be forced aloft over the cooler preceding air. The frontal cross section will show a very shallow slope because friction between the warm air and the surface tends to slow the movement of air in the lower levels, while the warm air aloft streams forward. (Picture a pencil being dragged across a tabletop. Now the friction of the eraser will tend to slow down that end of the pencil, shallowing its incline.)

As shown in Fig. 86, which depicts a cross section of a warm front, the associated weather will cover a wide area.

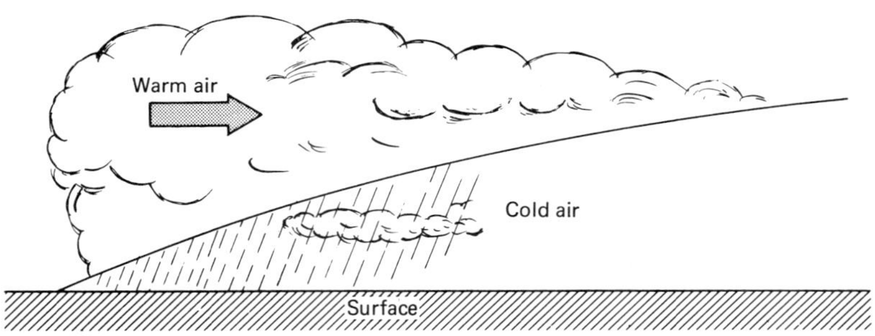


Fig. 86. The warm front.

### *Typical Sequence of Weather*

Let us now imagine a southwestward flight in a cold air mass, heading toward an approaching warm front. About 700 miles in advance of the front, high cirrus clouds will appear and gradually blend into cirrostratus and altostratus clouds. Precipitation will begin about 200 miles from the front and become steady as we enter the area of nimbostratus clouds. Precipitation fog may appear as the falling precipitation evaporates in the cold air. We will eventually notice a temperature increase and a clockwise shift of the wind to a southwesterly direction as we penetrate the front. Unlike the cold front, clearing weather does not immediately follow the front. Depending upon the moisture content and stability of the warm air, clearing weather may be 100 to 300 miles beyond the front.

The exact weather to be expected with warm front passage cannot be described. Similar to the cold front, there may be considerable variance from the typical conditions described here.

## The Stationary Front

If two air masses oppose each other in a manner so as to prevent any movement, the front becomes stationary. Weather conditions will be similar to warm front weather but will diminish in intensity as the front remains stationary. The wide band of weather may persist in one geographical area for many days, however.

## Frontal Waves and Cyclones

Quite often a frontal wave or low pressure area (cyclone) will form on a slow moving or stationary front.

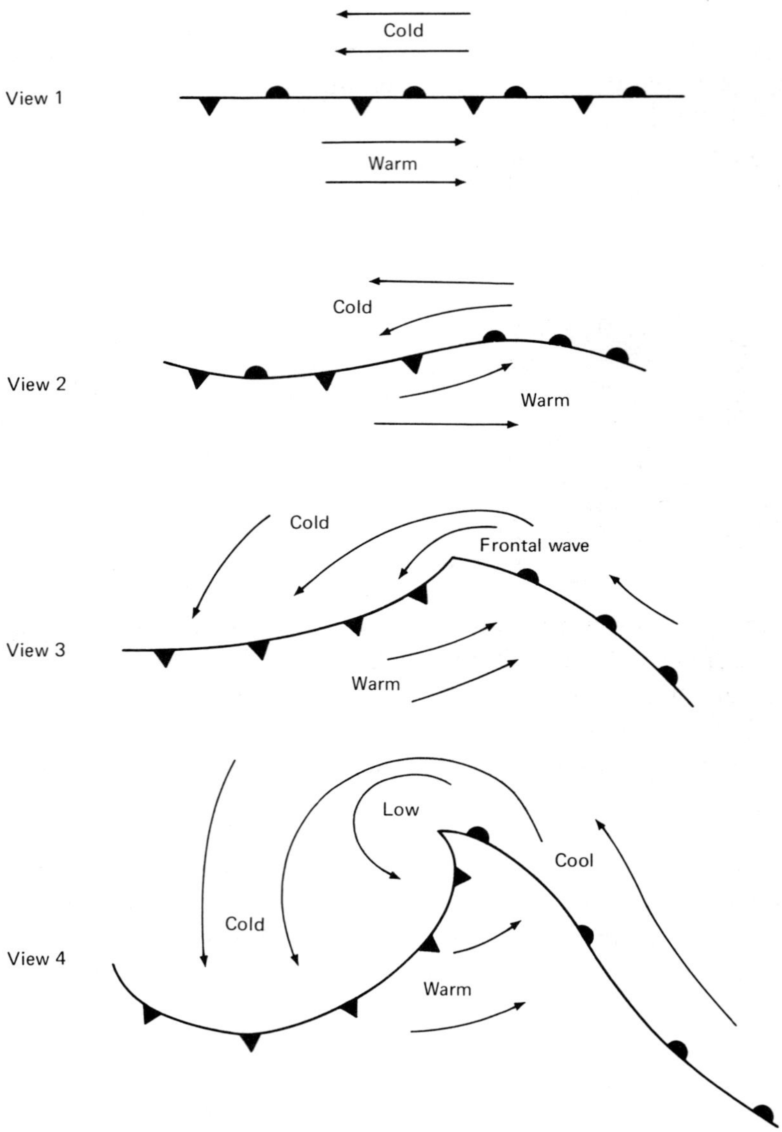


Fig. 87(a). Cyclonic circulation.

Figure 87 shows graphically the birth and death of a low pressure area.

Small disturbances (caused by irregular terrain or uneven surface heating) in the wind pattern of a stationary front (view 1) start a wave or bend in the frontal line (view 2). If the disturbance persists, the wave increases and cyclonic (counterclockwise) circulation begins. One section of the front begins to move as a cold front, while another section moves as a warm front (view 3).

A low pressure center forms at the peak of the wave which increases the cyclonic flow to a force strong enough to move both fronts (view 4). The cold front moves faster than the warm front, and eventually catches up with it, near the low pressure center. This section of the front (or combined fronts) is called an *occlusion* (view 5). At this point the cyclone is most intense. Figure 87, view 6 shows the occlusion growing in length. This causes the cyclonic flow to decrease, and the low pressure area weakens.

In the end (view 7), the low center disappears and the two fronts again become stationary.

View 5

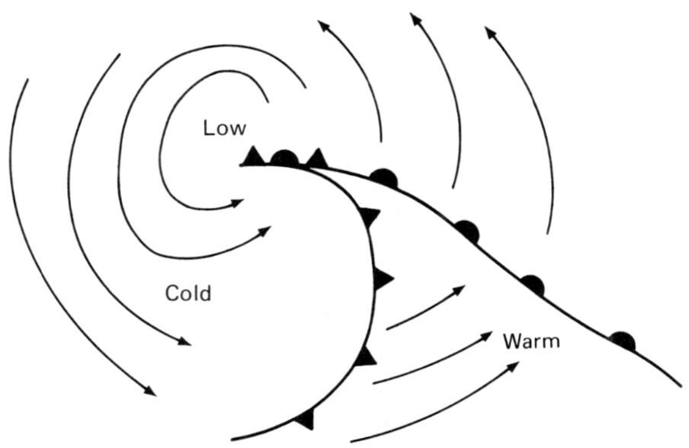


View 6

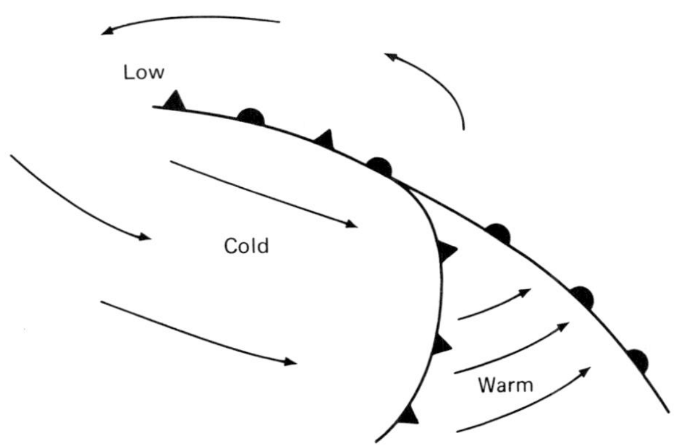


Figure 87 (b)

View 7

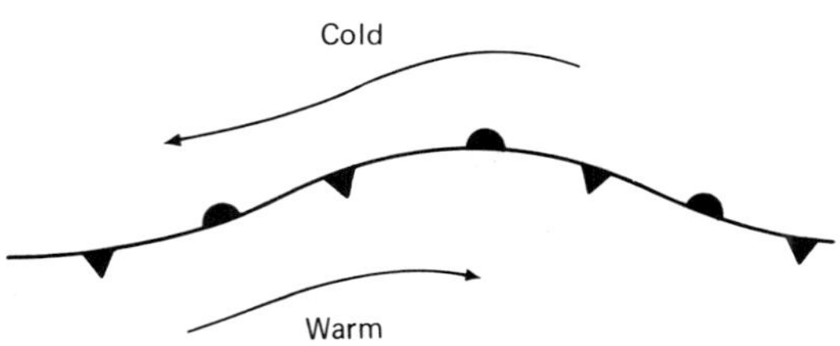

## Occlusions

When one front catches up with another it is called an occlusion. The weather in the occlusion will be characteristic of the cold and warm fronts combined, with one set of characteristics being more prominent.

*The Cold Front Occlusion*

In the cold front occlusion the weather most nearly resembles cold front weather. If we study Fig. 88 we will see that the cold air behind the occlusion is lower in temperature than either the warm air immediately preceding it or the air mass ahead of the warm front.

Therefore, the cold front remains on the surface, lifting both preceding air masses as it meets them.

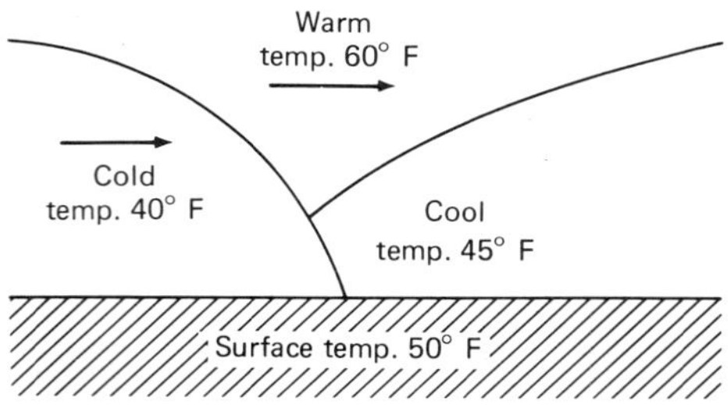


Fig. 88. The cold front occlusion.

*The Warm Front Occlusion*

The weather associated with a warm front occlusion more nearly resembles a warm front. Figure 89 shows this occlusion. The cold air mass is colder than the air immediately preceding it, but warmer than the air in front of the warm front. In this instance the cold air rides up over the warm front, leaving the warm front on the surface.

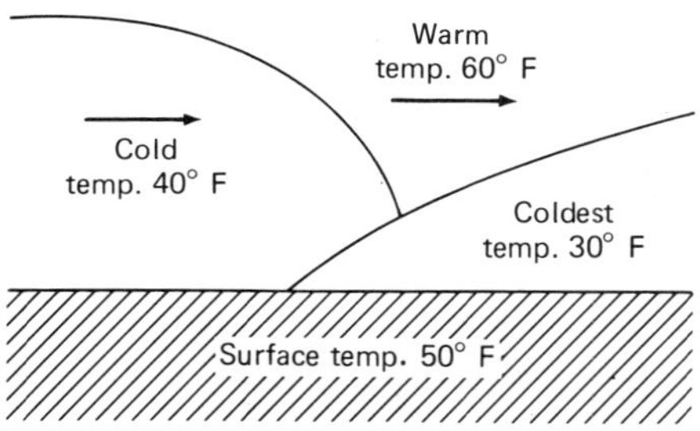


Fig. 89. The warm front occlusion.

## Upper Fronts

As we have seen, when an occlusion occurs, one front is forced aloft. This is called an *upper front*. When in flight, we may penetrate the surface front plus the upper front, which complicates our flight planning.

## Dry Fronts

Weather does not always accompany a front. Sometimes the air is too dry to support condensation even if lifting occurs. Usually a wind shift will still occur however.

## Frontolysis

As an air mass moves into a new geographical area, it becomes more and more modified. Eventually, the air mass becomes quite similar to its predecessor, so the zone of transition between the two becomes diffused. The front will gradually fade away. When a frontal zone diminishes in weather activity it is said to be in a process of *frontolysis*.

## Frontogenesis

When a front first forms and begins to intensify it is said to be in the process of *frontogenesis*.

## CHECK YOUR READING

1. What is a *front*?
2. At what altitudes will the most contrast between air masses be found?
3. List the differences in air masses which will be found when penetrating a front.
4. What is a pressure *trough*?
5. Are weather fronts always associated with major pressure troughs?
6. List the factors affecting frontal zone weather.
7. What is the definition of a cold front?
8. What is the definition of a warm front?
9. Describe the weather conditions found when flying in a warm air mass, heading toward a cold front.
10. Describe the weather conditions found when flying in a cold air mass, heading toward a warm front.
11. What determines whether an occlusion is designated warm or cold?
12. Describe the sequence of events in which a low pressure system forms on a stationary front.
13. Define *frontolysis*.
14. Define *frontogenesis*.

# 28

# THUNDERSTORMS

Thunderstorms, although local in nature, represent the ultimate in violent weather. They are very dangerous to pilots because of the severe atmospheric conditions which they contain. **Light aircraft pilots must avoid thunderstorms at all cost,** while pilots who fly heavier aircraft must comply with established penetration procedures.

Thunderstorms have destroyed aircraft in flight. We must learn to respect them, the giants of all cumulonimbus cloud formations.

## Formation

Specific atmospheric conditions must be present before thunderstorms may form. Those conditions are:

1. *Unstable Air.* The atmosphere must be either conditionally unstable or unstable and lifted until the air becomes warmer than its surroundings. (The altitude at which this occurs is called the *level of free convection.*) From this level upward, the air rises rapidly until at some altitude (as high as 70,000 feet) it will become cooled to the temperature of the surrounding air.

2. *Lifting Action.* The unstable air must be lifted until it reaches the level of free convection. The lifting action may be provided by any type of front, mountainous terrain, or convergence.

3. *Moisture.* The more moisture present in the air, the easier the level of free convection will be reached. As the warm, moist air rises, the water vapor is condensed which adds more heat to the air, increasing its instability.

## Types of Thunderstorms

There are two basic types of thunderstorms: (1) those formed along a front or squall line, and (2) those occuring within an air mass.

### *Frontal Thunderstorms*

During our discussion of frontal weather, we learned that thunderstorms frequently occur in association with a front. Cold fronts, because of their steeper slopes, lift the warm air more rapidly than do warm fronts. Therefore, thunderstorms are usually more frequent and more violent in cold frontal zones. Warm fronts may also contain thunderstorms imbedded in stratus clouds, making them difficult to avoid without airborne weather radar.

The squall line is composed of a near-solid line of thunderstorms which form along a cold front and move ahead of it. These storms are likely to be extremely severe.

### *Air Mass Thunderstorms*

Air mass thunderstorms are not associated with frontal activity, but rather occur within an air mass, due to localized disturbances which cause lifting. They are generally scattered over a wide area and isolated.

*Convective Thunderstorms.* Either heating from below, or convergence may cause thunderstorms if the air is warm, moist, and unstable. The usual time of day for active convective thunderstorms is between noon and sundown.

*Orographic Thunderstorms.* This type of thunderstorm is caused by the wind blowing warm, humid, unstable air up a mountainside. Again, the most common time of day for active orographic thunderstorms is between noon and sundown, because convective currents are added to the orographic lifting to produce the cumulonimbus cloud.

*Nocturnal Thunderstorms.* The nocturnal thunderstorm is most active in the late night or early morning hours. Meteorologists differ in their opinion as to the atmospheric phenomena causing them. They are quite common in the midwest section of the United States and can be very severe.

## The Life Cycle of a Thunderstorm

An individual thunderstorm rarely exceeds 10 miles in diameter and lasts from only 15 minutes to 1½ hours. However, thunderstorms may occur in clusters in which each storm cell is in a different stage of development. These clusters may extend over a 100 square mile area and last for six hours or more. Each cell within the cluster has a life cycle comprised of three stages, as shown in Fig. 90.

### *Cumulus Stage*

The thunderstorm always starts as a cumulus cloud. During the cumulus or building stage, most air currents are upward and may exceed 3,000 feet per

Fig. 90. Thunderstorms

minute. The cumulus cloud grows rapidly and becomes a cumulonimbus cloud. (One of the suprising factors which came to light with the advent of airborne weather radar was the great speed with which the cumulus cloud may grow to tremendous height.)

Little or no precipitation falls during the cumulus stage because of the strong updrafts within the cloud.

*Mature Stage*

When precipitation reaches the surface, the mature stage has begun. The updrafts are no longer able to support the growing water drops or frozen ice particles. As the rain, snow, or hail falls it drags air with it. This downward moving air is cooler than its surroundings. Most downdrafts are located within the center of the cell and spread outward as they plunge to earth at speeds in excess of 2,500 feet per minute. The horizontal spreading of the vertical currents when they meet the earth's surface causes extremely gusty surface winds.

The remaining updrafts increase their speed during this stage and may exceed 6,000 feet per minute. This is the period of greatest weather intensity within the storm.

*Dissipating Stage*

Ultimately the updrafts weaken, leaving only downdrafts, which produce heating and drying of the air within the storm. The upper levels of the cloud formation take on the appearance of a huge *anvil* which points to the direction of movement of the storm. Lower clouds will slowly change to stratiform. Precipitation gradually decreases as the storm dissipates. This is the longest of the three stages.

## Associated Weather

The thunderstorm is a weather *factory*. Most every conceivable combination of weather phenomena may be found within it.

### *Turbulence*

Thunderstorms are potential aircraft destroyers. Many official accident reports prove without doubt that the turbulence associated with a thunderstorm can destroy an aircraft.

Generally, severe or extreme turbulence will be found in the lower levels of a well-developed thunderstorm. It often will be impossible to maintain aircraft control due to the shearing action between gusts and drafts.

### *Surface Winds*

Aircraft attempting to land in the vicinity of a thunderstorm may be subjected to the hazardous *plow wind* which is the initial surge of wind resulting from the horizontal movement of the air when the downdrafts meet the surface of the earth. These winds may exceed 100 miles per hour while shifting direction 180 degrees.

### *Hail*

A hailstone is a lump of ice which forms in the interior of the thunderstorm and may fall to earth from it. (Severe vertical gusts often hurl the hail from the upper levels of the storm cloud, causing it to fall through the clear air 5 miles from the cell.) Hailstones may be as small as a pea or *as large as a grapefruit*, depending on the strength of the updrafts, the liquid content of the cloud, and the height of the storm.

Significant damage may be inflicted on the aircraft in just a few seconds if we fly through hail of only $\frac{1}{4}$-to $\frac{1}{2}$-inch diameter. Aircraft damage may include deformed leading edges, smashed windshields, shredded fabric, broken radio masts and antennas, and removed paint.

### *Lightning*

Although lightning is the most visual electrical charge produced by a thunderstorm, it does not often damage aircraft. The skin may be perforated, and electrical equipment may be damaged. In rare cases fuel vapors in the aircraft storage tanks may be ignited.

### *Precipitation Static*

When we fly through precipitation, our aircraft may accumulate a charge of static electricity. Although discharge of this electrical energy causes noise on the low frequency radio, it does not damage the aircraft. At night the discharge may be seen as a blueish wisp, or brush-like flame, emanating from the edges of our aircraft components. It is commonly called Saint Elmo's fire.

## Thunderstorm Flying

**All of us must avoid thunderstorms whenever possible.** We must also remember that hail, turbulence, and gusty surface winds may extend for many miles from the cloud formation, and plan our flight to remain well clear of the storm. Every aircraft has limitations. All of us have our own limitations, depending on experience, knowledge, and ability. *We must know, and abide by, these limitations.*

## CHECK YOUR READING

1. List the atmospheric conditions required in order to form thunderstorms.
2. What are the two main types of thunderstorms?
3. What are the major differences between the two types of thunderstorms?
4. What are the common kinds of air mass thunderstorms?
5. What are the three stages of the life cycle of a thunderstorm?
6. Which stage is the most severe?
7. Which stage is the longest lived?
8. List the weather hazards which may be present in and around thunderstorms.
9. What is Saint Elmo's fire? What damage may it inflict on an aircraft?
10. State your thoughts on thunderstorm flying.

# 29

# STRUCTURAL ICE

One of the major weather hazards affecting flight is the formation of structural ice. We must carefully study the conditions conducive to ice formation, the avoidance of icing conditions, and the removal of ice from the aircraft structure.

## Conditions Conducive to Ice Formation

Unlike carburetor ice, structural ice will form only when the outside air temperature is at or below freezing and visible moisture is present. The most severe icing occurs when the temperature is between 0 degrees Celsius and −10 degrees Celsius. Icing is not uncommon, however, at temperatures between −10 degrees Celsius and −25 degrees Celsius.

During specific meteorological conditions, it is possible to find liquid moisture falling through (or suspended in) below freezing air. For instance, we have seen in our discussion of frontal weather conditions, that it is possible to find a layer of warm air above a layer of colder air. Water drops may fall from the warm air into the colder air below. When an aircraft is flown through the cold air sector, the water drops will freeze immediately upon impact, covering the leading edges of all surfaces with ice.

## Types of Structural Ice

There are two main types of ice. Normally ice accumulation on the aircraft will be comprised of a mixture of both types, with one kind being more prominent.

*Clear Ice*

Clear ice resembles the ice in the freezer tray of a refrigerator. It is almost transparent, extremely hard, and adheres firmly to the surface. It is formed when large water drops splatter upon impact with the aircraft. After the drop covers the surface it freezes. Since large water drops will stay aloft only in the updrafts of turbulent air, clear ice will usually be found in unstable air.

Because of the spreading of the water drops before freezing, clear ice will conform to the shape of the leading edge of the aircraft components. As shown in Fig. 91, it does not change the shape of the component or cause much additional drag. It is heavy, however, and can build up at a rate in excess of two inches per minute if meteorological conditions are at their worst.

Clear ice is extremely difficult to remove with conventional deicing equipment.

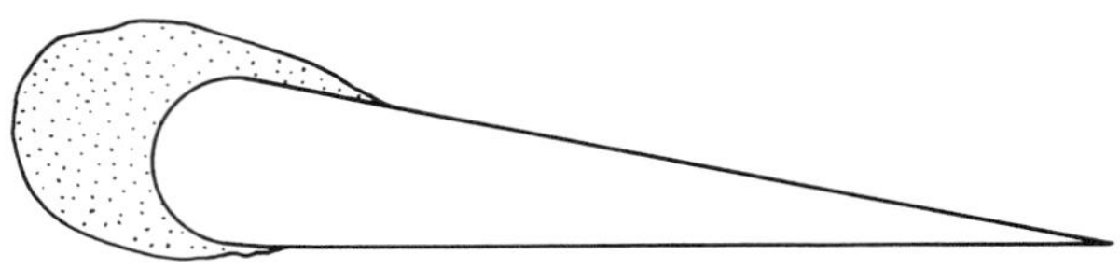

Fig. 91. Clear ice conforms to the shape of the airfoil.

*Rime Ice*

Rime ice resembles the white flaky ice on the sides of the freezer compartment in a refrigerator. It is soft and lacks the adherence of clear ice. Formation is caused by the freezing of small water droplets immediately upon impact without splattering. Small water droplets remain aloft with little or no vertical lifting, so rime ice is more often found in stable air.

Rime ice will build up in very irregular shapes, not conforming to the shape of the airfoil, but sticking out into the relative wind (Fig. 92). Although it is lighter than clear ice, the irregular shape will increase drag rapidly. Normally, conventional deicing equipment will remove it easily.

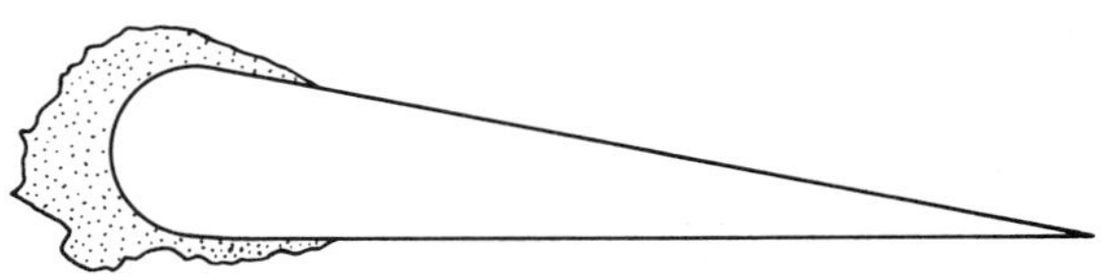

Fig. 92. Rime ice forms in irregular shapes.

## Effects of Structural Ice

When structural ice forms, loss of lift will begin almost immediately, due to a decrease in airspeed, an increase in drag, and the change of shape of the airfoil.

When the aircraft begins to descend due to insufficient lift, we must increase power (increasing fuel consumption) and increase the angle of attack in order to maintain altitude. This in turn will present a larger undersurface of the airfoil to the relative wind, allowing ice to form well aft of the leading edge.

The whirling propeller blades will gather ice more rapidly than the leading edges of the airfoil. This is because the propeller is traveling much faster in its circular path than the rest of the aircraft. It hits more raindrops in a given period of time, thus gathering more ice. Thrust will decrease due to the deforming of the shape of the propeller blades. Severe icing may cause the blades to become unbalanced, causing possible engine damage.

The weight of the structural ice, plus the increased drag, may cause small components, such as radio antenna masts, to break away from the aircraft.

Because of shape deforming, an airfoil on which structural ice has formed will stall at a higher than normal airspeed and at a lower than normal angle of attack. Therefore, we must exercise extreme caution and maintain higher than normal airspeeds, refraining from violent control actions or unusual aircraft attitudes when carrying ice.

The pitot–static system will be immediately affected if structural ice forms on the pitot mast or static ports. Ice will restrict the flow of ram air (in the pitot tube) or atmospheric bleed air (in the static ports) which will cause erroneous airspeed, altitude, and vertical speed indications.

Picture an aircraft which is being assaulted by structural ice. It is heavier than normal, developing less thrust than normal, slower than normal, and is reporting false instrument indications. We must have, and effectively use, anti-ice and deice equipment in order to sustain flight in these trying conditions.

## Avoidance of Icing Conditions

The best method of avoiding structural icing conditions is to perform a 180-degree turn when first encountering them. If a reversal of direction is not practical, we should initiate a *climb* as soon as possible. We will see why a climb is much preferred to a descent after reviewing the atmospheric conditions which are conducive to ice formation (Figure 93).

We must fly through air which is at or below freezing temperature. There must also be visible moisture present. If the moisture through which we fly is in liquid form, it must have fallen from an area of above freezing temperature. If it fell very far through below freezing air, it would have been in the frozen state prior to striking the aircraft. Therefore, we can legitimately deduce that warmer air must exist above and quite near our flight altitude, and we should climb immediately to the warmer air.

A descent should be made to avoid ice *only* if we have **definite** knowledge that a layer of warmer than freezing air lies *well above* the surface.

If structural icing conditions prevail or are forecasted, we must have an alternate plan of action foremost in mind, and we must use that plan before the aircraft becomes overburdened with ice. Detailed knowledge of air temperature above and below the flight altitude is a must. Pilot reports, inflight advisories, and weather bureau forecasts must be consulted.

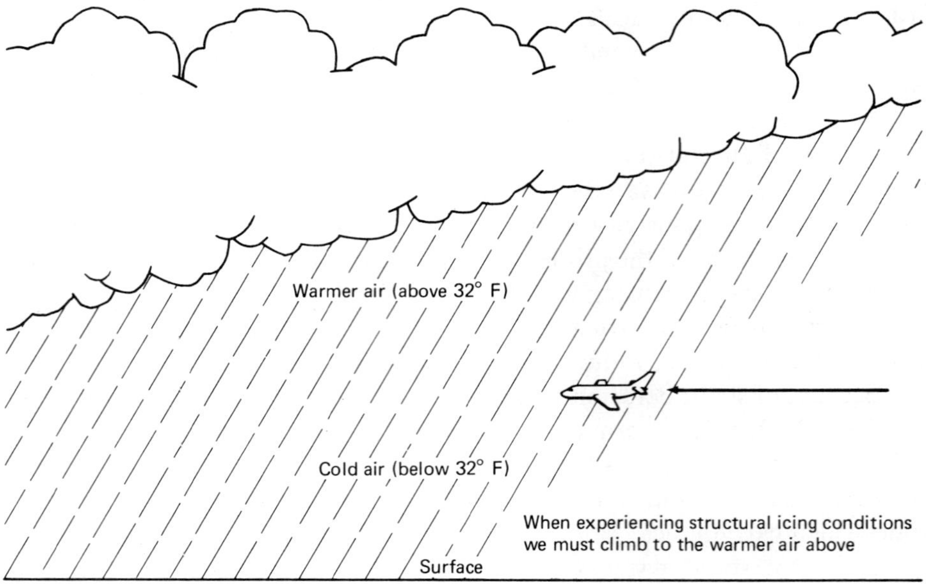


Fig. 93. Structural ice conditions.

## Control of Structural Ice

More sophisticated aircraft are equipped with deice and anti-ice equipment to aid us in the control of ice accumulation.

### *Deicing Equipment*

Deicing equipment is used to rid the aircraft of ice *after* it has formed. This equipment includes deicer *boots* on the leading edges of all lifting surfaces and on the propeller blades. The boot is simply a rubber tube which is inflated with air at will. Figure 94 shows an airfoil with the deflated boot flattened against it. After ice has formed, the boot is inflated and expands, cracking away the ice. This device must be used only *after* ice has formed, and then only for brief intervals. If the boot is used constantly, a shell of ice will form over the extended boot, leaving an airspace under it, in which the boot will expand and contract, never touching the ice.

An airfoil with deflated deicer boot.

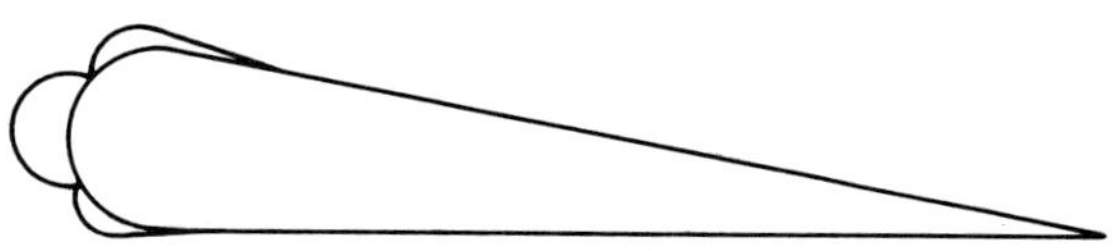

An airfoil with one section of the deicer boot expanded.

Fig. 94. Deicer boots.

### *Anti-icing Equipment*

Anti-icing equipment is used *prior to* entering icing conditions, and should be used constantly when in or near actual or suspected icing conditions. Anti-icing equipment includes heated wings, alcohol, pitot mast heaters, and propeller blade heaters.

*Heated Wings.* Some aircraft are equipped with heated wings. The leading edges of the lifting surfaces are fitted with either hot air ducts or electrical heating elements, which provide enough heat to keep the leading edge at a temperature above freezing. However, the leading edge *only* is warmed, so water tends to run rearward and freeze on the unheated portion of the wing. Another disadvantage is the fact that most aircraft use the wings for fuel storage, and either hot air ducts or electrical heating elements increase the hazard of fire in flight because of their close proximity to the fuel tanks.

*Alcohol.* Alcohol may be piped to the leading edges of the lifting surfaces and propeller blades and sprayed through tiny holes in the surface. Extreme care must be taken in order to keep the holes clean, however, and alcohol must be carried at all times, even though it may not be used for a particular flight. Also, the weight of alcohol must be deducted from the useful load of the aircraft.

*Pitot-Heaters.* Pitot heat is provided by electrical heating units in the pitot head. Enough heat is provided to keep the head free of all ice. It is most imperative that we use pitot heat *only* while in flight, as the heater element may burn out if there is insufficient ram air passing by it.

All anti-ice equipment is designed to prevent ice formation, **not** to remove already formed ice deposits.

## Frost

Frost is structural ice which forms on the top surfaces of an aircraft while on the ground. If the surface temperature drops to below freezing, and condensation occurs, a layer of frost will cover the aircraft.

All frost must be removed from the aircraft prior to flight. Although the thickness of the frost layer may be minute, it does add considerable drag, which may actually prevent flight. When we remove frost, we should pay particular attention to cleaning hinges, static ports, and moveable control surfaces, being sure to remove *all* frost.

## Solid Precipitation

Frozen precipitation will not accumulate on the aircraft in flight. Snow will not affect the structure of the aircraft, but pitot heat must be applied early to prevent the pitot head from packing with snow. Sleet or hail may damage the aircraft skin, remove paint, or break the windshield. Flight through this precipitation is hazardous (see Chapter 28, Thunderstorms).

## CHECK YOUR READING

1. What atmospheric conditions must be present in order for structural ice to form?
2. What are the two types of structural ice?
3. List the main differences between the two types of structural ice.
4. Which type can be expected to form in turbulent air?
5. Which type can be expected to form in smooth air?
6. List the effects of structural ice on the flight characteristics of an aircraft.
7. What methods are available to avoid structural ice?
8. What are the common methods of deicing an aircraft?
9. What anti-ice equipment might be installed on an aircraft?
10. What atmospheric conditions must be present in order to expect frost formation?
11. What preflight action is necessary if frost covers the aircraft?

# 30

# THE UNITED STATES AVIATION WEATHER SYSTEM

Because of the importance of weather to all aviation interests, information concerning current and forecasted weather must be made readily available.

All over the world, weather observers keep constant watch on the weather, forecasters analyze it and predict its future development, and communications equipment distributes pertinent information around the globe.

Through the years, tremendous cooperative efforts have developed between the U.S. Weather Bureau, Federal Aviation Administration, and the military services to economically gather and disseminate weather information.

## Observation

The weather observer may estimate or measure the existing weather on the surface, from the surface, and aloft.

### *Surface Observations*

Aviation weather surface observations are made by personnel of the Weather Bureau, Federal Aviation Administration, military services, and some scheduled airlines. Automatic observation stations are being added to the current network of sites.

### *Radar Observations*

At many locations, Weather Bureau and military personnel use radar to look at cloud and precipitation areas.

*Observations Aloft*

Many methods of observing weather aloft are now in use.

*Rawinsonde.* A balloon is sent aloft, which carries atmospheric measuring devices and a radio transmitter. Information concerning temperature, humidity, pressure, and wind is gathered to heights in excess of 100,000 feet.

*Pibals.* A balloon (without radio transmitter) is sent aloft and tracked visually to indicate winds aloft.

*Rocket.* The Department of Defense and NASA cooperate to gather atmospheric data by rocket.

*Pireps.* Pilots relay information (pilot reports) to the meteorologist concerning cloud tops, turbulence, icing conditions, and thunderstorms.

*Satellite.* A constant flow of photographs is transmitted from satellites at altitudes of 400 to 600 miles. When used in conjunction with surface observations, the photos show details of weather patterns.

*Reconnaissance Flights.* Military aircraft are used to penetrate, measure, and photograph meteorological phenomena over inaccessible geographic areas.

## Analysis and Forecasting

All of the data which is collected by the various methods of observation listed previously is carefully analyzed and used to predict future weather. A number of specific forecasting duties are performed, each by a specialized forecast center.

*Basic Forecasting Center*

All weather data in the Northern Hemisphere is collected at the Washington D.C. Meteorological Center. Prognostic charts are prepared, manually and electronically, which forecast general weather for about 18 hours after observation time. These charts are primarily used for long-term operational planning.

*Flight Advisory Weather Service (FAWS)*

Aviation weather specialists man the *FAWS* and devote all of their time to the preparation of forecasts for the aviation industry. These forecasts cover a small area (two or three states) and present great detail.

*Airport Stations*

Local weathermen stationed at the airport prepare very detailed forecasts for their home base and nearby airports. Often, the local weatherman will refine the area forecasts issued by the *FAWS* taking into consideration the local conditions which affect the weather. We will normally spend more time with the airport weatherman than with the men at the *FAWS* or Basic Forecast Center.

## *Special Centers*

Special weather forecasting is done in stations around the country. Primarily, they forecast severe or unusual weather conditions.

*Severe Local Storm Forecast Centers.* Located in Kansas City, Missouri, this weather station concerns itself with forecasting severe thunderstorms and associated weather, including hail, turbulence, and gusty surface winds. The forecast is issued for the entire contiguous 48 states for a period of 24 hours, with almost instant revision if required.

*The National Hurricane Center.* All tropical storms which may affect the eastern and southern seaboard of the United States are tracked and forecasted from the Hurricane Center in Miami, Florida. As the storm moves into a new area, a hurricane subcenter may be established at New Orleans, Louisiana; Washington D.C.; Boston, Massachusetts; or San Juan, Puerto Rico; as conditions demand.

The weather bureau office in San Francisco, California acts as the Hurricane Center for tropical storms or typhoons in the Pacific Ocean.

*High Altitude Forecast Center.* One unit of the National Meteorological Center in Washington D.C. is designated as the High Altitude Forecast Center. Forecasts are prepared for high altitude flight, and include wind, temperature, enroute weather, and terminal weather information.

*The National Weather Satellite Center (NWSC).* Located at the National Meteorological Center, the *NWSC* prepares cloud analyses, called *nephanalysis,* for distribution to selected stations. Television pictures transmitted from satellites in space are used to depict large weather patterns.

# Distribution

All of the collecting and analysing of weather data would be useless without rapid distribution. The information may be transmitted by teletypewriter, facsimile networks, long-line circuitry, or local systems.

## *Teletypewriter*

Recent improvements in the teletypewriter systems have increased the speed with which weather information can be distributed. However, due to the limited time available on each circuit, codes and symbols must be used in order to fit all of the data into the available time period. It is the interpretation of these codes and symbols to which we will devote many hours in future chapters.

## *Facsimile*

Weather analyses and forecasts in graphic form can be distributed rapidly on the facsimile network. A photo or drawing is *transmitted* electronically to the receiving station and then reassembled into graphic form.

*Long-Line*

The long-line system of communications is similar to the regular telephone system. Because of the expense and limited capability of this system the long-line is not extensively used to transmit weather information.

*Local Systems*

Air terminals are constantly expanding. The weather data which is received at the weather station is needed by many people throughout the terminal. Local distribution methods are used to rapidly transmit weather data to all interested persons in the terminal complex. These systems include the TelAutograph, Local Facsimile, Interphone, Closed-circuit Television, and Local Teletypewriter.

## Presentation of Data

After the weather data has been distributed, it must be presented for use by all interested persons. We must receive a complete briefing of the weather, both present and forecasted, in order to safely plan our flight. A number of pilot briefing methods are available.

*Recorded Weather*

We may receive a continuous broadcast of weather information for any area in a 250-mile radius by either tuning in many Federal Aviation Administration navigational radio aids or by dialing published telephone numbers (published in the Airman's Information Manual [AIM]).

*Person-to-Person*

We may speak directly to the weatherman by calling published (in AIM) telephone numbers.

*Closed-Circuit Television*

Closed-circuit television is being used at a number of briefing centers to show the pilot the weather charts while recorded comments are transmitted.

*Federal Aviation Administration Request-Reply Service*

We may contact any F.A.A. Flight Service Station in flight and request detailed weather information for local or distant stations.

## Summary

We have seen that hundreds of men across the country are constantly gathering, analyzing, forecasting, and disseminating weather information. These men are dedicated to serving us. We should make full use of these services and keep posted on all pertinent weather information.

## CHECK YOUR READING

1. List the government and private agencies which are concerned with the observation of weather.
2. List the methods of weather observation.
3. With which of the aviation weather forecast centers will we, as pilots, deal most regularly?
4. List the methods of distributing weather data.
5. List the methods used to present weather data to interested parties.
6. Why is the aviation industry the prime user of weather data?

# 31

# AVIATION WEATHER REPORTS

Of all the weather data available to us, we will probably use the Aviation Weather Report most often for a number of reasons.

The Aviation Weather Report is the only up-to-date local weather information available. It is published hourly, or more often if the weather changes rapidly. The information presented is in great detail for a very local area.

Due to the limited time available on the teletypewriter circuits, the large volume of information included in the report, and the number of stations reporting, the data is coded and symbolized. As we learn the symbols found in each segment of the report, we will realize the great volume of information contained in the Aviation Weather Report.

## Interpretation of Aviation Weather Reports

We must make every effort to form a mental picture of existing weather as well as the *trend* when interpreting the Aviation Weather Report. It is not sufficient to simply decode the message. We must know what the weather actually is.

## A Sample Aviation Weather Report

The following is a sample of the Aviation Weather Report.

023SA23221100
DCA 21 ⦶ M31 ⊕ 1/2VSW-F 152/32/30/3112G30/996/
VR28 WSHFT 0542E/ ⊕ 65/ ⊕ V ⦷ SB06 DRK NW
VSBY 1/4V3/4 → DCA 9/5 QAPES ↘ 9/1

*The Heading*

The heading is printed at the top of a long list of Aviation Weather Reports and is important to us in determining the time and date of the observation.

023 SA 23 22 1100

023 = The circuit identifier (used for distribution only)
SA = The traffic designator (used to distinguish this report from other weather information)
23 = The circuit identifier (repeated)
22 = The date (the *day* of the month is given)
1100 = The time of the observation (Greenwich mean time, using the 24-hour clock. To convert for local time, subtract 5 hours for Eastern standard time, 6 hours for Central standard time, 7 hours for Mountain standard time, and 8 hours for Pacific standard time.) 1100 hours Greenwich is equal to 0600 Eastern standard time.

*Group I*

The first group of symbols in each Aviation Weather Report identifies the reporting station and reports the sky condition, visibility, and restrictions to the visibility, if any.

DCA 21 ⦶ M31 ⊕ 1/2V SW—F

DCA = Station designator, Washington D.C. (A three-letter identifier is used for stations in the United States from which a report is sent. Canadian stations are identified by two letters. Normally, the three-letter identifier suggests the name of the station. For example: BOS, Boston, Massachusetts; OKC, Oklahoma City, Oklahoma; JFK, Kennedy International Airport, New York, N.Y.; BAL, Baltimore, Maryland; SFO, San Francisco, California.)
21 ⦶ M31 ⊕ = Sky cover. Scattered clouds at 2100 feet, overcast at 3100 feet. (The digits represent the height of the cloud base or bases *above the ground* in hundreds of feet. Each digit group is followed by a sky cover symbol.)

*Sky Cover Symbols.*
○ = Clear sky: less than 0.1 sky cover
⦶ = Scattered clouds: 0.1 to 0.5 sky cover
⦷ = Broken clouds: 0.6 to 0.9 sky cover (constitutes a *ceiling* unless preceded by "—")
⊕ = Overcast sky: more than 0.9 sky cover (constitutes a *ceiling* unless preceded by "—")
— = "Thin" when prefixed to above symbols.

—X = Partial obscuration: 0.1 to less than 1.0 of the sky hidden by precipitation or other obstruction to visibility (bases at the surface).

X = Obscuration: 1.0 of the sky hidden by precipitation or other obstruction to visibility (bases at the surface). Constitutes a *ceiling.*

*Ceiling.* The word *ceiling* becomes important to us in determining the legality of flight. (See Chapter 53, Federal Air Regulations, Part 91.) Therefore, we must remember that "broken," "overcast," or "obscuration" constitute a ceiling. The ceiling figure will always be preceded by one of the following letters to indicate how the ceiling height *above the ground* was determined.

| | |
|---|---|
| E = Estimated | A = Reported by aircraft |
| M = Measured | R = Radiosonde or radar |
| W = Indefinite | D = Estimated height of cirriform clouds |
| B = Balloon | |

U = Cirriform ceiling height unknown. When cirriform clouds do not constitute a ceiling and their height is unknown, the sky symbol will be preceded by a slant bar (/).

NOTE: If the ceiling height is lower than 3,000 feet (above the ground) and is variable, the letter V will follow the ceiling symbol. The remarks section of the report will state the range of ceiling height.

1/2V = Visibility one-half statute mile and variable: If the visibility is less than 3 statute miles and variable, it will be followed by the letter V. If the visibility is 6 statute miles or less, the restriction to visibility is always given following the visibility figure.

SW— = Precipitation: Light Snow Shower. Any precipitation, thunderstorm, or tornado will be noted here using the following symbols:

| | |
|---|---|
| A = hAil | E = slEet |
| L = drizzLe | S = Snow |
| R = Rain | W = shoWers |
| AP = small hail | IC = Ice Crystals |
| SG = Snow Grains | SP = Snow Pellets |
| T = Thunderstorms | Z = freeZing |

TORNADO will always be spelled out in full.

The intensity of precipitation is indicated with the following symbols which follow the precipitation symbol.

| | |
|---|---|
| – – = Very light | — = Light |
| no symbol = Moderate | + = Heavy |

F = Other restrictions to visibility: Fog. The following symbols will be used to designate other restrictions to visibility:

| | | |
|---|---|---|
| D = Dust | F = Fog | GF = Ground Fog |
| K = smoKe | H = Haze | B = Blowing (snow or dust) |

*Group II* The second grouping in the Aviation Weather Report is comprised of numbers, separated by slant lines (/).

152/32/30/3112G30/996

152 = Barometric pressure 1015.2 millibars. (The number 9 or 10 and the decimal point must be added to the code number. The result must fall between 950.0 and 1050.0 millibars.)

32 = The temperature is 32 degrees Fahrenheit.

30 = The dew point is 30 degrees Fahrenheit.

31 = The wind direction is from 310 degrees. (Wind direction is measured to the nearest ten degrees beginning at *true north* and moving clockwise. The final digit (0) is omitted in the report.) A calm wind will be reported as 00.

12 = The wind speed is 12 *knots*. If the wind speed is estimated, the value will be followed by the letter E. A calm wind will be reported as 00.

G30 = The peak gusts are 30 *knots*. (The peak gust recorded within the fifteen minutes preceding the report will be given following the letter "G.") The letter "Q" instead of "G" indicates a squall line is present.

996 = The altimeter setting, or barometric pressure in inches of mercury is 29.96 inches. (The first digit is omitted along with the decimal point. We must add a "2" or a "3" preceding the code number, and add the decimal point, keeping the result between 29.50 and 30.50 inches.)

*Group III*

This is the *remarks* section of the report, and contains any pertinent information not included in the body of the report.

VR28 WSHFT 0542E/ ⊕ 65/ ⊕ V ⦶ SB06 DRK NW VSBY 1/4V3/4

| | |
|---|---|
| VR28 | Runway visual range is 2800 feet. (The visibility is measured electronically and reported in hundreds of feet.) |
| WSHFT | The surface wind shifted its direction. |
| 0542E | The wind shift occured at 0542 Eastern standard time. (The 24-hour clock is used, but the time is *local*, not Greenwich.) |
| ⊕ 65 | The top of the overcast is 6500 feet above sea level. (The height *above sea level* of the bases or tops of cloud layers not visible from the station is reported in hundreds of feet.) NOTE: The height of cloud bases reported in the main body of the report is given as height *above the ground*. |
| ⊕ V ⦶ | The overcast cloud layer is variable to a broken layer. |
| SB06 | Snow began at 06 minutes past the hour preceding the hour of the report. (If the report shows no precipitation, and it stopped within an hour of the observation, it will also be noted. For example: SE52 would indicate that the snow ended at 52 minutes past the previous hour.) |
| DRK NW | The sky is dark to the Northwest. |
| VSBY | Visibility. |
| 1/4V3/4 | The visibility is variable between $\frac{1}{4}$ and $\frac{3}{4}$ statute mile. |

*Group IV* The remaining information concerns Notices to Airmen (NOTAMS). Instructions for interpreting the NOTAMS are found in the Airman's Information Manual (AIM). They concern the operational status of runways and other airport facilities (see Chapter 48).

NOTAMS → DCA 9/5 QAPES ↘ 9/1

## Summary

With constant use, the codes and symbols used in the Aviation Weather Report will become a second language. We should keep the following points in mind:

1. Only broken, overcast, or obscurations constitute a ceiling.
2. Cloud heights in the main body of the report are given as *height above the ground.*
3. Visibility is given in *statute miles.*
4. Temperature and dew point are in degrees Fahrenheit.
5. Wind direction is given in reference to *true* north.
6. Wind speed is given in *knots.*

Now, for practice, we should decode the following Aviation Weather Reports.

1. MKC 50 ⦶ 75 ⦶ M100 ⊕ 12 132/65/45/0220/992

Decoded: Kansas City, Missouri reports scattered clouds at 5,000 feet and 7,500 feet, with an overcast which was measured at 10,000 feet. The visibility is 12 statute miles. Barometric pressure is 1013.2 millibars, the temperature is 65 degrees Fahrenheit and the dew point is 45 degrees Fahrenheit. The surface wind is blowing from 020 degrees (true) at 20 knots. The altimeter setting, or barometric pressure in inches of mercury, is 29.92 inches.

2. TUL M3 ⊕ 1/2 RF 130/72/71/1410/991

Decoded: Tulsa, Oklahoma reports a measured ceiling of 300 feet and visibility of $\frac{1}{2}$ statute mile in moderate rain and fog. The barometric pressure is 1013.0 millibars. The temperature is 72 degrees Fahrenheit and the dew point is 71 degrees Fahrenheit. The surface wind is blowing from 140 degrees (true) at 10 knots. The altimeter setting or barometric pressure in inches of mercury is 29.91 inches.

FEDERAL AVIATION AGENCY
VFR PILOT EXAM-O-GRAM * No. 20

CEILING & VISIBILITY

An analysis of common mistakes on written examinations indicates that many applicants are improperly interpreting the ceilings and visibilities reported on hourly weather reports (sequence reports).

How about you? Do you really know what a ceiling is - what visibility is - and how both are reported??

A. CEILING:

1. Is defined as the "lowest layer of clouds or obscuring phenomena aloft that is reported as 'broken' (.6 to .9 coverage), or 'overcast' (more than .9 coverage) and not classified as 'thin'; or the height ascribed to surface-based obscuring phenomena not classified as 'partial'." This simply means that -

   a. Cloud coverage reported as broken or overcast constitutes a ceiling except when a minus sign precedes the cloud layer symbol ( -⦶, -⊕). When this occurs, that particular layer does not constitute a ceiling.

   b. If the sky is partially obscured (-X) it does not constitute a ceiling and no height will be given for this partial obscuration.
   Example: OKC -X18⦶M35⊕, ETC.

   c. Scattered clouds (⦶) do not constitute a ceiling.

2. For practical purposes, ceiling is the lowest height above the surface at which the total cloudiness between that level and the surface (as seen by a ground observer) covers more than half the sky.

*Exam-O-Grams are non-directive in nature and are issued solely as an information service to individuals interested in Airman Written Examinations.

3. Ceiling may be classified in several ways. This classification is shown by a letter which precedes the ceiling height. Some of the more important of these letters are:

   M = measured  E = estimated  A = aircraft  W = indefinite  B = balloon

   If one of these letter symbols does not precede the cloud symbol or if thin broken or overcast clouds, or a partial obscuration exists, there is no ceiling.
   Example: CBI -X10⦶14-⦶140-⊕5H, etc. means that no official ceiling exists at Columbia, Missouri.

B. CLOUDS:

1. In sequence reports the heights of cloud-base levels are given in feet above the ground, not above sea level. Note that the clouds are at approximately 1000 feet at station "A" and 6000 feet at station "B" in the illustration on the first page.

2. The figure for the height of the cloud base above the surface precedes each sky coverage symbol.

3. The last two digits of the cloud height are omitted; i. e., 1 means 100 feet, 14 means 1400 feet, and 140 means 14, 000 feet.

4. Clouds are reported in ascending order of height whether or not they constitute a ceiling.

5. Surface-based total obscurations (phenomena other than clouds such as fog, precipitation, dust, smoke, or haze) are not reported as clouds since this would be misleading but they are reported with a height value.
   Example: BUF W15X3/4S-F, etc., means that at Buffalo there is a vertical visibility of 1500 feet into a total obscuration. A definite cloud base cannot be seen from the ground.

C. VISIBILITY:

1. Visibility is the greatest distance on the earth's surface at which prominent objects can be seen and identified. This distance is not always the same in all directions. Therefore, the value for prevailing visibility (which is a ground visibility only) is based on surface observations and is stated in the hourly sequence reports. Prevailing visibility is always reported in statute miles and is the greatest surface visibility attained or surpassed throughout at least one-half of the horizon circle, but not necessarily continuous or for all of the horizon circle.

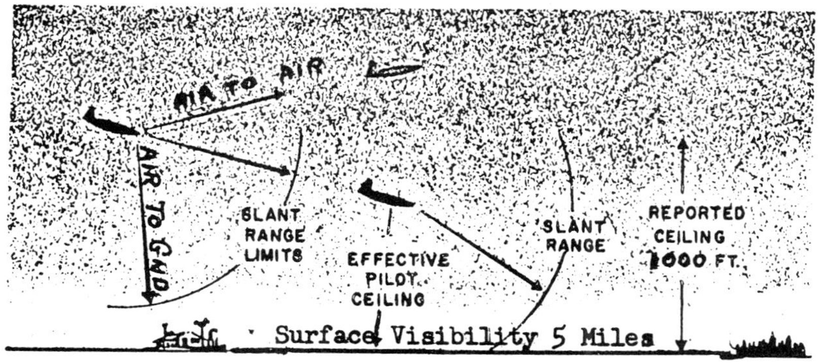


Types of Visibility in Aviation

All pilots should clearly understand that sequence report visibility may be much greater than either air to air, or slant range visibility, particularly when low ceilings and/or obscurations prevail.

2. Fractional values for visibility such as $1\frac{1}{4}$, $1\frac{1}{2}$, $2\frac{1}{2}$, etc., appear as 11/4, 11/2, 21/2, on the hourly sequence report because the teletype machines do not have fractions.
   Example: CLE M12⌀20⌀⊕21/2R-F ∅71/39/37/∅∅∅∅/941

3. It is not always possible to look at the visibility as reported in the main body of the teletype report and obtain complete information concerning surface visibility at the station because:

   a. If, in some direction from the station, there is a significant variation from the prevailing visibility, this variation will be found only in the remarks section.
      Example: PIT E18⊕4S-K ∅31/32/3∅/∅∅∅∅/957/VSBY N 1 means that the reported visibility is reduced to 1 mile to the north.

4. Weather occurring at the time of observation is also reported through use of letter symbols. These symbols follow the visibility as reported in statute miles. Some of the more common of these symbols are:

   R=Rain; S=Snow; T=Thunderstorm; Rw=Rain showers; A=Hail; E= Sleet

5. Except for thunderstorms, hail, and ice crystals, the intensity of weather is shown by:

   a. A plus sign (+) following the symbol to indicate "heavy".
   b. A minus sign (-) following the symbol to indicate "light".
   c. A double minus sign (--) following the symbol to indicate "very light".
   d. The absence of any sign indicates the intensity is "moderate".

   Example: STL 5⌀E10⊕2R--S-K etc.,which indicates that St. Louis has very light rain, light snow, and smoke. Note that these signs apply only to the weather and not to the obstructions to vision.

## QUESTIONS AND ANSWERS

LAX -X4-⦶14-⦶/-⊕4HK 196/66/63/ØØØØ/989/VSBY N 2

A. Based on the above report --

1. IS THERE A PARTIAL OBSCURATION AT 400 FEET OR SCATTERED CLOUDS AT 400 FEET? Ans. No height value is assigned to partial obscurations, and the figures for cloud heights above the ground precede the sky coverage symbol for those clouds; therefore, the figure 4 must refer to the scattered clouds.

2. WHAT IS THE CEILING? Ans. There is no official ceiling. No letter precedes the cloud symbols and all cloud coverage is reported as thin (-).

3. WHAT IS THE VISIBILITY? Ans. Prevailing surface visibility is 4 statute miles with haze and smoke, but to the north visibility is only 2 statute miles.

4. SHOULD A PILOT EXPECT HIS SLANT RANGE AIR TO GROUND VISIBILITY TO BE 4 MILES AT LOS ANGELES? Ans. No. It could be this or even more, but under the circumstances of obscuration, partial haze and smoke, it is probably less.

DEN 1Ø⦶M3Ø⦶8Ø⊕2VFK Ø31/75/65/1105/Ø1Ø/VSBY 1V3

B. Based on the above report --

1. WHAT IS THE CEILING? Ans. Measured 3000 feet above the surface.

2. AT 7500 FEET MEAN SEA LEVEL OVER DENVER, COLORADO, WOULD THE PILOT BE ABOVE OR BELOW THE CLOUDS REPORTED AT 3000 FEET? Ans. Below the clouds. At 7500 feet MSL he is approximately 2200 feet above the ground at Denver.

3. SHOULD A PILOT INTERPRET THIS REPORT TO MEAN THAT HE WILL FIND .9 OR MORE CLOUD COVER AT 8000 FEET ABOVE THE SURFACE? Ans. No. While it is possible that this situation actually exists, it is also true that such may not be the case at all. If there actually exists, visible from the ground, .5 coverage at 1000 feet, .4 coverage at 3000 feet, and .1 coverage at 8000 feet, it would be reported as 10⦶M30⦶80⊕. The combination of clouds at various levels may make it impossible for the ground observer to determine the actual percentage of cloud cover for all except the lowest level.

C. WHY SHOULD A PILOT BE EXTREMELY CAREFUL IN INTERPRETING CEILING VALUES, PARTICULARLY IN MOUNTAINOUS AREAS? Ans. If he does not he may "booby trap" himself into expecting an adequate ceiling when it does not exist! He must always relate surface elevation at the reporting station to terrain elevation along his flight route. The diagram on Page 1 illustrates what can happen to a 6000 foot ceiling when the surface elevation changes. He must also be aware that there are wide enroute variations from the reported ceilings (and visibility) even if there is no significant change in terrain elevation.

FAA AC 64 4843

FAA Aeronautical Center
Flight Standards Technical Division, AC-700
Operations Branch, AC-740
P. O. Box 25082
Oklahoma City, Oklahoma 73125

Exam-O-Grams available free of charge in limited quantities from this office.

FEDERAL AVIATION AGENCY

# VFR EXAM-O-GRAM* NO. 21

## FLYING INTO UNFAVORABLE WEATHER

How many times have you overheard or perhaps made similar statements yourself - "Let's go, we don't need weather, we'll make it okay." All too often a departing pilot merely glances at or completely ignores weather reports and forecasts because of a biased opinion that "weathermen never hit it right anyway". True, ceiling and visibility reports are sometimes estimated and a forecast is for conditions likely to occur; but if you don't utilize this information, your flight may be full of unhappy surprises. Accident investigation statistics continue to reveal "flight into unfavorable weather" as the chief cause of VFR fatal accidents. Results of FAA written examinations confirm that many pilots lack an adequate understanding of weather information.

WHY DO PILOTS FLY INTO UNFAVORABLE WEATHER?

1. GO-ITUS -- "I gotta get there . . ."; "I don't have time to wait." This is a condition that usually converts HOT PILOTS into COLD BODIES, and is a most difficult "disease" to cure. This attitude can be controlled only through sound reasoning and judgment by the individual.
2. MISINTERPRETATION OF FORECASTS AND REPORTS -- "It looks like VFR . . ."; "aw, it's good enough." Applicant performance on FAA written examinations indicate that the problem lies not in reading the data, but in knowing just what it means in terms of expected weather conditions.
3. FAILURE TO KEEP ABREAST OF WEATHER CHANGES -- Weather conditions do change, and the best way to keep informed en route is to listen to in-flight advisories and scheduled broadcasts 15 and 45 minutes past each hour.
4. IGNORING IN-FLIGHT WEATHER SIGNS -- "It's just a little shower . . ."; "just a few puffs of clouds." Rarely does weather suddenly go bad with no warning. Signs of deteriorating weather should be learned and observed by the VFR pilot.

## ARE YOU "WEATHER WISE" OR OTHERWISE?

DOES A STATION REPORT OF VFR CEILING MEAN EN ROUTE VFR? NO, the ceiling reported is the height above the reporting point only. It must also be related to the surrounding and en route terrain to determine if adequate VFR separation can be maintained between stations. (See Exam-O-Gram 20.) Additionally, unreported conditions between stations may be lower than those reported at the stations.

IS REPORTED VISIBILITY THE SAME AS VISIBILITY ALOFT? NO, the reported visibility is the visibility at the surface only. Conditions aloft may restrict flight visibility more or less than that reported. (See Exam-O-Gram 20.) Cockpit visibility in precipitation is further reduced by rain, drizzle, or snow spreading over the windshield. Forward visibility in a light snowfall may be zero due to the relative horizontal movement of the snow. Sunlight reflecting off haze or dust aloft reduces the visibility considerably.

WHAT CAN BE LEARNED FROM TEMPERATURE REPORTS? High temperatures reduce takeoff and landing performance. (See Exam-O-Gram 11.) Low temperatures reflect the approximate freezing level and the areas of possible icing in precipitation. Sudden temperature changes reveal the relative position of a front and its associated weather.

WHAT IS THE SIGNIFICANCE OF DEW POINT? Specifically, a dew point value relatively close (2° - 5°) to the air temperature is indicative of the probability of fog, low clouds, or precipitation.

*Exam-O-Grams are non-directive in nature and are issued solely as an information service to individuals interested in Airman Written Examinations.

WHY SHOULD THE REPORTED WIND DATA BE NOTED? The velocity and direction of the surface wind should be related to the runway at the point of intended landing to determine the degree of cross wind. Wind data also reflects the degree of turbulence to expect. A sudden shift in direction often reveals the position of frontal weather relative to a station.

OF WHAT VALUE IS THE ALTIMETER SETTING? Correct cruising altitudes and adequate vertical clearance are dependent on the application of altimeter settings. A rapid and continual drop in pressure (altimeter setting) forewarns of approaching inclement weather.

WHAT IS A PIREP AND WHERE IS IT FOUND? A PIREP is a report of weather conditions at flight altitude, particularly between stations, seen by the pilot instead of the ground observer. Reports are often broadcast, and a pilot report summary is disseminated hourly to stations by teletype. Cloud base and top reports are found in the Remarks section of sequence reports.

ARE YOU GETTING THE REAL PICTURE FROM FORECASTS AND REPORTS? Only when the above are considered in analyzing forecasts and reports will you have the full story.

WHAT ARE SOME OF THE WEATHER SIGNPOSTS AND THEIR WARNINGS?

Blowing Dust -- turbulence, poor visibility at low levels, particularly into the sun.

Low Layer of Haze -- possible fog or stratus cloud in early morning or late evening; poor visibility, particularly into the sun.

Light Puffs of Clouds at Low Levels -- probable fog or stratus cloud, particularly in early morning or late evening.

Ragged Cloud Base -- turbulence, erratic visibility, possible precipitation.

Bulbous Cloud Base -- turbulence, possible precipitation, conducive to TORNADOES.

Roll-Type Clouds -- DANGEROUS turbulence, dust and poor visibility, hazardous landing conditions, subsequent precipitation.

Line of Heavy Dark Clouds -- SEVERE turbulence, dust and poor visibility, hazardous landing conditions, precipitation, hail.

Opening in Wall of Dark Clouds (SUCKER HOLE) -- DANGEROUS turbulence, possible precipitation and poor visibility as the hole is entered.

Gradual Lowering and Thickening of the Ceiling -- inadequate terrain clearance, possible widespread precipitation, and fog.

Near Freezing Temperature -- poor visibility in precipitation with ice forming on the windshield as well as the aircraft structure.

THE 180° TURN IS AVIATION'S BEST SAFETY DEVICE -- IF USED PRIOR TO BEING ENVELOPED BY ADVERSE WEATHER. DON'T BE A "PUSHER" IN THE hope THAT THE WEATHER WILL GET BETTER!

## CHECK YOUR READING

1. What is the most current weather information available?
2. What is the time and day of the month of the Aviation Weather Reports following this heading?

   023SA23182300

3. List as many station identifiers as you can.

4. List the sky conditions and their symbols which indicate a *ceiling*.
5. How much of the sky may be covered if the report says *scattered*?
6. When will the method of determining cloud height be included in the report?
7. List the symbols and their meaning which are used to designate the method of measuring cloud height.
8. The height of clouds found in the main body of the report is in reference to what base?
9. List and define the symbols used for restrictions to visibility and precipitation.
10. State the barometric pressure in millibars for the following code numbers:

    520 483 234 765 222 123

11. List the wind directions indicated by the following code numbers:

    22 34 35 03 10 01

12. Where can we find the interpretation of NOTAMS?

# 32

# WEATHER CHARTS

Many weather charts are produced for our use, each depicting different weather data and weather at specific altitudes. We will learn the details of those charts we will use the most, and discuss generally those charts not used regularly.

## The Weather Depiction Chart

The weather depiction chart consists of plotted and analyzed data on cloud heights and visibility for the contiguous United States and southern Canada. We can, at a glance, receive information which will aid us in the planning of our flight. A sample is shown in Fig. 95.

Plotted data on the weather depiction chart includes:

1. Precipitation and/or other important weather phenomena.
2. Sky cover.
3. Visibility (only when 6 statute miles or less).
4. Cloud base (in hundreds of feet *above the ground* up to 20,000 feet).

Analysis consists of:

1. Areas in which the cloud ceilings are below 1,000 feet and/or visibilities are below 3 statute miles are enclosed by solid black lines and are shaded red by the weatherman.

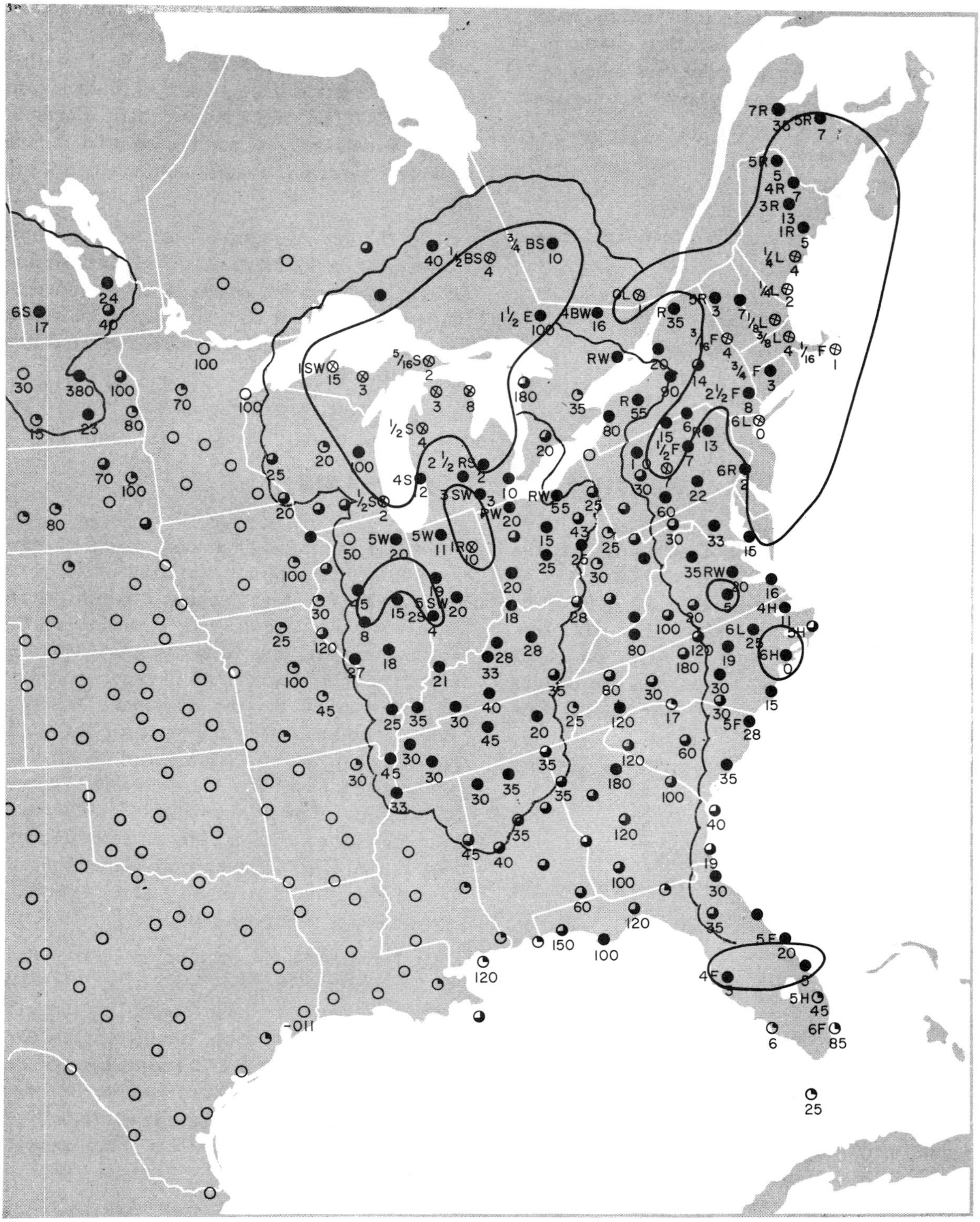

Fig. 95. Weather depiction chart.

2. Areas in which the cloud ceiling is below 5,000 feet but not below 1,000 feet and visibilities are above 3 miles are enclosed by a black scalloped line, and are shaded blue by weatherman.

Weather depiction charts are prepared every three hours beginning with the data observed at 0100 Greenwich mean time. By the time the chart reaches the local weather station, the information it contains is at least one hour and twenty minutes old. Therefore, we must consult the latest Aviation Weather Reports to update this information.

The Weather Depiction Chart should be used in conjunction with the Surface Weather Chart to properly form an accurate picture of the weather situation.

## The Surface Weather Charts

The Surface Weather Chart depicts plotted weather, analyzed pressure patterns at mean sea level, and surface location of fronts. A sample is shown in Fig. 96.

Although we will concentrate on pressure patterns and fronts more than on plotted data on the chart, we should be familiar with the more common symbols which are clustered around the reporting station (the station model).

### *The Station Model*

Figure 97 shows a sample station model with pertinent information as it will be plotted.

*A* is the station location. The amount of total cloud cover is indicated by the percentage of the station circle which is shaded in. For example:

○ = Clear sky
◔ = Partly cloudy—scattered clouds
◕ = Mostly cloudy—broken clouds
⊗ = Sky cover obscured (due to fog, haze, precipitation, etc.).

*B* is the surface wind direction. The arrow flies with the wind. For example:

 = East wind
 = Southeast wind
 = Southwest wind

*C* is the surface wind speed. Each barb on the wind arrow represents 10 knots of wind. A half barb represents 5 knots of wind. A pennant represents 50 knots of wind. For example:

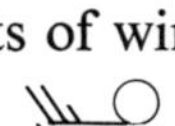 = West wind of 25 knots
 = Northwest wind of 15 knots
 = Southeast wind of 65 knots

*D* is the current barometric pressure in millibars. We must add a "9" or a "10" to the number shown so that the result falls between 950 and 1050 millibars. For example:

08 = <u>10</u>08 millibars
00 = <u>10</u>00 millibars

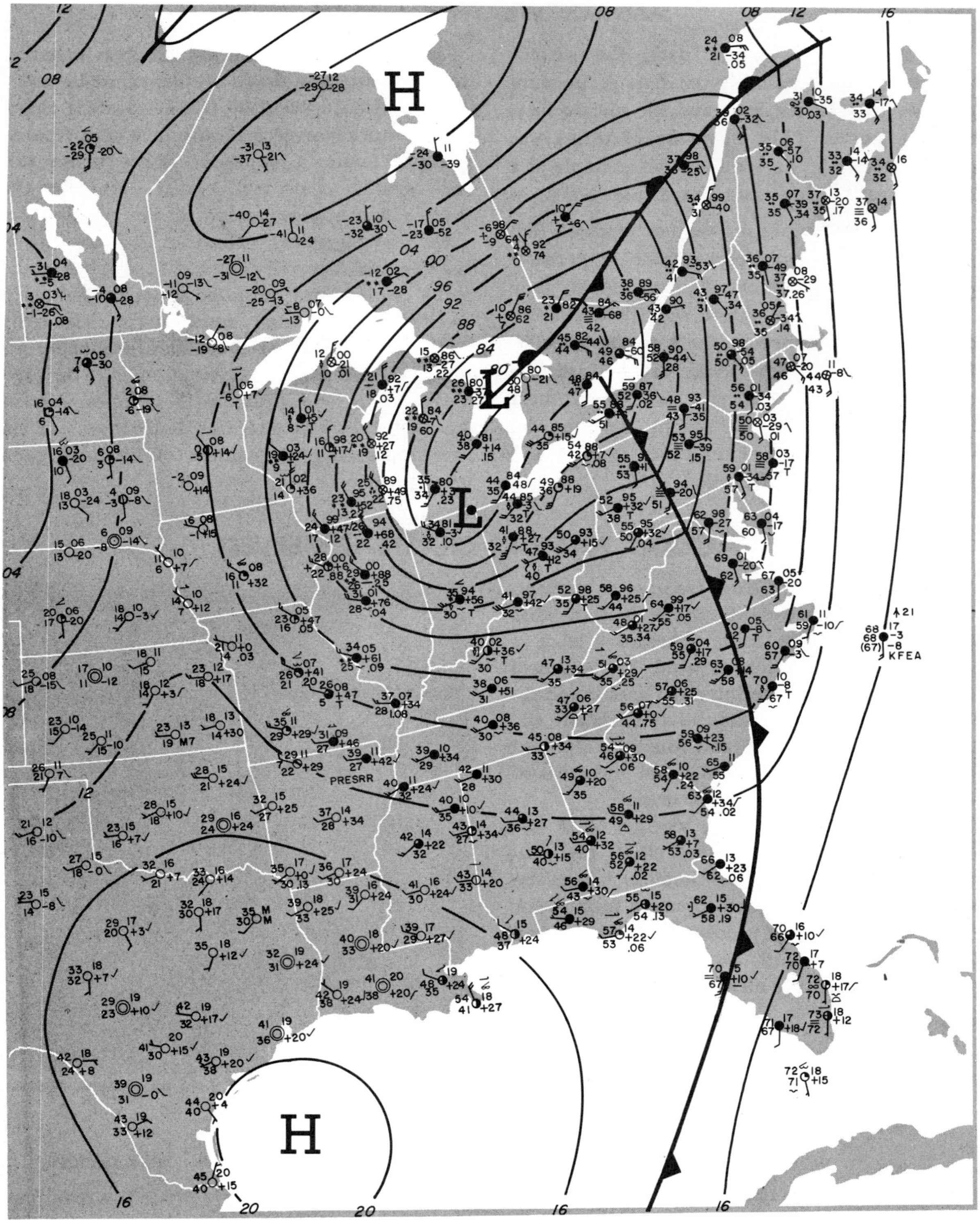

Fig. 96. Surface weather chart.

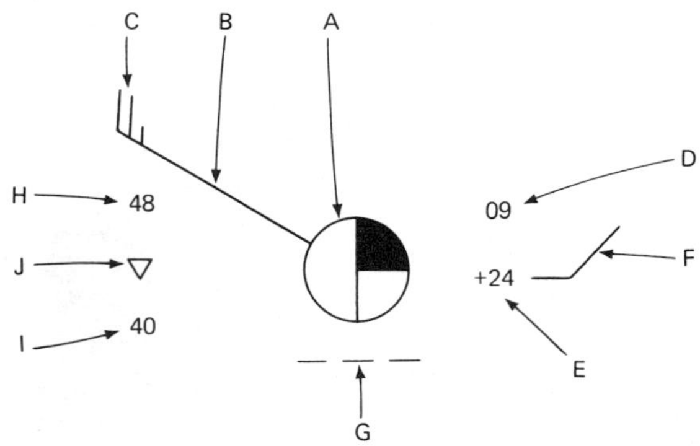


Fig. 97. The station model.

*53 = 953 millibars
*48 = 1048 millibars

*NOTE: If we added 9 or 10 erroneously, the result would not fall within 950 to 1050 millibars.

*E* is the amount of pressure change within the preceding three hours. We must add a decimal point. For example:

14 = 1.4 millibar change
02 = 0.2 millibar change
22 = 2.2 millibar change

*F* is the direction and manner of the pressure change, shown graphically. For example:

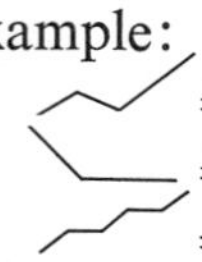

= slight climb, slight fall, rapid climb
= rapid fall, then steady
= rising unsteadily

*G* is the type of low clouds. For example:

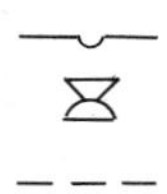

= low cumulus
= towering or building cumulus
= scud or low broken cloud

*H* is the temperature in degrees Fahrenheit. For example:

14 = 14 degrees Fahrenheit

*I* is the dew point in degrees Fahrenheit. For example:

8 = 8 degrees Fahrenheit

*J* is the precipitation. For example:

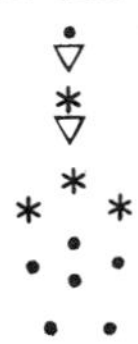

= rain shower
= snow shower
= snow (moderate)
= rain (heavy)
= rain (light)

*Major Depictions*

The more important major depictions on the surface weather chart include the following:

*Fronts:*

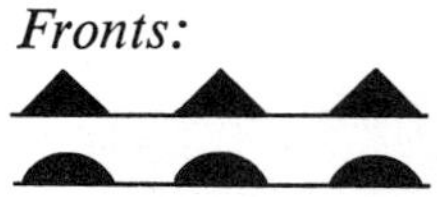

Cold Front (black or blue)
Warm Front (black or red)

Occluded Front (black or purple)
Stationary Front (black or alternating red and blue)
Front Aloft (black or in
appropriate color, but not fully shaded)

*Pressure Centers:*

Low Pressure: shown by large letter *L* in black or red
High Pressure: shown by large letter *H* in black or red

*Isobars:*

Lines connecting points of equal atmospheric pressure (at sea level) are called *isobars*. The wind on the surface normally blows parallel to the isobars or slightly across them (toward the low center or away from the high center).

*Precipitation:*

Areas of precipitation are shaded in gray or green.

The Surface Weather Chart for the contiguous 48 states is distributed every three hours beginning at 0000 Greenwich mean time. The Surface Weather Chart for the Northern Hemisphere is distributed every 12 hours.

We may compare previous charts with the current one to predict storm and front movement. However, we must consult current Aviation Weather Reports to update the Surface Weather Chart information.

## Prognostic Surface Charts

Prognostic charts for periods of 12 hours to three days are prepared to aid us in long-term flight planning. Pressure systems and fronts are shown as they are *expected* to appear at the posted future time. (See Fig. 98.)

## Winds Aloft Charts

Wind data observed by PIBAL and RAWINSONDE are plotted on the Winds Aloft Chart and published every 6 hours beginning at 0000 Greenwich mean time. Three sets of wind charts are published.

*Low Level Winds*

Winds for the second level above the surface; 5,000 feet; 8,000 feet; and 10,000 feet.

*Intermediate Winds*

14,000 feet; 20,000 feet; 25,000 feet; and 30,000 feet.

*Upper Level Winds*

35,000 feet; 40,000 feet; 50,000 feet; and 60,000 feet.

The Winds Aloft Chart shows *observed* (not forecast) winds and can be

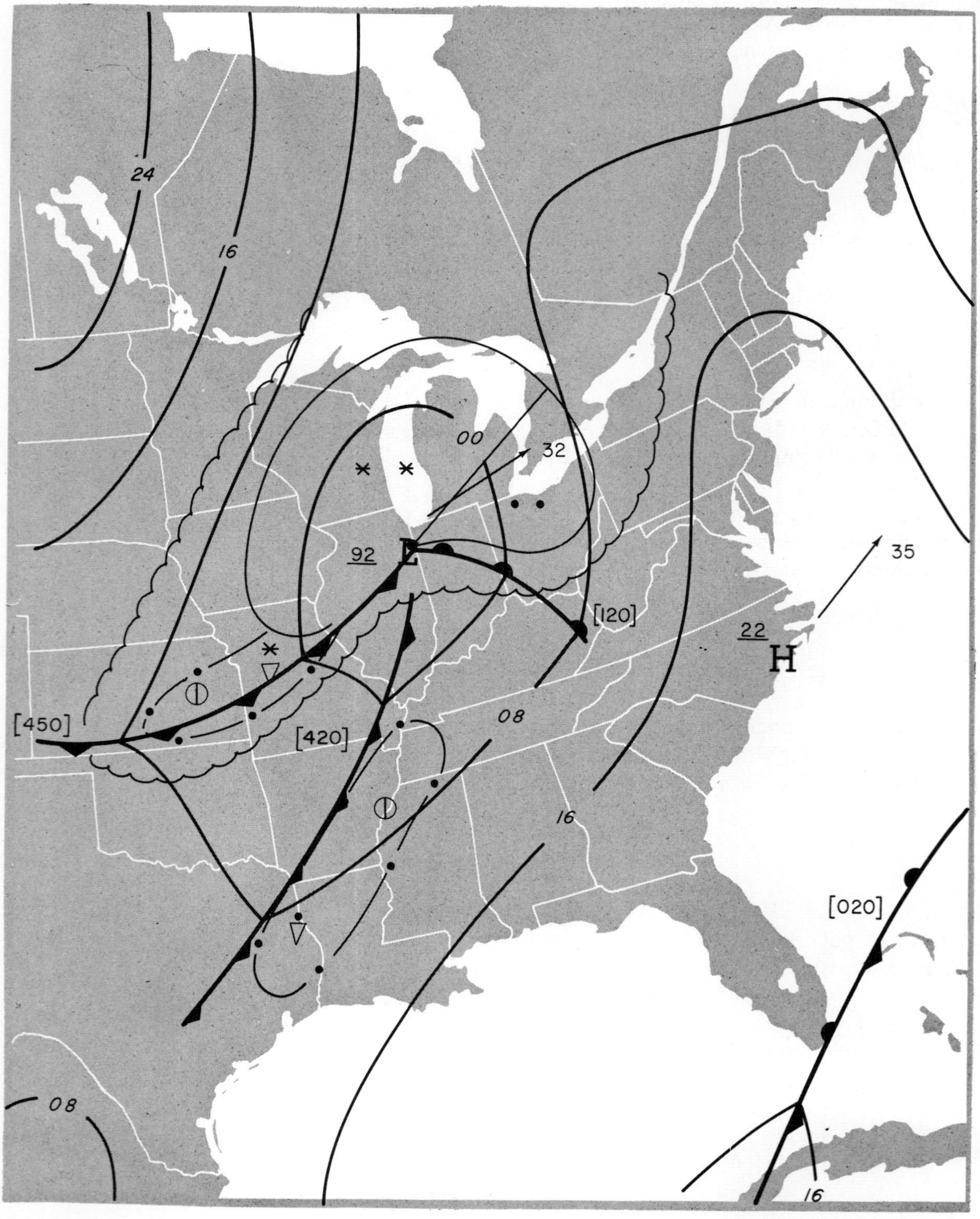

Fig. 98. Prognostic surface weather chart.

up to 9 hours old. Therefore, we should use it in conjunction with Winds Aloft Forecasts (see Chapter 34, Winds Aloft Forecast).

## Radar Summary Charts

The Radar Summary Chart is published every three hours, with plotted radar echoes taken from all radar observation sites in the contiguous United States. The approximate outline of all radar echoes are designated by scalloped black lines. (See Fig. 99.)

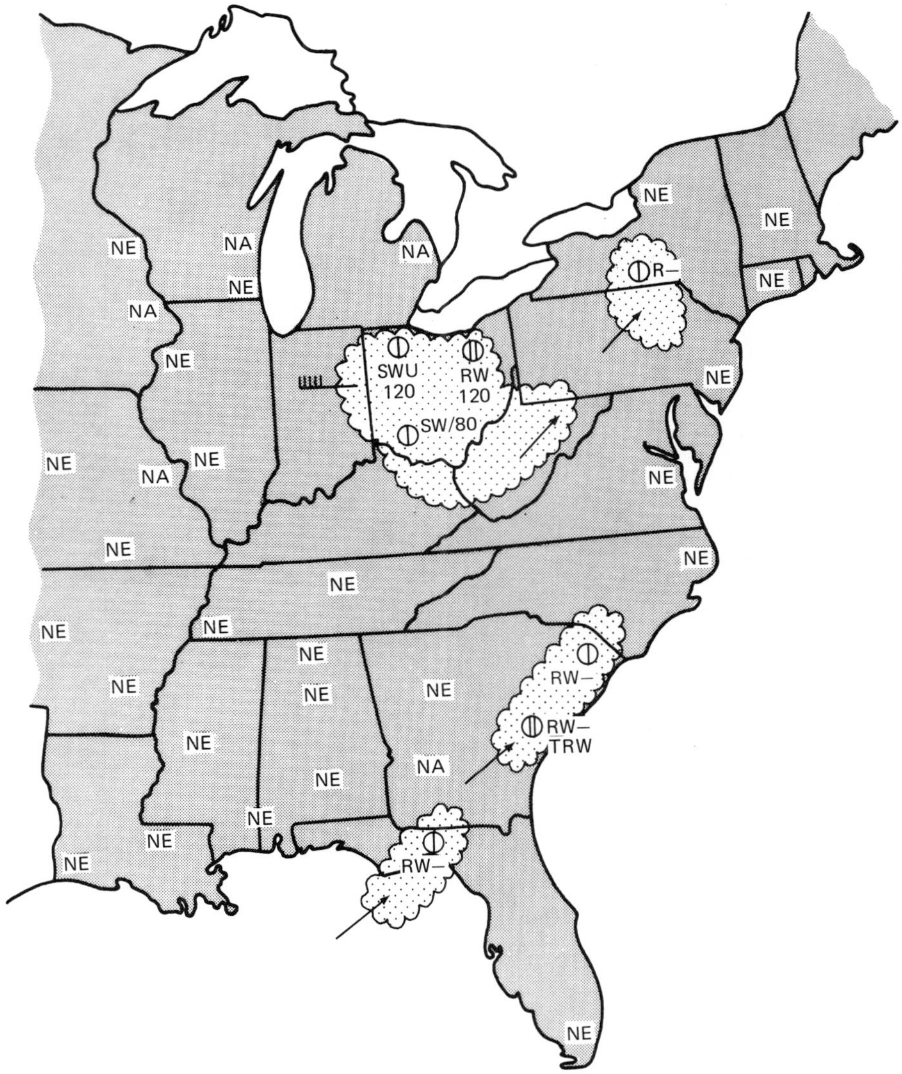


Fig. 99. Radar summary chart.

## CHECK YOUR READING

1. List and describe the data to be found on the Weather Depiction Chart.
2. How often are Weather Depiction Charts printed?
3. List the data to be found on the Surface Weather Chart.
4. How often is the Surface Weather Chart printed?
5. Draw the symbol used in the station model for a wind from 270 degrees at 45 knots.
6. Draw the symbols used on the Surface Weather Chart to depict a cold front, warm front, and occluded front.
7. What is an "isobar"?
8. List the altitudes for which a wind is given on the Winds Aloft Chart.
9. Does the Winds Aloft Chart depict actual or forecasted winds?
10. What information is available to update the Surface Weather Chart?

# 33
# THE AREA FORECAST

The area forecast for a specific region is used primarily for flight planning. If no terminal forecast is issued for a particular station, the area forecast may be used for that station weather.

The area forecast is prepared every 6 hours and presents a forecast for a 12-hour period, with an outlook for an additional 12 hours.

### *The Heading*

The heading identifies the forecasting center, the date and time of the forecast, and the time period for which it is valid (local time is used).

### *The Forecast Area*

The geographic area covered by the forecast is defined, normally using state lines, rivers, mountain ranges, or portions of states as boundaries.

### *Clouds and Weather*

In this section, expected flying conditions are described for the first 12-hour period. Abbreviated plain language is normally used, but sometimes weather symbols are used to describe sky cover, visibility, and restrictions to visibility.

### *Sky Cover*

Each layer of cloud is described except for broken or overcast layers with bases 5,000 feet or more above the ground and with less than 3,000 feet vertical separation (which are treated as one layer).

*Cloud Heights*

The height of cloud bases is given *in reference to the ground* and if the cloud constitutes a ceiling, the height will be preceded by the letter *C*. Heights are given in intervals of 100 feet up to 3,000 feet above the ground. Above 3,000 feet, 500-or 1,000- foot intervals are given. The height of cloud tops will be *in reference to mean sea level.* The word "TOPS" will identify the cloud top forecast.

*Visibility*

If the surface visibility is expected to be 8 statute miles or better, it will be omitted from the forecast. If the visibility is expected to be 6 statute miles or less, the obstruction to visibility will be given.

*Weather and Obstructions to Visibility*

A contracted form is used to describe weather and obstructions to vision (SNW, DRZL, etc.). If they are written as part of a complete symbolic group, standard symbols (S, L, etc.) are used.

*Surface Wind*

The contraction "SFC WND" precedes the wind direction and speed. The symbolic form will be used for any forecast of sustained wind of 25 knots or more (25G40). Wind direction is in reference to True North, and the speed is in knots. Gusty winds included in the forecast will be preceded by the word "GUSTY" or "GUSTS TO. . . ."

*Icing*

The section of the forecast in which expected icing is listed, will be preceded by the contraction "ICG." The type of ice will be described by the abbreviations "CLR" (clear), "RIME" (Rime), or "MXD" (Mixed). Ice intensity will be described as "LGT" (light), "MDT" (moderate), or "HVY" (heavy). If icing is expected to be in the clouds, the contraction "ICGIC" will be used. Icing in precipitation will be denoted by "ICGIP."

The height of the freezing level, whether or not icing is expected, will be given in hundreds of feet above mean sea level.

*Turbulence*

The contraction used for turbulence is "TURBC" and will be used when turbulence is expected which may affect flight safety. The heights mentioned are given in hundreds of feet above mean sea level.

*Outlook*

This section gives a brief statement of expected conditions in the 12-hour period following the forecast.

**Sample Weather Forecast**

FA ATL 071245
08E-20E WED

WRN NC WRN SC NRN GA ALA EXCP MOBILE AREA

CLDS AND WX.     OVER SRN AND CNTRL ALA

OVER NRN ALA AND NRN GA C15-20 ⦷ V ⦶ TOPS 40–60 BCMG
MSTLY C30 ⦷ V ⦶ TOPS 40 BY 12E AND 35 ⦶ BY 14E AND
○ BY 19E

OVR WRN SC 40 ⦶ V ⦷ MSL TOPS 90 BCMG ○ BY 19E

OVR WRN NC C20–40 ⦷ TOPS 100–120 LCLY C10 ⊕ 7 IN THE
MTNS WITH OCNL C10X2S- BCMG C40 ⦷ V ⦶ TOPS 80 BY 12E
WITH A FEW SNW FLRYS PRSTG IN THE MTNS. CLRG BY 20E.

ICG. LGT TO MDT RIME ICGIC WITH CHANCE OF MDT MXD
ICGIP WRN NC TIL 12E. FRZG LVL SFC OVER THE WRN
CAROLINAS AND NRN GA SLPG TO 30–40 CNTRL AND SRN ALA
LFTG TO 25 WRN NC AND 50 SRN ALA BY AFTN.

TURBC. MDT BLO 80–120 OVR WRN NC WRN SC NRN GA WITH
OCNL SVR TURBC OVR AND ALG E SLPS OF MTNS. TURBC WL
DMSH LATE AFTN. MOST PRBL AREA OF MDT CAT 250–350
OVER THE WRN CAROLINAS AND NRN GA ESPECIALLY OVR THE
MTNS.

OTLK. 20E WED-08E THU. ○ OVR AREA BCMG 200 ⦷ OVR
WRN ALA BY DABRK. VSBY UNRSTD LWRG LCLY TO 4KH BY 06E

As with the Aviation Weather Report, we must make an effort to picture the weather as forecast, not just interpret the abbreviations and contractions.

## CHECK YOUR READING

THE FOLLOWING QUESTIONS REFER TO THE SAMPLE AREA FORECAST IN THIS CHAPTER.

1. When will the weather be clear over northern Alabama?
2. What type of weather can we expect when flying through the mountain valleys in western North Carolina prior to 1200 Eastern standard time?
3. When will the mountain weather clear?
4. What will the height of the freezing level be over the western Carolinas?
5. Where might we expect to penetrate occasional severe turbulence prior to the late afternoon?

# 34

# MISCELLANEOUS WEATHER INFORMATION

The United States Weather Bureau issues reports and forecasts for special use by pilots, including high-altitude weather, transoceanic forecasts, and flight profile charts. Only those special reports which are used by most light aircraft pilots are included here.

## Terminal Forecasts

The Flight Advisory Weather Service prepares forecasts of special interest to aviation for a 12-hour period and a 24-hour period. These reports are concerned with the weather at specific terminal areas.

### *The 12-Hour Terminal Forecast*

The 12-hour forecast is prepared every six hours and transmitted by teletype to all interested parties.

*Content.* The terminal forecast gives the height and amount of expected sky cover, the visibility expected, weather or obstruction to visibility which is forecast, expected surface wind, and any appropriate remarks. Icing and cloud tops as well as turbulence are not included in the forecast.

*Time.* The filing time, as well as the valid time, is given in Greewich mean time. All times mentioned in the forecast itself are in reference to local time.

*Symbols and Codes.* The same symbols and codes that are used in the Aviation Weather Report are used here.

*Ceiling Identifier.* If any cloud layer is expected to constitute a ceiling, the cloud height figure will be preceded by the letter *C*.

*Cloud Height.* The height of the cloud is always given in reference to the ground.

*Visibility.* Only when the visibility is expected to be 8 statute miles or less will it be included in the terminal forecast. If the visibility is expected to be 6 statute miles or less, the restriction to visibility will be noted immediately following the visibility value.

*Surface Wind.* If the wind speed is forecasted to be less than 10 knots, the entire wind report will be omitted from the report. When the wind is given, its direction will be in reference to True North and the speed will be in knots. Gusty wind will be indicated by the letter *G*.

*Changes in Weather Conditions.* The forecast will include any expected significant changes in the weather. The time (local) of the change will precede the new weather.

*Amendments.* When the safety and/or efficiency of the flight may be affected by unexpected weather, an amended forecast will be issued.

### *The 24-Hour Terminal Forecast*

The 24-hour terminal forecast is issued every six hours for major terminal areas only. The format used is identical to the 12-hour terminal forecast.

### *Sample Terminal Forecasts*

FT1 121045
11Z–23Z WED
066

AMA 250 – ⦶ 2010. 1030C 250 – ⦷ 2018. 1630C COLD
FROPA C50 ⊕ 3RW – 3320G35

JBB 250 – ⦶ 2010. 1000C C250 ⦷ 2018.

PNC ○ 1612. 1230C 40 ⦶ 1815 ⦶ V ⦷. 1500C C35
2020G CHANCE BRF C10XTRW+ 2225G45

## Winds Aloft Forecast

Every 6 hours a forecast of the winds aloft is prepared and transmitted to interested parties. Each wind aloft forecast includes the forecasted wind direction in relation to True North and the wind speed in knots for the following altitudes.

3,000 feet (if station elevation is 2,000 feet or lower)

5,000 feet (if station elevation is 4,000 feet or lower)
10,000 feet
15,000 feet
25,000 feet

*Wind Speed*

The wind speed is given to the nearest whole knot. If the speed is forecasted to be less than 5 knots, it will be noted as 9900 (calm wind).

Forecast winds of 100 to 199 knots are indicated after subtracting 100 from the speed and adding 50 to the direction (which alerts the reader of the high wind). For example, a wind from 280 degrees at 125 knots will be noted as 7825.

*Temperature*

The expected temperature for each level is also included in the wind aloft forecast. It will be given in degrees Celsius. No temperature will be given for the 3,000-foot level, or for the 5,000-foot level if this is the lowest level for which the wind is forecasted.

*Sample Winds Aloft Forecast*

FD MEM 242350
00 — 12Z FRI

| LVL | 3000 | 5000 ft. | 10000 ft. | 15000 ft. | 20000 ft. | 25000 ft. |
|---|---|---|---|---|---|---|
| HRO | 2430 | 1945 + 11 | 2140 + 01 | 2445 — 11 | 2650 — 16 | 2665 — 25 |
| DYR | 2616 | 2430 + 09 | 2325 + 01 | 2540 — 12 | 2645 — 15 | 2660 — 25 |

*Interpolation*

When our flight altitude is between the levels for which a report is given, we must interpolate to find the forecasted wind for our particular altitude.

Suppose, for example, that the wind at 5,000 feet is forecasted to be from 270 degrees at 20 knots, and the wind at 10,000 feet is forecasted to be from 320 degrees at 45 knots. We plan to fly at an altitude of 8,000 feet.

There are five thousand-foot altitudes from 5,000 feet to 10,000 feet, and the wind direction is forecasted to change 50 degrees between 5,000 feet and 10,000 feet. Therefore, for each 1,000 feet of altitude change it is reasonable to assume that the wind will change 10 degrees. At 8,000 feet, the wind direction should be from 300 degrees. The wind speed increased 25 knots between 5,000 feet and 10,000 feet, or about 5 knots per 1,000 feet of altitude. Hence, the wind speed at 8,000 feet should be 35 knots. The wind we will use for flight planning will be 300 degrees at 35 knots.

## PIREPS

The U.S. Weather Bureau and the Federal Aviation Administration are interested in receiving reports of unexpected weather directly from pilots in flight. The *PIREPS* are pilot reports which describe upper cloud layers, turbulence, icing, or other weather phenomena which are not forecasted and may be useful to other pilots. We should make every effort to broadcast PIREPS knowing that they will be well-received by other pilots.

The PIREP will be transmitted by teletype in normal English. The type of aircraft issuing the report, the location of the flight, and the description of the weather or other phenomena will be given.

## AIRMETS

Weather considered potentially hazardous for light aircraft (single-engine and small twin-engine aircraft) is described in the *AIRMET* which is distributed via teletypewriter to all interested parties. AIRMETS will be broadcast by the Federal Aviation Administration over navigational radio facilities upon receipt.

The AIRMET will be issued if the following conditions exist or are expected to exist within two hours.

Moderate icing
Moderate turbulence
Winds of 40 knots or more within 2,000 feet of the surface
Ceilings below 1,000 feet and/or visibilities less than 2 statute miles

## SIGMETS

The *SIGMET* advisory concerns weather which is significant to the safety of large aircraft as well as light aircraft. The SIGMET will be issued if the following conditions exist or are expected to exist within 2 hours.

Tornadoes
Squall lines or lines of thunderstorms
Hail of $\frac{3}{4}$-inch diameter or larger
Heavy icing
Severe or extreme turbulence
Widespread sand or dust storms which reduce the visibility to 2 statute miles or less

## Aviation Weather Broadcasts

Aviation Weather Reports will be broadcast over Federal Aviation Administration navigational facilities at 15 minutes past every hour. In addition, we may request specific weather information at any time by contacting the nearest Flight Service Station.

Aviation weather within a radius of 150 miles of the broadcast station will be included in the broadcast at 15 minutes past the hour.

## CHECK YOUR READING

1. How often will the 12-hour terminal forecast be issued?
2. If a ceiling is forecast, how will it be designated in the 12-hour terminal forecast?
3. If the visibility is not included in the terminal forecast, it must be at least what value?
4. When will the surface wind be included in the terminal forecast?
5. For what altitudes will the winds aloft forecast be issued?
6. Write the code number for a wind of 300 degrees and 140 knots.
7. Will the temperature forecast in the winds aloft forecast be in degrees Celsius or Fahrenheit?
8. What is the difference between the AIRMET and the SIGMET?

PART III

# NAVIGATION

# 35

# NAVIGATION

Navigation is the art of finding our way from one point to another. We learn to navigate at a very early age. As toddlers we managed to find our way from one room to another by remembering familiar scenery enroute. As our world broadened to that outside our home, our navigation became more sophisticated, and we found our way to school, to work, and to distant cities. Therefore, navigation is nothing new to us.

## Methods of Navigation

There are four basic methods of navigation. We may use each one separately or in any combination.

### *Pilotage*

When we follow a road map to find a distant city, we use *pilotage*. We simply use a picture or chart of the route, and move from one landmark to the next, picking our way toward the destination. The prominent landmarks we use may be intersections of roads, lakes, large buildings, etc. Pilotage in flight is quite similar, in that we need only to draw a line on our chart from point of departure to the destination, make note of the prominent landmarks along that line, and search for them as we fly along our course. Naturally, from our vantage point aloft we may see greater distances than when on the ground, but we cannot see small landmarks.

### Dead Reckoning

If we combine distance, speed, time, and direction in mathematical computations in order to find our way, we use *dead reckoning* navigation. We have used this method many times prior to learning how to fly. For instance, when we give an automobile driver directions to "drive south for 10 miles, and then turn east and drive for about 5 minutes" or "at 50 miles per hour, it will take 20 minutes to reach Smithville," we have issued dead reckoning instructions for travel from one place to another. No landmarks need be mentioned, as we rely solely on the direction of travel, the distance, the speed, and the time required to reach the destination. This method of navigation is very useful when flying where no landmarks can be seen, such as at night, over clouds, or over large bodies of water.

### Radio Navigation

The chances are that we have never used radio navigation prior to beginning flight training. This method is now the most popular of all navigation procedures because of its accuracy and simplicity. A radio signal is received in the aircraft cockpit and presented to us either visually or aurally. We must direct the aircraft in the direction indicated to fly from one radio broadcast station to the next, or to any predetermined location.

### Celestial Navigation

The heavenly bodies (stars, sun, moon) may be used to pinpoint our position with great accuracy. This method of navigation is called *celestial navigation*. If we know the position of the heavenly bodies, and measure the angle between us and two or three of them, we can compute our location. Celestial navigation is no longer in popular use in aviation because of its complexity and the time consumed in calculations.

## Modern Aerial Navigation

In modern times, pilotage, dead reckoning, and radio navigation are combined, resulting in easy, accurate navigation. It is wise to use all three methods together whenever possible as added insurance against becoming lost in flight.

### Aids To Navigation

The Federal Government provides many aids to aerial navigation. We must become familiar with them so that we can derive maximum use from them.

*Aeronautical Charts.* Very detailed drawings of the earth's surface are printed in a number of different scales to provide us with a picture of the earth's prominent features and man-made landmarks.

*Radio Aids.* A broad network of radio broadcast stations covers the United States and Canada. Most of these stations are usable 24 hours a day, free of all direct cost, for navigation or communications.

*Federal Airways.* Radio signals from broadcast stations are used to define airways which cross the land. Although an airway may not exist to every desired destination, we will normally find that an airway will take us close to our point of intended landing, and, therefore, we should use airways whenever possible.

*Coast and Geodetic Charts.* The United States Government publishes maps and charts for use in radio navigation and terminal approaches. These publications are revised whenever any significant change takes place. The revisions are mailed to subscribers.

*Airman's Information Manual.* The Airman's Information Manual is published by the United States Government and revised as needed. Its six volumes describe pertinent information concerning navigational facilities, airports, procedures, and flight safety (see Chapter 48).

## Summary

Navigation is made quite simple by the abundant facilities and maps available. A firm foundation of navigational knowledge will aid us in the proper use of facilities and in using all of the available facilities to our best advantage.

## CHECK YOUR READING

1. List and describe the four basic methods of navigation.
2. What is the most popular method of aerial navigation?
3. List and describe the various aids to navigation.
4. Give an example of each type of navigation you have used prior to beginning flight training.

# 36

# NAVIGATION MAPS

Many special-use maps are published for pilots. We will discuss here only the most common charts in use.

Depending on the speed of the aircraft we will fly and the type of navigation we plan to use, one of the following maps will suit our specific need.

*Local Aeronautical Chart*

The *local* chart depicts large metropolitan areas, such as Boston, Chicago, and Los Angeles, in great detail. The scale of the local chart is one inch equals 250,000 inches, or one inch equals 4 statute miles. Only pilots of very slow aircraft will use the local chart because of the limited area depicted on it.

*Sectional Aeronautical Chart*

The *sectional* chart is the most popular of all pilotage maps in use by pilots of light aircraft. Although it does not show as much detail as the local chart, it depicts the surface quite accurately and completely. The scale is one inch equals 500,000 inches, or one inch equals 8 statute miles. A total of 37 sheets, printed back to back, cover the entire conterminous United States.

*World Aeronautical Chart (WAC) and Operational Navigation Chart (ONC)*

All of the land areas of the world are depicted on the World Aeronautical Chart (inside the U.S. except Hawaii) or the Operational Navigation Chart (outside the U.S. and Hawaii). The scale of these charts is one inch equals 1,000,000 inches, or one inch equals 16 statute miles. Some of the details printed on the sectional chart will be omitted from the *WAC* or *ONC* charts, but their scales make them ideally suited for moderate speed aircraft.

*Aeronautical Planning Chart*

The scale of this chart is one inch equals 5,000,000 inches, or one inch equals 80 statute miles, which restricts its use to long range flight planning.

*Avigation Charts*

Very little surface detail is shown on *avigation charts* for they are used primarily for radio navigation.

## Important Items on the Navigation Charts

The following items constitute the most important items on the sectional chart, the WAC chart, and the Aeronautical Planning Chart. The best way to become familiar with the charts is to use them often, looking up unknown symbols on the side panels, or back of the chart, or in the Airman's Information Manual.

*Date*

Naturally, it would be foolhardy to refer to a map which is out of date. As we will see in our years of flying, the landscape changes constantly. New highways are built, electric power lines erected, lakes created, and cities grow. The date of publication of each chart is listed in the Airman's Information Manual. We should always keep a current chart handy for use in orientation even when flying over familiar terrain. We must check the publication date often, for the charts are revised quite regularly.

*Radio Facilities*

Although the pilotage charts are not used primarily for radio navigation, the radio aids to enroute navigation and airways are shown on the charts. All very-high frequency transmitters and airways are printed in blue ink; low frequency radio facilities and airways are printed in magenta (maroon) ink.

*Topographical Information*

The topography, or height of the terrain is shown on the charts by relief shading, contours, elevation tints, and spot elevations. All elevations of terrain, airports, and obstructions are given in reference to their *height above sea level.*

*Restricted Areas*

Certain areas of the nation's airspace are restricted to flight, except with permission of the controlling agency. The restricted area is outlined in blue hatch lines and is assigned a number. The Airman's Information Manual will list all restricted areas by number and will describe the particular restrictions for each area.

*Prohibited Areas*

All flight is prohibited, except with prior permission of the controlling agency, in any *prohibited area.* These air spaces will be designated by a blue outline, with hatch marks covering the area. Again, the Airman's Information Manual will list and describe the prohibited areas.

*Airport Information*

Every airport on the chart will be accompanied by a description of the airport, including the type of surface of the runways, the length of the longest runway, availability of runway lights, and the radio facilities available. We should refer to the Airman's Information Manual and/or the side panels or back of the charts to become familiar with the symbols and codes used to describe airports.

*Chart Changes*

Any change in the charts between publications will be listed in the Airman's Information Manual. Prior to flight we should refer to this list to make sure that a new obstruction has not been erected, nor any other significant change taken place.

FEDERAL AVIATION AGENCY
VFR EXAM-O-GRAM* NO. 23
INTERPRETING SECTIONAL CHARTS (Series 1)

**For a flight from Little Rock, Arkansas, to Abilene, Texas, you select the current Little Rock and Dallas Sectional Charts. As you study these charts, you will notice a variety of airport symbols. A closer examination will reveal that there are differences in the symbols used between the two charts and a different method of representing some of the information. Depicted below are airport symbols selected from each chart. Each of these symbols, as well as the airport information, appears on the charts in a magenta (purplish red) color, indicating no control tower or FSS airport advisory service available.**

**Dallas Chart**

**Little Rock Chart**

**An analysis of Private Pilot Written Examinations indicates that applicants are having difficulty understanding and interpreting the aeronautical symbols and legends on the sectional charts. Referring to the airport symbols above and a sectional chart legend if available, how many of the following questions can you answer? It is suggested that you answer each question to the best of your ability, then check on page 2 for the answers and explanations.**

1. **Which airports have emergency facilities only or no facilities at all?**
2. **Which airports have facilities?**
3. **Which airports have hard-surfaced runways?**
4. **Which airports have no hard-surfaced runways?**
5. **Which airports have more than one hard-surfaced runway 1, 500 feet or longer?**
6. **Which airport is located at the highest elevation?**
7. **What is the elevation of this airport?**
8. **Which airport has the shortest landing area available?**

9. What is the length of this landing area?
10. Which airports have UNICOM available?
11. Which airports have a rotating light?
12. Which airports have lighting facilities?
13. Which airports have lighting available only upon prior request by the pilot?
14. What is the length of the longest hard-surfaced runway at the Gainesville Airport and which one is it?
15. Which airport has no hard-surfaced runways but has a rotating light?
16. Which airports would be easiest to find at night?
17. Which airports have hard-surfaced runways but have no facilities or, at best, have emergency facilities?

NOTE: The symbols and explanations used in this Exam-O-Gram reflect the current practices on most Sectional Charts. Remember, chart legends are always subject to change. Use the legend (front and back) for each particular chart.

| Q | Answer | Explanation | |
|---|---|---|---|
| 1 | Whaley, Caddo Mills, Weiner, Petit Jean | Those airport symbols without the 4 projections indicate airports with emergency or no facilities |  |
| 2 | Decatur, Cleburne Mun., Gainesville, Carlisle, Searcy | Those airport symbols with the 4 projections indicate airports with facilities |  |
| 3 | Caddo Mills, Cleburne Mun., Gainesville, Searcy, Petit Jean | Hard-surfaced runways are outlined in the air port symbol on some charts such as Dallas. On other charts such as Little Rock, an "H" in the airport information block indicates hard-surfaced runways. | 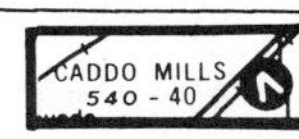  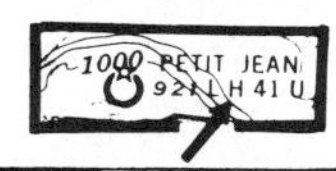  |
| 4 | Whaley, Decatur, Weiner, Carlisle | The absence of an "H" or runway outlines indicates the absence of hard-surfaced runways. |     |
| 5 | Caddo Mills, Gainesville, and possibly Petit Jean and Searcy | All hard-surfaced runways 1500' or longer are outlined on the Dallas-type charts. You are unable to determine the number of hard-surfaced runways on Little Rock-type charts. | |
| 6 | Decatur | The series of numbers on the left of the airport information block gives the elevation of the airport in feet. |   |
| 7 | 1059 feet | | |
| 8 | Whaley | The series of numbers on the right side in the information block gives the length of the longest runway or landing strip in hundreds of feet. |   |
| 9 | 1800 feet | | |
| 10 | Cleburne Mun., Petit Jean, Carlisle, Searcy | UNICOM availability on 122. 8 mc at airports without control towers is indicated by a "U" at the far right end of the information block. | 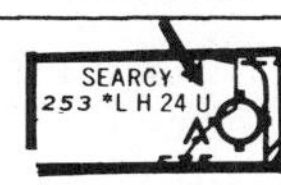  |
| 11 | Decatur, Petit Jean, Cleburne Mun. | A star at the top of the airport symbol indicates a rotating light. |   |
| 12 | Cleburne Mun., Gainesville, Petit Jean, Searcy | An "L" in the airport information line following the elevation indicates lighting facilities. |   |
| 13 | Gainesville Searcy | Airport lighting only on prior request is indicated by an asterisk (*) preceding the "L". |   |

| | | | |
|---|---|---|---|
| 14 | 4700 feet | For those airports having hard-surfaced runways the length given in the airport information line is that of the longest hard-surfaced runway. However, there is nothing to tell you which one is the longest unless you can determine this from the relative lengths of the runway outlines. |  |
| 15 | Decatur | See answers and explanations for Q 4 and Q 11. |  |
| 16 | Decatur, Petit Jean, Cleburne Mun. | The rotating light would point them out; however, the airport at Decatur has no lighting facilities. | |
| 17 | Caddo Mills, Petit Jean | See answers and explanations to Q 1 and Q 3. |  |

Other areas of difficulty in reading and interpreting sectional charts are obstruction and terrain elevations. Using the two excerpts, see how many of the following questions you can answer. Answers and explanations are given below.

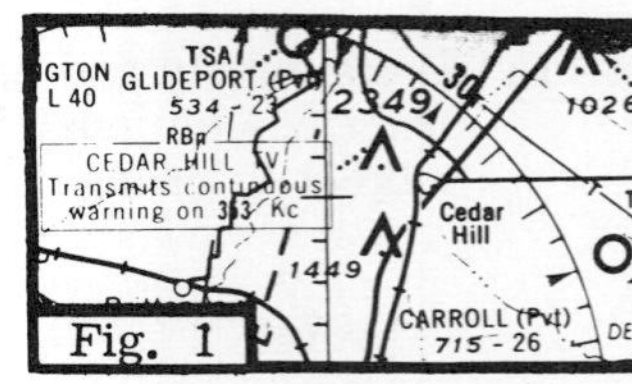

Fig. 1

- - - - - - - - - - - - - - -

1. What is the elevation of the highest obstruction in Fig. 1?
2. How can you determine the approximate height of an obstruction above the ground (Fig. 1)?
3. What is the elevation of the highest terrain (Fig. 2)?
4. At what elevation intervals are contour lines shown on sectional charts (Fig. 2)?
5. What is the elevation of the highest contour line (Fig. 2)?

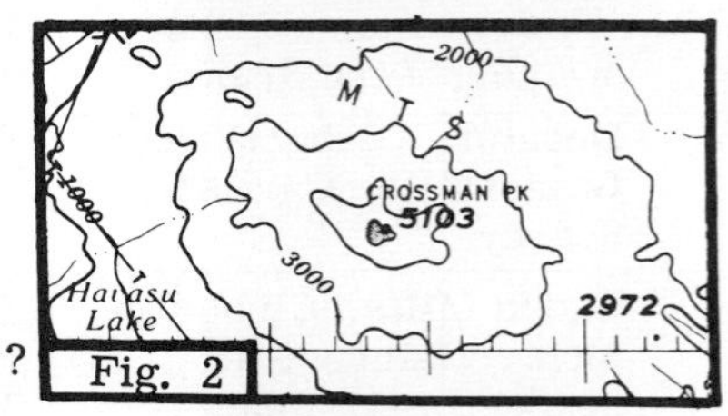

Fig. 2

6. What is the significance of contour lines being close together? Far apart?
7. What elevation ranges are indicated by the various color shadings on sectional charts?

---

## ANSWERS AND EXPLANATIONS TO QUESTIONS 1 THROUGH 7

1. The top of the highest obstruction in Fig. 1 is 2,349 feet above sea level. The number shown by an obstruction symbol is the elevation of the obstruction top above sea level.
2. If color shading or contour lines indicate surrounding terrain is fairly level, you could compare the height of the obstruction with the elevation of airports in the immediate area. A comparison of the obstruction height with the elevation of the two airports in Fig. 1 indicates that it probably extends 1,600 to 1,800 feet above the ground -- a formidable obstruction when letting down or climbing out in that area.
3. The elevation of the highest terrain in Fig. 2 is 5,103 feet (Crossman Peak). This is shown by the black dot with the elevation above sea level written beside it. 
4. Contour lines are shown at 1,000-foot intervals. You may have to follow a contour line for some distance before you can find that elevation which it represents. Sometimes there will be no elevation indication on the contour line; however, you can generally determine its elevation through nearby contours.
5. The elevation of the highest contour line in Fig. 2 is 5,000 feet. The small enclosed area on the edge of which Crossman Peak appears is the 5,000-foot contour line. This is determined by noting the 3,000-foot contour line which is labeled, then the next one which is the unlabeled 4,000-foot line, which would make the small one the 5,000-foot contour line.
6. The closer together the contour lines, the steeper the slope of the terrain; the farther apart they are, the shallower the slope.

7. A different color shading is used for each 1,000-foot interval up to 3,000 feet; for each 2,000-foot interval from 3,000 to 9,000 feet; and for each 3,000-foot interval above 9,000 feet. The elevation color shading chart is found along the lower right edge corner of a sectional chart. At the top of this color shading chart, the highest terrain and obstruction elevations are shown for the particular sectional chart.

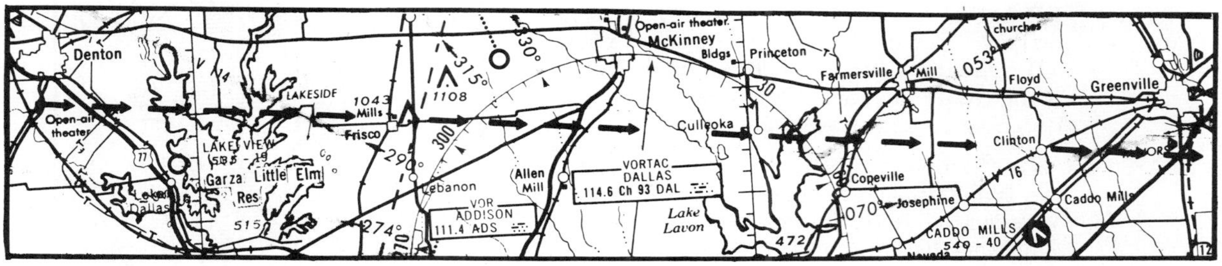


Which group represents the best set of checkpoints along the route indicated by the arrows?

| I | II | III |
|---|---|---|
| Denton | Garza-Little Elm Res | Denton |
| Garza-Little Elm Res | Frisco | Frisco |
| McKinney | Culleoka | McKinney |
| Lake Lavon | Lake Lavon | Culleoka |

Group I represents the best set of checkpoints since it includes only prominent landmarks -- large lakes and large towns. Even though the large towns do not lie exactly on route, they are close enough to be easily identified. Groups II and III represent the poorer groups because of the inclusion of the town of Culleoka which is represented by the symbol ○ . This symbol often represents small towns or villages that cannot be relied on even though they do fall right on course. In sparsely populated areas, towns represented by this symbol become more important as checkpoints provided you know your approximate whereabouts. If you are lost, however, they may be very difficult to identify. The best checkpoints are represented by combinations of prominent checkpoints such as the town of Denton and Garza-Little Elm Res along with the roads and railroads. Large rivers are also excellent checkpoints when they have prominent loops or bends or in combination with other prominent checkpoints.

## FEDERAL AVIATION AGENCY
## VFR EXAM-O-GRAM* NO. 24

## INTERPRETING SECTIONAL CHARTS (Series 2)

Someone once said: "A sectional map tells you everything except how to fold it up after you have used it." Most pilots know that there is a lot of information printed on the front and back of the sectional charts, but do many of them really know what is there?

This is the second in a series of three Exam-O-Grams dealing with interpreting and understanding the aeronautical symbols and legend of the sectional charts. These Exam-O-Grams are intended to help clarify certain areas of misunderstanding for written examination applicants.

The task of selecting the proper communications frequencies is often a very real problem for the pilot who is not accustomed to using the charts and other publications that list this data. Let's take a look at the manner in which tower and Flight Service Station (FSS) radio frequencies, Control Zones. . . etc. appear on the charts.

IF A PILOT PLANS TO LAND AT HAWKINS AIRPORT (JACKSON, MISS.), WHAT TOWER FREQUENCY INFORMATION CAN HE FIND IN THE BOX TO THE RIGHT? The tower frequencies listed are:

278 kc - LF/MF tower frequency shown first is in magenta (a purplish shade of red). Your ADF receiver tuned to 278 kc could be used if the tower personnel knew that you were listening on this frequency.

118.7 mc - is the primary VHF frequency. You can transmit and receive on this frequency if your equipment permits. This and the remaining frequencies listed are printed in blue on the chart.

126.2 mc - assigned primarily as a VHF frequency for military aircraft.

257.8 mc - a UHF military frequency.

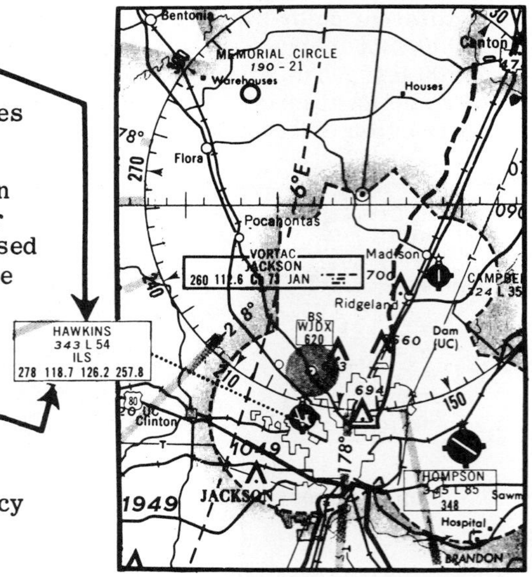


> NOTE: If your radio equipment is such that you are unable to transmit and receive on the same frequency (see 118.7 mc above), you can transmit to Hawkins tower on the Standard tower frequency 122.5 mc (explanation on page 2 of this Exam-O-Gram). In transmitting to other towers around the country you will find that all FAA towers receive on one of these four frequencies (122.4, 122.5, 122.6 or 122.7 mc), along with their other assigned frequencies. Normally only one of these four frequencies is assigned to an individual tower and the other three are not monitored by that facility. Notice how easy these frequencies are to remember.
>
> All towers and FSS transmit and receive on Emergency Frequency 121.5 mc. This is the universal VHF channel for emergency and distress communication and its use should be restricted to such calls. It is generally available at all FAA and military stations through either the tower or the FSS.

WHAT IS THE MEANING OF THE STATEMENT PRINTED IN THE LEGEND ON THE BACK OF ALL SECTIONAL CHARTS THAT READS: "122.5 MC STANDARD TOWER GUARDING FREQUENCY FOR PRIVATE AIRCRAFT"? It means that most FAA towers listen (receive you) on 122.5 mc. For that reason the airport box on the chart will not list this frequency since it is the Standard tower frequency. See example to the right:

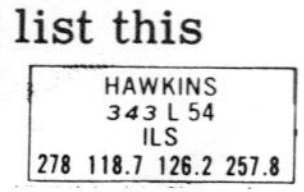


IN ADDITION TO THE PRIMARY VHF FREQUENCY 119.7 MC SHOWN IN THE WILEY POST AIRPORT BOX, ON WHAT OTHER VHF FREQUENCY CAN THE TOWER RECEIVE? On the Standard frequency 122.5 mc.

IF AN FAA TOWER DOES NOT RECEIVE ON THE STANDARD FREQUENCY 122.5 MC., HOW WILL THIS BE INDICATED IN THE AIRPORT BOX ON THE CHART? One of the other FAA tower frequencies (122.4, 122.6, 122.7 mc) - known as Nonstandard tower frequencies - will be the last frequency printed in the box and it will be followed by the letter "G." This is illustrated in the airport box to the right which indicates that the Will Rogers tower receives on the Nonstandard frequency 122.7. (Note: As new charts are published, the letter "R" will be substituted for the letter "G"; e.g., 122.7R.)

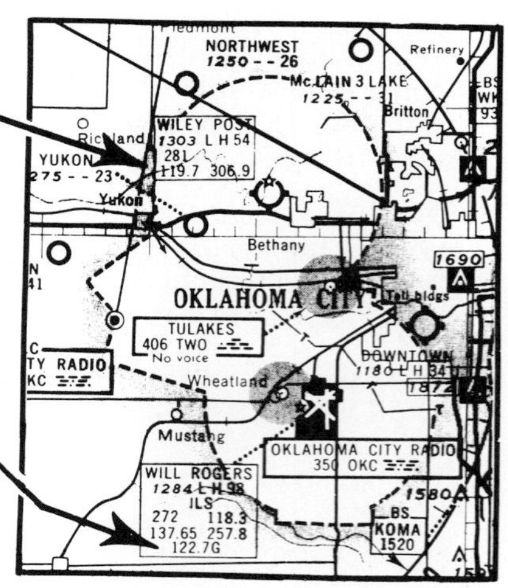

WHAT WOULD BE THE DISADVANTAGE OF HAVING BOTH WILEY POST TOWER AND WILL ROGERS TOWER RECEIVING ON THE STANDARD FREQUENCY 122. 5 MC? The two airports are located less than 10 miles apart and a great deal of confusion would result if pilots were calling each tower on the same frequency. To avoid such confusion, Will Rogers tower has been assigned one of the nonstandard frequencies (122. 7 mc).

☆ ☆ ☆ ☆ ☆

WHAT IS A CONTROL ZONE AND HOW IS IT SHOWN ON THE SECTIONAL CHART? The Control Zone is outlined by a broken line. It is an area extending upward from the surface of the earth which may include one or more airports. It is normally a circular area of five statute miles in radius with extensions where necessary to include instrument approach and departure paths.

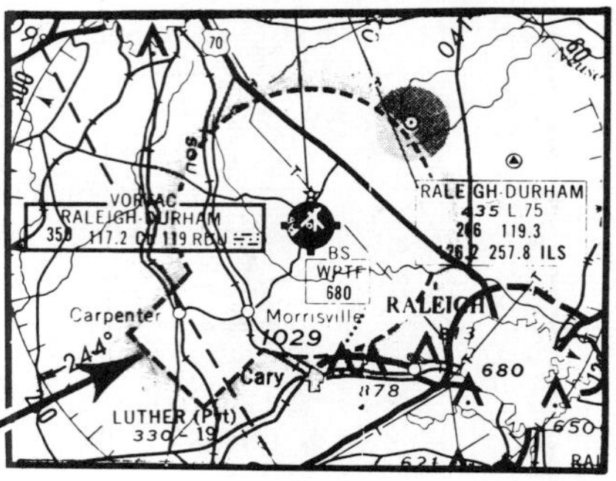


☆ ☆ ☆ ☆ ☆

A PILOT ON A VFR CROSS-COUNTRY DESIRES TO CONTACT AN FSS STATION. ON WHAT FREQUENCY CAN HE EXPECT A REPLY? The legend on the back of the chart states: "Reply from FSS to private aircraft will normally be made on the VOR (VORTAC) voice channel unless 122. 2 mc is requested. . . etc." This applies to enroute aircraft and not to those attempting to land at an airport where an FSS provides airport advisory service. (See example.) If the pilot was flying over the Waco VORTAC he would receive the FSS on the VORTAC frequency 115. 3 mc unless he requested a reply on 122. 2 mc or one of the other FSS frequencies.

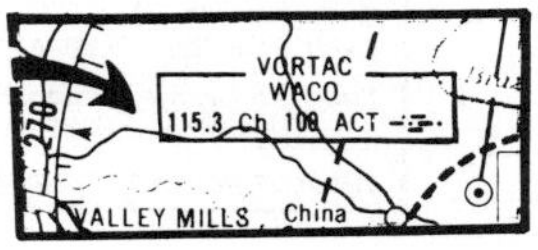


**DO ALL FLIGHT SERVICE STATIONS TRANSMIT AND RECEIVE ON THE SAME FREQUENCIES?** Yes, there are no variations in FSS (VHF) frequencies except communications on VOR (VORTAC) facility frequency. Remember these frequencies and save time in the future! ! (1) You can transmit to all FSSs on 122.1 mc, which is their standard listening frequency. (2) You can receive all FSSs on 122.2 mc since this is the FSS transmitting frequency for private aircraft. (3) In addition to the above frequencies, you can also transmit to and receive all FSSs on 126.7 and 123. 6 mc plus the Emergency Frequency 121.5 mc.

> NOTE: In communicating with an FSS or tower, the pilot should specify, in the initial call, the frequency on which he expects a reply, if other than primary.

**WHAT IS THE PRIMARY TRANSMITTING FREQUENCY OF THE FSS WHERE AIRPORT ADVISORY SERVICE (AAS) IS PROVIDED?** The legend on the front of the chart states that the transmitting frequency 122.2 mc has been designated for this service. Pilot requests for AAS on other station frequencies will be honored -- but the FSS will then simultaneously transmit on the requested frequency and 122.2 mc, in order to keep other aircraft in the vicinity aware of the subject aircraft's position.

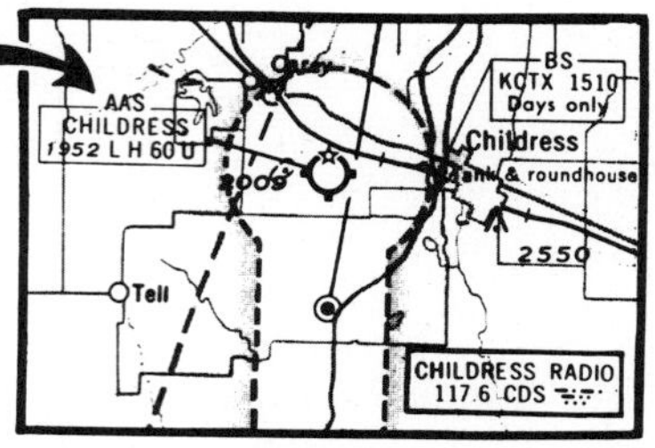


**WHY ARE SOME OF THE AIRPORT SYMBOLS BLUE AND OTHERS MAGENTA?** Sectional Charts show all airports with an operative airport traffic control tower or Airport Advisory Service (AAS), in dark blue. All other airports, large or small, that do not provide this service are shown on the charts in magenta.

★★★★★

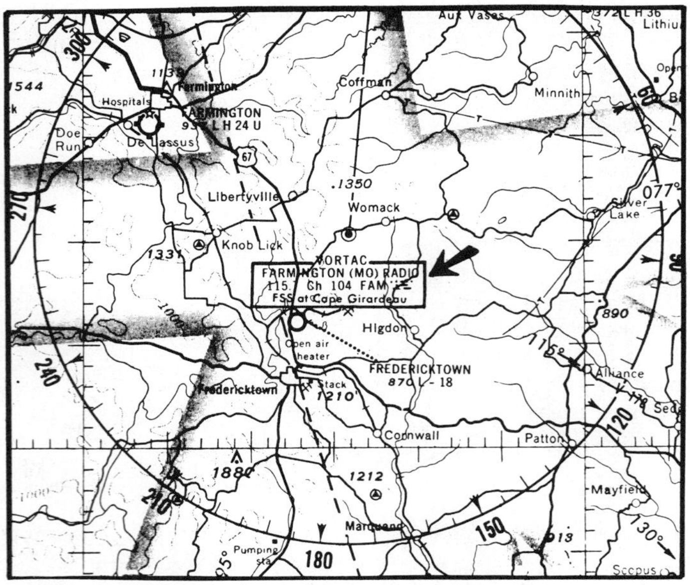


To the left appears the Farmington, Mo. VORTAC. You will notice that radio voice facilities at Farmington are remotely controlled by Cape Girardeau FSS (48 statute miles southeast of Farmington).

IF A PILOT FLYING STINSON N1122P CALLS FARMINGTON RADIO, HOW WILL HE BE ANSWERED? The Cape Girardeau FSS will reply: "STINSON ONE ONE TWO TWO P(apa), (THIS IS) FARMINGTON RADIO, GO AHEAD."

★ ★ ★

Scheduled broadcasts over Farmington Radio will begin with an announcement - THIS IS CAPE GIRARDEAU AREA RADIO.

FAA Aeronautical Center
Flight Standards Technical Division, AC-700
Operations Branch, AC-740
P. O. Box 25082
Oklahoma City, Oklahoma 73125
4/64
Exam-O-Grams available free of charge in limited quantities from this office.


## FEDERAL AVIATION AGENCY
## VFR EXAM-O-GRAM* NO. 25

### INTERPRETING SECTIONAL CHARTS (Series 3)

This is the last in a series of three Exam-O-Grams dealing with understanding and interpreting the aeronautical symbols and legend of sectional charts. See Exam-O-Grams No. 23 & 24.

### Military Climb Corridors

Pilots of Century Series aircraft (F-101, F-104, F-106, etc.) on active air defense missions are unable to see and safely avoid other aircraft during the climb phase of a scramble. Anyone who has observed the takeoff of an aircraft such as the F-106, can fully understand the pilot's limited forward visibility in a maximum power climb. In addition to maintaining steep pitch angles and airspeed control, the Century Series fighter pilot's attention is also on such tasks as check-in, identification, procedures, and spatial orientation.

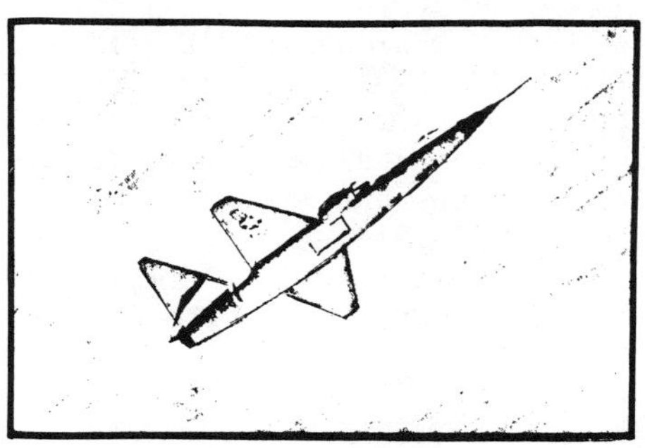

In the interest of safety, the Department of Defense and the FAA have agreed to establish restricted corridors to segregate such operations from other air traffic. All flights through these areas must obtain prior approval from the controlling agency. Positive control is enforced within the corridor airspace under the sole jurisdiction of the controlling agency which is usually the associated control tower or approach control.

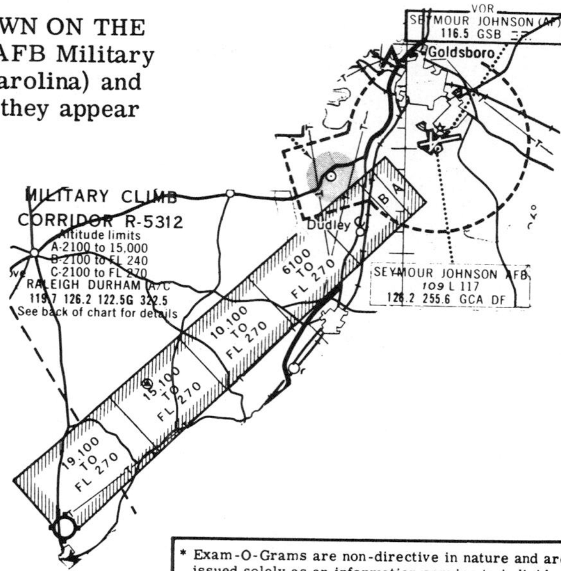


HOW ARE MILITARY CLIMB CORRIDORS SHOWN ON THE SECTIONAL CHARTS? The Seymour Johnson AFB Military Climb Corridor (located at Goldsboro, North Carolina) and shown to the right, is a typical example of how they appear on the front of the charts.

WHAT DOES THE "R-5312" MEAN? This denotes that the Seymour Johnson AFB Climb Corridor has been designated a Restricted Area (R-5312). By referring to the legend on the back of the chart you can locate and identify the data which applies to Restricted Area R-5312.

Note the location and lateral limits of the Climb Corridor in relation to terrain and aeronautical facilities. Also notice that the Seymour Johnson AFB elevation is 109 feet above sea level, but in order to stay in round figures 100 feet was used to establish the lower altitude limits of the corridor.

* Exam-O-Grams are non-directive in nature and are issued solely as an information service to individuals interested in Airman Written Examinations.

HOW LARGE IS THE CLIMB CORRIDOR R-5312? R-5312 is 3.45 statute miles wide at the starting point near Seymour Johnson AFB and it expands to a width of 4.5 statute miles at the 27 statute miles termination point.

IS IT EVER LEGAL TO FLY THROUGH R-5312? Notes on the back of the chart indicate that no person shall operate an aircraft within the Restricted Area between the designated altitudes - unless prior permission has been issued by the FAA Raleigh-Durham, N.C. Approach Control.

DO THE LOWER ALTITUDE LIMITS OF THE CLIMB CORRIDOR APPLY TO ALTITUDES ABOVE SEA LEVEL OR ALTITUDES ABOVE THE SURFACE? When climb corridors are established they are based on absolute altitudes, however, the lower altitude limits of the climb corridor printed on the sectional chart are designated MSL and represent heights above the average level of the sea (Mean Sea Level).

* * * * * *

IN THE EXAMPLE BELOW, CAN THE PILOT OF AIRCRAFT NO. 1 LEGALLY FLY VFR UNDER SECTOR "C" OF CLIMB CORRIDOR R-5312? Yes, but he must remain below 2,100 feet MSL since he will be flying under an area whose lower limit is 2,100 feet MSL.

LATERAL LIMITS OF MILITARY CLIMB CORRIDOR R-5312

NOTE: When a pilot flies under sectors A, B, or C of Restricted Area R-5312 he must remain below 2,100 feet MSL. Flying under the corridor at this point will require constant vigilance, since the pilot is confined to an area with everything else that flies - from canvas back ducks to another pilot in a "bug smasher".

CAN AIRCRAFT NO. 2 ABOVE LEGALLY FLY UNDER THE CLIMB CORRIDOR WITHOUT A CLEARANCE FROM FAA RALEIGH-DURHAM APPROACH CONTROL? Yes, provided he remains below 6,100 feet MSL.

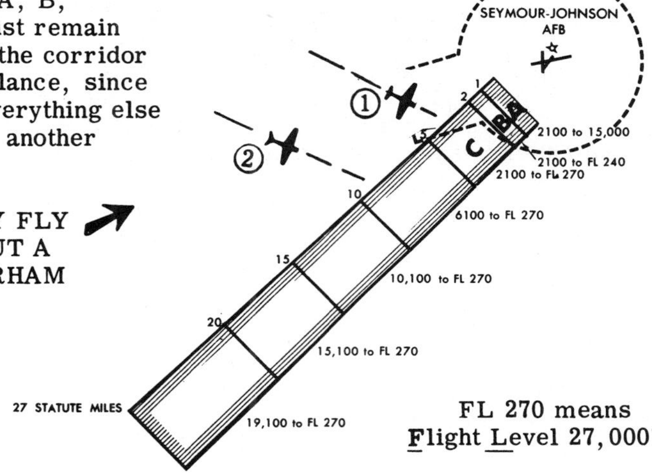


FL 270 means Flight Level 27,000'

PROFILE SHOWING UPPER AND LOWER LEVEL OF MILITARY CLIMB CORRIDOR

NOTE:
Most other type Restricted Areas start at the surface of the terrain and go upward. Notice, however, that the Seymour Johnson AFB Climb Corridor restricted airspace starts at 2,100 feet MSL and its lower limits are stair-stepped upward as distance from the control zone is increased. See illustration to the right.

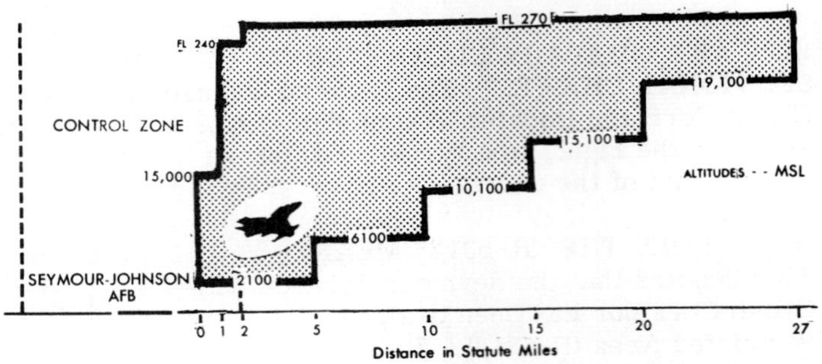


WHERE CAN A PILOT FIND INFORMATION CONCERNING THE RESTRICTED AREA CONTROLLING AGENCY? The back of the sectional chart contains a listing showing the "Appropriate Authority" that he must contact before flying through the Restricted Area. A listing of this and other data, such as altitude and time restrictions, can be found on the back of all sectional charts that display Restricted, Prohibited, Warning and Caution Areas (see examples below).

Excerpts

PROHIBITED, RESTRICTED, WARNING, AND CAUTION AREAS ON WASHINGTON SECTIONAL CHART

| NO. | NAME | ALTITUDE | TIME | APPROPRIATE AUTHORITY |
|---|---|---|---|---|
| P-56 | Washington, D. C. | Unlimited | Unlimited | |
| R-6612 | Dahlgren Complex, Va. | To 7000 | 0800-1700 Mon. through Friday | † FAA Washington ARTC Center or area FSS<br>Comdr. Naval Proving Grounds, Dahlgren, Va. |
| W-106 | Patchogue, N.Y. | Unlimited | Sunrise to sunset | C. O. NAS New York, Brooklyn 34, N.Y. |

P - Prohibited R - Restricted W - Warning C - Caution † - Controlling Agency

HOW ARE RESTRICTED AREAS OTHER THAN MILITARY CLIMB CORRIDORS SHOWN ON THE CHARTS? The Restricted Areas are depicted on the charts in many shapes and sizes and their boundaries are outlined by crosshatching as the two examples on the right illustrate.

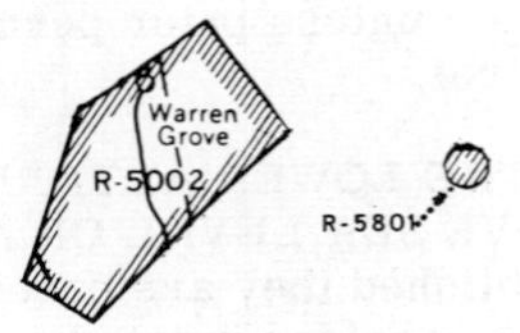


WHAT IS A CAUTION AREA AND HOW IS IT SHOWN ON THE CHARTS? It is an area in which there is a visible hazard to flight or navigation, such as aircraft maneuvers, parachute drops, etc. There is no restriction on flight, but they are to warn the pilot that activities may be in progress which will require him to be constantly on the alert. See example to the right.

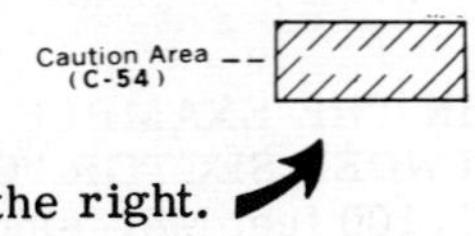


WHAT IS A WARNING AREA?... WHERE CAN THEY BE FOUND? It is a specified area over international waters within or over which there may exist activities constituting a potential danger to aircraft. They are established to permit military maneuvers and firing in certain off-shore areas as a necessary part of combat training. Warning Areas are found along the Atlantic coast, Pacific coast, and Gulf of Mexico coast. In flights along an uneven coastline pilots may attempt to take shortcuts over water and unintentionally fly in a Warning Area, since some Warning Areas lie within 3 statute miles of the coast. Be Alert and refer to the appropriate chart before attempting flight along any coastline!

WHAT IS A PROHIBITED AREA?... ARE THEY NUMEROUS?
It is airspace identified by an area on the surface of the earth within which the flight of aircraft is prohibited. No, Prohibited Areas are rare. The capitol in Washington, D. C. is a Prohibited Area. Notice how the wide spaced crosshatching covers the entire Prohibited Area in the example.

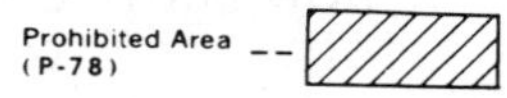


* * * * *

IS THE SAME INFORMATION FOUND ON THE BACK OF ALL SECTIONAL CHARTS?
No, there is some data common to all charts, but in general there is a great variation in the information listed for different parts of the country. For example, by comparing the New York and Albuquerque charts, we find (a) Half of the back of the New York sectional chart is devoted to airport data listing for those airports which appear on the front of the chart; (b) In the case of the Albuquerque chart, airports are fewer in number and only a small portion of the space on the back is needed for airport data listing. This space is utilized by including a Koch Chart and data pertinent to flying over mountainous terrain.

WHY ARE SECTIONAL CHARTS RECOMMENDED FOR VFR FLIGHT IN MOUNTAINOUS AREAS? When flying over unfamiliar mountainous terrain Sectional Charts are advantageous over other charts for locating, identifying by name, and establishing routes through specific mountain passes. To illustrate this point, refer to the chart excerpts to the left below, which cover identical areas.

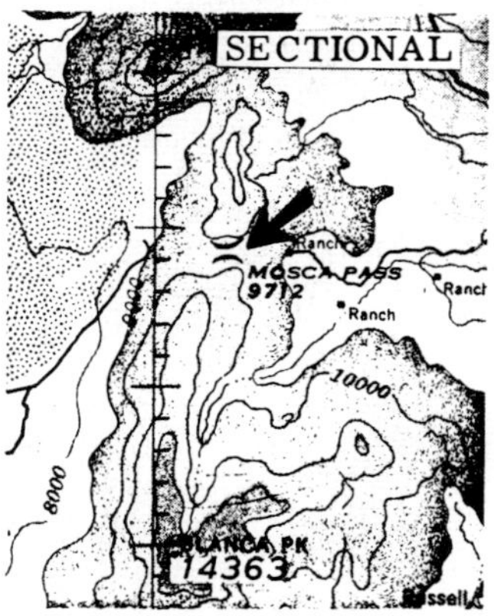


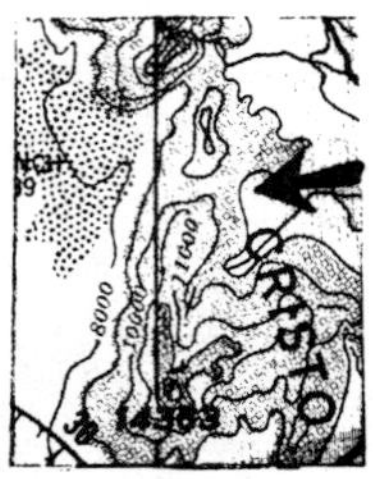

WAC

Assume that Pilot Reports indicate that VFR flight through certain passes is not possible due to snow showers or other weather, but that a pilot has just reported good VFR conditions through MOSCA PASS. Noting the excerpts to the left, MOSCA PASS can be identified by name and located only by reference to the Sectional Chart, and the elevations are more realistically pictured to determine the necessary flight altitudes.

* * * * *

Many other aeronautical symbols appear on Sectional Charts that have not been mentioned in this series of Exam-O-Grams. Some of the following topographical symbols are used on all charts while others may only be found on a few charts. Test your knowledge in the quiz below by filling in the spaces provided. Check your answers with those at bottom of the page.

## HOW MANY OF THESE SECTIONAL CHART SYMBOLS CAN YOU IDENTIFY?

(1) 

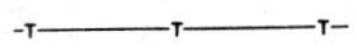

____________________

(2)  ____________________

(3) 

____________________

(4)  ____________________

(5)  ____________________

(6) 

____________________

(7)  ____________________

(8) 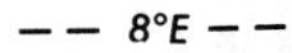

____________________

(9)  ____________________

(10)  ____________________

(11) 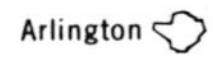

____________________

(12) 

____________________

(13) 

____________________

(14) 

____________________

## DO YOUR ANSWERS AGREE WITH THESE?

1. Prominent transmission line.
2. VFR routes through mountain passes.
3. U. S. Road Marker.
4. Oil tanks and oil fields.
5. Highway tunnel.
6. Race tracks or stadiums.
7. Swamps and marshes.
8. Isogonic line.
9. Railroad overpass.
10. Multiple track railroad.
11. Cities
12. Small cities and large towns.
13. Towns.
14. Small towns and villages.

## CHECK YOUR READING

1. What is the scale of the *local* chart?
2. What is the scale of the *sectional* chart?
3. Which of the charts is the most popular for light aircraft pilots?
4. What is the difference between the World Aeronautical Chart and the Operational Navigation Chart?
5. What is the scale of the WAC chart?
6. If you were planning a flight of 1,200 miles, which chart would you use for preliminary flight planning?
7. Where can you find the date of the latest published chart?
8. How are very-high frequency radio facilities noted on the chart?
9. List the methods used on the charts to show topographic features of the earth's surface.
10. What is the difference between the restricted area and the prohibited area?

# 37

# LATITUDE, LONGITUDE, AND DIRECTION

Let us imagine for a moment how difficult it would be to describe a meeting place in a large city, if there were no street signs. We might say something like, "We will meet in front of the building which is 20 paces east of the monument in the center of town, and 452 paces north of the water fountain." How much more simple and precise to say, "We will meet at the corner of Oak and Elm Streets."

If we multiply by 1,000 the complexity of finding a spot in one city with no reference grid of streets, we might realize the immensity of the problem of finding a spot on the earth's surface without the use of a reference grid. Fortunately, the earth is covered by a series of imaginary lines which form a standard grid, and to which everyone the world over can refer when giving instructions as to a meeting place, a destination, or a point of departure.

## Longitude and Latitude

### *Longitude*

The earth is divided into 360 equal parts, similar to the sections of an orange, as shown in Fig. 100.

The lines which form each section are called *longitude lines*, and converge at the True North and True South poles. The "prime" longitude line, or *prime meridian* lies at Greenwich, England, and all longitude is measured east or west from the prime meridian until, on the opposite side of the world, the 180 degree longitude line is reached.

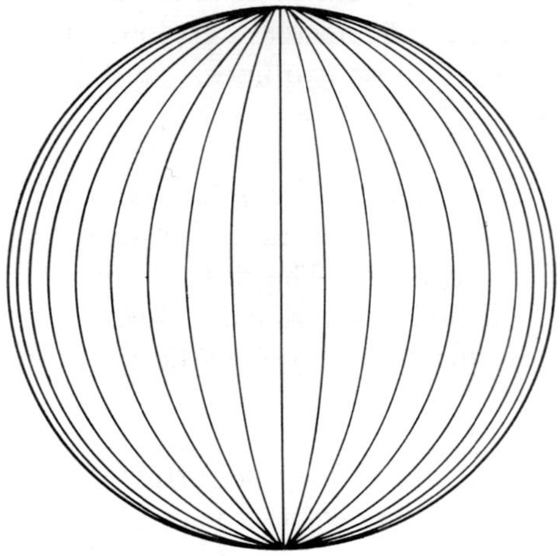

Fig. 100. Longitude lines (each line represents 10° of longitude).

Since the earth rotates through a full 360 degrees of longitude in 24 hours, each 15 degrees of longitude represents one hour of time (360 degrees divided by 24 hours equals 15 degrees of longitude).

*Latitude*

If we add to our earth another series of lines, the grid will be formed which we can use to pinpoint our location. This pattern of lines divides the earth into 180 equal parts, and these segments will look like slices of an orange, as shown in Fig. 101.

The lines which form these "slice-like" segments are called *latitude lines* or parallels. The latitude lines do not converge, but are parallel. The equator is the "prime" latitude line. Latitude is measured north or south from the equator until the pole is reached at the 90 degree point.

*Coordinates*

We need only to state the latitude and longitude of any spot on the earth to describe the location of that spot.

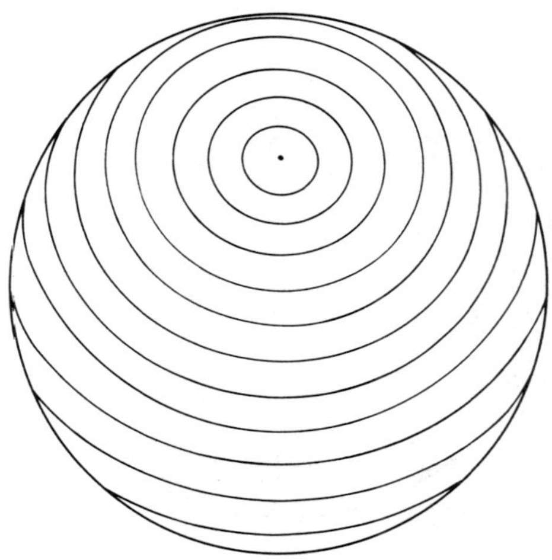

Fig. 101. Latitude lines (each line represents 10° of latitude).

Obviously, if we are at 45 degrees north latitude, and at the same time we are at 82 degrees west longitude, our only possible location is at the intersection of those two lines.

*Minutes of Longitude or Latitude*

Each degree of longitude and latitude is divided into 60 equal parts, called *minutes*. The distance between each minute of latitude is 6,080 feet, or one nautical mile. The distance between each minute of longitude is one nautical mile *only at the equator*. (Because the longitude lines converge at each pole, the distance between them diminishes as we move away from the equator.)

*Seconds of Longitude or Latitude*

Each minute may be further divided into 60 equal parts, called *seconds*. Normally, only degrees and minutes are used in aerial navigation.

*Coordinates*

To find a position on the earth's surface, we first find the coordinates (longitude and latitude) of that position in the appropriate publication. (All airport coordinates for instance, are published in the Airman's Information Manual.) Then we find either of the coordinates on the chart, and follow it until it crosses the other coordinate line.

For example, suppose we wish to find airport A which is located at 40 degrees, 30 minutes north latitude, and 95 degrees, 10 minutes west longitude. We simply locate the 40 degree, 30 minute north latitude line, and follow it until it crosses the 95 degree, 10 minute west longitude line, as shown in Fig. 102.

## Direction

If we stand in the center of a room and turn about until our eyes have scanned the entire room, we will have rotated 360 degrees, or a full circle. Imagine that we began when facing a picture on the wall which we will use as a reference point. When we have turned one quarter of the way around the circle we will have rotated 90 degrees. When our back is to the reference point we will have

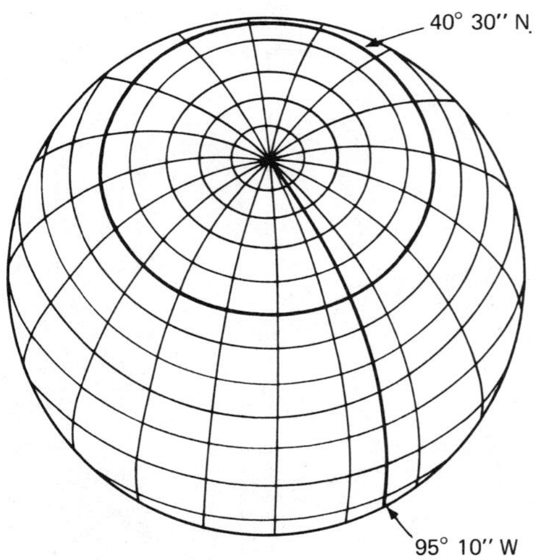


Fig. 102. Coordinates. Using latitude and longitude lines to pinpoint position (each line represents 10° of latitude or longitude).

turned 180 degrees, or one-half of the circle, and so forth until we again face the reference point.

Now imagine that a friend is in an adjoining room, and we wish to ask him to turn to a specific direction as we turn. We must be sure of two things. First, his beginning point of reference must be in an identical direction to ours. Secondly, he must rotate in the same direction as we rotate. We can ask our friend to turn 90 degrees to his right, as we turn 90 degrees to our right, and be sure that he will be facing the same direction as are we.

That is what *direction* is all about. It is simply a standardized method of describing the angle between the reference point and the line along which we point.

*The True North Pole*

For the purpose of aerial navigation in the Northern Hemisphere, we use the True North Pole as our standard reference point. (In the Southern Hemisphere the True South Pole is the standard reference point. We will confine our discussion to the Northern Hemisphere.)

The True North Pole is the point on the earth's surface where the converging longitude lines meet and is not related in any way to the Magnetic Pole. All of our charts are drawn with longitude and latitude lines as a standard grid. Therefore, True North is a standard point on all charts.

Direction is always measured *clockwise* from a line which connects our position with True North, as shown in Fig. 103.

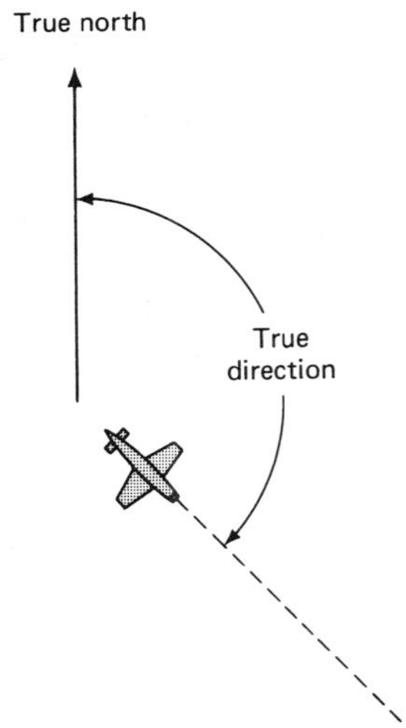


Fig. 103. Direction. Direction is measured *clockwise* from True North.

### CHECK YOUR READING

1. What is the purpose of building a grid of longitude and latitude?
2. What is the important difference between longitude and latitude?
3. What is the distance in nautical miles between each minute of latitude?
4. One hour of time is represented by how many degrees of longitude?

5. What is a *minute* of longitude?
6. Where is the *prime meridian* located?
7. What are the coordinates of Chicago, Illinois?
8. Why is the True North Pole used as a reference for all map makers?
9. What are the numerical equivalents for the four cardinal points on the compass? (North, East, West, and South)
10. Why is direction always measured clockwise from the reference point of True North?

# 38

# NAVIGATION COURSES

As we learned in the previous chapter, True North is the standard reference point from which direction is measured clockwise. There are, however, other reference points which are used in measuring direction.

## Course

A *course* is a straight line drawn on a navigation chart from the point of departure to the point of destination.

### *True Course*

The *true course* is the *angle* between True North and the course line. (Now we have assigned a *direction* to the course.)

Figure 104 shows a course line from A to B.

### *Plotting the True Course*

Before beginning our flight planning, we must determine the direction of the flight. We draw a course line from the point of departure to the point of destination using a straight edge. With a protractor or navigation plotter we measure the angle between True North and the course, as shown in Fig. 105. The longitude lines point to True North, so one of them will form one side of the angle and the course will form the other side of the angle. Since the longitude lines converge at the poles, we must use the longitude line nearest the midpoint of the course for the most accurate measurement.

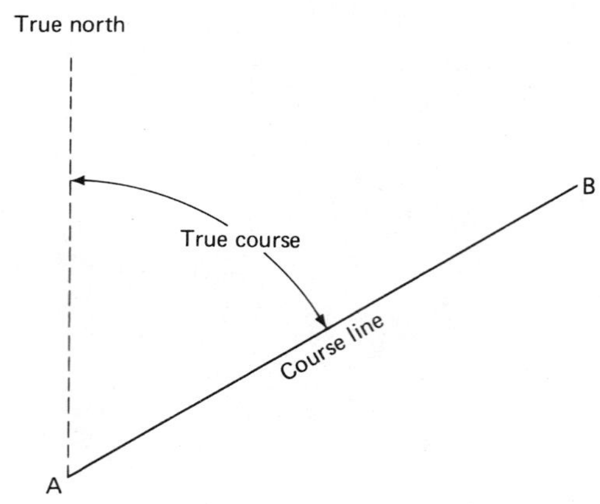


Fig. 104. True course: the angle between True North and the course line.

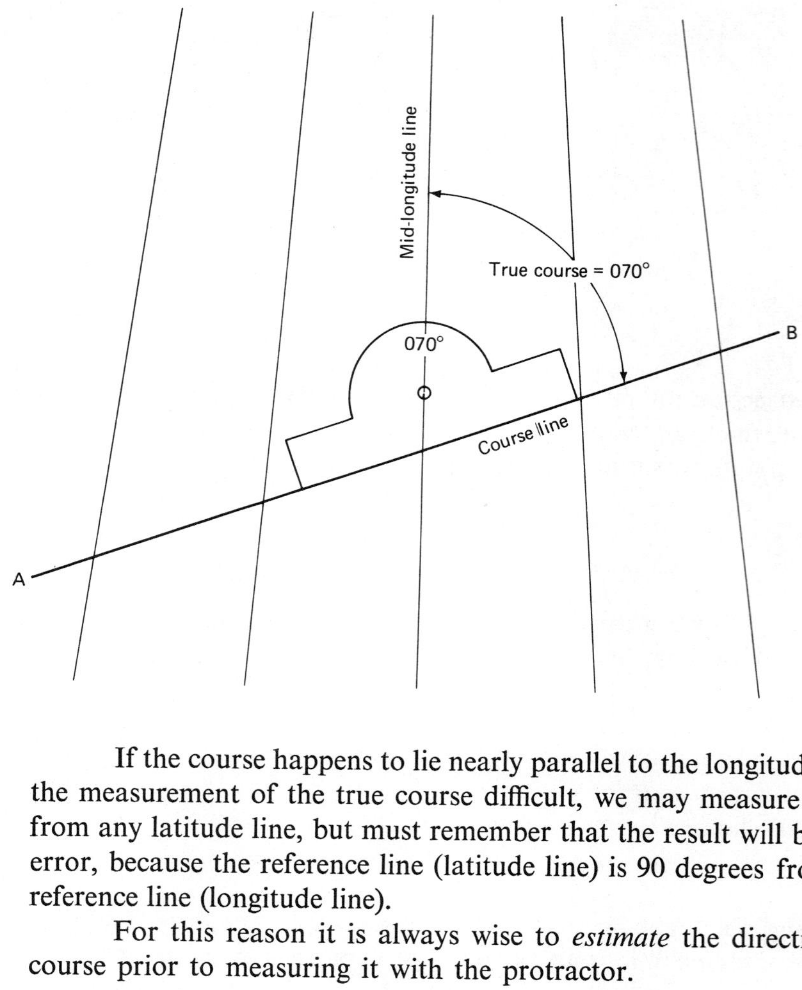


Fig. 105. Measure the true course from the mid-longitude line.

If the course happens to lie nearly parallel to the longitude lines, making the measurement of the true course difficult, we may measure the true course from any latitude line, but must remember that the result will be 90 degrees in error, because the reference line (latitude line) is 90 degrees from the normal reference line (longitude line).

For this reason it is always wise to *estimate* the direction of the true course prior to measuring it with the protractor.

### *The Magnetic North Pole*

The Magnetic North Pole is the point on the earth from which the magnetic field emanates and to which a perfect *compass* points. Natural phenomena cause drifting of the Magnetic North Pole, so it cannot be used as a reference point when drawing maps and charts.

*Variation*

Variation is the angle between a line drawn from our position to the True North Pole and a line from our position to the Magnetic North Pole, as shown in Figure 106.

This angle changes as our vantage point changes. If, from our point of view, Magnetic North is east of True North, variation is called *easterly*. If Magnetic North is west of True North, variation is called *westerly*.

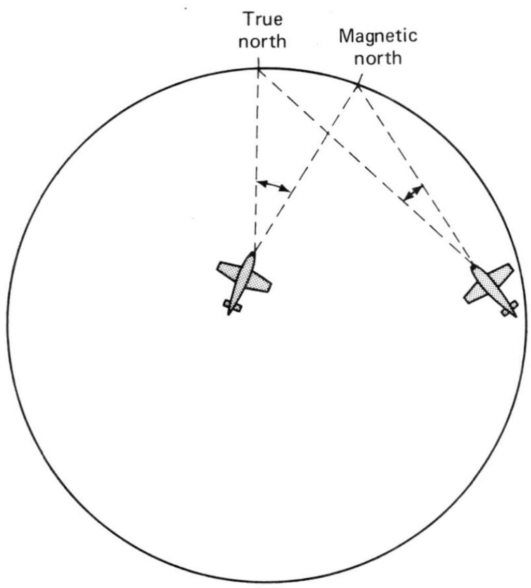


Fig. 106. Variation: the angle between True North and Magnetic North.

*Isogonic Lines*

Variation is depicted on navigation charts by a line which connects points of equal variation, called the *isogonic line*. The isogonic lines are irregular or "wavy" due to local ore deposits which give the effect of changing the position of the Magnetic North Pole. The lines on the chart which connect points of zero variation are called *agonic lines*.

Figure 107 shows a map of the United States with isogonic and agonic lines.

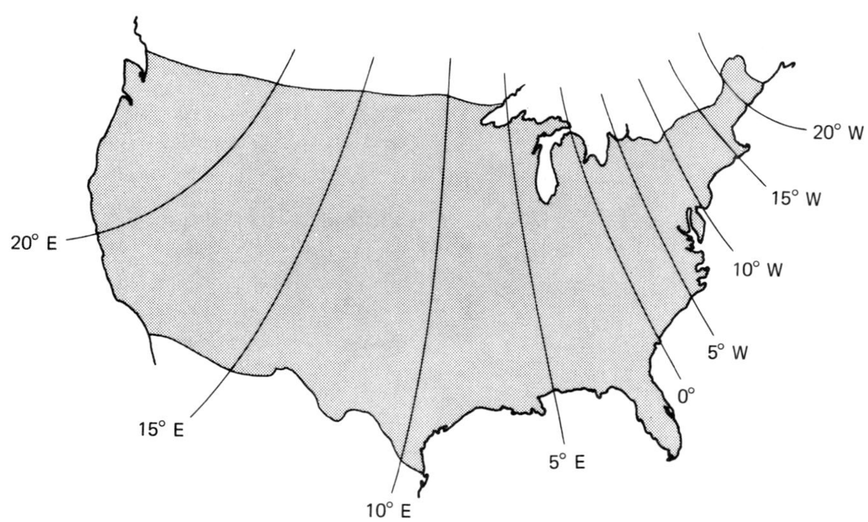


Fig. 107. Variation in the United States.

*The Magnetic Course*

If we refer to Fig. 108, which shows our true course A-B of 075 degrees and westerly variation, we will see that with westerly variation the angle between Magnetic North and the course is *greater than* the true course. Therefore, the westerly variation must have been *added* to the true course, resulting in a *magnetic course*. (Easterly variation would have to be subtracted from the true course.)

The magnetic course is, therefore, the angle between Magnetic North and the course.

Now we have assigned a "magnetic direction" to the course line.

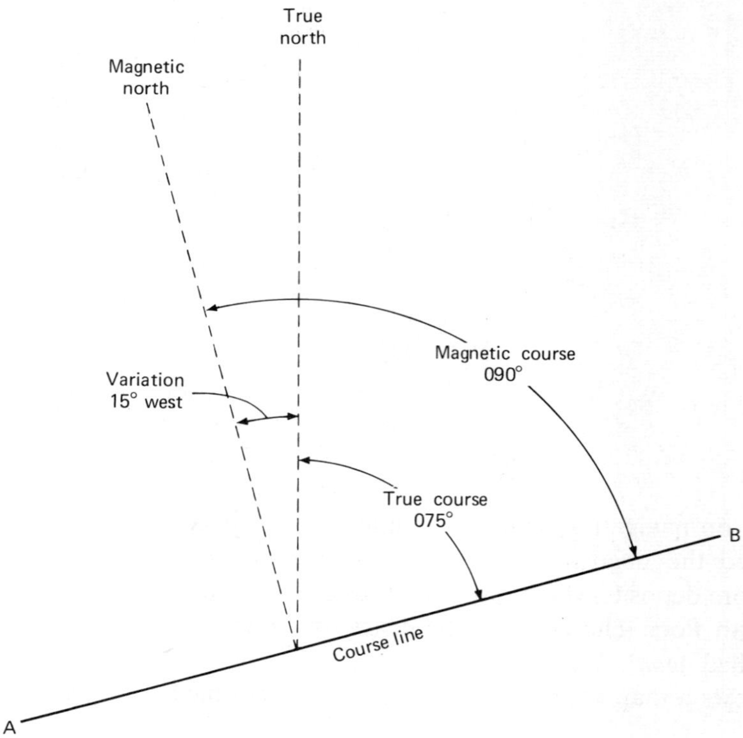


Fig. 108. The Magnetic Course.

*Deviation*

Unfortunately, the magnetic compass is affected by the aircraft electrical system, radios, and metal components. Hence, it does not always point to Magnetic North, but to what is called *Compass North.* This error is called *deviation.*

Every compass is accompanied by a *compass card* or *deviation card* which specifies the deviation on many points of the compass. A sample compass card is shown in Fig. 109.

Compass correction card

| For | 000 | 030 | 060 | 090 | 120 | 150 | 180 | 210 | 240 | 270 | 300 | 330 |
|---|---|---|---|---|---|---|---|---|---|---|---|---|
| Steer | 002 | 031 | 060 | 089 | 117 | 149 | 178 | 210 | 242 | 273 | 304 | 332 |

Fig. 109. The compass card.

After computing the magnetic course, we must refer to the compass card to find the deviation for the course and apply the correction to the magnetic course.

## *The Compass Course*

The application of deviation to the magnetic course results in the *Compass Course*. Compass Course then, is the angle between Compass North and the course. Figure 110 shows this angle.

Now we have assigned a direction to the course which is applicable to one specific compass.

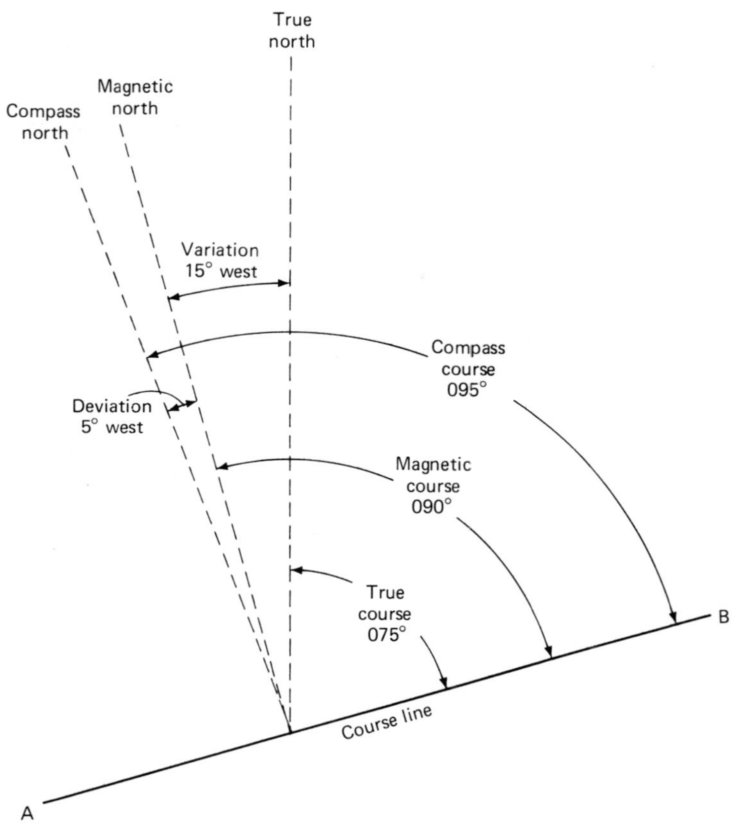


Fig. 110. The compass course.

## Review

The *course* is a straight line on a map.

The *true course* is the angle between True North and the course.

*Variation* is the angle between True North and Magnetic North.

The *magnetic course* is the angle between Magnetic North and the course.

*Deviation* is the angle between Compass North and the course (caused by aircraft components).

*Compass course* is the angle between Compass North and the course.

### Computation

Step 1. Draw the course line.
2. Find the true course. (Measure the angle between the course and the mid-longitude line.)
3. Apply variation (add westerly; subtract easterly).
4. Magnetic course (the result of step 3).
5. Apply deviation (from the compass card).
6. Compass course (the result of step 5).

## CHECK YOUR READING

1. What is the difference between a *course* and a *true course*?
2. Which longitude line must be used when measuring the true course?
3. If a latitude line is used to measure the true course, what correction must be applied to the angle?
4. Why isn't the Magnetic North Pole used as the standard reference on maps?
5. What is the True North Pole?
6. What is *variation*?
7. Where can we find variation for flight planning?
8. What is an *isogonic line*?
9. If variation is applied to the true course, what is the result called?
10. What is *deviation*? Where can we find it for each specific compass?
11. When deviation is applied to the magnetic course, what is the result called?
12. Using the compass card in this chapter solve for the compass course. True course = 330 degrees, Variation = 11 degrees East, Compass course = ?

# 39

# NAVIGATION HEADINGS

Thus far our discussion has been concerned only with *courses* from one point to another. However, our aircraft will maintain the desired course only if we fly a particular heading. Therefore, we must know the difference between a course and a heading, and how to compute the aircraft heading which will result in maintaining a course.

**Heading**

The *heading* is the direction in which the nose of the aircraft is pointing.

*True Heading*

The *true heading* is the angle between True North and the heading. (Now we have assigned a *direction* to the heading.) The true heading is always measured clockwise from True North.

*Computing the True Heading*

The difference between a course and a heading is *drift* which is caused by the wind. If there were no wind, we could simply point the aircraft along the course line and maintain that course. Unfortunately, the wind does blow and will cause the aircraft to drift to one side of the course or the other, unless the wind is perfectly aligned with the course. As shown in Fig. 111, we must actually point or head the aircraft upwind, and allow the aircraft to be drifted back to the course.

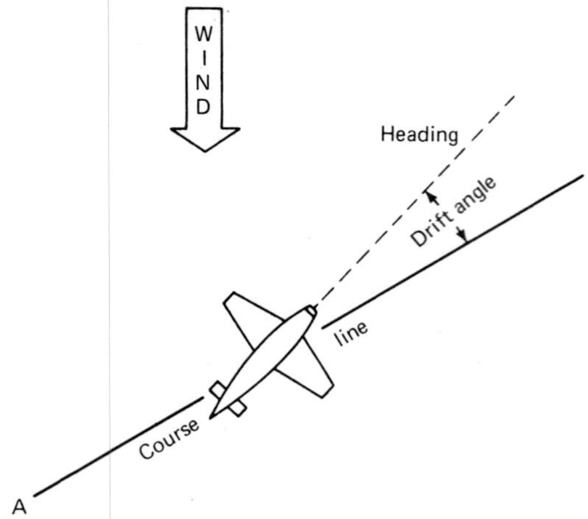


Fig. 111. Course ± drift = heading.

Since the wind aloft is always given in reference to *True* North, we must apply the wind to the *true* course in order to find the *true* heading. In other words, the true course will change to true heading by simply introducing a correction for the wind effect (drift). We will devote an entire chapter to the specific computations required in finding the true heading (see Chapter 40).

### *The Magnetic Heading*

After we compute our true heading, we must apply variation to arrive at a heading which is in reference to the Magnetic North Pole. True heading with variation applied is called magnetic heading, which is the angle between Magnetic North and the heading. Now we have assigned a value to the heading which can be found on a magnetic compass.

### *The Compass Heading*

We learned that no aircraft compass is perfect. It is affected by aircraft components and installation errors. Therefore, we must apply deviation to the magnetic heading to arrive at a compass heading. Now we have assigned a value to the original heading to which *one specific compass* can refer.

### Review

The *heading* is the direction in which the nose of the aircraft points.

The *true heading* is the angle between True North and the heading.

Wind effect (*drift*) is applied to the true course (rather than the magnetic course or compass course) because the wind direction is given in reference to True North.

The difference between a course and a heading is the wind effect (drift).

Variation is applied to the true heading, resulting in a magnetic heading.

The magnetic heading is the angle between Magnetic North and the heading.

Deviation is applied to the magnetic heading in order to adjust for the specific errors in an individual compass.

Compass heading is the angle between compass North and the heading. It is the result of applying deviation to the magnetic heading.

The end result of the computations above is a compass heading to which we must refer in order to maintain a desired course, with a specific wind, in a specific geographic location.

**Computation**

1. Draw the course line on the sectional chart.
2. Find the true course. (Measure the angle between the course line and the mid-longitude line.)
3. Apply wind drift correction (see Chapter 40). The result is true heading.
4. Apply variation (add westerly, subtract easterly) to find the magnetic heading.
5. Apply deviation (from the compass card) to find the compass heading.

## CHECK YOUR READING

1. What is the definition of *heading*?
2. What is the difference between the heading and the course?
3. What is the definition of *true heading*?
4. Why is the wind applied to the true course instead of the magnetic course?
5. What is *drift*?
6. What is the definition of *magnetic heading*?
7. Where do we find *deviation*?
8. What is the definition of *compass heading*?
9. List the required steps to compute compass heading, beginning with the course line.
10. List the required computations to find the true heading, if compass heading, variation, and deviation are given.

# 40

# THE WIND VECTOR DIAGRAM

We have learned that the natural wind has absolutely no effect on the speed of the aircraft or the performance of the aircraft *in relation to the air*, but the natural wind *does* have an important effect on the speed and direction of flight *in relation to the ground.*

Since the time required for a flight between two points depends on our speed *over the ground*, and the amount of drift affects our course *over the ground*, we must carefully compute the full effect of the natural wind prior to beginning a cross-country flight.

We can accomplish these computations very easily by drawing an accurate picture of the proposed flight conditions in any convenient scale. This picture is called a *wind vector diagram.*

## A Vector

A *vector* is a line which depicts two values. The specific direction in which the line points represents one value. The length of the vector depicts another value. We may assign the direction of flight through the air and the distance traveled through the air to one vector line. Another vector line will represent the direction of movement over the ground and the distance traveled over the ground. The third vector line will show the direction from which the wind is blowing and the distance the wind will travel.

All of the vector lines in one vector diagram must represent the identical time period. That is, if we draw a wind vector line representing one hour of wind, we must also draw vector lines representing one hour of flight through the air and over the ground.

Figure 112 shows a vector line.

The direction of the vector line is south. The length of the vector line is one inch. Suppose the scale of this drawing is one inch equals 50 nautical miles. This vector line may represent an automobile driving south for 50 nautical miles, an airplane flying south for 50 nautical miles, or a wind blowing from the north (to the south) at 50 knots (50 nautical miles per hour).

Since we commonly refer to the speed of the wind in relation to one hour of time (for example, 50 nautical miles *per hour*), we will draw our vector diagram to represent one hour of flight time. The results of one hour of flight time can then be applied to the duration of the flight.

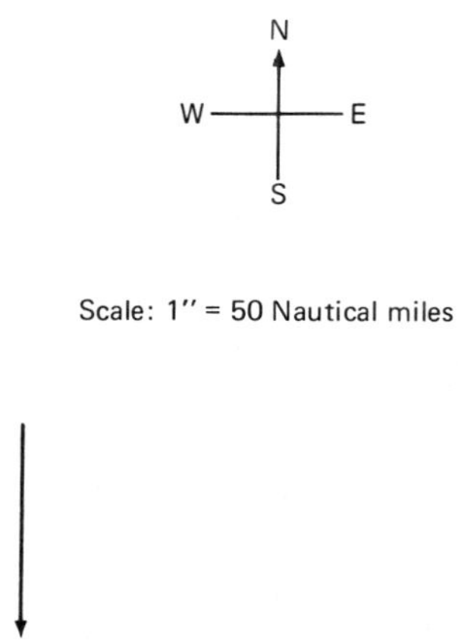


Fig. 112. A vector line.

*Vector Values*

Remember that every line in a vector diagram must depict *two* values. The direction of the vector line will represent the *direction of movement* of the wind or of the aircraft (either through the air or over the ground) and the length of the vector line will represent the speed of the wind or of the aircraft (either through the air or over the ground).

## Building the Wind Vector Diagram

We can construct a Wind Vector Diagram which will represent one hour of flight with a piece of graph paper, a protractor, a straightedge, and a pencil.

The resulting picture will be a triangle, showing the true course and groundspeed on one leg, the true heading and true airspeed on another leg, and the wind direction and wind speed on the third leg. There is a definite reason why each pair of values must go together on one vector line. Obviously, the wind speed and direction make a natural pair. The true course and groundspeed form a pair because they are both in reference to the *ground.* The true heading and true airspeed form a pair because they are both in reference to the *air*.

## Plotting the First Vector Line

Although each of the three legs of the wind vector diagram has two values, the only leg having two *known* values is the wind leg, so we shall plot it first.

Assume that the Weather Bureau has given us a wind at our planned altitude of 045 degrees at 40 knots.

This means that the wind is blowing *from* 045 degrees in relation to True North.

Figure 113 shows a wind vector line plotted from 045 degrees. The scale of our drawing is one inch equals 40 nautical miles, so the length of the wind vector line must be one inch (representing 40 knots).

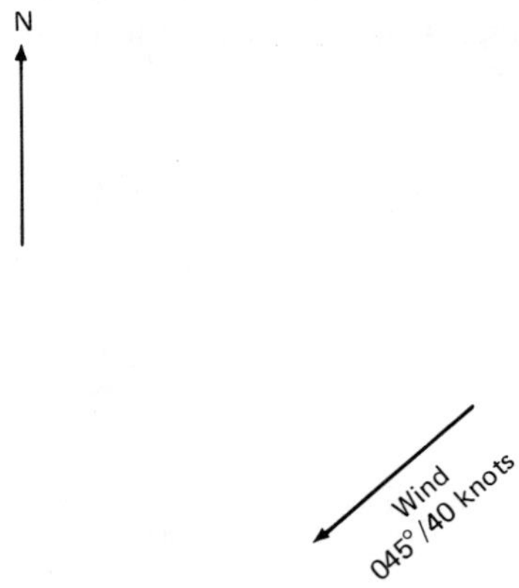


Fig. 113. The wind vector line. The wind vector line: representing a wind of 40 knots from 045°.

## Plotting the Second Vector Line

The direction of the true course line is known (we measured it on the sectional chart) so we shall plot this leg of the triangle next. Let us assume that the true course equals 090°. The true course leg of the triangle must be drawn oriented to 090 degrees, as shown in Fig. 114. The length of the true course line, representing groundspeed, is still unknown, so the length of the true course line will be indefinite.

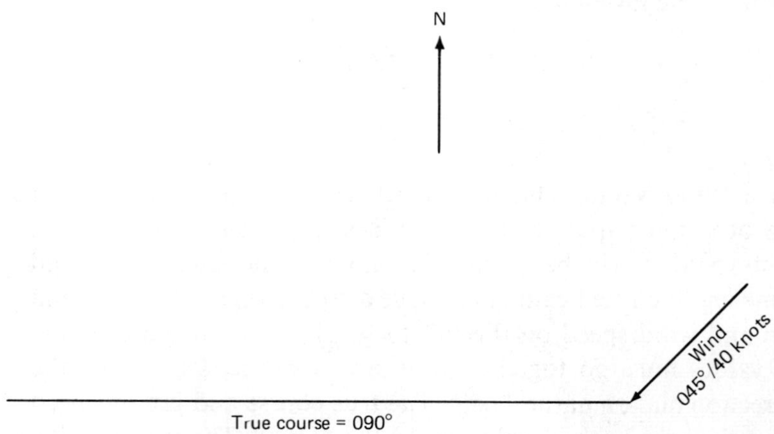


Fig. 114. The wind vector line and the true course. A wind vector line and a true course line (090°) of undetermined length.

Notice that the true course line was drawn so as to show the wind line blowing *toward* it. This is correct, as we want the wind to drift our aircraft *toward* the true course, *not* away from it.

*Plotting the Third Vector Line*

The third line on our wind vector diagram will represent the true heading and true airspeed. The *known* value of this leg is the true airspeed which in this example we will assume to be 180 knots.

We begin at the beginning of the wind line and strike an arc which is four and one-half inches in radius (representing 180 knots) until it intercepts the true course line, as shown in Fig. 115.

What we actually did in this step is back-track from the beginning of the wind line a distance equal to one hour of true airspeed, to find our position along the true course one hour ago.

After connecting the beginning of the wind line with the point of interception of the arc and the true course, our triangle will be completed, as shown in Fig. 116.

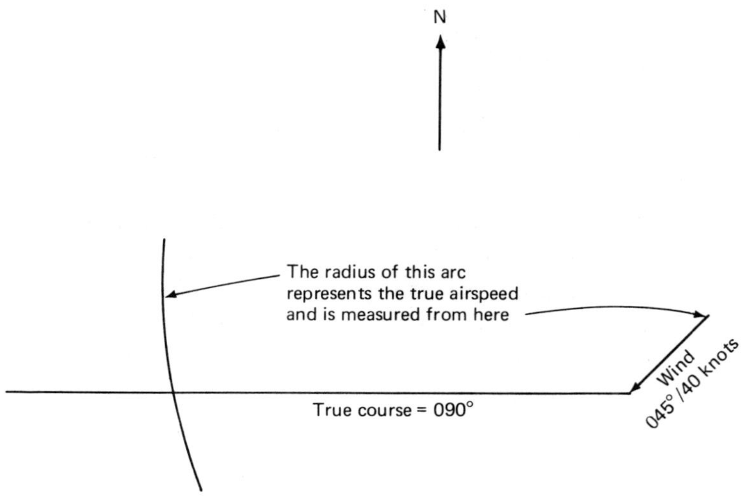


The true airspeed value is used to strike an arc which intercepts the true course line.

**Fig. 115. Plotting true airspeed.**

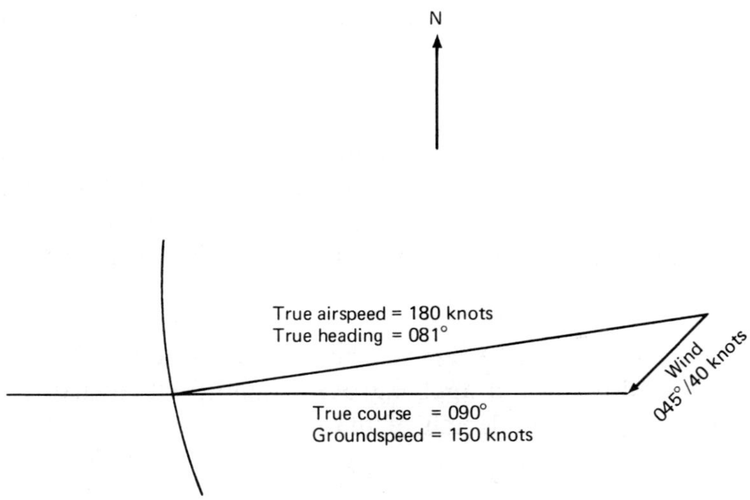


**Fig. 116. The completed wind vector triangle.**

*Finding the Unknowns*

The two unknown values of the six vector values were the groundspeed and the true heading.

*Groundspeed.* We can find the groundspeed by measuring the length of the true course vector. If we do so in Fig. 116, we will find the line to be $3\frac{3}{4}$ inches long which represents 150 knots on our scale.

*True Heading.* We can find the true heading in two ways. We can measure the angle between the true course and the true heading (the drift angle) and apply this correction to the true course, or we can measure the true heading directly with a protractor.

Using either method, we will find that the true heading is 081 degrees.

## Review

Each line of the wind vector diagram has two values:

Line No. 1 The wind vector line: wind direction and speed are represented. Both values are known.
Line No. 2 The true course—groundspeed line: true course is known; groundspeed is unknown.
Line No. 3 The true heading—true airspeed line: true heading is unknown; true airspeed is known.

The wind vector diagram represents a total of six values. If any four values are known, the other two can be found.

The true course and groundspeed values must be on the same vector line because they both refer to the *surface*. The true heading and true airspeed go on the same vector line because they both refer to the *air*.

## Application of Results to the Entire Flight

After finding the true heading and groundspeed for one hour, we must apply them to the total flight.

*True Heading Application*

If the wind remains constant, the true heading will remain the same for the entire flight. In some instances, the Weather Bureau will give us a variety of winds aloft, one for each altitude or geographic location. In that case, a new wind vector diagram will have to be constructed for each wind situation.

*Groundspeed Application*

We can find the total time enroute by measuring the length of the flight on the sectional chart and applying our groundspeed to that distance. Assume, for instance, that the length of the flight we have been planning is 250 nautical miles. At a groundspeed of 150 knots, our time enroute will be one hour and forty minutes.

## Using the Navigation Computer for the Wind Vector Diagram

Since it is awkward to carry graph paper and protractors in the cockpit, we must learn to complete the wind vector diagram on the navigation computer. One side of the computer is equipped with a moveable graph card and a transparent drawing area.

We will not need a protractor when drawing on the computer because the drawing area rotates 360 degrees, and we use a reference point or *true index* on the outer ring to indicate the desired direction, as shown in Fig. 117.

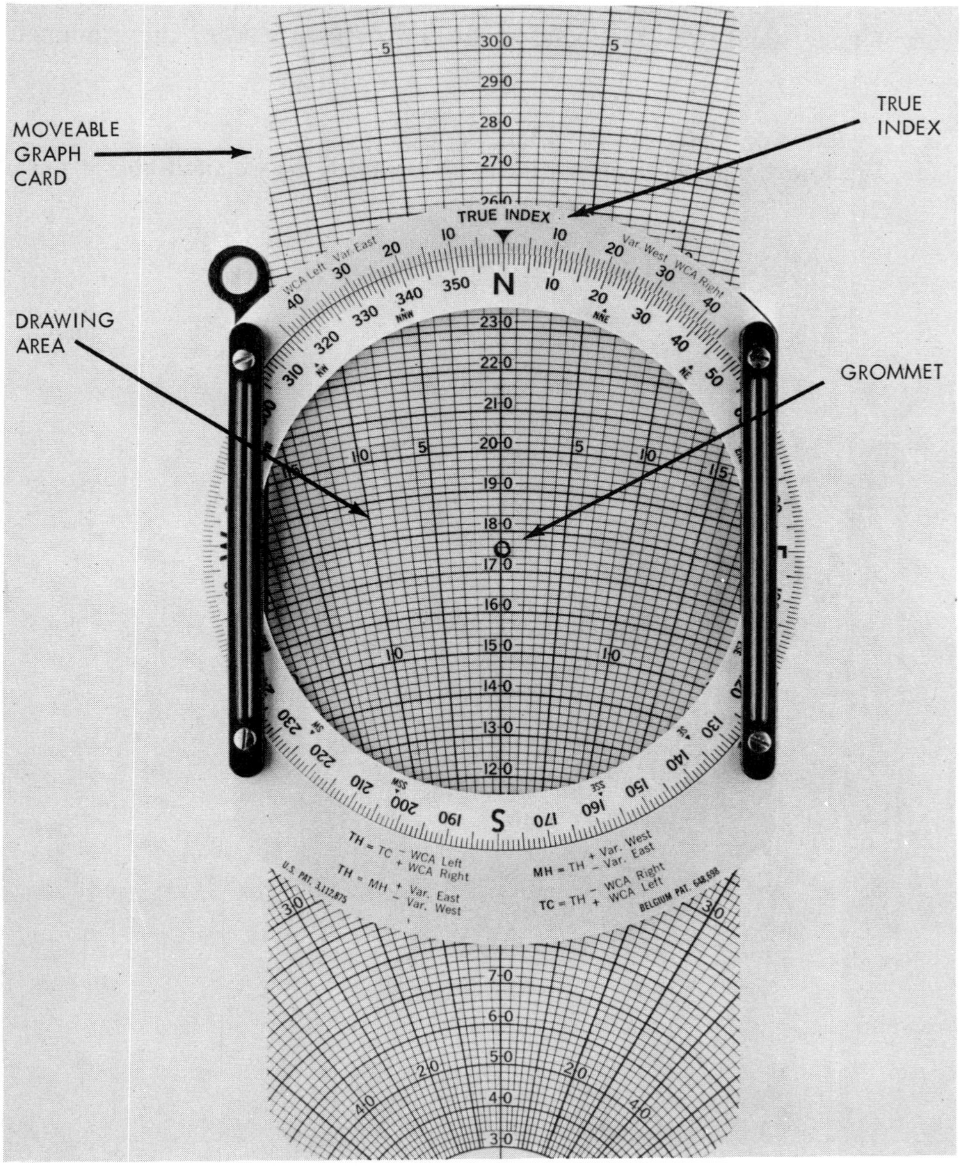


Fig. 117. The navigation computer.

When a point of the compass is placed under the true index, the centerline of the graph card represents a vector line pointing *from* that direction.

### *Plotting the First Vector Line (The Wind Line)*

We will begin with the wind vector line on the computer, just as we did when using the graph paper, because *both* the wind direction and speed are known. The wind is blowing *from* 045 degrees, so we will rotate the drawing wheel until 045 degrees lies under the true index. The centerline of the graph card now represents the wind vector line. The length of this vector is 40 knots, representing the wind speed. Therefore, we will limit the length of the wind vector line to 40 knots. Notice that the graph card is marked in units of two nautical miles, making it easy for us to draw a line over the centerline of the graph card, 40 units in length.

Figure 118 shows the wind vector line drawn on the computer. Notice that the arrow flies with the wind, pointing *from* 045 degrees, ending in the "grommet" in the center of the drawing area. After completing the next step, we will understand why the wind arrow must point *toward* the grommet.

### *Plotting the Second Vector Line*

The True Course—Groundspeed Line.

We know that the true course is 090 degrees, so we shall plot it next.

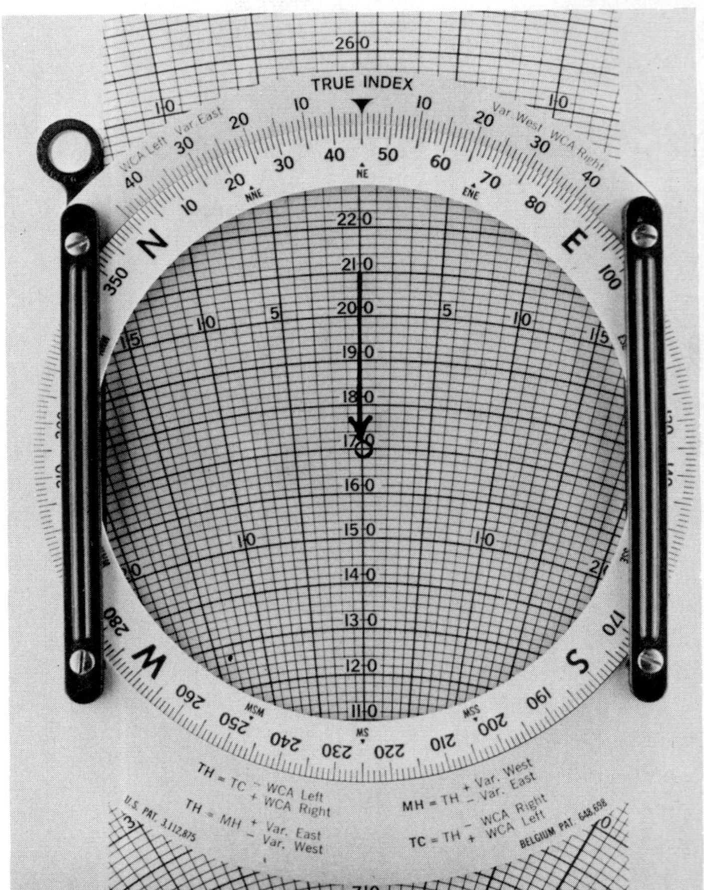


Fig. 118. The wind vector line.

We rotate the drawing wheel until 090 degrees lies under the true index as shown in Fig. 119.

The centerline of the graph card now represents the true course, and we should draw it in, over the centerline, on the drawing wheel. Now it becomes obvious that the wind vector line *had to be drawn into the grommet, in order for it to show the wind blowing toward the true course.* (Remember, the wind *always* blows the aircraft *toward* the true course.)

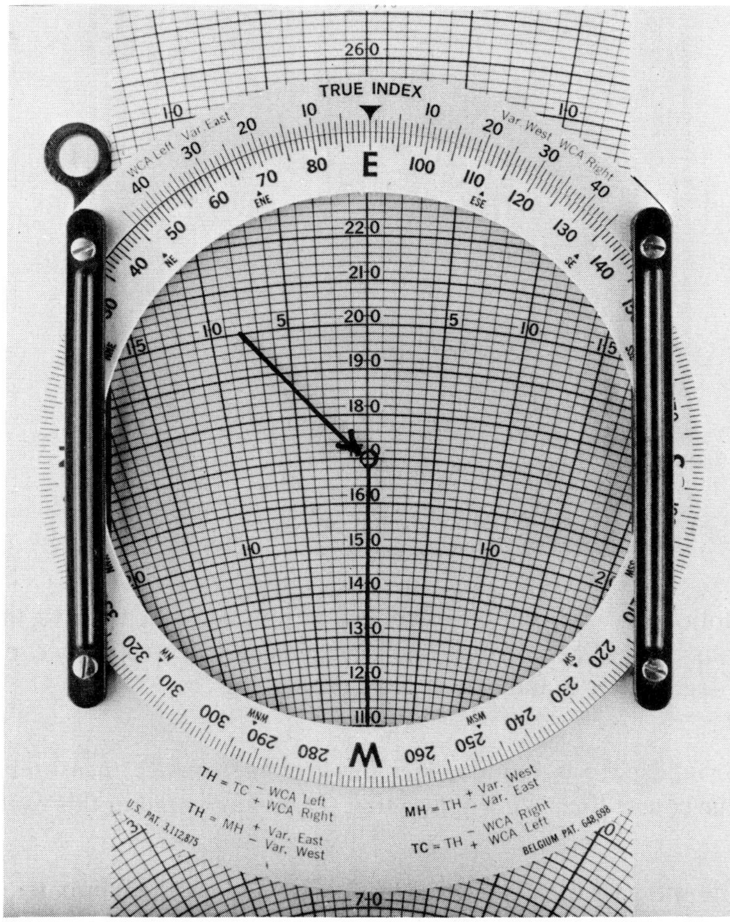


Fig. 119. The true course line.

### *Plotting the Third Vector Line*

The True Heading—True Airspeed Line.

When we plotted the wind vector diagram on graph paper, we measured back from the beginning of the wind line a distance equal to one hour of true airspeed. We will do exactly the same now, by sliding the graph card until 180 knots lies under the beginning of the wind arrow.

Now we will draw our true heading line, beginning at the 150 knot arc, down to the bottom of the drawing area, *making sure that our true heading line remains parallel* to the converging vertical lines on the graph card.
(Note: Although the true heading line and the true course line do not actually meet at the bottom of the drawing area, they do converge and would eventually meet if the area was large enough.)

Figure 120 shows the third line drawn on the computer.

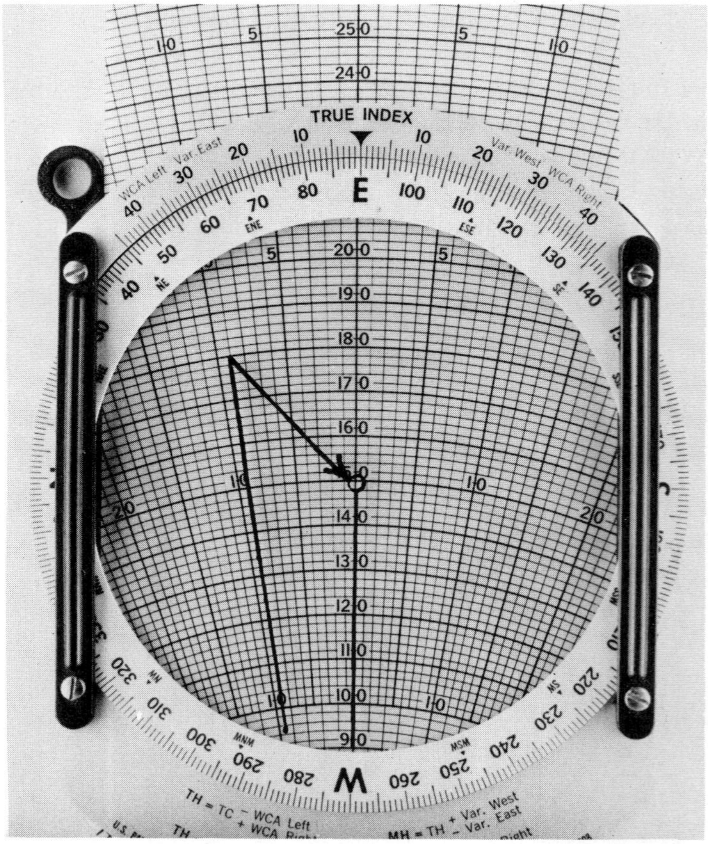


Fig. 120. The complete wind vector diagram.

*Solving for True Heading*

Notice that the true heading line is nine vertical lines to the left of the true course line. Each vertical line represents one degree. Hence, our true heading is 9 degrees left of the true course, or 081 degrees.

*Solving for Groundspeed*

Groundspeed is found under the grommet, which measures the length of the true course line. We will find that the groundspeed in this example is 150 knots.

*Final Computations*

The wind vector diagram, when completed on the computer, looks exactly the same as the diagram which we did on the graph paper. All we have done is transferred it to a more convenient drawing area.

We must now apply variation to the true heading in order to arrive at a magnetic heading and apply deviation to the magnetic heading to arrive at the compass heading.

The groundspeed must be applied to the total length of the true course to find the total time enroute.

## CHECK YOUR READING

1. When drawing the wind vector diagram, why is the wind line drawn first?
2. When using the navigation computer, why does the wind vector line point *to* the grommet?

3. Solve the following problems with graph paper and protractor:
   a) True course = 330 degrees
      Wind = 270 degrees at 40 knots
      True airspeed = 220 knots
      True heading = ?
      Groundspeed = ?
   b) True course = 190 degrees
      Wind = 245 degrees at 35 knots
      True airspeed = 105 knots
      True heading = ?
      Groundspeed = ?
4. Solve the following problems using the navigation computer:
   a) True course = 260 degrees
      Wind = 050 degrees at 42 knots
      True airspeed = 158 knots
      True heading = ?
      Groundspeed = ?
   b) True course = 045 degrees
      Wind = 090 degrees at 20 knots
      True airspeed = 90 knots
      True heading = ?
      Groundspeed = ?
5. With the following information given, solve the problem below with the navigation computer.

   *GIVEN:* True course = 060 degrees
   True airspeed = 145 knots
   Wind = 095 degrees at 34 knots
   Variation = 15 degrees westerly
   Deviation = 002 degrees easterly
   Distance = 440 nautical miles

   *FIND:* Compass course = ?
   True heading = ?
   Magnetic course = ?
   Magnetic heading = ?
   Groundspeed = ?
   Time enroute = ?
   Compass heading = ?

# 41

# THE NAVIGATION CALCULATOR

Contrary to popular belief, the mathematics involved in navigation is quite basic. In order to make the required calculations rapidly and accurately, we should learn to use the navigation calculator.

The calculator (shown in Fig. 121) is equipped with a rotating disk which is attached to the underlying base. Beginning with the outermost scale on the calculator, we must carefully study the engravings and divisions of each scale.

*The Outer Scale*

The outermost scale is often called the *miles scale*. Notice in Fig. 122 that the number 190 appears at the top of the scale and number 200 is to the immediate right. (Figure 122, arrow 1.) There are five spaces between the numbers 190 and 200, each space representing 2 units. However, we may assign a different value to the number 190 by moving the decimal point. We may call it 19.0, 1900, etc. If we assign a value of 19 to the number 190 on the calculator, then the number 200 will be assigned a value of 20, and each of the five spaces between 19 and 20 will equal 0.2.

On the bottom portion of the outer scale is the number 60. Notice in Fig. 123, arrow 2, that the number 55 is to the right of number 60, and there are 5 large spaces between number 60 and number 55. In this segment of the scale, each space would equal 1 if the values assigned were 60 and 55. If, however, we choose to assign values of 600 and 550 to the digits 60 and 55, each large space between would equal 10 units.

Obviously, we must use caution in interpreting the value represented by the digits and spaces on the scale.

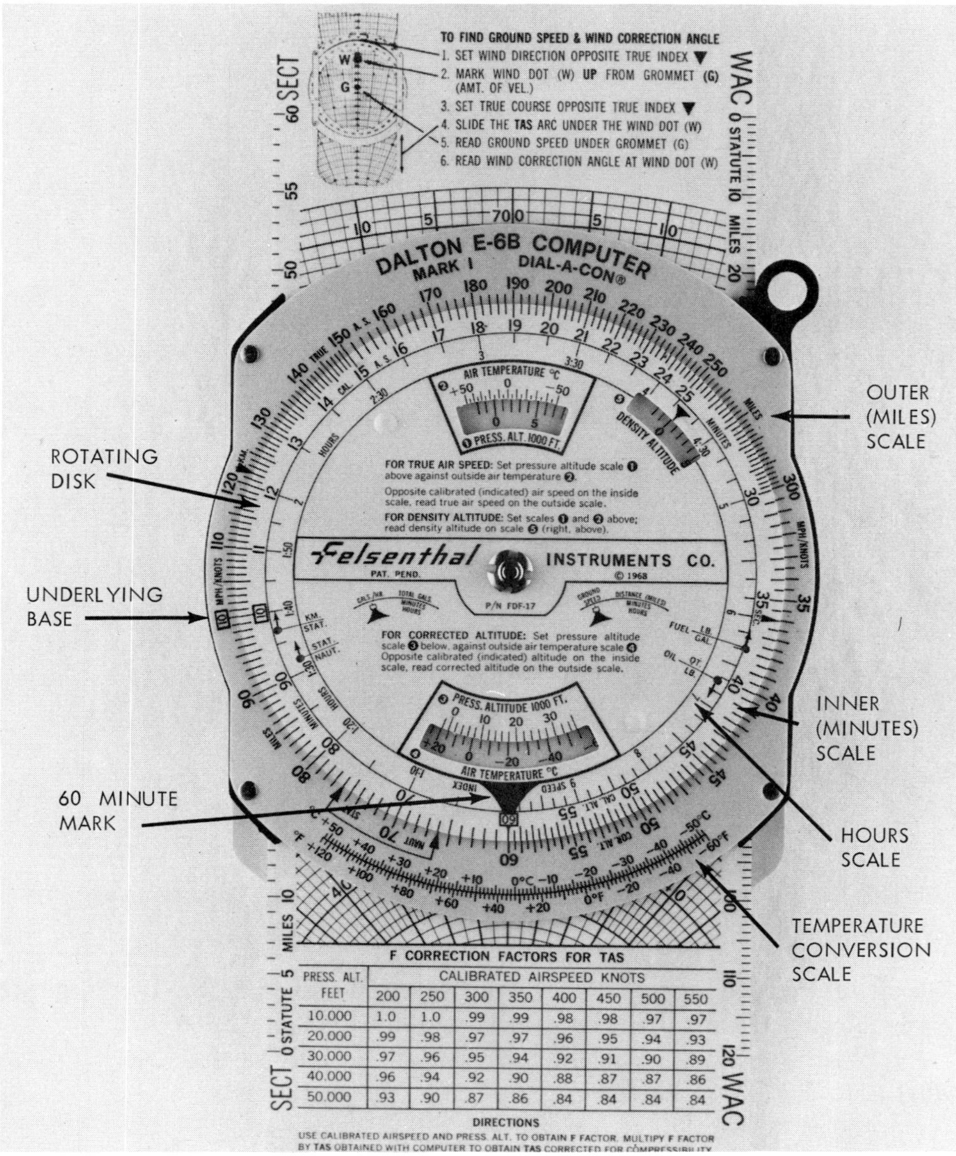


Fig. 121. The navigation calculator.

### *The Inner Scale*

The inner scale is also referred to as the *minutes scale*. The engraved digits and spaces are identical to those on the outer scale.

### *The Hours Scale*

Inside of the inner scale, an additional scale depicts hours and minutes. This is called the *hours scale*.

## Multiplication and Division

The calculator may be used for multiplication and division problems. Suppose, for instance, we wish to multiply 246 by 5.2. Before using the calculator, we should estimate the answer so as to avoid confusion concerning the placing of the decimal point in the result. We may easily estimate the result by rounding

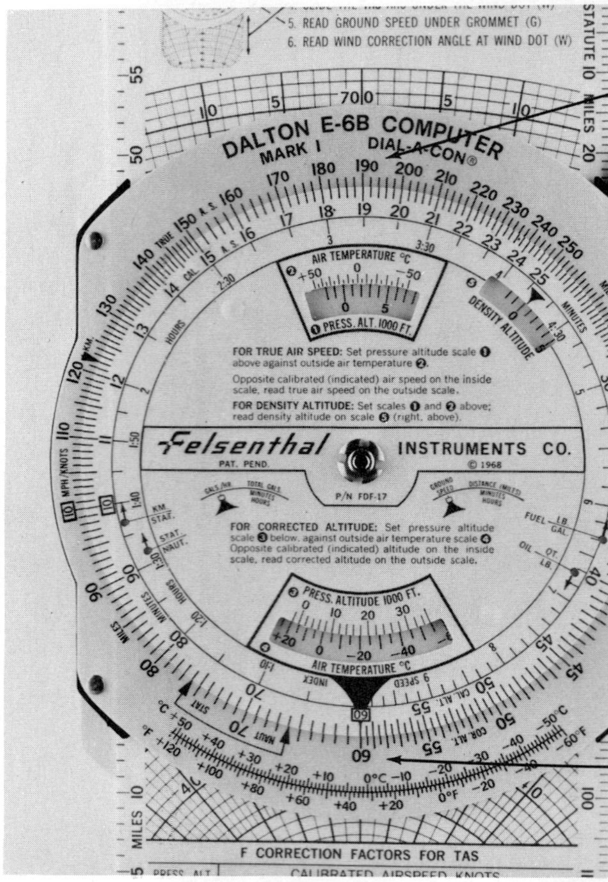


Fig. 122. Calculator engravings.

246 off to 250, and by rounding 5.2 off to 5 ($250 \times 5 = 1250$). Therefore, we know the answer should be about 1250, and not 125.0 or 12,500.

When we multiply two numbers we are actually saying "246 is to 1 as what is to 5.2." This statement, written mathematically is

$$\frac{246}{1} = \frac{?}{5.2}$$

When setting the calculator for this problem, we will set 246 on the outer scale, opposite 1 on the inner scale (Fig. 123, arrow 1). We must find 5.2 on the inner scale and find the answer above it, on the outer scale. As shown in Fig. 123, arrow 2, the outer scale reads 1278.

Because we already estimated the answer to be about 1250, we know the answer must be 1278 and not 127.8 or 12,780, etc.

Complete these practice problems by using the calculator. Be sure to estimate the answer first. (Answers are provided in the Appendix.)

1. $32 \times 17 =$
2. $142 \times 9 =$
3. $245 \times 52 =$

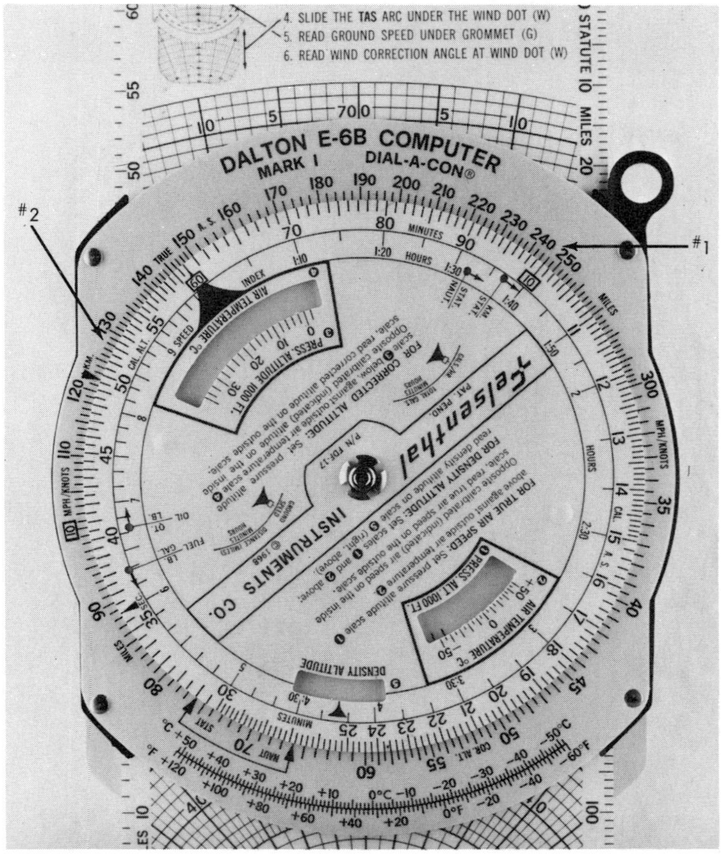

Fig. 123. Multiplication.

4. $7.8 \times 12.2 =$
5. $0.7 \times 378 =$
6. $1.25 \times 7870 =$

Division on the calculator is quite similar. Suppose, for instance, we wish to divide 730 by 17. Estimating the answer first will eliminate any doubt as to where to place the decimal point in the result. We can easily estimate the answer by changing 730 to 800, and by changing 17 to 20, before mentally dividing ($800 \div 20 = 40$). We know the answer must be about 40 and not 400 or 4.0, etc.

When we divide one number by another, we actually say "730 is to 17 as what is to 1." To express this sentence mathematically we write

$$\frac{730}{17} = \frac{?}{1}$$

The calculator should be set exactly the same, as shown in Fig. 124.

730 on the outer scale is placed over 17 on the inner scale (Fig. 124, arrow 1). The answer (43) is found on the outer scale above number 1 on the inner scale (Fig. 124, arrow 2). Since we estimated the answer to be about 40, we know the true answer is 43 and not 430 or 4.3, etc.

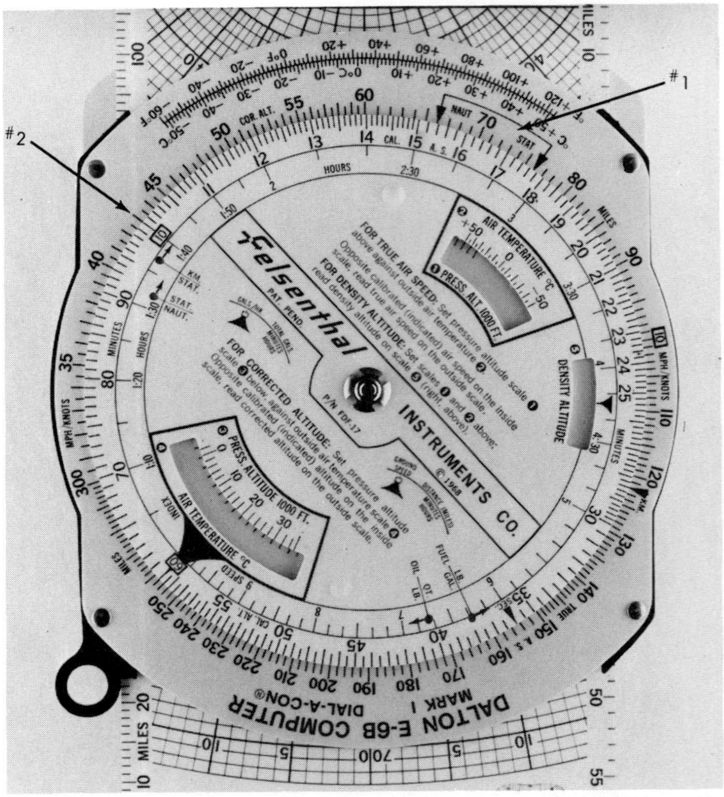

Fig. 124. Division.

Complete these practice problems using the calculator. Estimate each answer first. (Answers are provided in the Appendix.)

1. $300 \div 20 =$
2. $224 \div 16 =$
3. $9.2 \div 1.7 =$
4. $2240 \div 408 =$
5. $336 \div 250 =$
6. $1455 \div 70 =$

## Percentages

The calculator is easily used to compute percentages. For instance, suppose we wish to know what per cent of 75 gallons of fuel is 42 gallons. An estimate will reduce the chance of error in the answer. Since 42 gallons is a little more than one-half of 75 gallons, we know that the exact answer should be in the neighborhood of 50 per cent.

When solving this problem, we say, "75 gallons is 100% of the fuel. What per cent is 42 gallons?" Or, "75 gallons is to 100 per cent as 42 gallons is to what per cent?" To express this mathematically we write

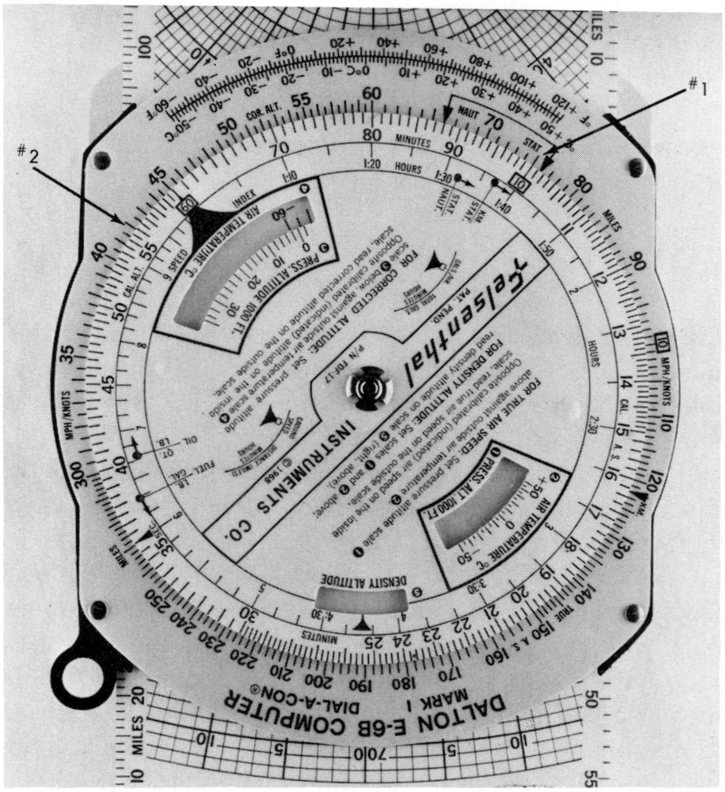

Fig. 125. Percentage.

$$\frac{75}{100} = \frac{42}{?}$$

The calculator is set exactly the same as shown in Fig. 125. 75 is placed over 100 (Fig. 125, arrow 1) and the answer (55.9%) is found under 42 (Fig. 125, arrow 2).

Complete the following problems with the calculator. Estimate the answer first. (Answers are provided in the Appendix.)

1. What per cent of 850 is 400?
2. What per cent of 45 is 9?
3. What per cent of 7.8 is 1.3?
4. What is 28% of 95?
5. What is 85% of 430?
6. What is 66% of 1040?

## Time—Speed—Distance

One of the most common problems of navigation is one which is easily solved with the calculator; that is, finding the time required to fly a specific distance at a specific groundspeed. Suppose, for instance, we have calculated the groundspeed to be 160 knots. The total length of the flight is 310 nautical miles. We wish to know how long it will take to complete the trip.

When we say our speed is 160 knots, we are actually saying that we expect to travel 160 nautical miles in one hour, or 60 minutes. To express this problem mathematically we write

$$\frac{160}{60} = \frac{310}{?}$$

or: 160 nautical miles is to 60 minutes as 310 nautical miles is to how many minutes? Before using the calculator we will estimate the answer. 310 nautical miles is about twice 160 nautical miles. Hence, the flight should take about 2 hours.

The calculator will be set up exactly the same as the mathematic equation above. Because the 60 minute mark is used so often, the calculator is engraved with a large black arrowhead at the 60 minute mark.

We must find 160 on the outer scale, and place 60 (or the black arrowhead) under it as shown in Fig. 126, arrow 1. If we then find 310 on the outer scale, we can read directly beneath it to find the answer (116 minutes, or one hour and 56 minutes), as shown in Fig. 126, arrow 2. Since we estimated the answer to be 2 hours, or 120 minutes, we know the true answer cannot be 11.6 minutes or 1160 minutes, etc. If we look below the minutes scale, we can also read the answer on the hours and minutes scale (1 : 56).

Complete the following problems using the calculator. Estimate each answer first. (Answers are provided in the Appendix.)

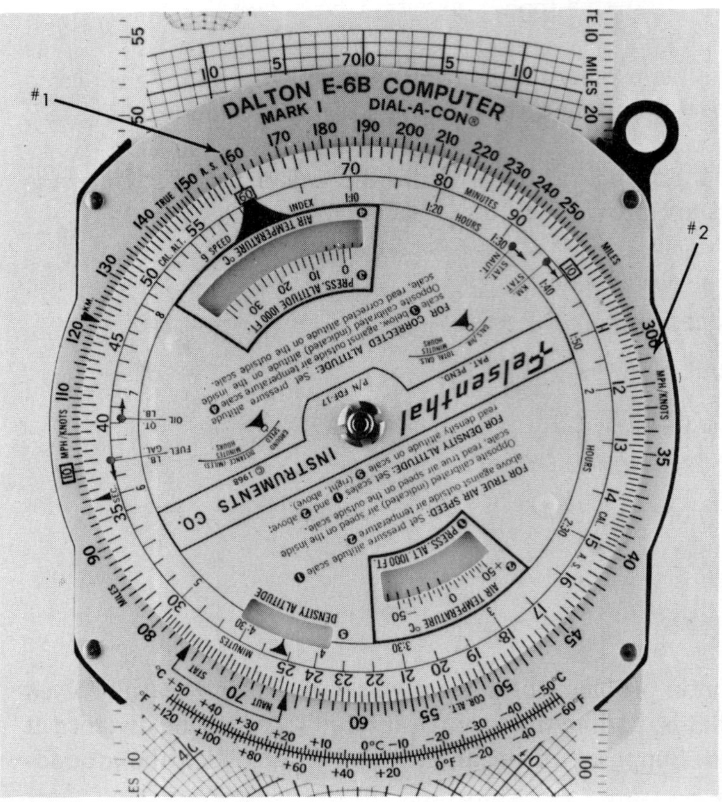


Fig. 126. Time/speed/distance.

1. Groundspeed = 220 knots, distance = 480 nautical miles, time = ?
2. 〃 = 104 knots, 〃 = 302 〃 〃 〃 = ?
3. 〃 = 85 knots, 〃 = 153 〃 〃 〃 = ?
4. 〃 = 420 knots, 〃 = 190 〃 〃 〃 = ?
5. 〃 = 125 knots, 〃 = 1105 〃 〃 〃 = ?
6. 〃 = 190 knots, 〃 = 70 〃 〃 〃 = ?

Sometimes, we know the distance we have flown and the time required to fly that distance. We wish to find out groundspeed. For example, suppose we timed our crossing of a prominent landmark and found that it was 5 minutes past the hour. After flying to the next prominent landmark, a distance of 72 nautical miles, we check the clock again, which reads 53 minutes past the hour. We flew 72 nautical miles in 48 minutes. How fast did we fly?

To express this problem mathematically, we write

$$\frac{72}{48} = \frac{?}{60}$$

We must estimate the answer in order to eliminate confusion as to where to place the decimal point in our true answer. 72 is about 50 per cent greater than 48, so the answer must be about 50 per cent greater than 60, or about 90.

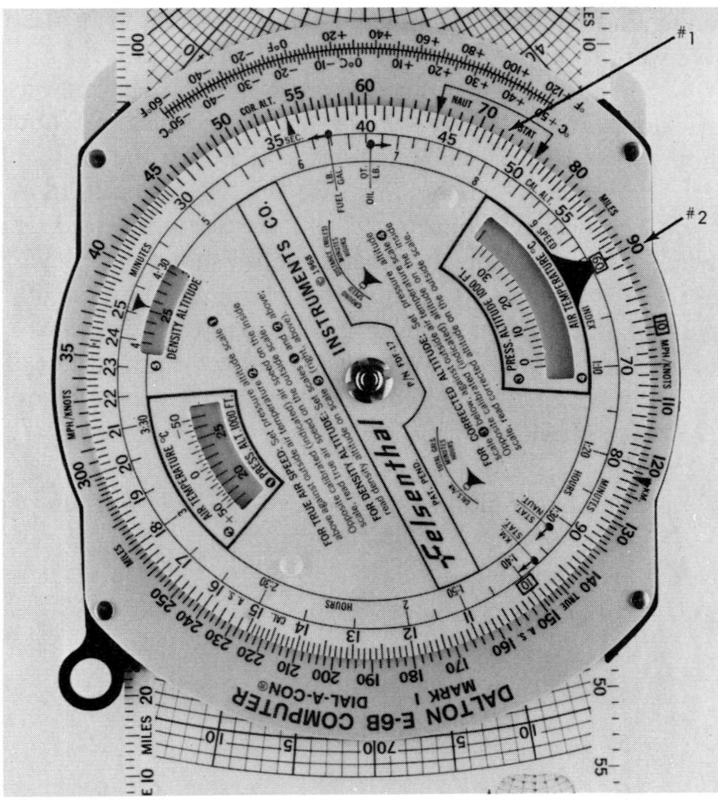


Fig. 127. Time/speed/distance.

We will set the calculator with 72 on the outer scale opposite 48 on the inner scale (Fig. 127, arrow 1). The result (90) will appear over the 60 minute mark (black arrowhead), Fig. 127, arrow 2.

Complete the following problems with the calculator, estimating each answer first. (Answers provided in Appendix.)

| | Distance | Time | Groundspeed |
|---|---|---|---|
| 1. | 405 nautical miles | 130 minutes | |
| 2. | 22 nautical miles | 14 minutes | |
| 3. | 86 nautical miles | 65 minutes | |
| 4. | 109 nautical miles | 42 minutes | |
| 5. | 1250 nautical miles | 550 minutes | |
| 6. | 14 nautical miles | 4 minutes | |

**Fuel Consumption**

Suppose we wish to compute the fuel which will be consumed during our flight. The cruise control (fuel consumption chart) in the Aircraft Owner's Handbook will specify the fuel consumption at specific power settings and altitudes. For instance, assume the chart shows a fuel consumption rate of 11.5 gallons per hour, and we have computed the time enroute to be 4 hours and 50 minutes. We want to know how much fuel will be consumed.

We can estimate the answer by calling 11.5 gallons per hour 12 gallons per hour, and rounding 4 hours and 50 minutes to 5 hours. At 12 gallons per hour, we will consume 60 gallons of fuel in 5 hours.

Figure 128 shows this problem as it should appear on the calculator.

Notice that 11.5 on the outer scale appears opposite 60 minutes (the black arrowhead) on the inner scale, as shown in Fig. 128, arrow 1. Looking above 4 hours and 50 minutes (or 290 minutes) on the inner scale, we find 55.6 gallons on the outer scale (Fig. 128, arrow 2). Since we estimated the answer to be about 60 gallons, we know the answer is not 5.56 or 556 gallons.

Practice the fuel consumption problems by completing the following. (Answers provided in the Appendix.)

| | Consumption Rate | Time Enroute | Fuel Used |
|---|---|---|---|
| 1. | 14 | 1 + 56 | |
| 2. | 22 | 2 + 25 | |
| 3. | 7.4 | 3 + 15 | |
| 4. | 9.8 | 0 + 47 | |
| 5. | 8.2 | 2 + 55 | |
| 6. | 40.2 | 3 + 52 | |

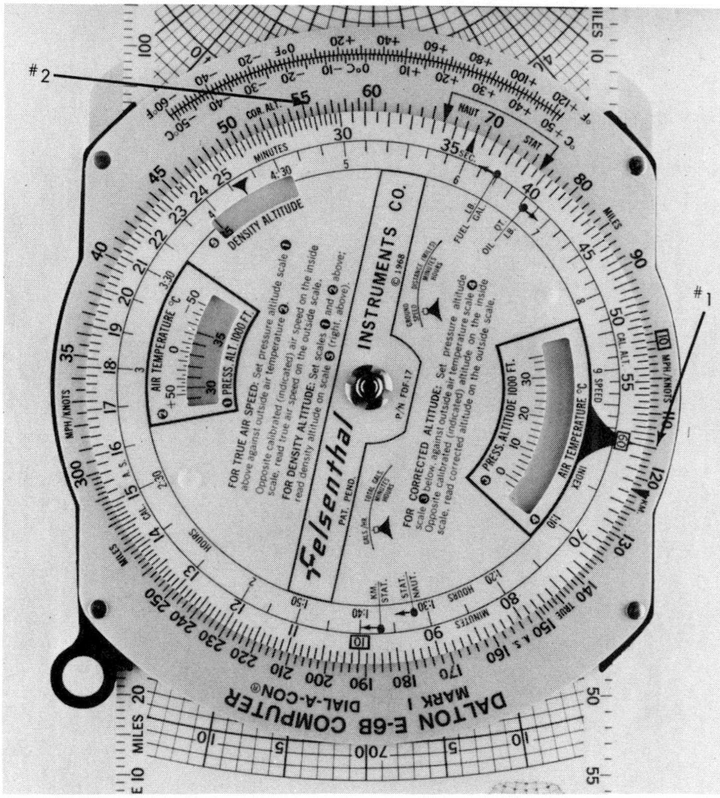


Fig. 128. Fuel consumption.

## Mileage Conversions

Although the aviation industry is converting to nautical miles as a standard (instead of statute miles), the conversion is quite slow. There are, therefore, many times when it is necessary to convert statute miles to nautical miles (or vice versa) or even to kilometers.

A statute mile is 5,280 feet.

A nautical mile is 6,080 feet.

A kilometer is 3,280 feet.

For the purpose of estimating our mileage conversions, we should remember that there are 15 per cent more statute than nautical miles in a given distance and about 85 per cent more kilometers than nautical miles in a given distance.

The calculator is engraved with reference marks in order to simplify mileage conversions. As shown in Fig. 129, on the outer scale at numbers 66 and 76 there are two arrows marked "Naut" and "Stat" respectively. At number 122 on the outer scale is the arrow marked "KM."

Suppose we wish to convert 245 nautical miles to statute miles. We know that the result should be about 15 per cent greater than 245, or about 280. We must set 245 on the inner scale under the NAUT arrow, and read 282 under the STAT arrow. Without moving the wheel, we can look under the KM arrow to find 453 kilometers. (Since we know that there are about 85 per cent more kilometers than nautical miles in a given distance, we know the answer should be 453 and not 45.3 or 4530.) Figure 129 depicts this problem as it should be set up on the calculator.

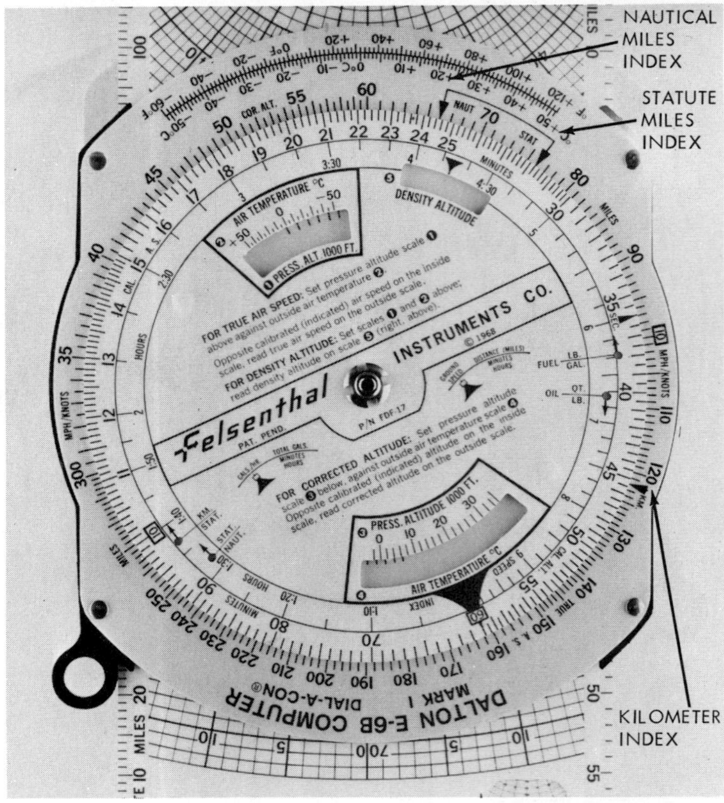


Fig.129. Mileage conversion engravings.

Complete the following conversions, estimating each answer first. (Answers provided in the Appendix.)

| | NAUT | STAT | KM |
|---|---|---|---|
| 1. | 80 | ? | ? |
| 2. | ? | 174 | ? |
| 3. | ? | ? | 3500 |
| 4. | 20 | ? | ? |
| 5. | ? | 810 | ? |
| 6. | ? | 12.50 | ? |

## Temperature Conversions

Converting degrees Fahrenheit to degrees Celsius is quite simple with the calculator, for a direct reading scale is engraved on the outer edge, as shown in Fig. 130.

We must only look at the scale to determine the temperature conversion.

Convert the following temperatures. (Answers provided in the Appendix.)

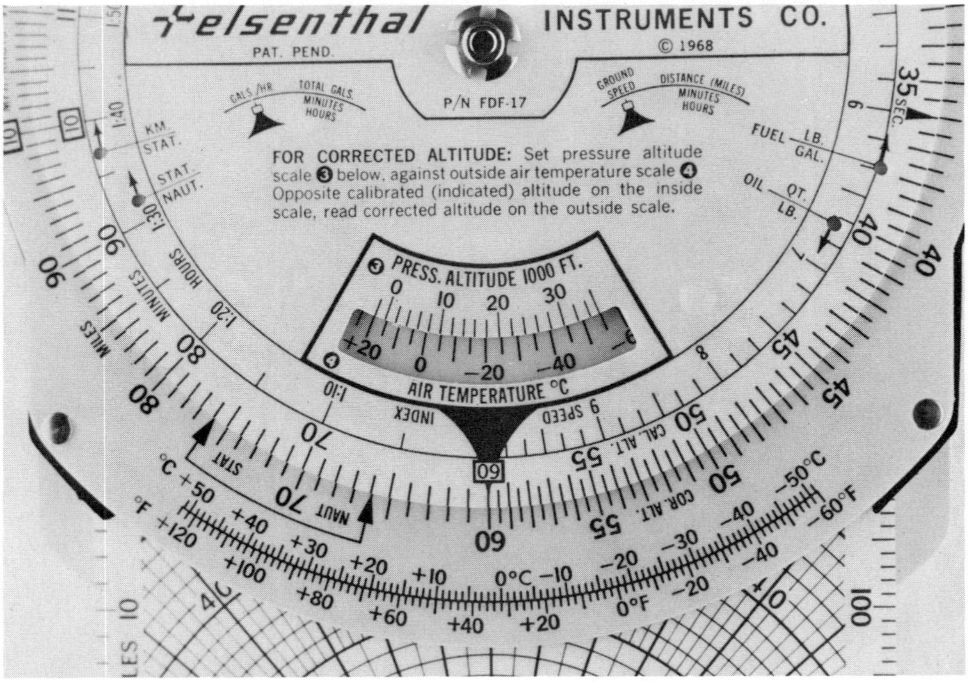


Fig. 130. Temperature conversion.

1. +30° F =
2. −30° C =
3. +80° F =
4. +15° C =
5. +25° C =
6. 0° F =

## Density Altitude

As we have learned, aircraft performance is dependent upon the density altitude. Therefore, we will often need to compute the density altitude on the calculator. We must first know the outside air temperature in degrees Celsius and the pressure altitude. (Pressure altitude is found by setting 29.92 in the altimeter barometric pressure window and reading pressure altitude directly from the altimeter.)

Suppose the outside air temperature is +30 degrees Celsius, and the pressure altitude is 10,500 feet. Using the cut-out windows on the calculator (as shown in Fig. 131), we set the pressure altitude opposite the temperature and read density altitude in the window.

The calculator in Fig. 131 is set for the problem described, and we see 14,500 feet in the density altitude window.

Compute the following problems. (Answers are provided in the Appendix.)

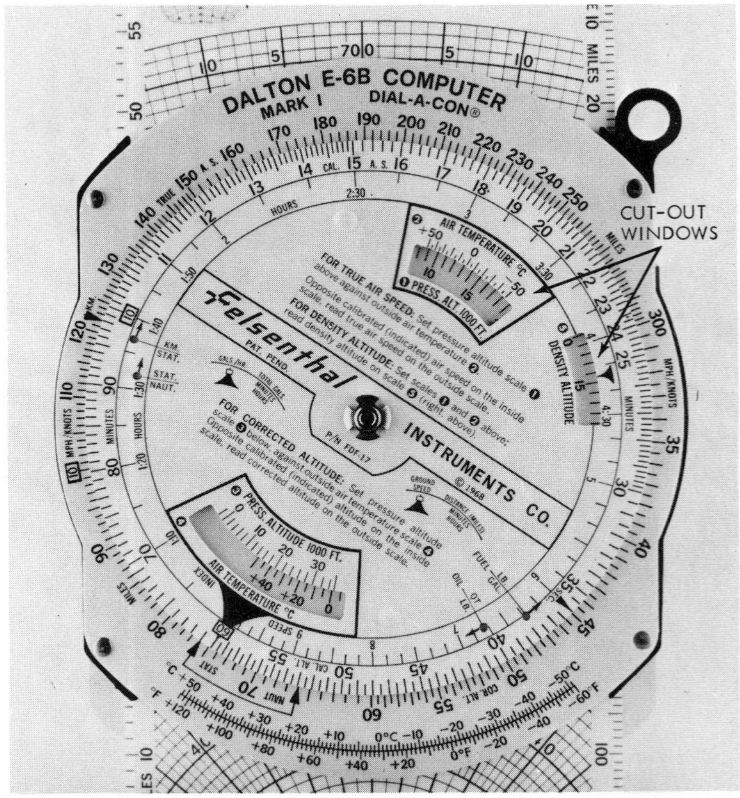


Fig. 131. Density altitude.

| | OAT | Pressure Altitude | Density Altitude |
|---|---|---|---|
| 1. | 0 degrees Celsius | 5,000 feet | |
| 2. | +13 degrees Celsius | 10,000 〃 | |
| 3. | −25 degrees Celsius | 10,000 〃 | |
| 4. | 0 degrees Celsius | 12,000 〃 | |
| 5. | +10 degrees Celsius | 3,500 〃 | |
| 6. | −10 degrees Celsius | 7,000 〃 | |

## Computing True Airspeed

Wind vector diagrams are computed using the true airspeed. To find true airspeed, we must know the outside air temperature, pressure altitude, and calibrated airspeed. Calibrated airspeed is indicated airspeed, corrected for installation errors, and is found in the Airspeed Correction Chart in the Aircraft Owner's Handbook.

Figure 132 depicts the calculator as it should be set for the following problem.

Outside air temperature = −20 degrees Celsius, pressure altitude = 9,500 feet, and calibrated airspeed is 120 knots.

The pressure altitude is set opposite the outside air temperature in the

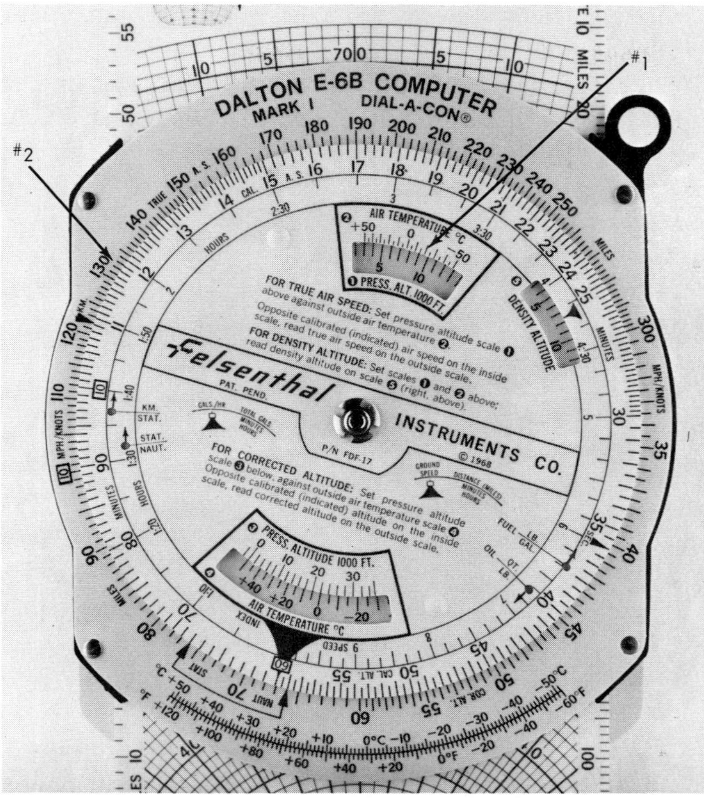


Fig. 132. True airspeed.

true airspeed cut-out window on the calculator (Fig. 132, arrow 1). True airspeed is found on the outer scale above the calibrated airspeed (Fig. 132, arrow 2). The true airspeed is 134 knots. Compute the following problems. (Answers are provided in the Appendix.)

| | OAT | Pressure Altitude | CAS | TAS |
|---|---|---|---|---|
| 1. | +10 degrees Celsius | 7,500 feet | 140 knots | |
| 2. | − 5 degrees Celsius | 8,000 feet | 105 knots | |
| 3. | −15 degrees Celsius | 9,500 feet | 202 knots | |
| 4. | + 7 degrees Celsius | 5,000 feet | 85 knots | |
| 5. | 0 degrees Celsius | 12,000 feet | 150 knots | |
| 6. | +20 degrees Celsius | 6,500 feet | 185 knots | |

## Off-Course Correction

In the event we find ourselves off course during a cross-country flight, we may use the calculator to compute a new heading to return to the destination.

Suppose the total length of our flight is 370 nautical miles. After flying 130 nautical miles we find ourselves 14 nautical miles to the left of course.

We wish to know how many degrees to turn right in order to intercept our course at the destination.

Our calculations are based on the fact that one nautical mile off course after flying 60 nautical miles is equal to one degree off course.

The problem is done in two parts. First, we compute the number of degrees it will be necessary to turn in order to parallel the course. Figure 133 shows the calculator set for this part of the problem. We have flown 14 miles off course after travelling 130 miles, so 14 on the outer scale is set above 130 on the inner scale (Fig. 133, arrow 1). The number of degrees to turn in order to parallel the course is found above the 60 (black arrow index) on the outer scale. Figure 133, arrow 2, shows this to be $6\frac{1}{2}$ degrees.

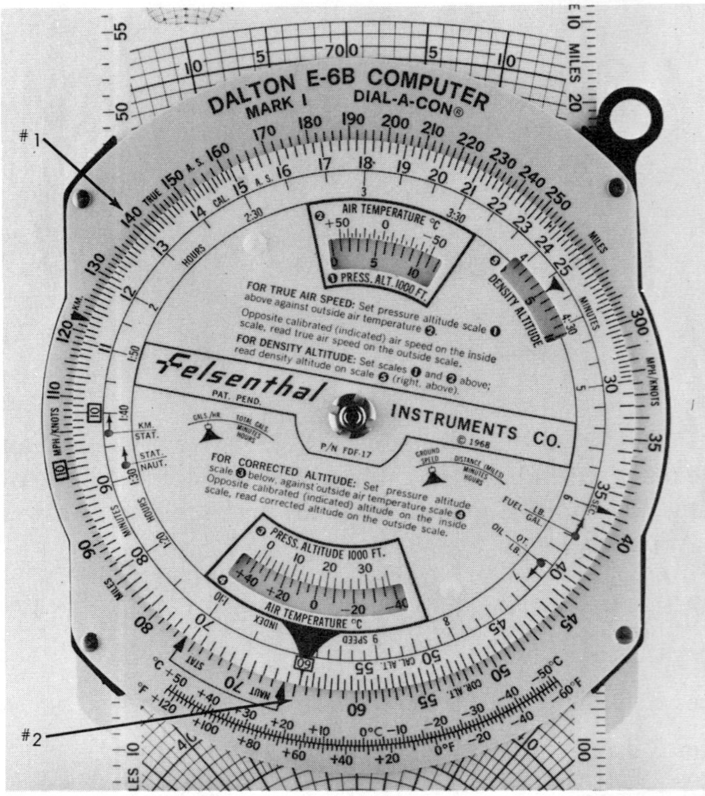


Fig. 133. Off-course correction (step 1).

In the second part of the problem, we compute the number of additional degrees to turn to intercept the course. Figure 134 shows the calculator set up for this part of the problem. We have 240 nautical miles to fly (370 nautical miles minus 130 nautical miles). Therefore, we set 14 on the outer scale over 240 on the inner scale (Fig. 134, arrow 1) and read $3\frac{1}{2}$ degrees additional turn over the 60 (Fig. 134, arrow 2). We know, therefore, that a total turn of 10 degrees right will return us to the course at the destination ($6\frac{1}{2}° + 3\frac{1}{2}°$).

Compute the following problems. (Answers are provided in the Appendix.)

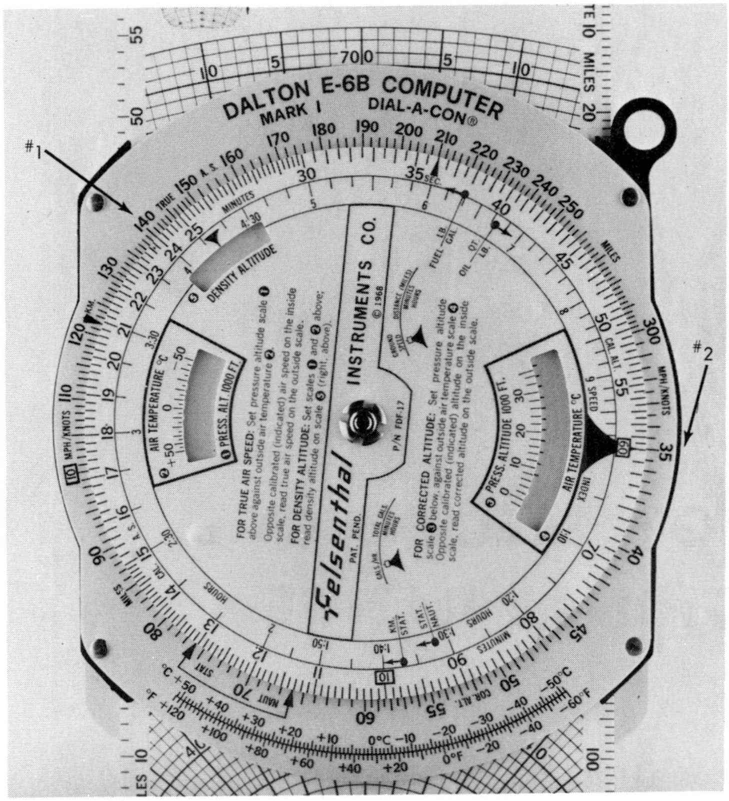


Fig. 134. Off-course correction (step 2).

| | Total Flight | Miles Off Course | Distance Flown | Total Heading Change |
|---|---|---|---|---|
| 1. | 440 | 12 | 200 | |
| 2. | 105 | 3 | 55 | |
| 3. | 90 | 5 | 30 | |
| 4. | 325 | 10 | 105 | |
| 5. | 182 | 12 | 102 | |
| 6. | 1500 | 25 | 500 | |

# 42

# THE VERY-HIGH FREQUENCY OMNI-DIRECTIONAL RANGE

The most popular and most accurate system of radio navigation in use today is the nationwide network of transmitters called very-high frequency omni-directional ranges (often called OMNI or VOR).

## Principles of VOR

The Federal Aviation Administration has installed VOR transmitters all across the United States. Even while flying in the most remote areas of the country we will normally be within reception distance of these helpful aids to navigation.

The VOR operates on the frequency band between 108.0 and 117.9 MHz. This frequency band, called very-high frequency (VHF) offers the advantage of static-free reception which is quite reliable in all atmospheric conditions. Certain propeller rpm settings may cause fluctuations in the readout of the VOR but a small change in rpm will normally remedy this problem. VHF reception is accurate only when the receiver is within *line-of-sight* of the station. Line-of-sight does not mean that the receiver must be within a short distance of the transmitter, but rather that there are no obstructions between the transmitter and the receiver. VHF signals will not bend around the curvature of the earth, or penetrate buildings or mountains. For this reason we should remember that we must increase our altitude as our distance from the transmitter increases, in order to enjoy loud and clear VHF reception. Figure 135 depicts the minimum reception altitudes at various distances from the VOR transmitter.

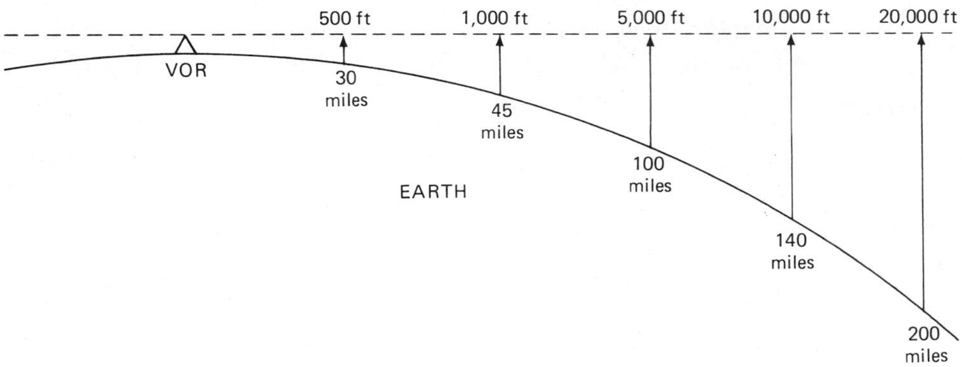


Fig. 135. VOR reception altitude.

The VOR transmitter constantly broadcasts signals which are received by the airborne radio and converted into *radials* which resemble spokes of a wheel emanating from the VOR station. Actually, there are an infinite number of these radials around each station, but only 360 of them are considered—one for every degree in a circle, as shown in Fig. 136.

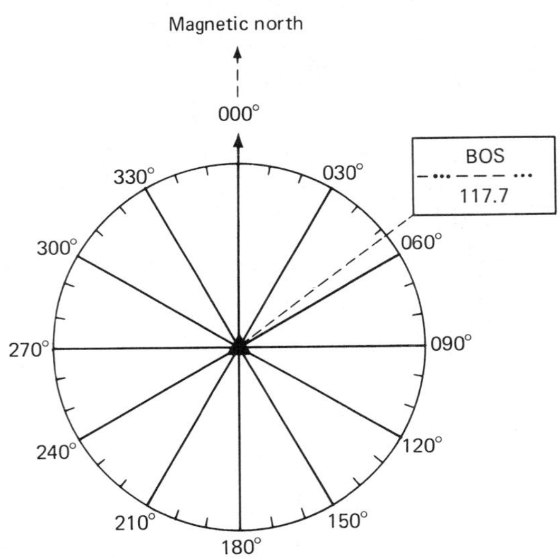


Fig. 136. Chart depiction of a VOR.

## Chart Depiction

OMNI stations are printed on aeronautical charts in blue ink and are surrounded by a large circle called the *compass rose*. Hatch marks on the compass rose indicate ten-degree increments, and the OMNI is oriented to *Magnetic North*. Therefore, all of the radials are oriented to Magnetic North also. A blue box is printed near the compass rose in which the name of the station appears, and the identifying Morse code signal. This identifying signal will be broadcast every 15 seconds, either in Morse code or by recorded voice.

## Voice Broadcast

Most OMNI stations are also utilized as a voice broadcast station over which transcribed weather information is broadcast at 15 minutes past every hour.

We may also contact the nearest Flight Service Station while in flight and request a reply on the OMNI station. Voice broadcasts do not interrupt navigation signals.

## Uses of the VOR

We may use the VOR for many navigation problems. We may follow a *magnetic radial* (so called because all radials are oriented to Magnetic North) outbound from a station to any destination, or we may follow a magnetic radial inbound toward any station.

While flying on a radial of one station, we may tune in a second station to note the radial we are crossing, thus pinpointing our position.

The federal airways are defined by radials of VOR stations. Hence, we may accurately complete cross-country flight by simply following radials from one station to the next.

When a VOR station is placed at or near a terminal airport, a radial may be used to align the aircraft with the final approach path of a specific runway. In this instance the VOR is used for terminal approaches during instrument flight conditions.

Using one OMNI station and a simple mathematical formula, we can determine our distance from the station and the time required to fly there.

Before discussing the various uses in detail, we must become familiar with the components of the airborne receiver and the function of these components.

## Components of the VOR Receiver

### *The On–Off–Volume Control*

Just as an auto radio has an on–off–volume control, so does the VOR receiver. The control should be turned on and the volume set at maximum. The receiver will normally require about 30 seconds to warm up. All components of the VOR receiver are shown in Fig. 137.

### *The Frequency Selector Control*

Two types of frequency selectors are now in production. The most popular and more sophisticated model is the crystal-controlled tuner which looks like a television channel selector. We simply turn the selector wheel until the proper published frequency for the desired station appears in the window of the selector.

The older tuner is a small hand crank, similar to a very old home radio tuner. The hand crank is rotated until a moving reference line appears at the proper position on the frequency band. This type of tuner is not as accurate, and we may need to hunt along the frequency band for the exact setting which results in the best reception.

After we tune the OMNI station, and adjust the volume control to a comfortable level, we *must identify* the station by listening to the coded or voice

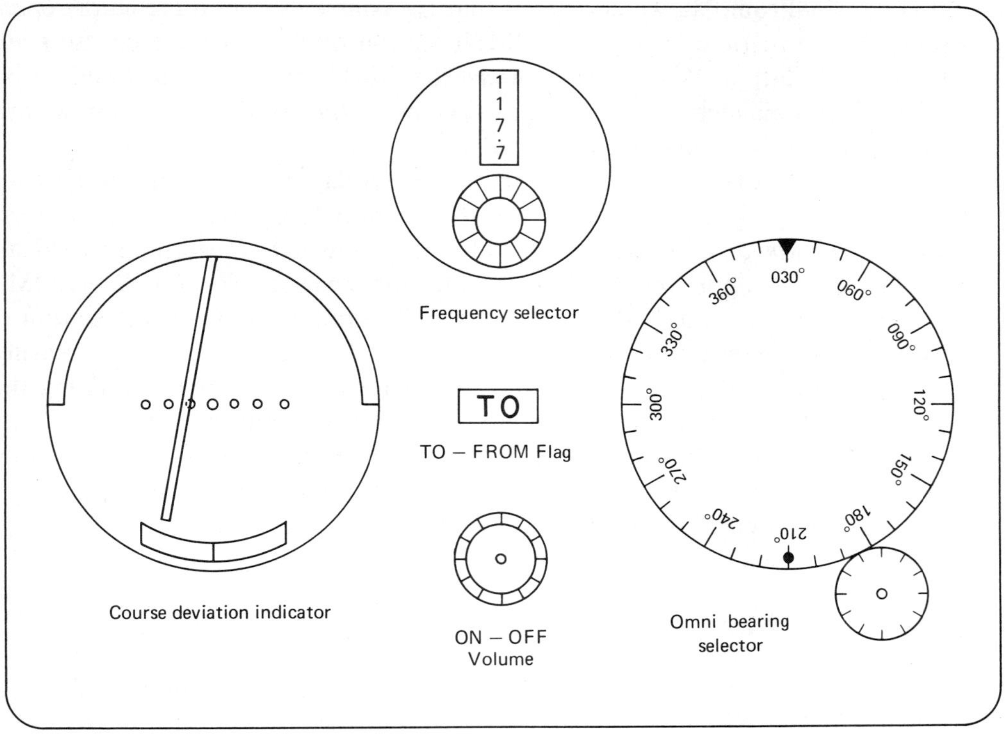


Fig. 137. Typical airborne VOR equipment.

identifier. Since some OMNI stations are used as remote voice transmitters for an FAA Flight Service Station many miles away, we must *not* identify the station by listening to *any* voice transmission except the recorded station identifier. It is imperative that we properly identify the OMNI station, for frequencies are sometimes changed, or, if the flight is at a high altitude, it is possible to receive distant stations on a common frequency. Many pilots have successfully navigated to a VOR station, only to find that they had tuned in the wrong one!

### *The Omni Bearing Selector (OBS)*

Again, current production models vary, but in most cases the omni bearing selector is comprised of a dial which is rotated by hand, bringing one of the 360 degrees of the circle under an index mark. The omni bearing selector will be used in conjunction with the course deviation indicator and the To-From flag to indicate our position relative to the OMNI station.

### *The Course Deviation Indicator (CDI)*

Sometimes called the *Left–Right Needle*, the course deviation indicator represents the magnetic course toward or away from the OMNI station which is selected on the omni bearing selector. For instance, if we select 090 degrees on the omni bearing selector, the course deviation indicator will represent *either* the 090 degree magnetic course *to* the OMNI station or the 090 degree magnetic course *from* the OMNI station.

*The To-From Flag* It is the To–From flag which tells us that the course deviation indicator represents the magnetic course *to* the OMNI station or the magnetic course *from* the OMNI station. When flying *toward* the OMNI station we are said to be *inbound* on a magnetic radial. When flying *away from* an OMNI station we are said to be *outbound* on a magnetic radial.

It is important to note that the To-From flag does *not* indicate that we are actually flying to or from the OMNI station. A "TO" indication *does* show us that we *would be* heading to the station *if* and *only if* we maintained an aircraft heading approximately equal to the omni bearing selection. A "FROM" indication shows us *only* that we would be heading away from the station *if* we maintained an aircraft heading approximately equal to the omni bearing selection. The To-From flag has no way of knowing if we *are* maintaining the proper heading. It is *not* affected in any way by the aircraft heading.

If a red flag appears in the To-From window, the receiver is not to be used for navigation. Either the receiver is not functioning properly, or the signal being received is too weak to be accurate.

**Finding Position**

In order to find our position using the VOR we simply tune and identify the nearest OMNI station. We will then rotate the omni bearing selector until the course deviation indicator centers in the face of the instrument, as shown in Fig. 138.

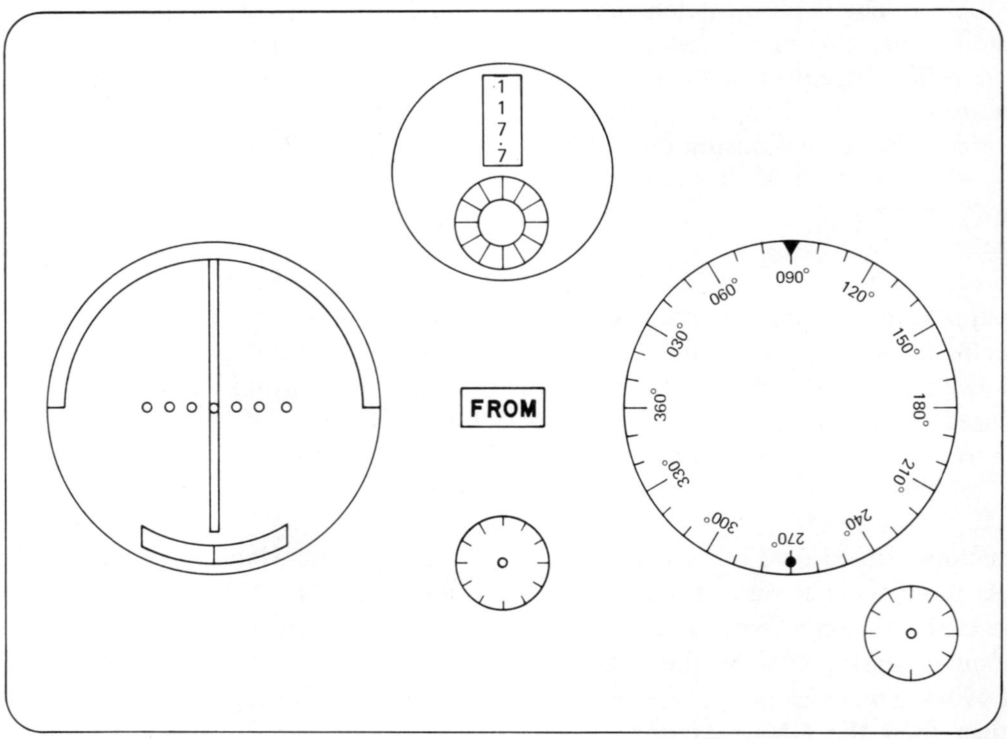


Fig. 138. Airborne equipment.

Assume that an omni bearing selection of 090 degrees centers the course deviation indicator, and a "FROM" indication appears on the To-From flag. These readings indicate that the aircraft is at a position 090 degrees **from** the station, or on the 090 magnetic radial, as shown in Fig. 139.

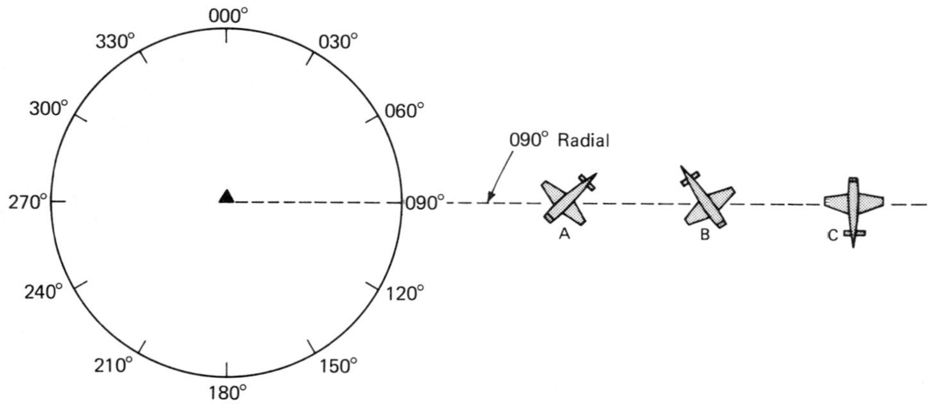


Fig. 139. Aircraft A, B, and C have identical VOR indications.

The OMNI receiver is *not* indicating the heading of the aircraft. The identical readings will result no matter what the aircraft heading. In Fig. 139, for instance, aircraft A, B, and C will all have the same OMNI readings.

The To-From flag does *not* indicate that the aircraft is heading away from the station. It is telling us that our current position is 090 degrees from the station and that we will be proceeding away from the station *if* we maintain a heading of about 090 degrees.

If we now rotated the omni bearing selector a full 180 degrees to a reading of 270 degrees, the course deviation indicator would again center, but the To-From flag would read "TO." Remember, we did not change the aircraft heading—only the omni bearing selector. Aircraft A, B, and C in Fig. 139 would all show the same indications. The OMNI is still telling us that our position relative to the station is on the 090 degree radial, or that an approximate heading of 270 degrees will take us ***to*** the station. But the aircraft radio receiver cannot tell if the aircraft is actually heading in that approximate direction.

Let us now assume that we have tuned and identified another VOR station. We center the course deviation indicator by rotating the omni bearing selector. The reading under the omni bearing selector index reads 345 degrees and the To-From flag says "TO." Referring to Fig. 140 at what position must the aircraft be?

The aircraft must be in position C (on the 165 degree radial) as that is the only place from which a heading of about 345 degrees would carry the flight ***to*** the VOR station. If the flight were in position B, (on the 345 degree radial) the course deviation indicator would also be centered with an omni bearing selection of 345 degrees, but the To-From flag would read "FROM" because a heading of about 345 degrees would take the aircraft in position B ***away from*** the station. In position A, D, E, and F the course deviation indicator would not be centered with an omni bearing selection of 345 degrees.

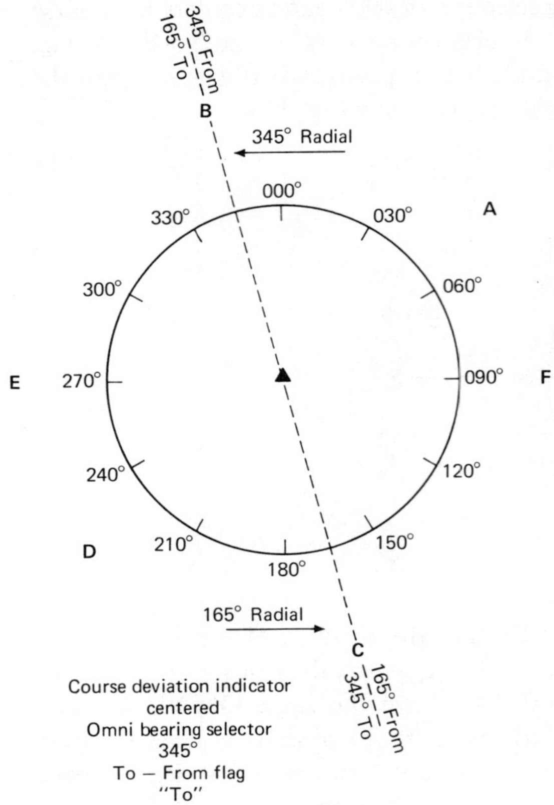

Figure 140

What can we tell about the aircraft heading from the given OMNI indications? Nothing! Remember, the OMNI receiver has nothing whatever to do with the heading of the aircraft.

## Enroute Navigation

Referring to Fig. 141, we see airport A which is our point of departure, and airport B which is our destination. Both airports are on the 220 degree radial of the Belle Isle VOR.

After takeoff from airport A we may tune in and identify the Belle Isle VOR and set 220 degrees on the omni bearing selector. This should center the course deviation indicator with a "FROM" reading on the To-From flag. Now, if we turn to a heading of 220 degrees, we can "follow the course deviation indicator" until we arrive at airport B. Following the course deviation indicator means this: The aircraft is assumed to be in the center of the OMNI instrument. If the course deviation indicator moves to the left of center, as shown in Fig. 141, view 1, it indicates that our desired course is to the left. We must turn left to return to the course. Conversely, if the course deviation indicator moves to the right (Fig. 141, view 2), we must turn right to return to the course. This is called *true sensing*. That is, the course deviation indicator truly represents the position of the course. We will shortly see *reverse sensing* in which

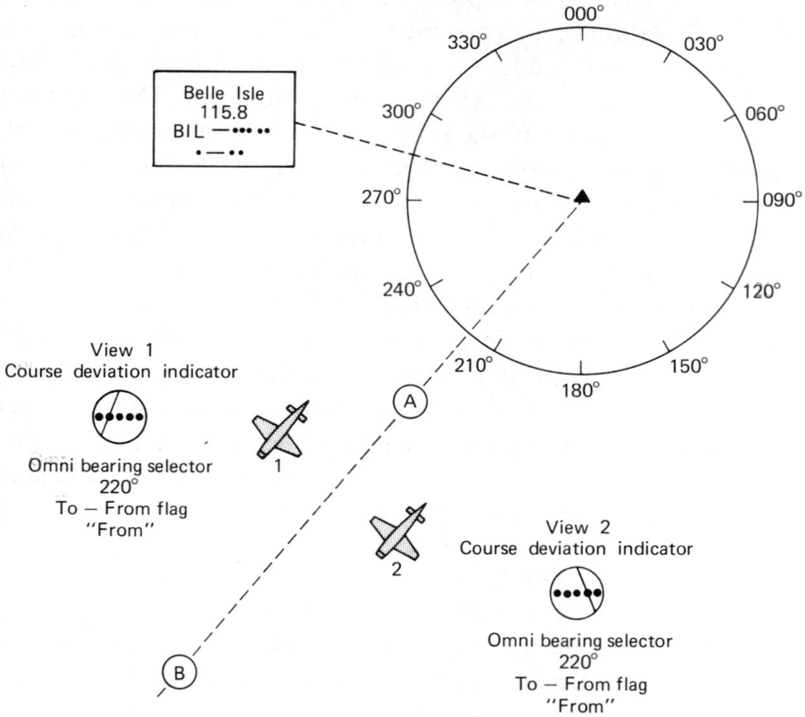


Figure 141

the course deviation indicator moves in a direction opposite to that which is desired.

## Station Passage

As our aircraft moves toward a VOR transmitter, the course deviation indicator becomes more sensitive because the radials are closer together at the station. Full deflection of the course deviation indicator represents 10 degrees off course. When the aircraft is 60 miles from the transmitter, a 10-degree off course indication would mean that we were 10 miles off course. When the aircraft is six miles from the transmitter, the same 10-degree off course indication would mean that we were only one mile off course.

With practice, we will sense our approach to a transmitter due to the rapid deflection of the course deviation indicator with very small aircraft heading changes. At the time of station passage, the course deviation indicator will oscillate violently from side to side and the To-From flag will switch from "TO" to "FROM" or vice versa. *The only reliable indication of VOR station passage is the reversal of the To-From flag.*

## Reverse Sensing

In reference to Fig. 141, when the flight was over airport A the course deviation indicator could also have been centered with the omni bearing selector set to

040 degrees. The To-From flag would have read "TO" however, telling us that an aircraft heading of approximately 040 degrees will take the aircraft *toward* the VOR station. Let us assume that the OMNI readings are as stated, but that the aircraft is still heading toward airport B—a heading of about 220 degrees. Under these conditions, reverse sensing will occur. That is, if the course deviation indicator moves to the right of center, we must turn the aircraft to the left to return to the course (opposite to the true sensing condition). This reverse sensing occurs because the aircraft is not doing what the OMNI is ordering. It is flying *away from* the VOR station on a heading of 220 degrees, but the OMNI is set for a flight on a heading of 040 degrees *to* the VOR station. In other words, the aircraft is "lying" to the OMNI by heading in a direction 180 degrees from that set on the omni bearing selector, so the OMNI is "lying" to the aircraft by giving reverse sensing indications.

To avoid reverse sensing we must make sure that the aircraft is heading in approximately the same direction as we have selected on the omni bearing selector. Actually, the aircraft heading must be within 089 degrees of the omni bearing selection for true sensing.

*Remember these simple rules:*

1. The To-From flag does *not* indicate that the aircraft is flying to or from the VOR station. It does tell us that the omni bearing selection *will* take the flight to or from the VOR *if* the aircraft is headed in approximately the same direction as shown on the omni bearing selector.
2. True sensing (fly into the needle) will result if the aircraft heading is within 089 degrees of the omni bearing selector.
3. The course deviation indicator represents the desired magnetic course. The aircraft is always pictured to be in the center of the OMNI instrument.

## Intercepting a Desired Radial

Suppose the desired radial of a VOR lies somewhere to the north of the departure airport A in Fig. 142. In this instance, we must take off and fly to the 090 degree radial before proceeding outbound on it. Normally, a good interception angle to the radial is about 45 degrees. Hence, we must turn the aircraft to a heading of 045 degrees while tuning in and identifying the VOR station. With 090 degrees set in the omni bearing selector, the To-From flag will read "FROM" (view 1). As we approach the desired radial of 090 degrees, the course deviation indicator will slowly move toward the center of the dial. As it approaches the center, we will turn the aircraft to a heading of 090 degrees to fly outbound on the radial (view 2).

## Wind Drift Correction

Now let us assume that our flight is in a strong north wind condition. When on the 090 degree radial outbound from the VOR station, the north wind will drift our aircraft south of the course, as shown in Fig. 143, view 1.

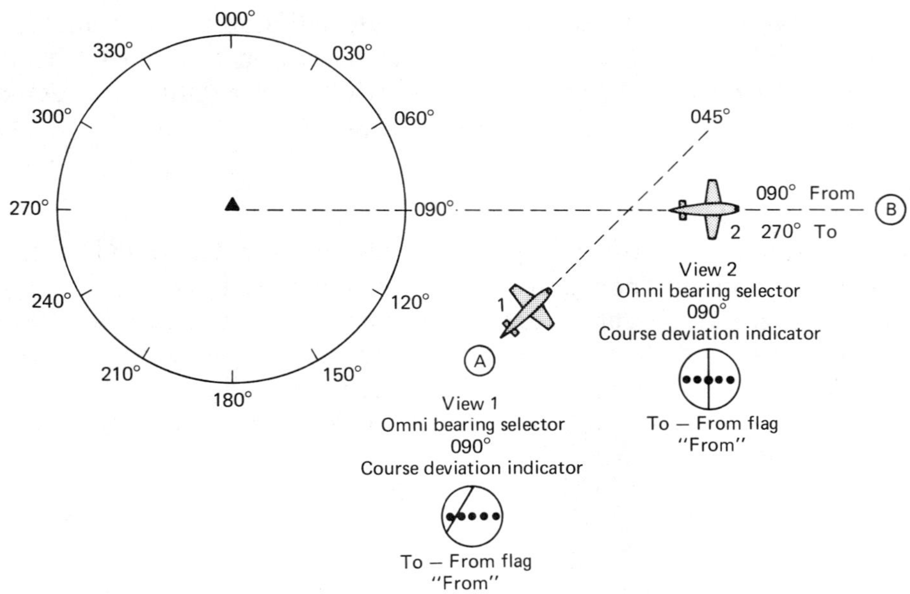


Figure 142

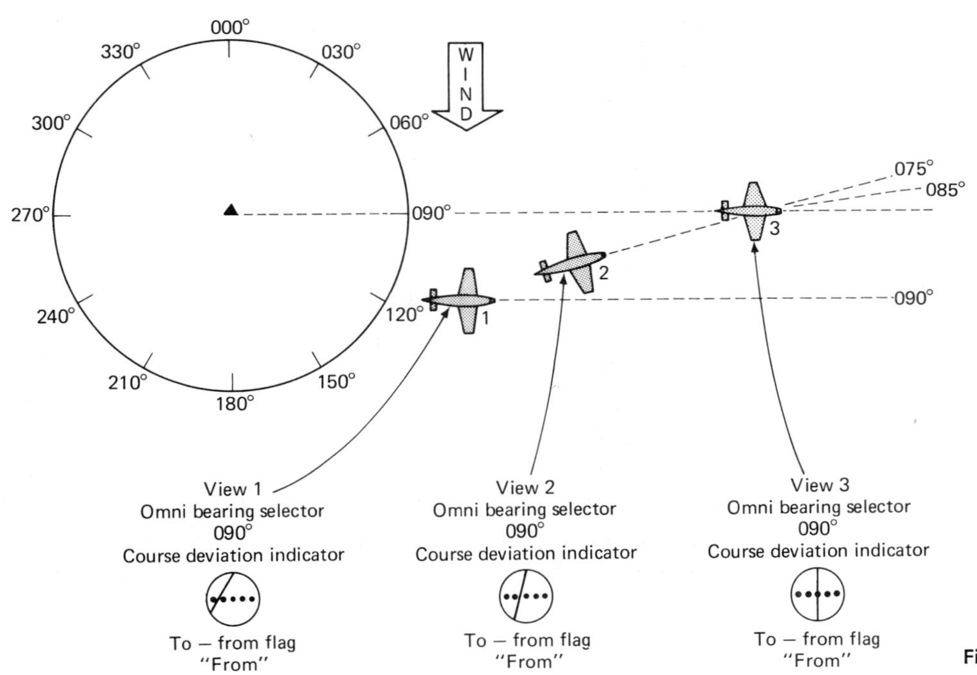


**Fig. 143. Drift correction.**

Since the OMNI is in a true sensing condition, the aircraft must be turned left, into the needle, to return to the course (Fig. 143, view 2). Normally, a turn of 10–15 degrees will suffice to reconverge on the course, unless we allowed the aircraft to drift a great distance before making a correction. As we approach the course, the course deviation indicator will again center.

If, on the original heading of 090 degrees, the aircraft drifted off course to the right, it would not be sensible to return to the same heading again. We must therefore, maintain a heading of about 085 degrees to provide a drift correction angle of about five degrees (Fig. 143, view 3).

We will watch the course deviation indicator to see if another small heading correction is needed in order to remain on course. (A left turn will correct for additional drift; a right turn will lessen the drift correction angle.)

## Position Plotting

The nationwide system of OMNI stations comes in handy indeed for quick and easy position finding. We first decide which VOR stations to use for the *fix*. An accurate fix can be determined with two OMNI stations. Preferably, one station should be dead ahead or directly aft of our aircraft, and the second station should be off either wing tip. These ideal conditions will not always be available, but we should make every effort to pick two transmitters as close to the described position as possible.

If two OMNI receivers are available, each should be tuned to a VOR station. The course deviation indicators are then centered, while reading a "FROM" on the To-From flag.

If only one OMNI receiver is available, we will first tune and identify the station from which the bearing is changing the slowest (the station ahead of or aft of our aircraft). After noting the radial and drawing it outbound from the first OMNI station, we must tune and identify the second station. Again, we note the radial which centers the course deviation indicator and plot it on our chart from the OMNI station. This procedure will result in the most accurate fix because the aircraft is moving during the time spent in tuning the second OMNI station, but the change in bearing to the first station will be minimal.

For example, referring to Fig. 144, we will first tune and identify Silver VOR, centering the course deviation indicator with the omni bearing selector, making sure we read a "FROM" indication on the To-From flag. We will then draw a line on our chart from the center of the Silver VOR compass rose at the same bearing as registered on the omni bearing selector (190 degrees). Our aircraft must be somewhere along that radial.

Then we will tune and identify the Tarmac VOR, repeating the process of centering the course deviation indicator with a "FROM" reading on the To-From flag. We shall then draw the new radial (275 degrees) from the Tarmac VOR. The interception of the two radials pinpoints our location.

## Computing the Time to Fly to the VOR

By using a simple formula and making one turn in our aircraft, we may compute quite accurately the time required to fly from our present position to the VOR station.

As shown in Fig. 145, the position of the aircraft relative to the station is determined as previously described. Then a turn is made, so that the VOR station is located directly off either wing tip (Fig. 145, view 2).

During the turn the aircraft will naturally move to another radial, so

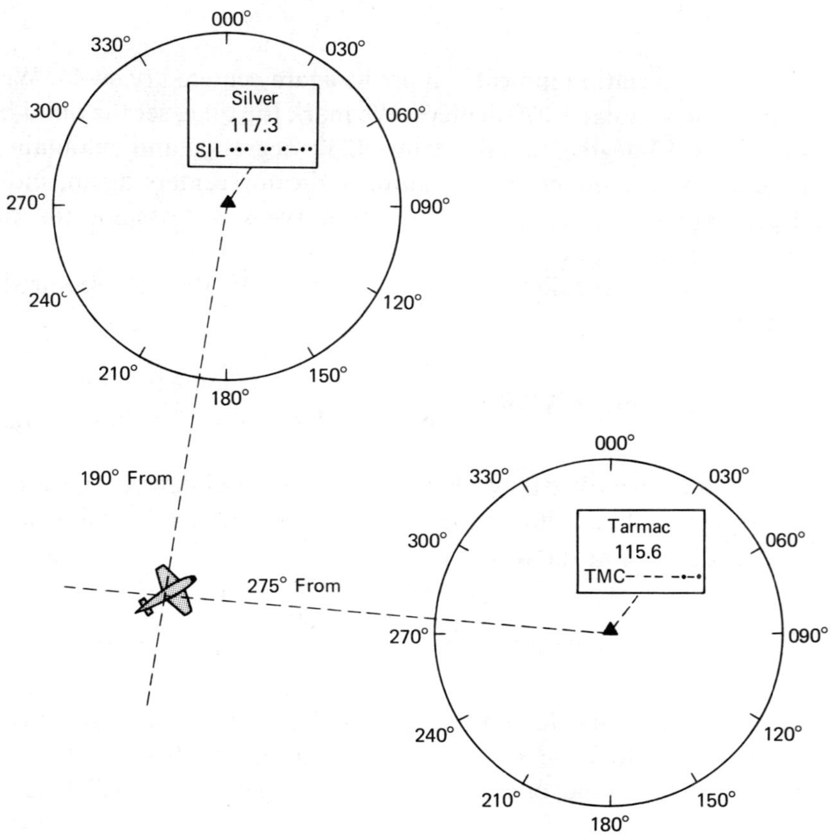


Fig. 144. Two-station position fix.

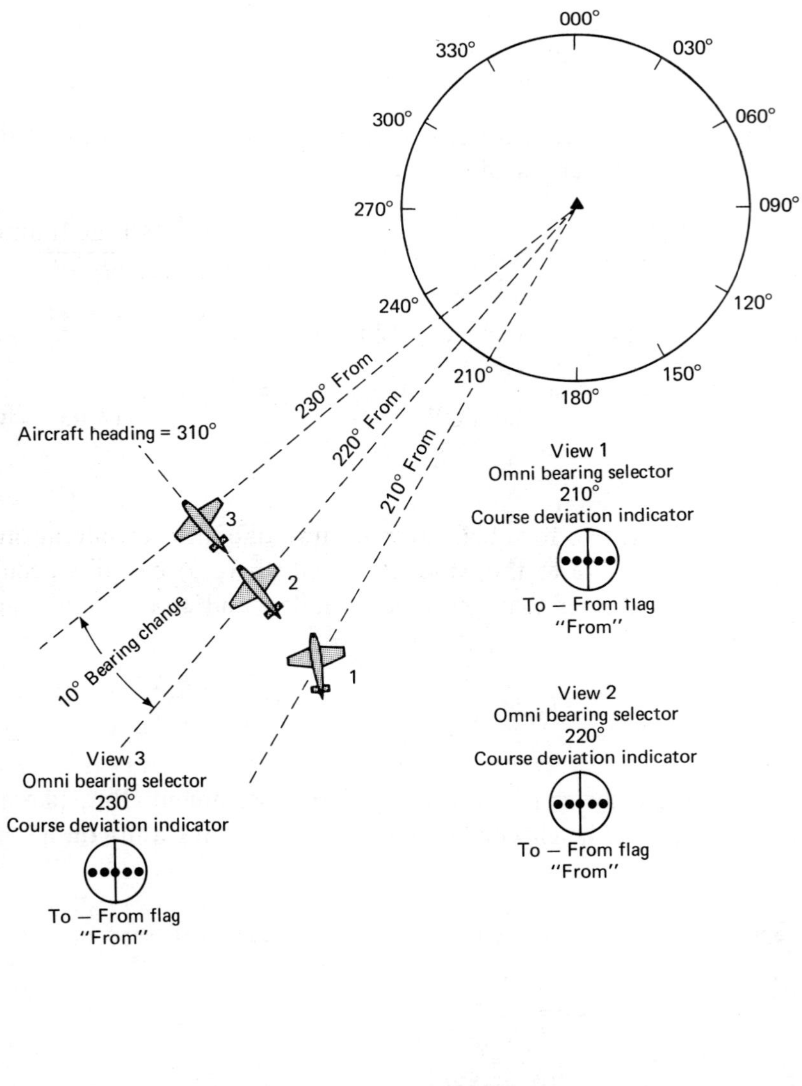


Fig. 145. Time/distance problem.

the course deviation indicator must be again centered (view 2). We will assume that the new radial is 220 degrees. We mark the time, set the omni bearing selector for a 10-degree radial change (230 degrees), and maintain the aircraft heading. When the course deviation indicator centers again, indicating a 10-degree radial change, we note the time (view 3). Assume the time was one and one-half minutes.

We use the following formula to compute the time required to fly to the station.

$$\text{Time to VOR} = \frac{\text{Time (in minutes)} \times 60}{\text{Number of degrees of radial change}}$$

This formula will work with any number of degrees of radial change, but 10 degrees will provide enough change for accurate timing and will make mental division in the formula easy.

$$\text{Time to VOR} = \frac{1.5 \times 60}{10} = \frac{90}{10} = 9 \text{ minutes}$$

When using the ten degree radial change, we can simplify our figuring more by counting the seconds required to accomplish the radial change, and moving the decimal one place to the left. The result will be minutes to the station.

### Distance to The VOR Station

The exact same procedure is used to compute the distance to the station. The formula, however, is different.

$$\text{Distance to the VOR} = \frac{\text{True airspeed} \times \text{time (minutes)}}{\text{Number of degrees of radial change}}$$

Assume our airspeed is 120 knots.

$$\text{Distance to the VOR} = \frac{120 \times 1.5}{10} = \frac{180}{10} = 18 \text{ nautical miles}$$

#### *Wind Effect on Time/Distance Computations*

Although the two formulas given do not take into consideration any change in groundspeed due to the wind, they will prove to be quite accurate unless the distance is greater than 60 nautical miles, and/or the wind speed is in excess of 30 knots.

### OMNI Receiver Checks

The VOR receiver is a precise, reliable instrument. But, like any mechanical device, it must be checked carefully before betting a life on it.

At numerous airports around the country, special test VOR transmitters have been installed, enabling us to perform a simple check of our OMNI receiver. Their locations and frequencies are listed in the Airman's Information Manual (see Chapter 48). When the test station is tuned in, the course deviation indicator should center with an omni bearing selection of 180 or 360 degrees (plus or minus 004 degrees). The 180 degree selection should result in a "TO" indication, and the 360 degree selection should result in a "FROM" indication on the To-From flag. Some airports (also listed in the Airman's Information Manual) have a designated spot painted on the surface to which the aircraft must be taxied. The VOR test station is tuned in, and the course deviation indicator centered. The omni bearing selection must be within 004 degrees of the published test radial.

A third method (much less accurate) is to fly over a designated landmark, which is described in the Airman's Information Manual, with a specific radial selected on the omni bearing selector. Again, the course deviation indicator should center when the omni bearing selector is within 004 degrees of the published radial.

### Use the VOR

The only way to become proficient in the use of the VOR is to use it often. With constant practice, we shall develop the trust and confidence deserved by this marvelous navigation equipment.

FEDERAL AVIATION AGENCY

VFR EXAM-O-GRAM *NO. 15

HOW TO USE VOR

(Series 1)

The VHF omnirange, commonly referred to as VOR, has largely replaced the low frequency, four-course radio range as an aid to navigation. VOR offers many advantages over the low frequency range -- (1) it is less susceptible to static; (2) an unlimited number of courses are available; and (3) navigational information is visual rather than aural. Since most modern aircraft are equipped with VOR receiving equipment, it behooves each pilot to know and understand how to use this equipment properly. The VFR pilot should use radio navigation along with other methods of navigation to maintain his desired course. Always double-check your position by aeronautical chart.

WHAT ARE THE VOR RECEIVER COMPONENTS USED BY THE PILOT? (1) COURSE (or omnibearing) SELECTOR; (2) LEFT-RIGHT (or vertical) needle; and (3) TO-FROM indicator.

WHAT IS THE FUNCTION OF EACH OF THESE COMPONENTS?

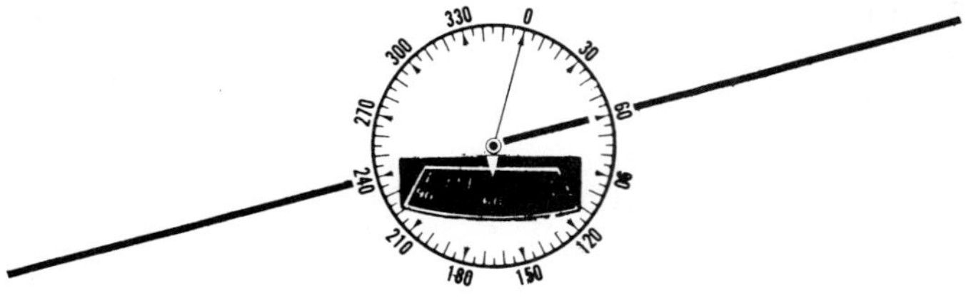


(1) COURSE SELECTOR (CS) -- permits the selection of any course.

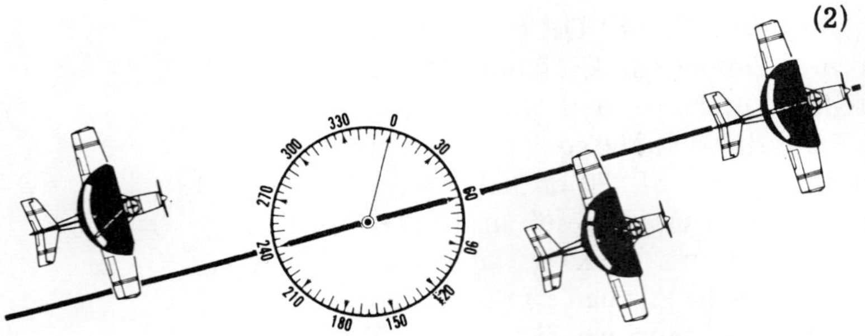


(2) LEFT-RIGHT (L-R) needle -- shows the position of the aircraft in relation to the course selected. If the course line is drawn on the chart, passing through the VOR station to which tuned, the L-R needle indicates on which side of the aircraft the desired course lies.

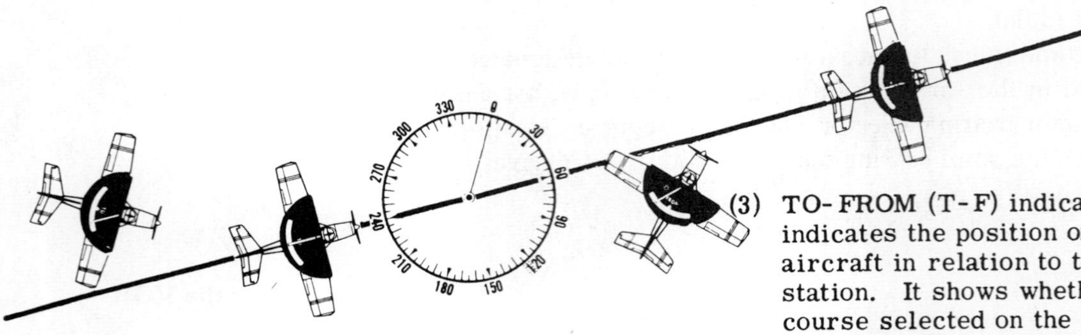


(3) TO-FROM (T-F) indicator -- indicates the position of the aircraft in relation to the VOR station. It shows whether the course selected on the CS, if intercepted and flown, will take you to or from the station.

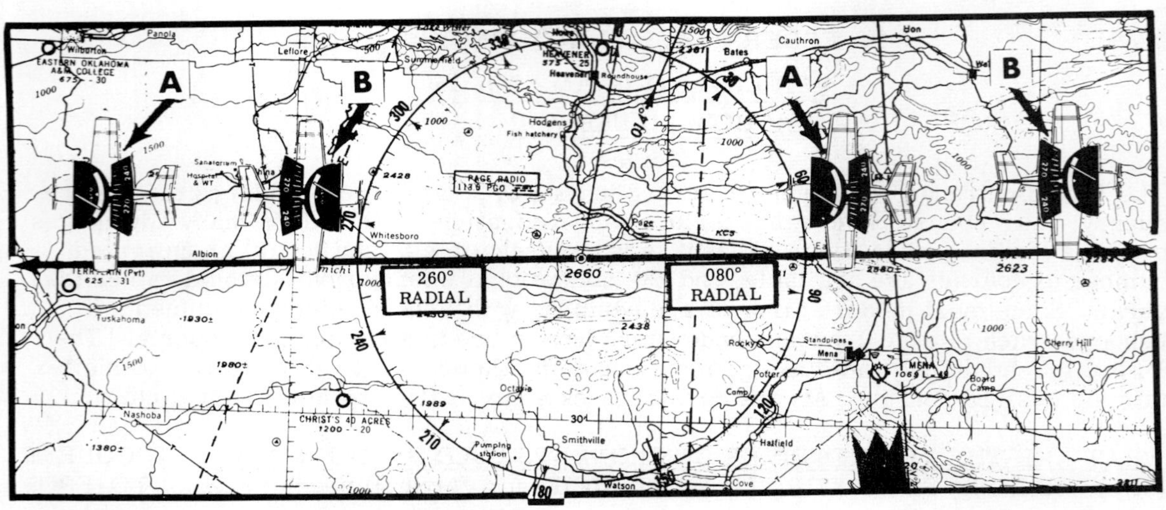


WHAT IS A RADIAL? It is a line of magnetic bearing extending from the station. Note the easterly-westerly line in the illustration. East of the VOR station, this line is the 080° radial; west of the station, this line is the 260° radial.

WHAT IS PROPER SENSING? Simply this -- if the L-R needle is to the right, the desired course is to your right; if the L-R needle is to the left, the desired course is to your left. In other words, the desired course is on the same side as the L-R needle. (NOTE: With reverse or opposite sensing, the course is on the opposite side from the L-R needle.)

HOW CAN YOU BE SURE THAT THE LEFT-RIGHT NEEDLE IS GIVING PROPER SENSING? By ensuring that the heading of the aircraft is approximately the same as the course selected on the CS. Assume you wish to maintain a course of 260° which passes over PAGE VOR station. Your heading naturally will be approximately 260° depending on wind direction and speed. In this case your CS should be adjusted to 260° regardless of whether you are east or west of the VOR station. In other words, always set the CS on the course you are flying -- not the reciprocal of your course. Note the L-R needle in both airplanes above labeled A. They give proper sensing.

WHEN DOES THE LEFT-RIGHT NEEDLE GIVE REVERSE SENSING? When the aircraft heading and course selected on the CS are approximately reciprocals (actually, anytime the angle between heading and radial selected is greater than 90°). For example, both airplanes labeled B above are trying to maintain an easterly course of 080°; however, the CS is set on 260° (the reciprocal of 080°). Note that the L-R needle gives reverse sensing in both airplanes labeled B. In this case, the CS should have been set on 080° to get proper sensing.

IS THE INDICATION ON THE TO-FROM INDICATOR DEPENDENT ON THE HEADING OF THE AIRCRAFT? NO! It is dependent only on the setting of the CS and the direction of the aircraft from the station. Note that in the 2 airplanes east of the VOR station the TO-FROM indicates TO; west of the station, the TO-FROM indicates FROM. You could pivot any of the 4 aircraft in the illustration through 360° and there would be no change in the indication on the T-F indicator.

WHEN DOES THE TO-FROM INDICATOR GIVE A NEUTRAL INDICATION? (1) When an unreliable signal is being received (you are either too far from the station or at too low an altitude, or the station is not properly tuned in); or (2) when you pass directly over the station; or (3) when you cross the radials perpendicular to the course selected on the CS. For example, as the airplanes above cross the 350° or 170° radials (and a short distance on either side), the TO-FROM would have a neutral indication.

## FEDERAL AVIATION AGENCY
## VFR EXAM-O-GRAM *NO. 16
## HOW TO USE VOR
### (Series 2)

IN WHAT THREE GENERAL WAYS MAY A VOR STATION BE USED AS A NAVIGATIONAL AID?

1- To fly a course directly to a station (Airplane 1).

2- To fly a course directly away from a station (Airplane 2).

3- To determine the direction or bearing of your aircraft from a particular station or stations (Airplane 3).

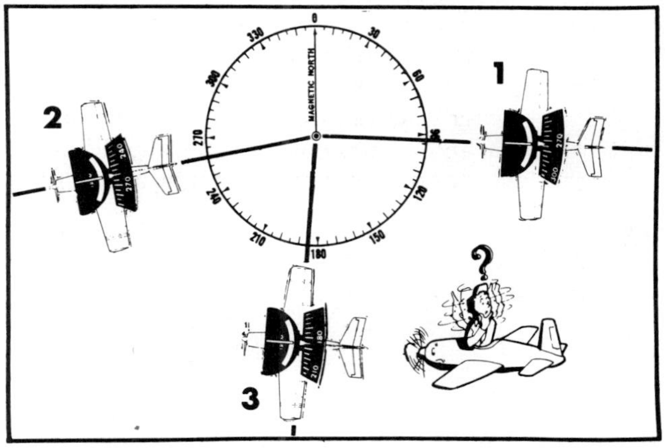


WHAT PROCEDURE SHOULD YOU FOLLOW TO FLY DIRECTLY TO A STATION WHEN YOU ALREADY KNOW YOUR APPROXIMATE DIRECTION FROM THE STATION? (Assume you are flying from Stephan Airport to Hartington Airport; en route you decide to fly to O'Neill Airport. Visualizing your position, you know you are east of the O'Neill VORTAC.)

1- Tune in and aurally identify the station (O'Neill VORTAC).
2- Turn CS (Course Selector) until the L-R (Left-Right) Needle is centered and TO-FROM indicates TO (Airplane 1).

3- Turn to a heading approximately the same as the setting on the CS (Airplane 2).
4- The L-R Needle now has "proper sensing" (i. e. it is displaced in the direction of the desired course). Keep it centered by making turn corrections toward the needle.

(NOTE: After completing Step 3, you could re-center the L-R needle by readjusting the CS, then follow this new course inbound.)

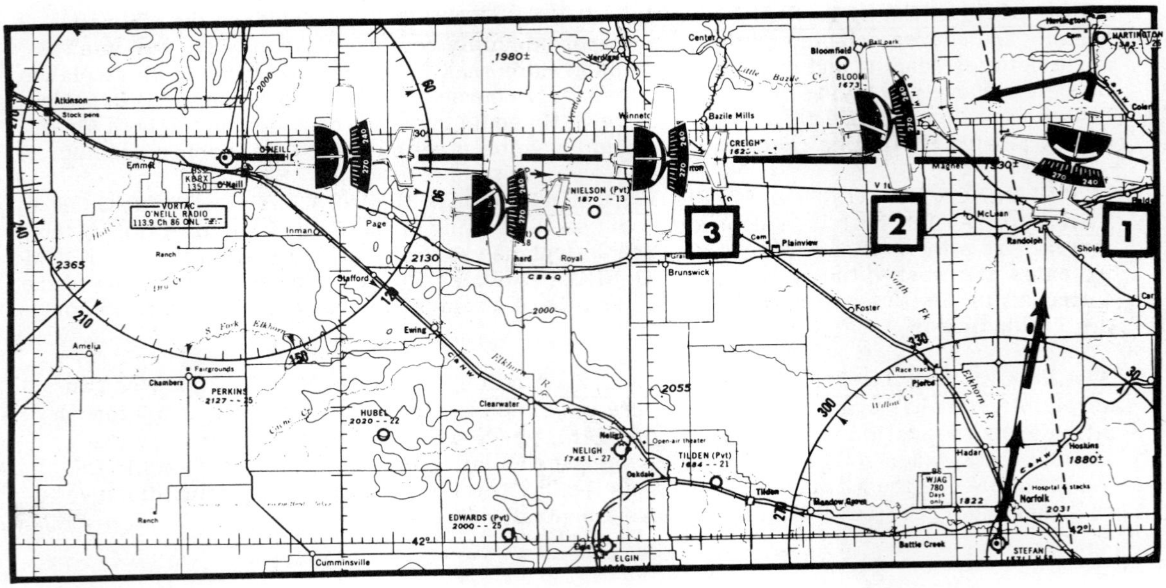

AFTER ARRIVING OVER A STATION, WHAT PROCEDURE SHOULD YOU FOLLOW TO FLY DIRECTLY FROM THE STATION? (Illustration below).

1- As you pass over the station TO will change to FROM. (Airplanes 1, 2A and 2B).
2- At this time turn the CS to the radial which you wish to follow outbound (Airplane 2B).
3- Turn the aircraft to a heading which approximates the new setting of the CS.
4- The L-R Needle now has "proper sensing". Keep it centered by making correcting turns toward the needle. If the needle is displaced to the left, the desired course is to the left and a correcting turn should be made to the left; if the needle is displaced right, the desired course is right and a turn correction should be made to the right.

(NOTE: If your outbound course from the station is the same as the inbound course, then make no change in the CS in Step 2 and continue on your same general heading -- Airplanes 1 and 2A.)

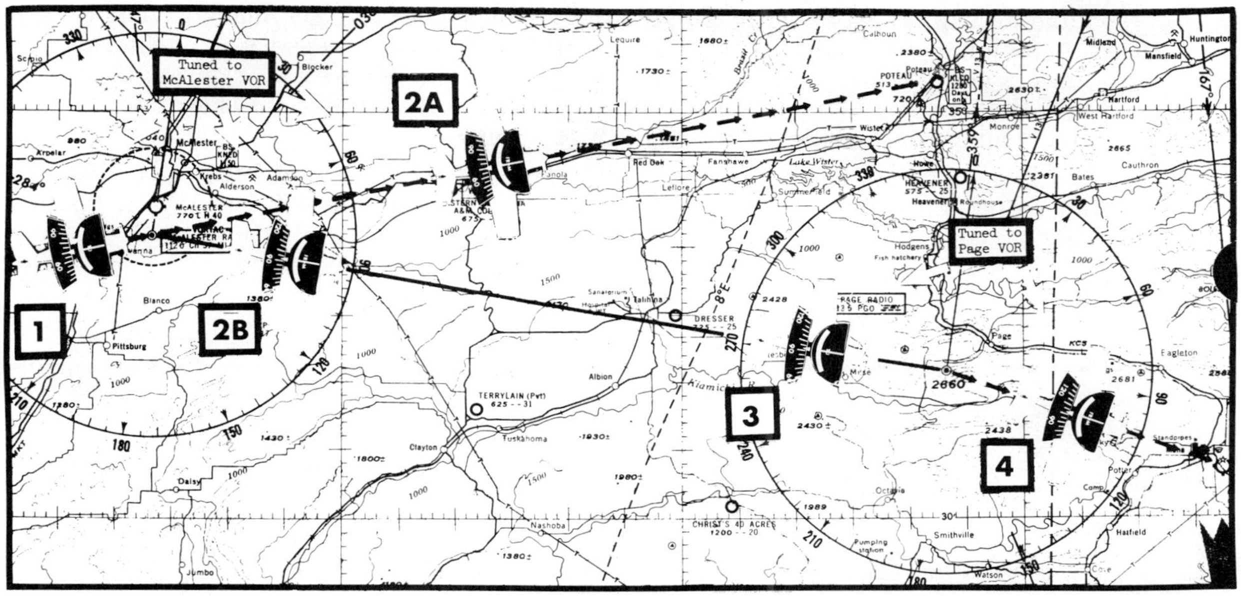


UNDER WHAT SITUATIONS WOULD YOU BE FLYING DIRECTLY FROM A VOR STATION? (Illustration above).

1- The first half of a flight between VOR stations (Airplane 2B).
2- To help you find an airport located in the vicinity of a VOR station by flying outbound from the station along the radial on which the airport is located (Airplane 2A or 4).

WHAT PROCEDURE SHOULD YOU FOLLOW TO DETERMINE YOUR DIRECTION OR BEARING FROM A STATION? (Illustration below).

1- Tune in and aurally identify the station (Columbia VOR).
2- Rotate the CS until the L-R Needle is centered and the TO-FROM indicator indicates FROM (Airplanes 2 and 3).
3- The setting of the CS represents the radial on which you are located (185° for #2 and 220° for #3). Draw the radial on the chart and you will have your line of position.
4- Visualize your position -- you are south and southwest respectively from the station. You do not know how far south or southwest of the station without additional information.

IN WHAT WAY WILL IT BE HELPFUL TO YOU TO KNOW YOUR BEARING FROM A VOR STATION? (Illustration above).

1- If you are flying a course from Vance VOR to the Augusta VOR station (Airplanes 1, 2, 3, 4), you can determine your position along this route by finding your bearing from a VOR station on either side of your route (Airplanes 2 and 3).
2- If you are uncertain of your position, you may determine your bearing from two or more VOR stations. Draw these radials on the chart and your position will be where they intersect. Actually, you are doing this in the illustration above --

you are maintaining a specific radial FROM Vance VOR and TO Augusta VOR. You determine your radial from the Columbia VOR. The intersection of the two radials is your position. You would keep your radio tuned to Vance during the first half of the flight and to Augusta during the second half except when determining your radial from the Columbia VOR.

<u>(NOTE:</u> Knowledge of your exact position during the second half of this flight becomes very important so that you may be sure of avoiding the Prohibited Area, P-378, southwest of Airplane 3).

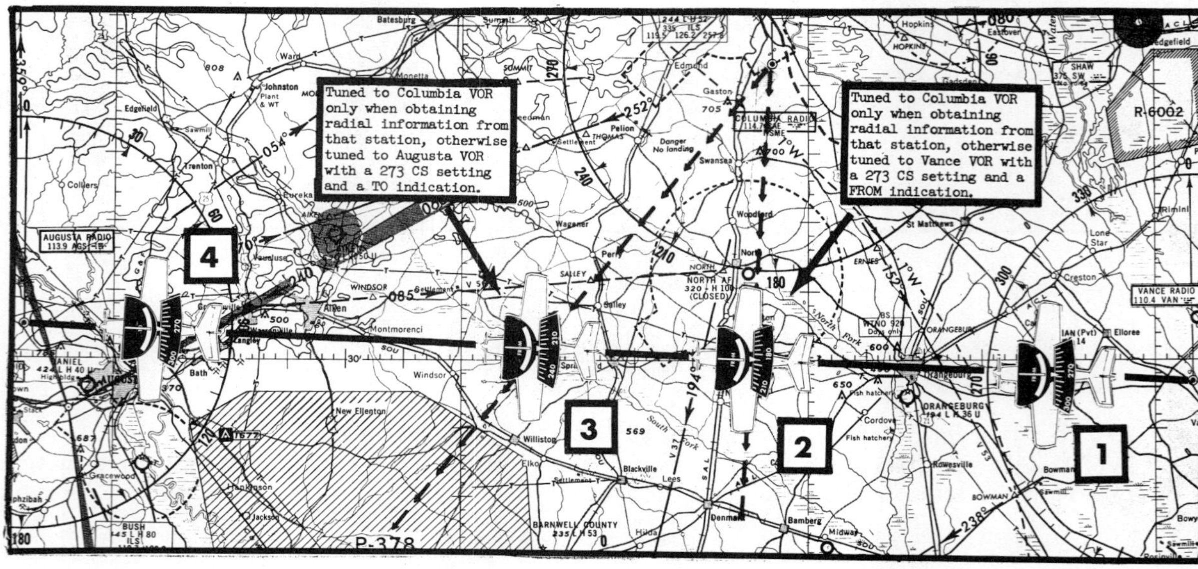


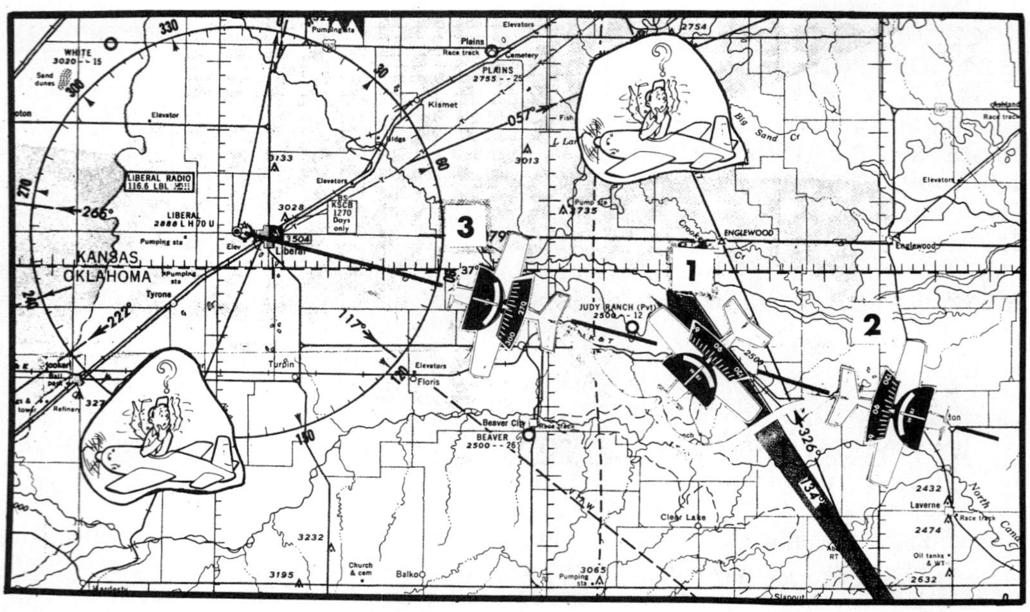

WHAT ARE THE SUGGESTED STEPS TO FOLLOW WHEN YOUR POSITION RELATIVE TO A VOR STATION IS UNKNOWN?

1- Tune in and aurally identify the station (Liberal VOR in the illustration above).
2- Turn the CS until the L-R Needle is centered and FROM appears on the TO-FROM Indicator (Airplane 1). (Remember, the heading of your aircraft will not affect the reading of the TO-FROM Indicator). The resulting setting of the CS tells you the radial on which the aircraft is located or, in other words, your bearing or direction from the station. For example: If the CS reads 225°, you are southwest of the station; 090°, you are east, etc. In the illustration, you are on the 093° radial; however, you do not know how far east of the station you are.
3- Visualize your position relative to the station -- always do this! After determining your bearing and visualizing your position, if you wish to fly directly away from the station along the radial on which you are located, you merely turn to a heading approximately the same as the setting on the CS (Airplane 2). The L-R Needle will have "proper sensing" and you should make corrections (toward the needle) to keep the L-R Needle centered.
4- After determining your bearing and visualizing your position, if you wish to fly directly to the station from your present position, you rotate the CS (approximately 180°) until the L-R Needle is centered and the TO-FROM indicates TO. Turn the aircraft to a heading approximately the same as the setting on the CS and make corrections (toward the needle) to keep the L-R Needle centered (Airplane 3). Since the heading of the aircraft and the setting on the CS are approximately the same, the L-R Needle will have "proper sensing".

IT MAKES GOOD SENSE TO HAVE PROPER SENSING

GET SHARP ON VOR - IT CAN TAKE THE HEADACHE OUT OF NAVIGATION

## CHECK YOUR READING

1. What is the frequency band used for VOR transmission?
2. List the advantages and disadvantages of very-high frequency transmission.
3. After turning on the VOR receiver and selecting the published VOR frequency, what important step must be accomplished?
4. What is the outbound course from a VOR called? What is the inbound course to a VOR called? Are these courses relative to True or Magnetic North?
5. How are OMNI stations identified? How often?
6. What voice transmissions may be heard on most VOR stations?
7. List the various uses to which we may put the VOR.
8. What does the course deviation indicator represent?
9. For what do we use the omni bearing selector?
10. If our aircraft is 5,000 feet above the surface, from what distance may we expect to receive a usable signal?
11. What does the To-From flag tell us?
12. List the steps required to find our position relative to one VOR station.
13. Assume our aircraft is on a heading of 040 degrees. The omni bearing selector is

set to 220 degrees, the course deviation indicator is centered, and the To-From flag reads "TO." What will change when the aircraft heading is changed from 040 degrees?

14. When using one VOR receiver to plot a fix from two VOR stations, which station should be tuned and identified first?
15. How do we recognize VOR station passage?
16. Describe the situation in which the VOR receiver will give *reverse sensing* indications.

# 43

# THE AUTOMATIC DIRECTION FINDER

Although it is not as popular as the very-high frequency Omni directional range, the automatic direction finder is useful as a backup system of radio navigation, as a means of monitoring aircraft position while enroute, for plotting fixes, and for receiving radio communications while on the ground or when VHF reception is impossible.

## Principles of Operation

The automatic direction finder (ADF) operates on the low-frequency band between 190 and 1,750 kilocycles. The wavelength of the radio signal in this frequency band is such that the signal can be received even when the transmitter is not within line-of-sight of the receiver. Therefore, we can receive the low-frequency signal while on the surface, or while behind mountains or buildings. Unfortunately, the low-frequency band is very susceptible to static produced by thunderstorms, precipitation, or other electrical impulses. Depending on the power of the transmitter, the low-frequency signal may be received from great distances (1,000 miles is not uncommon).

### *The Loop Antenna*

The automatic direction finder is dependent upon a particular receiving antenna called a *loop* which is shown in Fig. 146.

As the loop antenna receives a signal, it electronically measures the signal strength. The closer the antenna is to the transmitter, the stronger will

Fig. 146. The ADF loop antenna.

be the signal. The loop rotates until the signal strength on each side of the loop is equal. In other words, the loop will rotate until it lies perpendicular to the transmitted signal, as shown in Fig. 147.

A servomechanism moves an arrow on the face of the cockpit instrument as the loop rotates. The arrow will always point *to* the station.

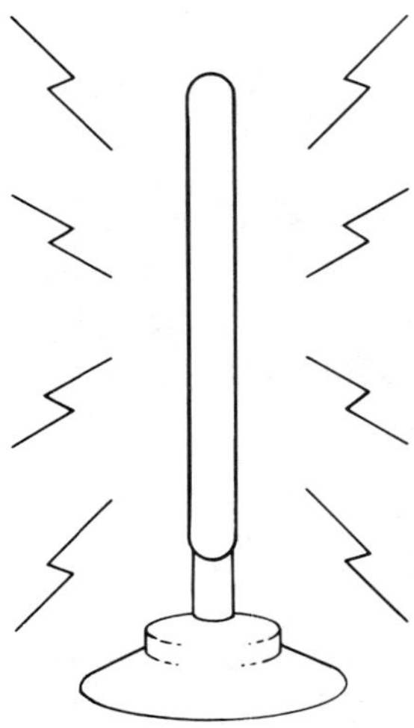

Fig. 147. The ADF loop antenna perpendicular to the transmitted signal.

*Airborne Equipment*

The automatic direction finder is quite simple and inexpensive. Although production models vary in design, the controls and components on all models will be similar to those described here and depicted in Fig. 148.

*On–Off–Volume Control.* As with any airborne radio receiver, the ADF is equipped with an On–Off–Volume control. The receiver requires about 30 seconds to warm up after we turn it on. The volume control should be turned at least halfway toward the full volume position.

*Tuner.* A hand crank or knob is used to scan the frequency band for the published radio frequency. (Frequencies are published in the Airman's Information

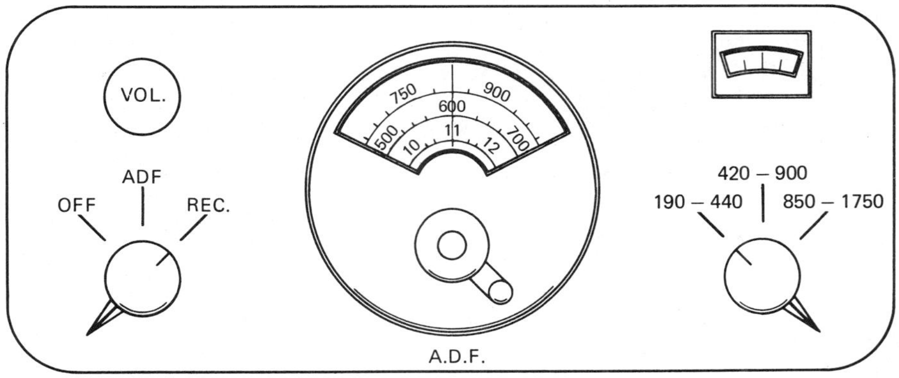


Fig. 148. A typical ADF airborne receiver.

Manual and on navigation charts.) On less sophisticated models, the frequency index may not be accurate, and we may need to "hunt" along the dial in the vicinity of the published frequency to locate it.

*The Function Switch.* A function switch is provided to prevent the ADF loop from constantly *homing in* on each station we receive as we scan the frequency dial. While we tune the ADF, we should place the function switch in the *receive* position. The function switch must be in the ADF or *automatic* position in order to enjoy automatic homing.

*Tuning Meter.* More sophisticated models will be equipped with a tuning meter. This device resembles an ammeter. As we tune the station, we should watch the tuning meter and tune to the highest reception level.

*Station Identification.* As we learned when discussing the VOR, it is extremely important to properly identify any station we intend to use for navigation. A low frequency broadcast station which is specifically designed for navigational use will be identified every 30 seconds by a three-letter Morse code identifier. Standard radio broadcast stations may also be used for automatic direction finding, but normally are identified only once in 30 minutes.

*The ADF Instrument.* On the instrument panel, we will find the ADF indicator. As shown in Fig. 149, the indicator is a round dial which is engraved with the 360 degrees of a circle. Zero, or North, is always aligned with the nose of the aircraft. Therefore, East (090 degrees) will be aligned with the right wing tip, South (180 degrees) will be aligned with the aircraft tail, and West (270 degrees) will be aligned with the left wingtip. The ADF needle is mounted on a pivot in the center of the instrument face and is free to swing in either direction the full 360 degrees.

## Bearings

Many pilots find the word "bearing" confusing. *Bearing* simply means the direction of a straight line between the nose of the aircraft and the transmitting

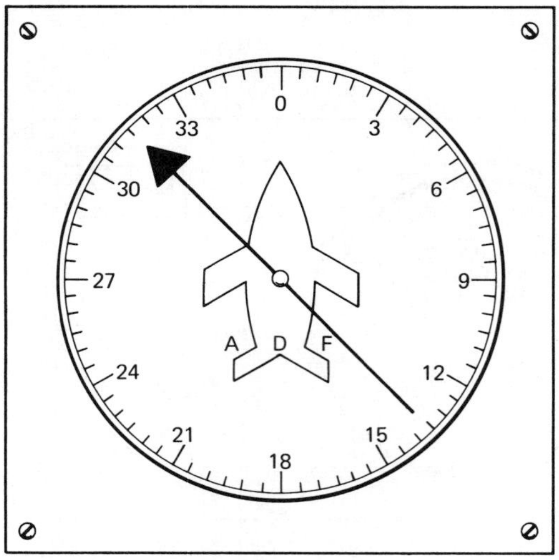


Fig. 149. A typical ADF airborne indicator.

station. If we find the word bearing confusing, we may substitute the word "direction" which should dispel any confusion.

*Relative Bearing*

The relative bearing is the angle between the nose of the aircraft and a line drawn to the station. In other words, relative bearing is the direction in which the station lies, *in relation to the nose of the aircraft*, as shown in Fig. 150.

If the aircraft is pointing directly at the station, the relative bearing is zero, for there is no angle between the nose and the station. If the station lies off the right wingtip the relative bearing is 090 degrees, and so forth, all around the points of the compass. Notice that we have made no mention of the aircraft heading in our discussion of relative bearing, for aircraft heading has no connection with relative bearing.

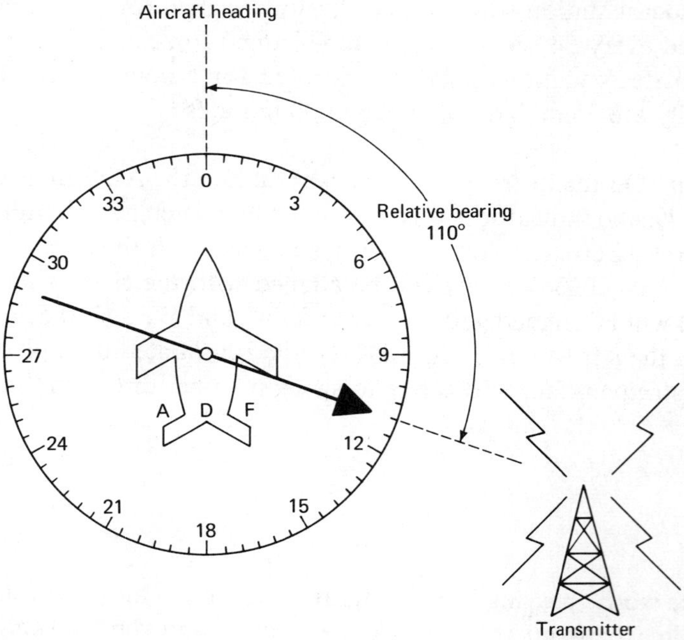


Fig. 150. Relative bearing.

*True Bearing*

The true bearing is the angle between True North and a line drawn from our position to the station. True bearing is the bearing or direction in which the station lies from our position, as measured in reference to True North. We find the true bearing by using the following formula:

$$\text{True bearing} = \text{true heading} + \text{relative bearing}$$

Figure 151 shows how this formula works. Suppose the true heading is 040 degrees, and the relative bearing is 110 degrees. (The ADF needle reads 110 degrees.) We add the true heading (040) to the relative bearing (110) and find that the station lies in a true direction of 150 degrees from our position.

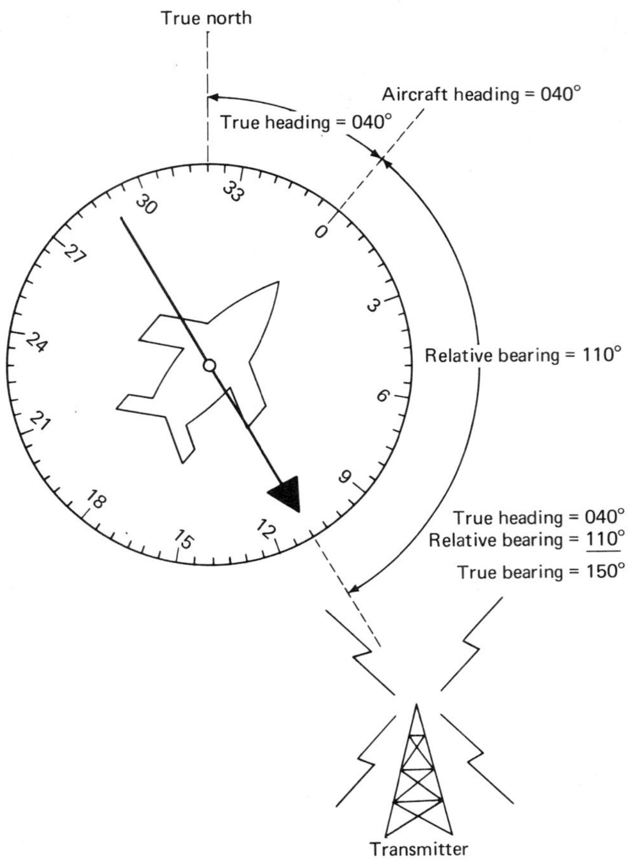


Fig. 151. True bearing = true heading + relative bearing.

*Magnetic Bearing*

The magnetic bearing is the angle between Magnetic North and a line drawn from the present position to the station, as shown in Fig. 152. We can find the magnetic bearing by using the following formula:

$$\text{Magnetic bearing} = \text{magnetic heading} + \text{relative bearing}$$

For example, if the magnetic heading is 028 degrees, and the relative bearing is 110 degrees, the magnetic bearing will be 138 degrees (028 + 110).

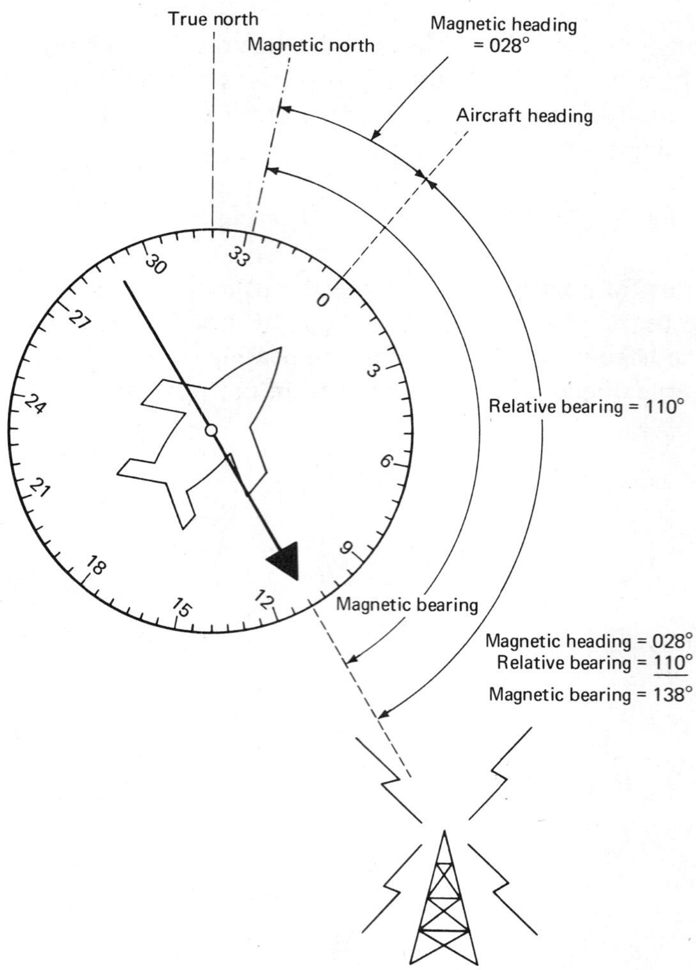


Fig. 152. Magnetic bearing = magnetic heading + relative bearing.

The station lies in a direction of 138 degrees from our position, as measured in reference to Magnetic North.

*Compass Bearing*

The compass bearing is the angle between Compass North and a straight line drawn from the present position to the station, as shown in Fig. 153. We can find the compass bearing by using the following formula:

$$\text{Compass bearing} = \text{compass heading} + \text{relative bearing}$$

For example, if our compass heading is 200 degrees, and our relative bearing is 100 degrees, the compass bearing is 300 degrees (200 + 100). The station lies in a direction of 300 degrees from our position as measured in reference to Compass North.

## Homing

Homing or "bird dogging" to a station is the simplest way to navigate using the automatic direction finder, but it is not always the most efficient method. If we

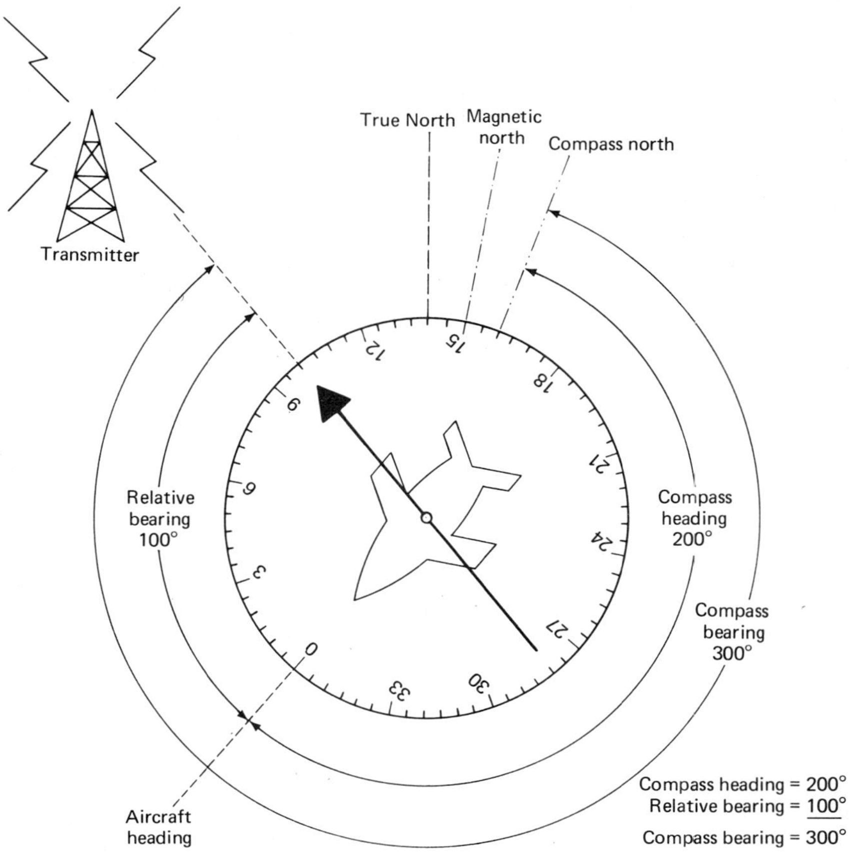


**Fig. 153. Compass bearing = compass heading + relative bearing.**

desire to home on a station, we tune and identify it, and place the function switch on the ADF position. After the ADF needle settles down, pointing to the station, we simply turn the aircraft until the relative bearing is zero. That is, the ADF needle points to the nose of the aircraft.

If there is no crosswind to drift the aircraft, we can maintain the original heading and the ADF needle will remain pointing straight ahead, to the station. However, if any crosswind exists, the aircraft will drift from the original flight path, as shown in Fig. 154.

If, as we drift to the left of the straight line to the station, we maintain a constant aircraft heading, the ADF needle will begin to point to the right of the nose. If we are to proceed to the station, we must turn right until the ADF needle again points to zero, or to the nose, as depicted in Fig. 154, view 3. Again the right crosswind will cause the aircraft to drift to the left, and we must turn further right to continue pointing directly at the station. As we can see, homing with any crosswind will result in a curved course to the station.

## Tracking

Tracking with the automatic direction finder entails the establishing of a proper drift correction angle to prevent a curved course to the station.

Assuming the same wind conditions as previously existed, let us suppose

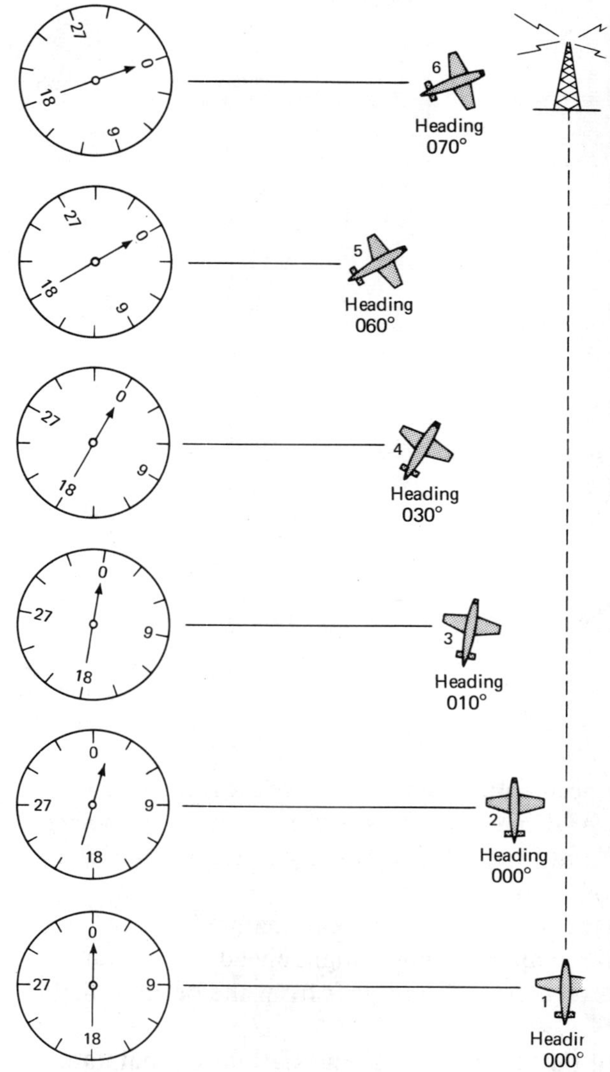


Fig. 154. Homing with ADF.

that we wish to track on a straight course to the station. We tune and identify the station, place the function switch on the ADF position, and allow the needle to settle down, pointing to the station. We then turn the aircraft until the ADF needle points to the nose or zero, and note that the magnetic heading is 050 degrees.

We maintain that heading until the crosswind causes the aircraft to drift left, which will be indicated by the ADF needle moving to the right. As illustrated in Fig. 155, view 2, we should allow the aircraft to drift about five degrees from the original track.

From this position, a right turn of 5 degrees would again point the aircraft to the station, but we would not return to the original track. So, an addi-

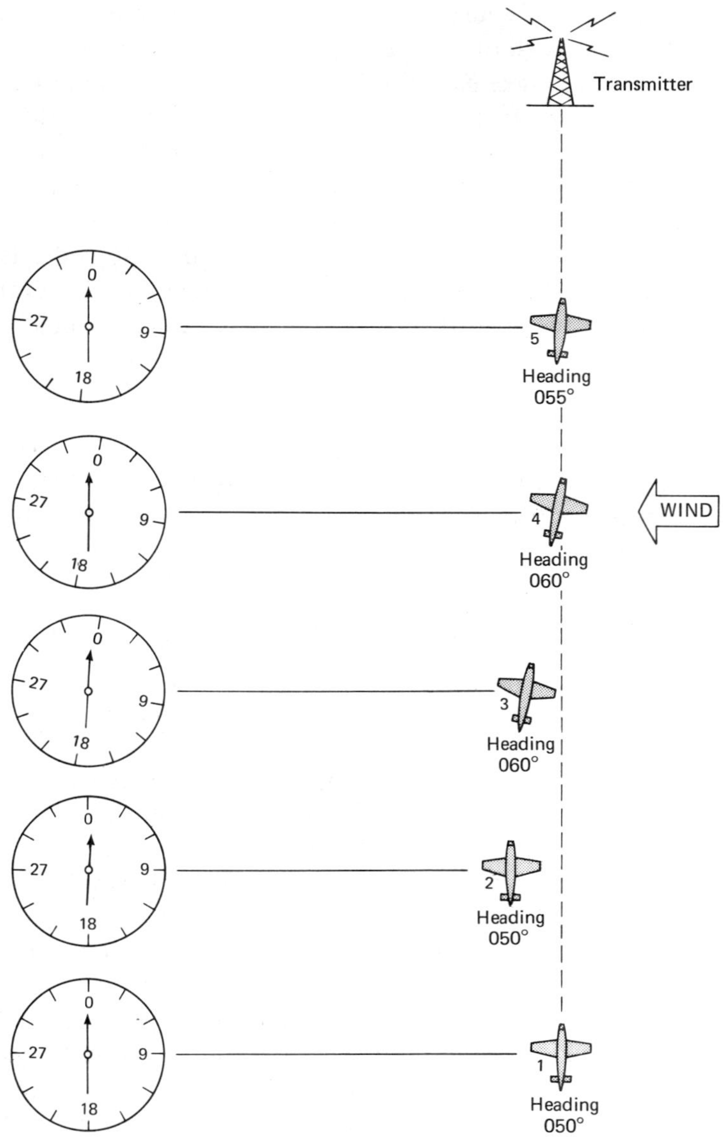


Fig. 155. Tracking with ADF.

tional 5-degree right turn will be in order, to return us to the track. As shown in Figure 155, view 3, after we complete a total of 10 degrees of right turn, the ADF needle will move to a position 5 degrees to the left of the nose. (Actually the ADF needle does not move, for it is pinned to the station. In reality the instrument case and the aircraft move under the needle.)

Assume that our new heading of 060 degrees will overcome the drift and return us to the original course. We will know that we are on that course when the ADF needle shows a displacement to the *left* of the nose in the same magnitude as our new heading is displaced to the *right* from the original heading (10 degrees). This is shown in Fig. 155, view 4.

We can see at this point if we returned to our original aircraft heading of 050 degrees, the ADF needle would point to the nose or zero on the ADF instrument. However, if we were to do this, the crosswind would only drift the aircraft left of course again. Our new heading of 060 degrees took us from an *off course* position to an *on course* position. Any heading which takes us ***to*** a course will take us ***through*** that course if we maintain it. Therefore, a compromise heading, halfway between the original heading (050 degrees) and the new heading (060 degrees) should keep the aircraft on course. Therefore, we should turn left to 055 degrees. As shown in Fig. 155, view 5, the ADF needle will remain displaced 5 degrees to the left of the nose. As long as the ADF needle shows a *left* displacement which is equal to the *right* displacement of the aircraft nose (from the original heading), we know that the aircraft is maintaining the correct amount of drift correction to remain on course.

### Station Passage

When the aircraft proceeds over the station, the ADF needle will rotate 180 degrees, pointing to the station, which will then lie to the rear of the aircraft.

### Tracking From a Station

We may also track away from any low-frequency transmitter using the automatic direction finder. As we leave the station, the ADF needle will settle down on 180 degrees, pointing to the tail of the aircraft. As depicted in Fig. 156, assume that we wish to track outbound on a magnetic bearing of 050 degrees.

The crosswind from our right will drift the aircraft to the left of course, as shown in Fig. 156, view 2. The ADF needle will move to the right of the tail as the left drift occurs. Suppose the ADF needle moves 5 degrees right.

A 10-degree right turn, to a heading of 060 degrees will overcome the drift and return the aircraft to the course. As we complete this turn the ADF needle will indicate an additional 10 degrees to the right of the tail, or a relative bearing of 165 degrees. (This is shown in Fig. 156, view 3.)

When we return to the desired course, the ADF needle will show the station as being 10 degrees to the right of the tail, as shown in Fig. 156, view 4. We can always tell if the aircraft is on course by mentally turning the aircraft to the original heading which would have carried the aircraft along the course with no crosswind. For instance, in this case, we would mentally turn the aircraft 10 degrees left, to a heading of 050 degrees. As we did so, the ADF needle would return to 180 degrees, indicating that the station again is directly behind the aircraft.

In other words, when tracking outbound, we know we are on course when the ADF needle is displaced from the tail in the same direction and in the same degree as the nose is displaced from the desired bearing.

However, if we did return to a 050 degree magnetic heading, the aircraft would again drift left of course. Our new heading of 060 degrees returned the aircraft to the course, so naturally it will bring us through the course if we maintain it. Therefore, a compromise heading of 055 degrees should be maintained to keep the aircraft on course (Fig. 156, view 5).

As long as the displacement of the nose from the desired course equals

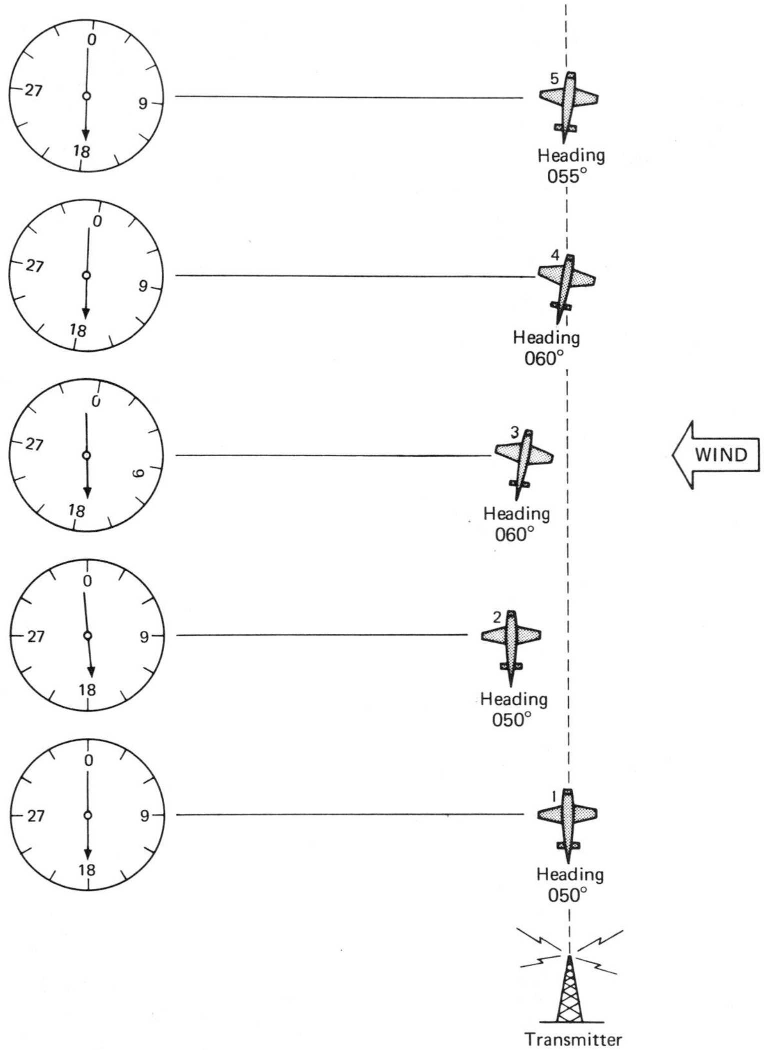


Fig. 156. Tracking outbound with ADF.

the needle displacement from 180 degrees (the tail), we will know that we are maintaining the course.

## Plotting Position

The automatic direction finder may also be used to compute a position fix. As we learned when discussing the VOR, two stations should be used for an accurate fix. Ideally, one station should be directly ahead of the aircraft, and the other station should be abeam of the aircraft. We first tune and identify the station which is most nearly in line with the path of flight, because the bearing from that station will change more slowly than the bearing from the abeam station.

If we intend to plot our position on a sectional chart, we must compute a true bearing, because we will be using the longitude lines on the sectional chart to measure angles, and they are aligned with True North.

Figure 157 shows this sample fix.

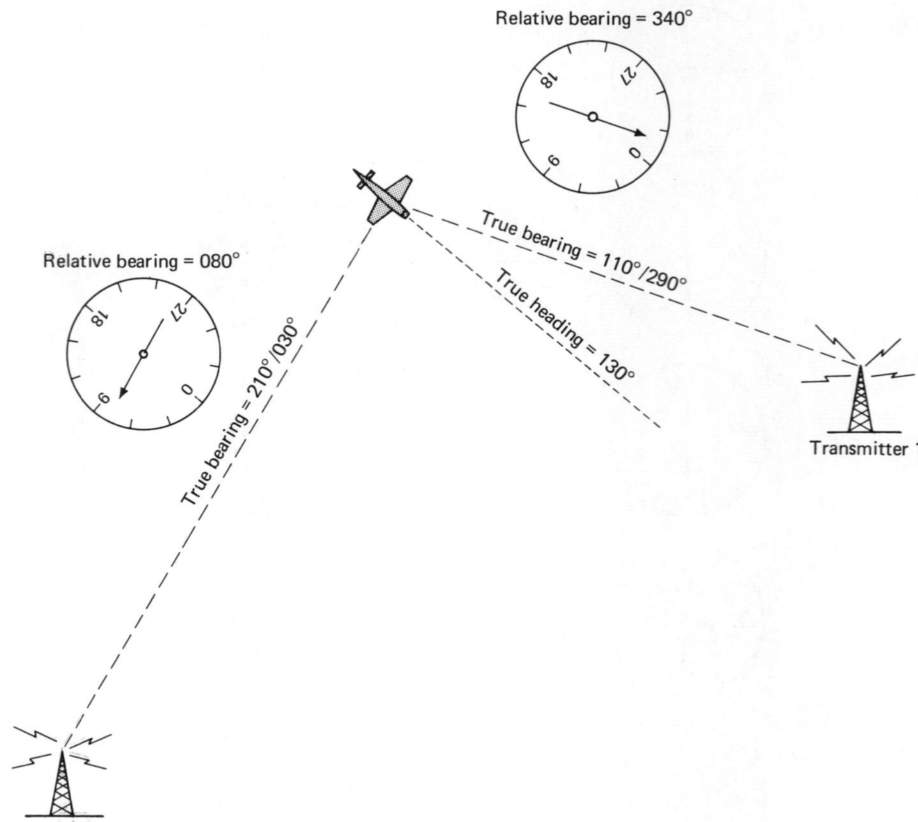


Fig. 157. Position fix.

Suppose the true heading is 130 degrees, while maintaining a true course of 120 degrees. We receive a relative bearing of 340 degrees from station number 1. Using the formula: True bearing = true heading + relative bearing, we find that the true bearing equals 130 degrees plus 340 degrees, or 470 degrees. We must deduct 360 degrees, for there are only 360 degrees in a circle. 470 degrees minus 360 degrees = 110 degrees. Therefore, station number 1 lies in a true direction of 110 degrees from the aircraft. In order to plot our position from the station, we must find the reciprocal of 110 degrees, by adding or subtracting 180 degrees. We may then draw a line from station number 1 in a direction of 290 degrees (in relation to True North) on which we are located.

We will accomplish the same procedure using station number 2. Assume that the relative bearing is 080 degrees. The true bearing equals 130 plus 080 degrees, or 210 degrees. The reciprocal of 210 degrees, or 030 degrees will then be plotted on the sectional chart in relation to True North.

The point of intersection of the two true bearing lines will represent the current position of the aircraft.

## Time/Distance Checks

The automatic direction finder may also be used to compute the time or distance from a station. The formulas are the same as were used when we learned time/distance checks using the VOR.

$$\text{Time to station} = \frac{\text{Time for bearing change (minutes)} \times 60}{\text{Degrees of change}}$$

$$\text{Distance to station} = \frac{\text{Time in minutes for bearing change} \times \text{airspeed}}{\text{Degrees of change}}$$

For either computation, we tune and identify the station, and turn the aircraft until the ADF needle is 5 degrees in front of either wing tip, as shown in Fig. 158, view 1.

We then maintain the aircraft heading until the ADF needle points 5 degrees aft of the wingtip, as shown in Fig. 158, view 2. We note the time required for the 10-degree bearing change, and using either formula, compute the time and/or distance to the station.

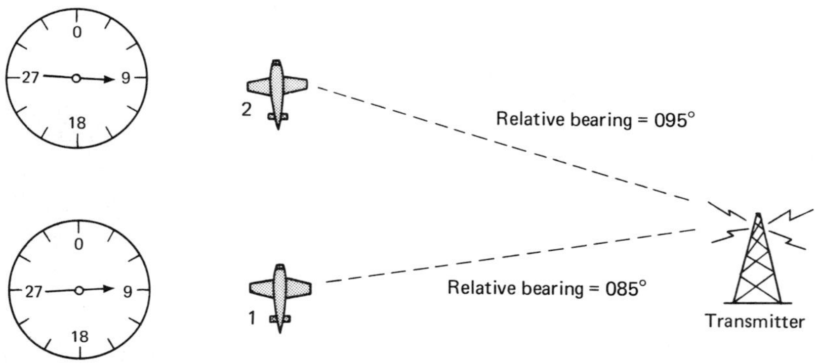


Fig. 158. Time/distance checks.

Department of Transportation

FEDERAL AVIATION ADMINISTRATION

## VFR PILOT EXAM-O-GRAM° NO. 39

### SIMPLE ADF FOR VFR NAVIGATION

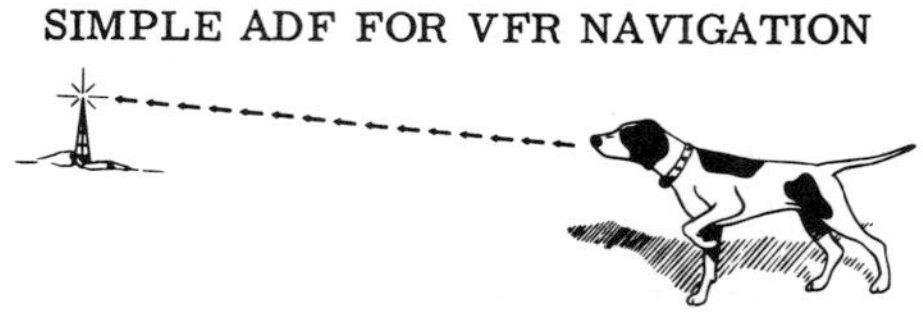

To test the applicant's knowledge of the practical aspects of cross-country flying, FAA written examinations contain test items on the use of radio aids to VFR navigation. In analyzing applicant performance on written examinations, it is apparent that a high percentage of applicants lack an adequate understanding of Automatic Direction Finding (ADF) procedures. Since a large number of modern aircraft are ADF equipped, it is essential that VFR pilots have a reasonable knowledge of the use that can be made of this important navigational aid. Since there is very little instructional material available to the VFR pilot in this area of knowledge, the material which follows is pertinent, timely, and promotes aviation safety.

The methods presented here are only the basic and more common uses of ADF suitable for VFR flights where "pin point" precision is not a vital factor. These methods include how to (1) fly to the station, (2) determine the direction of your position from the station, and (3) determine the ADF indication that signifies arrival at a desired direction from the station.

WHAT TYPE OF RADIO STATIONS CAN BE USED FOR ADF? Most ADF receivers receive signals in the frequency spectrum of 190 KHz (Kilo Hertz) to 1750 KHz, which includes LF (low frequency) and MF (medium frequency) navigation facilities, and the AM (amplitude modulation) commercial broadcast stations. Primarily for air navigation, the LF/MF stations are FAA and private non-directional radio beacons (Rbn), ILS compass locators (LOM), and four-course radio ranges (MRA). Marine radio beacons can be used but present difficulties in air navigation because they transmit only at brief scheduled periods and are arranged in groups of three or more along the coast, with each in the group transmitting on the same frequency. AM stations, or standard broadcast stations, are useful for air navigation; but remember that they can be identified only when the broadcast is interrupted for "station identification," some operate only during daylight hours, and many of the low-powered stations transmit on identical frequencies causing erratic ADF indications.

HOW DOES AN ADF INSTRUMENT INDICATE DIRECTION TO THE STATION? As implied by its nickname "Bird Dog," the ADF has automatic direction qualities which result in the indicator always pointing to the station to which it is tuned. This action is presented to the pilot on the vertically mounted face of the instrument (Fig. 1 and 2); that is, when the pointer is straight up (nose position), the station is ahead of the aircraft; when the pointer is straight down (tail position), the station is behind the aircraft; and when pointing 90° to either side (wing tip position), the station is off the respective wing tip.

Fig. 1

WHAT DOES THE ADF INDICATION MEAN IN TERMS OF BEARINGS TO THE STATION? The more commonly used ADF instrument (sometimes termed Radio Compass), to which this Exam-O-Gram relates, has a stationary azimuth dial graduated up to 360° (with 360 or 0 at the top of the instrument representing the aircraft nose). The bearing pointer shows only the relative bearing, or angle from the nose of the aircraft to the station; i.e., the Relative Bearing in Fig. 1 is 060°. A more sophisticated instrument called a Radio Magnetic Indicator (RMI) uses a 360° azimuth dial which, being slaved to a gyro compass, turns with the aircraft to continually show the Magnetic Heading of the aircraft at the top of the instrument. With this rotating azimuth thus referenced to a magnetic direction, the bearing pointer superimposed on the azimuth, then indicates the Magnetic Bearing to the station. For example, in Fig. 2, the Magnetic Heading is 240°; the Magnetic Bearing to the station shown by the No. 1 needle, is 104°. The No. 2 needle shows a Magnetic Bearing of 040° to another station.

Fig. 2

HOW CAN ADF BE USED TO FLY TO THE STATION? The easiest, and perhaps the most common method, is to "home" to the station. Since the ADF pointer always points to the station, simply head the aircraft so that the pointer is on the 0° or nose position. The station, then, will be directly ahead of the aircraft. With a crosswind, however, the aircraft would continually drift and, unless a change in heading is made, would no longer be headed straight to the station. This would be indicated by the pointer moving to the windward side of the nose position. By turning into the wind (toward the pointer) so as to continually return the pointer to the 0° position, the aircraft is flown to the station, although in a curving flight path, as shown in Fig. 3 inbound. (As this curving flight path deviates from the direct course, use caution to avoid drifting into unanticipated obstructions or terrain.) The lighter the crosswind and the shorter the distance, the less the flight path curves. Upon arrival at and passing the station, the pointer will swing from a nose position to a tail position.

IS IT POSSIBLE TO "HOME" AWAY FROM THE STATION? For all practical purposes--NO! "Homing" away from the station can be accomplished only if there is no crosswind. Attempting to keep the station directly behind the aircraft in a crosswind by turning to keep the pointer on the tail or 180° position, requires that the aircraft be turned more and more to a downwind heading. This, of course, results in the aircraft getting further and further from the desired course (Fig. 3 outbound).

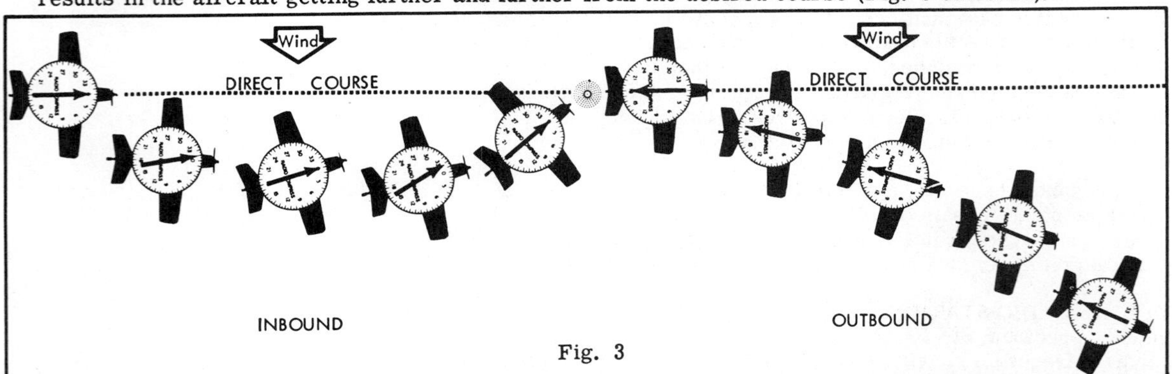

Fig. 3

OF WHAT IMPORTANCE IS THE MAGNETIC HEADING OF THE AIRCRAFT WHEN USING ADF? ADF should be considered as a moving, "fluid" thing. The number to which the indicator points on the fixed azimuth dial does not mean anything directionally until it is related to the aircraft's heading. Due to this relationship, the heading must be observed carefully when reading the Relative Bearing-to-the-station. Anytime the heading is changed, the Relative Bearing will be changed an equal number of degrees.

HOW CAN THE MAGNETIC BEARING TO A STATION BE DETERMINED ON A FIXED ADF AZIMUTH DIAL? Looking at the ADF instrument, imagine yourself as being in the center of the fixed azimuth, with the nose of the aircraft at the 0° position, the tail at the 180° position, and the left and right wing tips at the 270° and 090° positions respectively. When the pointer is on the nose position you are heading to the station and the Magnetic Bearing can be read directly from your compass (plus or minus deviation). If the pointer is left or right of the nose, note the direction and number of degrees of turn that would (if you were to head to the station) move the pointer to the nose position, and mentally apply this to your heading. For example, in Figure 4, a turn 60° to the left would place the pointer on the nose position. 60° left of the 090° Magnetic Heading is a Magnetic Bearing of 030° to the station. Your location, then, is southwest of the station, and if you were to head to the station, your heading would be 030°.

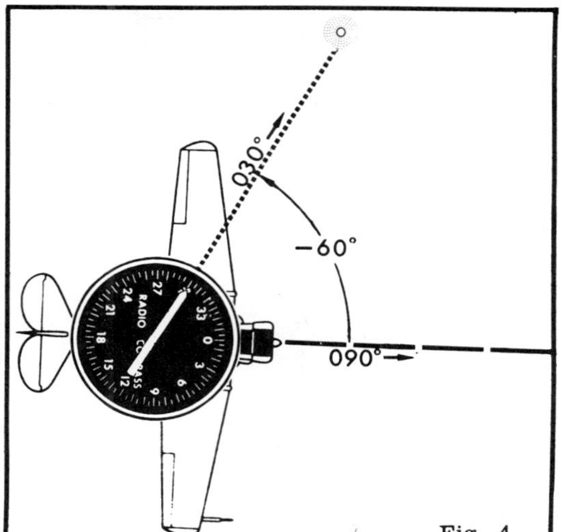

Fig. 4

HOW CAN THE MAGNETIC BEARING FROM THE STATION BE DETERMINED? Since the direction from the station is the opposite of the direction to the station, it can be determined by following the steps discussed in the preceding paragraph and add or subtract 180°, as appropriate, to the aircraft-to-station bearing. In other words, find the reciprocal of that bearing. If the ADF pointer happens to be behind the wing-tip position (flying away from the station), an alternate method is to note the number of degrees and the direction of turn that would move the pointer to the tail position, and apply to the heading. For example, in Figure 5, a turn 45° to the right would place the pointer on the tail position. 45° to the right of the 030° Magnetic Heading is a Magnetic Bearing of 075° from the station (east-northeast of the station). Just as "radials" always extend outward from a VOR station in a magnetic direction, the Magnetic Bearings from an ADF station should be thought of as "radials" of that ADF station. It is important to determine the radial (bearing from the station) because to locate your position, the line of position must be plotted from the known station location, as similarly done in VOR orientation.

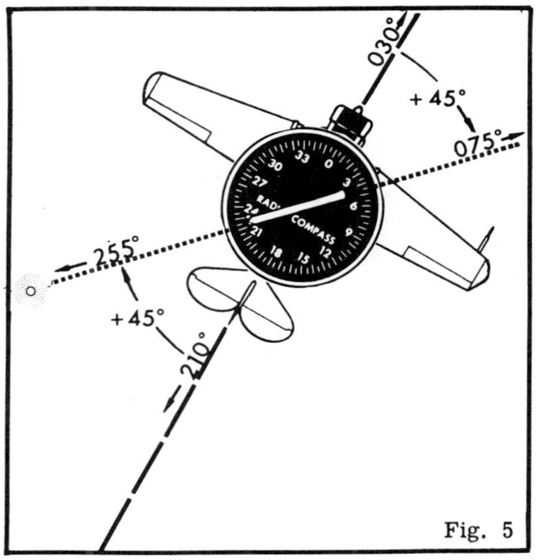

Fig. 5

CAN THE ADF INDICATION WHICH SIGNIFIES ARRIVAL AT A SPECIFIC BEARING-TO-THE-STATION BE PREDETERMINED? YES! From the aeronautical chart, first ascertain whether the station is left or right of the course being flown. Then, after selecting the Magnetic Bearing-to-the-station that you desire to intercept, determine the angular difference between that bearing and your Magnetic Heading (angle of intercept). With the aircraft headed to the 0° position of the ADF azimuth, the bearing indicator, in pointing to the station, will show the relative angle between the aircraft's nose and the station. As you continue on course this angle will gradually change, since the position of the aircraft relative to the station is changing. Arrival at the preselected bearing-to-the-station will be indicated when the pointer shows the difference between the heading and that bearing (angle of intercept). For example, as shown in Figure 6, if your Magnetic Heading is to be 315° and the selected bearing-to-the-station is 225°, the angular difference is 90° left. Arrival on this bearing then, will be indicated when the pointer is 90° left of the nose position. If the station is to the right of the course on a bearing of 045°, the 90° angle between the heading and the bearing would be shown to the right of the nose position, as in Figure 7.

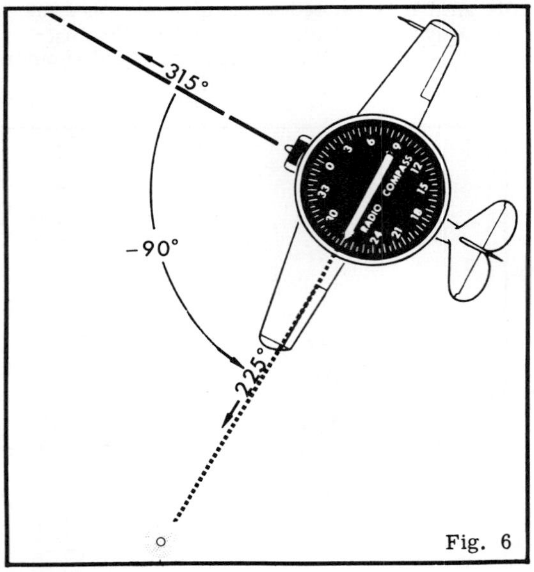


Fig. 6

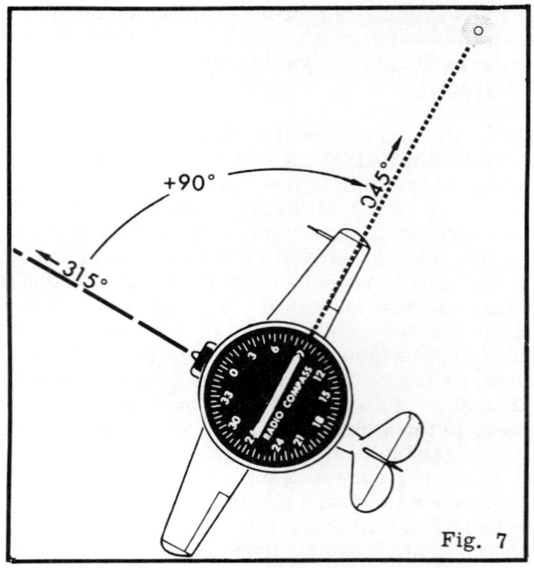


Fig. 7

HOW CAN MAGNETIC BEARINGS TO OR FROM A STATION BE DETERMINED ON AERONAUTICAL CHARTS? Since a compass rose or azimuth is not shown at LF/MF or AM stations on VFR-type charts, the most accurate way to obtain a Magnetic Bearing is to measure the direction with a plotter, taking into account the local magnetic variation. However, the many VOR azimuths and airway courses (already oriented to magnetic direction) printed on the chart can be used satisfactorily for approximation of ADF bearings on VFR flights. This approximation can be made on the basis of the direction of the nearest VOR radial or airway that most closely parallels the bearing of the ADF station. Remember, though, that the VOR radial or printed airway direction is outbound from the station. To find the bearing to the station, simply determine the reciprocal of the parallel radial or airway.

HOW CAN ADF BE USED TO SUPPLEMENT VOR NAVIGATION? One of the most valuable uses of ADF is the determination of your position along the course being flown. Even though you are following a course along a VOR radial, obtaining an ADF bearing that crosses the course will establish your "fix" or position along that course. This is particularly advantageous when an off-course VOR is not available for a cross bearing or when the only VOR receiver must be used as the primary tracking system.

WHAT CAN CAUSE AN ERRONEOUS ADF INDICATION? As mentioned earlier in this Exam-O-Gram, two or more standard broadcast stations may transmit on the same (or close to the same) frequency and interfere with each other's signal. This causes the ADF pointer to oscillate in an attempt to discriminate between stations. Whenever possible, choose stations of higher power and lower frequencies or wait until you are closer to the station before using it. Another source of erroneous bearings is improper tuning, or tuning in the fringes of a signal. Always tune to the center of the signal, which may be a few KHz (Kilo Hertz) on either side of the published frequency. And, of course, always make an absolute identification of the station before using it for navigation purposes. In the vicinity of electrical storms, the ADF pointer tends to swing from the radio station to the center of the storm at every flash of lightning. This makes it difficult to obtain reliable bearings. Erroneous or fluctuating bearings may also result from the deflection of radio waves from the surface of mountains. Use caution when taking bearings over mountainous terrain.

## CHECK YOUR READING

1. What frequency band does the automatic direction finder receive?
2. List the advantages offered by low-frequency transmission.
3. List the disadvantages of low-frequency transmission.
4. How does the *loop* antenna work?
5. Why is a function switch a desirable feature of an airborne ADF receiver?

6. What does *bearing* mean in respect to navigation?
7. What is *relative bearing*?
8. What is *true bearing*?
9. What is *magnetic bearing*?
10. Why is *homing* with the ADF sometimes inefficient?
11. What is the difference between homing and tracking with ADF?
12. Assume we are tracking inbound on the 170-degree magnetic bearing from the ADF station. With no wind, what will be the magnetic heading?
13. Assume the true heading is 045 degrees. The ADF needle points to a station directly off our left wing tip. What is the true bearing to the station?
14. Assume the magnetic heading is 240 degrees. The ADF needle points to a station and reads 240 degrees on the ADF instrument. To what magnetic heading must we turn to point the aircraft to the station?

# 44

# OTHER AIDS TO RADIO NAVIGATION

Recent advances in radio aids include the *Radar Beacon* or *Transponder* and *Distance Measuring Equipment*. Since these aids can be put to very good use in VFR flying, and because many light aircraft are equipped with one or both of these electronic devices, we should have at least a working knowledge of them.

## The Transponder

The transponder is an airborne electronic device which is used to positively identify traffic on a surface based radar scope. The transponder emits a select signal which, if the surface radar is controlled to interpret that select signal, shows a bright return on the radar screen. Specific signals or *codes* have been designated to indicate aircraft in specific conditions. For instance, all VFR traffic which are transponder equipped may be requested by Air Traffic Control to emit "code 3300." The surface radar equipment may then be adjusted so as to receive a special identifying return signal from only those aircraft which are operating on transponder code 3300. In this manner, air traffic may be segregated on the radar scope.

In addition to this selectivity, each individual transponder may be activated to emit a powerful signal which will appear on the radar screen as a bright slash mark. This enables the air traffic controller to positively identify any one aircraft on the scope, simply by requesting the pilot to identify himself.

Reception of transponders is line-of-sight, just as the signals of the very-high frequency omni directional ranges are. Low altitude, or antenna shielding by the aircraft itself may therefore result in reduced range.

When operating VFR with radar advisory service of any kind, we should advise the controller if our aircraft is transponder equipped. The controller will instruct us as to what code he wishes and when to identify ourselves.

*Transponder Phraseology*

Standard phraseology has been instituted to remove confusion during congested communications conditions.

*Squawk* (number): operate the transponder on designated code.

*Ident:* engage the identification feature of the transponder. Most transponders are equipped with a small push button for identification use.

*Squawk standby:* switch the transponder to the standby position.

*Squawk low/normal:* operate the transponder on low or normal sensitivity as specified.

*Stop squawk:* switch off the transponder.

*Squawk mayday:* operate the transponder in the emergency position. (Code 7700 has been designated for emergency use. The transmission of a code 7700 signal is instantly received on the surface radar.)

*Restriction*

Under no circumstances should we operate a transponder on code 0000.

When selecting the code number, we should *avoid switching through code 7700* as it might cause a false alarm in the ground facilities.

## Distance Measuring Equipment (DME)

Distance Measuring Equipment enables us to tell our distance instantly from a VOR station which is DME equipped.

An interrogating signal is transmitted from the airborne equipment which is received at the ground station (VOR). The ground station equipment is triggered and a second signal is transmitted to the aircraft. The airborne equipment measures the time between the interrogation and the return signal, and converts this time measurement into a mileage reading on our instrument panel, on either a dial or digital counter.

The mileage indicated on the DME will be *nautical* and slant range. The difference between the actual miles to the station and the slant range mileage is not critical, particularly if the slant range distance is more than the aircraft's altitude divided by 1,000. For example, if the aircraft is at 8,500 feet, slant range is accurate enough for precise computations until within 8.5 miles from the transmitter.

Groundspeed checks may be made by noting the time in minutes required to fly any convenient distance. The following formula is applied to compute the groundspeed.

$$\frac{\text{Distance flown} \times 60}{\text{Elapsed time in minutes}} = \text{Groundspeed}$$

To simplify computation, we may use a three-minute time period and multiply the distance travelled by 20.

## CHECK YOUR READING

1. Of what use is a transponder?
2. What is the transponder code used to indicate an emergency condition?
3. Who designates the transponder code to be used for each specific flight?
4. What is the meaning of *squawk ident*?
5. What transponder code must never be used by civilian pilots?
6. What radio facility is combined with DME?
7. Is the mileage indicated by DME statute or nautical?
8. What would be our groundspeed if we took 3 minutes to fly 9 miles?

# PART IV

# COMMUNICATIONS

# 45

# COMMUNICATIONS

As pilots we will spend a great deal of time communicating to surface-based radio stations, particularly as we progress to commercial and instrument flying. Tests have shown conclusively that the pilot who learns to speak effectively early in his flight training will have less difficulty in using the radio, and will more efficiently converse with others during the remainder of his flying career. If we begin with poor communication habits we will tend to keep them forever, but if we begin properly we will learn and build good speech habits.

The voice which may be natural and easy to listen to in normal conversation may not be acceptable for flight communications. Just as the artist uses his brush on canvas differently than he would if he painted a house, so must we use our voice in a special manner when using our radio communication equipment.

## Elocution

Recent studies by the Federal Aviation Administration have shown certain speech problems to be the most common. We must make an effort to correct these problems in our speech.

### *Rate Too Fast*

If we try to get too much information into a short time span, we simply confuse the listener. This will lead to constant repeating, the tendency to speak before keying the transmitter, or mouth movement void of sound.

*Articulation*

We must pronounce each word carefully and avoid running words together if we expect others to understand every word. Remember, inflight communications are normally brief and to the point. Since every word counts, every word must be clearly understood. We will discuss common problems of articulation later.

*Loudness*

The tendency to begin each sentence loudly and "trail off" or gasp at the end of the sentence is common. A "singsong" voice tends to rise and fall in volume as well as tone. We should try to maintain one volume throughout our conversation, which should be the normal conversational volume.

If we are hesitant or lack confidence in our performance, we may tend to begin our conversation quite loudly, but the volume will decrease as our confidence wavers.

Many pilots experience the problem of increasing volume in their conversation, particularly when they become impatient. Radio transmitters overmodulate if the voice becomes too loud, which results in a "garbled" conversation.

*Phraseology*

The meaning of a sentence may be made difficult to understand if we change only the phraseology. For instance, if we ask a friend how to pronounce the word BA-SEB-ALL he is likely to stumble in confusion before recognizing the word BASE-BALL as he is more used to hearing it spelled. "I went to bed . . . . . . . . . . . . early last night." is difficult to recognize because of the long pause after the word "bed." Therefore, we must learn and use proper phraseology when speaking on the radio.

*High Pitch*

A note of urgency or exasperation is conveyed by a high pitched voice. For instance, say aloud, "What are you doing?" in a normal voice, and repeat it in a high pitched scream.

*Anger*

It is surprising how easily anger can be transmitted by the *tone* of a voice. The old expression, "It's not what you say, but how you say it," is quite true. We must use care not to convey anger in our voices, even when angered, because anger is infectious. If the receiving party senses anger in the tone of our message, he will undoubtedly react with anger. Confusion will lead to danger if the pilot and the controller are busy transmitting anger instead of expediting the flow of traffic.

*Vocal Support*

When speaking, we must give our voices all of the support we can. We cannot do this while smoking, eating, or drinking. Fatigue also shows in the voice. A broadcast by a voice which lacks proper support is difficult if not impossible to interpret.

Although a "singsong" voice is certainly not desirable, neither is a monotone. The voice is capable of changing tone, and our ear is used to hearing tone change. Vocal inflection in the proper place aids in the interpretation of our message.

*Monotone*

An accent or dialect adds to the problem of interpretation. English is the international standard language of flight, but many pilots and communications specialists speak a foreign language when off the job. Their tendency is to slur their English, particularly when tense. This problem is easily overcome once we are aware of it.

*Dialect*

If we continually pause during a sentence, we give the impression that we are thinking while we are talking instead of thinking before we speak. The introduction of the sound "a-a-a-ah" is one of the most common and disconcerting habits, and unfortunately it is not limited to inexperienced communicators.

*Pause*

We cannot use too many words in our transmission if we expect others to understand us. We must say what we wish and nothing more. The use of frills or eight syllable words should be avoided.

*Excessive Words*

Most of us are lazy with our pronunciation. It becomes a habit, for instance, to say "Juh eat?" instead of "Did you eat?" Mispronounced words become difficult to understand when broadcast.

*Pronunciation*

Some of us will undoubtedly become overconscious of our speech and will, therefore, try too hard. We can easily confuse the listener by overemphasizing our words.

*Overemphasis*

## Self-Improvement

We can best determine our vocal qualities by listening to them. Unfortunately, we cannot hear our voices as others do. We become used to hearing our own bad habits, and until they are pointed out, they go unnoticed. Our sinus cavities affect the tone of our voices. Our own ears do not hear the same tone when we speak as do the ears of another. Therefore, if we wish to listen to ourselves properly, we should use a recording device. Practice of common phraseology, tone, volume, and speed will soon improve our ability to communicate.

## Articulation

Proper articulation demands that all of the sounds which comprise a message are clear and distinct. We must not omit sounds from words, slight or distort any important sound, or substitute one sound for another. The intelligibility of a message depends upon the articulation of each syllable of each word.

Good articulation requires work. We must establish proper habits by consciously striving to improve. If we listen to a popular news broadcaster on television or radio, we will normally enjoy hearing proper articulation. Many commercial messages, particularly those voiced by women, go too far with the pronunciation of every letter and as a result sound theatrical if not ridiculous. Therefore, we can learn by listening to and criticizing others.

### *Assimilation*

The change of one sound to another because of the influence of a neighboring sound, is called *assimilation*. For instance, the phrase, "the dish shines" is normally pronounced "the di shines." This pronunciation is perfectly acceptable. But we must use care not to go to extremes in our assimilation.

### *Telescoping*

If we slur two or more words, we *telescope* them. For instance, "must be" is commonly mispronounced as "muzzbe." Another common example of telescoping is the phrase "no, jew." instead of "No, did you?"

## Self-Improvement in Articulation

We must develop distinct, clear-cut pronunciation. We must avoid being overly precise, or lazy. Improvement will come after thinking how sounds are made, feeling each word, and listening to others and ourselves.

## Microphone Technique

Engineers, designers, and manufacturers of microphones devote a great amount of time and money to produce an instrument which will carry our voice vibrations to a transmitter. Unfortunately, many pilots abuse the microphone to such an extent as to nullify its design features. Just as with all mechanical devices, the radio microphone must be used properly if it is to perform within design standards.

### *Common Errors in Microphone Use*

A few of the more common errors in the use of the microphone are discussed here. An awareness of these problems will help us to improve our technique.

*Turning Away From The Microphone.* Most microphones are *noise cancelling*, which means that outside noises are cancelled out. Therefore, we must speak directly into the microphone.

*Microphone Too Far Away.* Our lips must be no more than one-quarter of an inch from the microphone. If we speak from a greater distance, the bass tones will disappear, and greater intelligibility will result.

*Microphone Too Close.* If we "swallow" the microphone, breath blast will cause excess noise in the broadcast.

*Poor Timing.* Many pilots tend to activate the "push to talk" button after they begin talking. Others tend to release it before finishing their sentence. These practices result in "clipped" messages which are difficult or impossible to comprehend.

## Terminology and Phraseology

Specific terms and phrases are used when communicating by aeronautical radio, each having a particular meaning. Some of the more common ones are presented here for study. The best way to become familiar with all of them is to listen to radio transmissions often.

### *Aircraft Call Sign*

When we initiate a call to a ground station of any kind, we will use the complete aircraft call sign, preceded by the make or type of aircraft.

Example: "Cessna Two Four Seven Three Alfa."

An abbreviated aircraft call sign may be used, but *only* after being initiated by the ground station. (This precaution is observed because the ground station may be in contact with two or more aircraft having some identical digits in their call signs.) The shortened call sign will consist of the aircraft type followed by the last three characters of the identification number.

Example: "Cessna Seven Three Alfa."

### *Ground Station Call Sign*

When contacting a ground station of any kind, we will use the name of the location of the airport, followed by the appropriate indication of the type of station.

Example: "Boston Tower" (airport traffic control tower)
"Miami Ground Control" (ground control position in a tower)
"Chicago Approach Control" (tower radar or nonradar position)
"Los Angeles Radio" (F.A.A. Flight Service Station serving Los Angeles)

## Contact Procedure

Each radio contact is comprised of four parts. Usually the omission of any part only serves to confuse the contact, requiring repeats.

*The Call-up*

The call sign of the station *being called* is given first. This alerts the particular station operator to the fact that a message is about to be transmitted for his attention. The words "this is" precede the full call sign of the caller. The word "over" signifies that the message is ended and a reply is expected.

Example: "Kansas City radio, this is Piper six seven four two Zulu, over."

*The Reply*

The reply is composed of the call sign of the station to which the reply is directed, the words "this is", the call sign of the replier, and the word "over."

Example: "Piper six seven four two Zulu, this is Kansas City radio, over." (Notice that the reply from the ground station did not contain a shortened aircraft call sign. Therefore, the pilot will not shorten the call sign either.)

*The Message*

The message is that portion of the broadcast which conveys or requests information. The message is preceded by the call sign of the ground station, the words "this is" (may be omitted), and the full aircraft call sign. The message is ended by the word "over."

Example: "Kansas City radio, Piper six seven four two Zulu, requesting Kansas City surface wind, over."

*The Acknowledgement and/or Ending*

The end of a conversation consists of the call sign, the word "Roger" or "Wilco" and the word "Out." "Roger" means "I have received all of your transmission." "Wilco" means "I will comply with your request." "Out" means, "This conversation is ended. I expect no reply."

Example: "Piper four two Zulu, Kansas City radio, surface wind at Kansas City, three zero zero degrees, two five knots, over."
"Kansas City radio, Piper four two Zulu, Roger, Out."

**Broadcasting Technique**

Our transmission should be concise, and in a normal conversational tone.

When we initiate a call-up to any facility, we must indicate the frequency on which a reply is expected *if other than normal.* The Chicago Flight Service Station, for example, has the ability to transmit over many different VOR stations. If we wish a reply on the Des Plaines VOR we must say "request reply Des Plaines VOR."

We should attempt to contact the *nearest* facility which can fill our need, not the most distant.

Because flight service stations broadcast weather information at 15 minutes past every hour, we should avoid contacting them at that time, unless our message is urgent. It is good operating procedure to listen to the station for a moment or two prior to calling it, to determine that the operator is not in the midst of a conversation with someone else.

When making a position report, the name of the reporting point should be used in lieu of the phrase "over your station" because the station from which the communicator is broadcasting may be many miles from the station over which we receive his broadcast.

## Standard Words and Phrases

Whenever practical, we should use the standard words and phrases listed here. However, at any time, we may feel free to use common language over the radio. We must not refrain from communicating simply because we do not know the appropriate terminology. The use of plain, simple language is certainly proper if these terms do not come to mind. *Caution*! It is unlawful to curse or swear on the radio.

| WORD OR PHRASE | MEANING |
|---|---|
| Acknowledge | "Let me know if you have received and understood this message." |
| Affirmative | "Yes." |
| Correction | "An error has been made in this transmission. The correct version is . . ." |
| Go ahead | "Proceed with your message." |
| How do you hear me | Self-explanatory |
| I say again | Self-explanatory |
| Negative | "That is not correct." |
| Out | "This conversation has ended. No response is expected." |
| Over | "My transmission is ended and I expect a reply from you." |

Note. Since "Out" means that no reply is expected, and "Over" means that a reply is expected, we cannot use the two words concurrently.

| | |
|---|---|
| Read back | "Repeat all of this message back to me." |
| Roger | "I have received all of your last transmission." |

Note: "Roger" does *not* mean "affirmative." It simply indicates that the complete message was received.

| | |
|---|---|
| Say again | Self-explanatory |
| Speak slower | Self-explanatory |
| Stand by | If used alone: "I must pause a few seconds." If the pause is longer than a few seconds, or if "Stand by" is used to prevent another station from transmitting, it must be followed by the ending "Out." |

| | |
|---|---|
| That is correct | Self-explanatory |
| Verify | "Check with the originator." |
| Wilco | "I have received your message, I understand it, and will comply with it." |

Note: The use of the words "Roger—Wilco" is redundant. That is, "Roger" indicates that the message has been received, but in addition to that, "Wilco" means that the message is understood and will be complied with.

| | |
|---|---|
| Words twice | (a) As a request: "Communication is difficult. please say every phrase twice."<br>(b) As information: "Since communication is difficult, every phrase in this message will be spoken twice. |

## Time

The Federal Aviation Administration uses Greenwich Mean Time for all operations. We should use GMT (Zulu) time for aeronautical purposes, but local time is acceptable. The following table shows the conversion factors from local time to Greenwich Mean Time.

| TO CONVERT FROM | TO GMT (ZULU) |
|---|---|
| Eastern Standard Time | add five hours |
| Eastern Daylight Time | add four hours |
| Central Standard Time | add six hours |
| Central Daylight Time | add five hours |
| Mountain Standard Time | add seven hours |
| Mountain Daylight Time | add six hours |
| Pacific Standard Time | add eight hours |
| Pacific Daylight Time | add seven hours |

All radio transmissions will refer to time in respect to the twenty-four hour clock. The hour is indicated by the first two digits, and the minutes past the hour is indicated by the last two digits.

Example: 0142 Forty-two minutes past the hour of one (AM)
1735 Thirty-five minutes past the hour of five (PM)

When time is transmitted, each digit will be spoken.

Example: 0142 "Zero one four two"
1735 "One seven three five"

Time may be transmitted in minutes past the hour only, but only when no misunderstanding is likely to occur.

Example: 0142 "Four two" (Forty-two minutes past the current hour, or past the last hour if the current time is earlier than forty-two minutes past the hour.)

## Figures

When speaking numbers which indicate either ceiling heights or upper wind levels up to 9,900 feet, we should state hundreds and thousands in round numbers.

| | | |
|---|---|---|
| Example: | 500 | Five hundred |
| | 1,400 | One thousand, four hundred |
| | 4,800 | Four thousand, eight hundred |
| | 9,900 | Niner thousand, niner hundred |

Numbers above 9,900 shall be spoken by separating the digits preceding the word "thousand."

| | | |
|---|---|---|
| Example: | 11,000 | One one thousand |
| | 14,000 | One four thousand |
| | 17,500 | One seven thousand, five hundred |
| | 22,100 | Twenty-two thousand, one hundred |

Transmit airway or jet route numbers as follows:

| | | |
|---|---|---|
| Example: | V-16 | Victor Sixteen |
| | J-538 | J Five thirty-eight |

All other numbers will be transmitted by pronouncing each digit.

| | | |
|---|---|---|
| Example: | 10 | One zero |
| | 85 | Eight five |
| | 584 | Five eight four |
| | 1,870 | One eight seven zero |
| | 14,213 | One four two one three |
| | 9 | Niner |
| | 5 | Fife |

If a radio frequency contains a decimal point, it is spoken as "point."

| | | |
|---|---|---|
| Example: | 121.5 | One two one point five |

## Speeds

When transmitting aircraft speeds, each digit is spoken separately, followed by the word "knots."

| | | |
|---|---|---|
| Example: | 95 knots | Niner fife knots |
| | 125 knots | One two fife knots |
| | 320 knots | Three two zero knots |

## Phonetic Alphabet

When conditions are adverse for communications, such as during frequency congestion, the phonetic alphabet is used for single letters or to spell out difficult words. We should use the phonetic alphabet when using our aircraft call sign, particularly on the intitial call-up.

If we cannot remember the appropriate word to be used in the phonetic alphabet, we may substitute phonetic equivalents.

| | | | |
|---|---|---|---|
| A | Alfa (AL-FAH) | N | November (NO-VEM-BER) |
| B | Bravo (BRAH-VOH) | O | Oscar (OSS-CAR) |
| C | Charlie (CHAR-LEE) | P | Papa (PAH-PAH) |
| D | Delta (DELL-TAH) | Q | Quebec (KEH-BECK) |
| E | Echo (ECK-OH) | R | Romeo (ROW-ME-OH) |
| F | Foxtrot (FOKS-TROT) | S | Sierra (SEE-AIR-RAH) |
| G | Golf (GOLF) | T | Tango (TANG-GO) |
| H | Hotel (HOH-TEL) | U | Uniform (YOU-NEE-FORM) |
| I | India (IN-DEE-AH) | V | Victor (VIC-TAH) |
| J | Juliett (JEW-LEE-ETT) | W | Whiskey (WISS-KEY) |
| K | Kilo (KEY-LOH) | X | Xray (ECKS-RAY) |
| L | Lima (LEE-MAH) | Y | Yankee (YANG-KEY) |
| M | Mike (MIKE) | Z | Zulu (ZOO-LOO) |

| | | | | | | | | | |
|---|---|---|---|---|---|---|---|---|---|
| 1 | WUN | 2 | TOO | 3 | TREE | 4 | FOW-ER | 5 | FIFE |
| 6 | SIX | 7 | SEV-EN | 8 | AIT | 9 | NINER | 0 | ZERO |

## Frequency Tuning

No attempt is made to provide specific frequencies in this text as they are changed quite often. We should always refer to the Airman's Information Manual or navigation chart for current frequencies. We will, however, discuss the proper procedure to be used when tuning a frequency for transmission or reception.

As we gain experience in flight, it will become obvious that many pilots do not think before acting. This is particularly true as far as communications is concerned. Just as we must dial the correct combination of digits on the telephone in order to contact the correct party, so must we select the proper frequency if we expect to reach the correct ground station with our airborne radio equipment. And yet, if we listen to the radio for very long, we will constantly hear pilots calling a facility on the wrong frequency.

### *Initial Call-up*

When we initially call a ground station we should state the frequency on which the reply is expected if any doubt whatever exists as to the frequency to be used. This procedure serves two purposes. First, it will remind us to check to see that our receiver is on the proper frequency. Secondly, it will inform the ground station as to what frequency to use for a reply. Remember, many ground stations have dozens of frequencies at their disposal. There is not much sense in using all of them to answer one call, particularly during periods of communications congestion.

It is a good idea to listen for a moment or two for a reply before initiating another call. The volume control should be on high. The ground communicator may be talking to another aircraft on another frequency and, therefore, be unable to answer our call. We should give him time to finish the conversation before repeating our call.

If, after three calls, we receive no reply, there is a good chance that we have done something wrong. Instead of repeating our call time and again, we should make a systematic check of our equipment and procedures.

The On–Off–Volume control must be turned on and the volume placed in the high position.

The proper frequency for transmission must be selected. A check of the Airman's Information Manual may show a second frequency which we might try.

The proper reception frequency must be selected. We should check the navigation chart or the Airman's Information Manual for the proper frequency.

The microphone must be activated to key the transmitter. Normally we will hear the microphone click on the transmitter if it is "hot." Sometimes the microphone plug is not connected to the jack, and although the radio is operative, we are unable to transmit.

Some pilots, particularly the inexperienced, in their haste to transmit, attempt to speak into the back of the microphone. We must always speak directly into the front of the instrument. (A pattern of tiny holes or mesh will mark the front of the microphone.) If all of the above checks prove negative, we may attempt another call.

It is astounding how the microphone causes "stage fright." We should think of it as a telephone. We "dial" a number, speak distinctly, and listen for a reply. It is as simple as that. The only way to become proficient in the use of the radio is to use it often. Don't hesitate to transmit often.

## Clearances

A clearance issued by Air Traffic Control, which includes the tower, ground control, approach control, departure control, and enroute control is predicated on *known* traffic and *known* airport conditions. An Air Traffic Control clearance is authorization to proceed under specified conditions in controlled airspace issued for the purpose of *preventing collision between known aircraft*. An Air Traffic Control clearance is *not* an authorization to deviate from any rule or regulation, nor to conduct any unsafe operation of the aircraft.

Federal Air Regulation 91.3 (a) states: "The pilot in command of an aircraft is directly responsible for, and the final authority as to the operation of, that aircraft." In other words, no matter what the Air Traffic Control clearance directs, it is *our* responsibility to conduct a safe and proper operation. If a clearance is not acceptable, we must simply request an amended clearance.

If we request a clearance for a different course of action, we are expected to act accordingly. (For example, we may request a different runway, takeoff from an intersection, a different taxi route, etc.)

### *Use of Runways*

At an airport where a runway-use plan is not established, Air Traffic Control clearances may specify:

1. The runway most nearly aligned with the wind when it is five knots or more,
2. The "calm wind" runway when it is less than five knots,

3. Another runway if operationally advantageous.

The controller need not inquire if we desire to use a runway different than that specified, or offer a choice of runways. If we desire to use a runway different than that specified, we must inform the controller.

At airports having runway-use plans (for noise abatement), the runway assignment still does not preclude our having a choice of runway. We should, however, cooperate with Air Traffic Control in the assignment of a runway if at all possible so that traffic flow is not unduly disturbed.

*Taxi Clearance*

When an airport traffic control tower is in operation, we must receive a clearance prior to taxiing on a runway, taking off, or landing. Authorization to "taxi to" a specific runway is authorization to cross all runways that intersect the taxi route, unless we receive instructions to the contrary. Authorization to "taxi to" a specific runway does *not* constitute a clearance to taxi on that runway.

In order to more efficiently control the flow of traffic at an airport, the controller may initiate intersection takeoffs, or may approve them at our request. If, however, we desire the use of a different intersection, or the full length of the runway, we must simply advise Air Traffic Control accordingly. We may also request the available runway length from any particular intersection.

*Takeoff Clearance*

When Air Traffic Control issues a clearance, we are expected to comply with it after acceptance. If the clearance execution time is at our discretion, the clearance will include the phrase "at pilot's discretion."

Air Traffic Control will sometimes include or imply the word "immediately" in an instruction, to denote urgency in a situation. Our expeditious compliance is expected and necessary for safety. However, we must remember that the *final authority* and ultimate responsibility for our flight is the *pilot in command.* For example, a common expression used by Air Traffic Control communicators, particularly at busy airports, is "You are cleared to land; make first available turn off the runway . . . traffic on one-half mile final behind you." The inexperienced pilot hears that traffic is on one-half mile final and makes every effort to clear the runway fast, using the first available turnoff, even if the aircraft is still going 100 miles per hour. As a result, the aircraft groundloops at the turnoff.

What the controller really meant, although he did not express it was, "Make the first available turnoff *if you, the pilot in command, think that it is safe to do so.*"

Another common instruction is "Cleared for immediate takeoff." We know that a takeoff in a light aircraft immediately behind a four-engine jet transport is likely to be hazardous due to jet wash or wing-tip vortices. Therefore, if we are placed in that situation, it is our responsibility to request an amended clearance. We cannot expect the tower controller to read our mind, however, so we must *inform him* of our inability to accept the clearance as issued.

*Adherence to a Clearance*

When an air traffic clearance has been obtained, we shall not deviate from it or the provisions contained in it unless an amended clearance has been obtained. The last clearance we receive has precedence over the related items contained in a previous clearance.

## Communications Facilities

The network of communications facilities which covers the globe was built with tax money for use by taxpayers. Our life aloft will be made much easier if we use this marvelous system to its fullest extent. Whether we simply wish a taxi to meet us at our destination or an emergency steer to a safe airport, there is a facility to help us. In this chapter, we will discuss facilities and their use during nonemergency conditions. Emergency aid will be discussed in Chapter 49 (Emergency Procedures).

## Departure

The communications available upon departure from an airport depend on the particular airport. Some have control tower facilities; some do not. Others have a Flight Service Station located on the field; others are equipped with advisory services only. The small outlying airport may not be equipped with any radio facilities.

*Unicom Facility*

When we are about to begin taxiing, we should contact the Unicom operator and furnish him with our position on the airport and our intentions. The Unicom operator will provide the runway in use, an estimate of the surface wind, and traffic information (at his discretion).

*Flight Service Station Facility*

The Flight Service Station at an airport which is not equipped with an operating control tower will provide *advisory* information to departing aircraft. Traffic control is *not* exercised, however. The Airport Traffic Advisory provided by the Flight Service Station will include the wind direction and speed, favored or designated runway, altimeter setting, *known traffic* (caution: aircraft operating in the vicinity may not be *known*, and therefore will not be reported), notices to airmen, airport taxi routes, and traffic pattern information.

When ready to taxi, we should notify the Flight Service Station of our intentions, including the aircraft identification, type, and location on the airport, and the destination and type of flight plan.

*Control Tower—Not Radio Equipped*

If the airport has a control tower, but that tower is not radio equipped, the tower personnel will use a high-intensity, narrow beam signal light to control traffic. Although this system offers some opportunity to control traffic, it is at best a poor replacement for radio communications because we may not be looking at the tower when the light signal is shown, and because directions issued are very limited.

The following table shows light gun signals and their meaning when received in flight and on the surface.

| Color and Type Signal | Meaning on Surface | Meaning in Flight |
|---|---|---|
| Steady Green | Cleared for takeoff | Cleared to land |
| Flashing Green | Cleared to taxi | Return for landing (to be followed by steady green at proper time) |
| Steady Red | Stop | Give way to other aircraft and continue circling. |
| Flashing Red | Taxi clear of landing area (runway) in use | Airport unsafe; do not land |
| Flashing White | Return to starting point on airport | Not applicable |
| Alternating Red and Green | General warning signal, exercise extreme caution. | |

Between the hours of sunset and sunrise, if we wish to attract the attention of the tower operator, we should turn on the taxi or landing lights and point the aircraft toward the tower. We should leave the light on until an appropriate signal is received.

We can acknowledge the light signal by blinking the taxi or landing lights at night, or by moving the ailerons or rudder during daylight.

*Control Tower—Operational Radio*

Before taxiing, we must contact the control tower on the appropriate ground control frequency. We should remain on that frequency until we are ready to request clearance for takeoff. (Ground control frequencies are limited for use by the tower and aircraft or vehicles on the surface.)

*Radar Service for Departure*

Pilot participation in Radar Departure Assistance is urged. The safe, orderly flow of departure traffic can be assisted by the use of radar at those facilities which are properly equipped.

To participate, we must request "Traffic Information Service" on the initial contact with the ground controller and advise him of our initial direction of intended flight. After takeoff, the tower will advise us as to what frequency to use to contact Departure Control, and will tell us when to turn to that frequency.

## Enroute

In the contiguous 48 states of the United States, two route systems have been established for air navigational purposes; the VOR and L/MF System, and the Jet Route System. We will discuss in this text only the VOR/LMF System, because we will not be flying in the high altitudes of the Jet Route System until we become more experienced.

The VOR and L/MF System is comprised of airways designated from 700 feet *above the surface* (or in some cases higher) up to but not including 18,000 feet *mean sea level.*

The VOR airways depend solely on VOR/VORTAC aids, and are called *vector airways.* The vector airways are numbered much like the surface highways. When we file our flight plan (see Chapter 56) we need only to name the airways on which we intend to fly in order to designate our intended route of flight.

The L/MF portion of the airway system is obsolete and has been replaced almost in its entirety by the VOR/VORTAC system.

While enroute, we may take advantage of the many communications facilities which are available.

### *Flight Service Stations*

The Federal Aviation Administration operates Flight Service Stations to provide exactly what the name implies, flight services. When in flight we may contact the Flight Service Station to file a flight plan, receive a weather briefing, check on NOTAMS, or simply ask a question concerning any aspect of our flight.

The Flight Service Station normally transmits over a nearby VOR station, as well as other specified frequencies. (Refer to the Airman's Information Manual for frequencies.)

### *Lake, Island, and Mountain Reporting Service*

When we fly over large bodies of water, swamps, or mountains, we should take advantage of the special flight monitoring services provided by certain Flight Service Stations. The areas covered by Lake, Island, and Mountain Reporting Services are listed in the Airman's Information Manual, excerpts from which are shown in Fig. 159 and 160. If we desire monitoring, we must contact the appropriate Flight Service Station by telephone or radio and give our present position, type of aircraft, altitude, indicated airspeed, proposed route of flight, and mainland estimate.

After making these arrangements, we must make a radio contact every 10 minutes with a Flight Service Station while enroute. If contact is lost for 15 minutes, search and rescue facilities will be immediately notified. This simple procedure will save important time for search and rescue missions in the case of an emergency.

### *Radar Assistance*

If we are able to communicate with an Air Traffic Control radar equipped facility, are within their radar coverage, and can be radar identified, we may

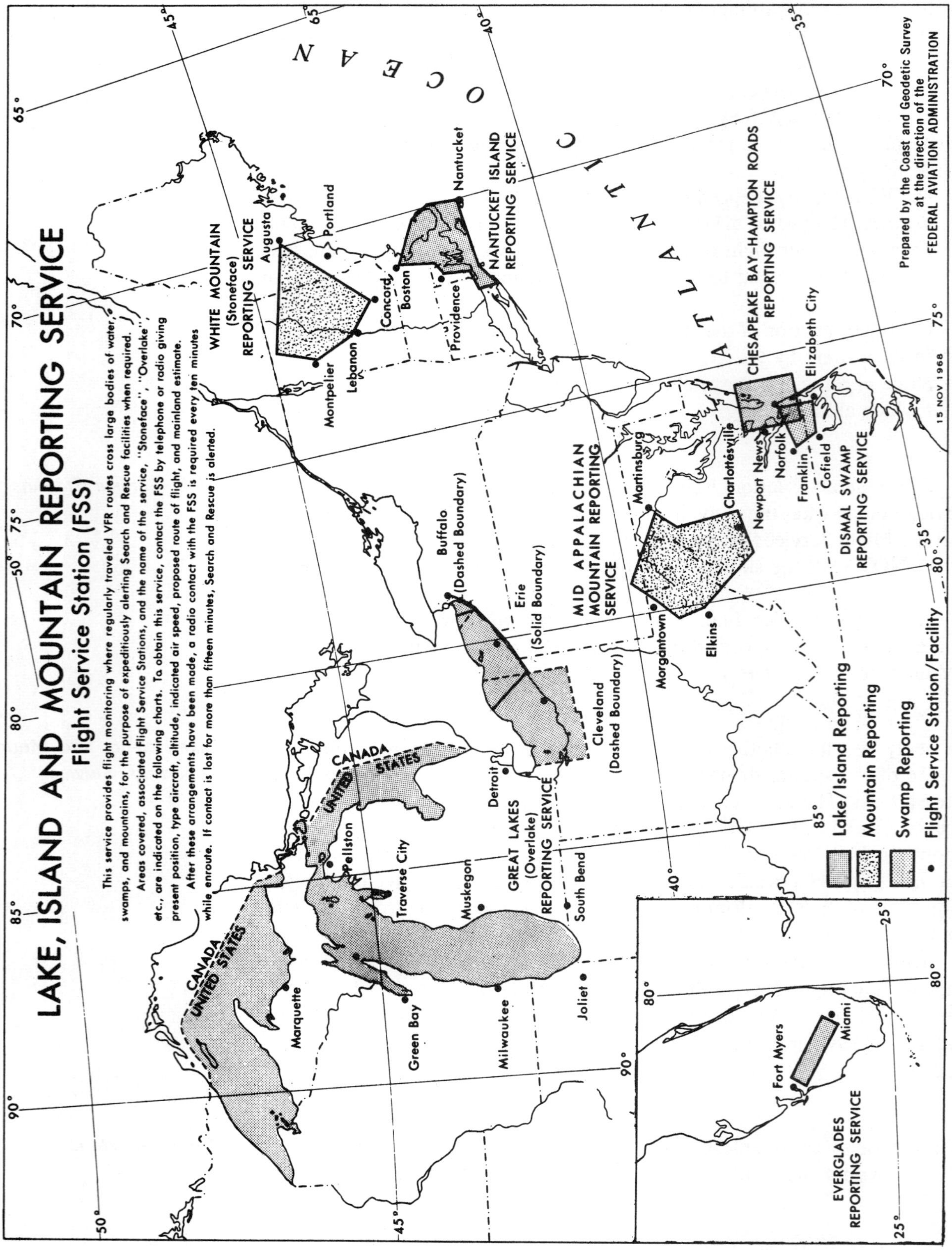
LAKE, ISLAND AND MOUNTAIN REPORTING SERVICE
Flight Service Station (FSS)
This service provides flight monitoring where regularly traveled VFR routes cross large bodies of water, swamps, and mountains, for the purpose of expeditiously alerting Search and Rescue facilities when required.
Areas covered, associated Flight Service Stations, and the name of the service, "Stoneface", "Overlake" etc., are indicated on the following charts. To obtain this service, contact the FSS by telephone or radio giving present position, type aircraft, altitude, indicated air speed, proposed route of flight, and mainland estimate.
After these arrangements have been made, a radio contact with the FSS is required every ten minutes while enroute. If contact is lost for more than fifteen minutes, Search and Rescue is alerted.
WHITE MOUNTAIN (Stoneface) REPORTING SERVICE
NANTUCKET ISLAND REPORTING SERVICE
CHESAPEAKE BAY–HAMPTON ROADS REPORTING SERVICE
MID APPALACHIAN MOUNTAIN REPORTING SERVICE
DISMAL SWAMP REPORTING SERVICE
GREAT LAKES (Overlake) REPORTING SERVICE
EVERGLADES REPORTING SERVICE
ATLANTIC OCEAN
CANADA
UNITED STATES
Augusta
Portland
Concord
Boston
Nantucket
Providence
Lebanon
Montpelier
Buffalo (Dashed Boundary)
Erie (Solid Boundary)
Cleveland (Dashed Boundary)
Martinsburg
Morgantown
Elkins
Charlottesville
Newport News
Norfolk
Franklin
Cofield
Elizabeth City
Detroit
Pellston
Traverse City
Muskegon
South Bend
Joliet
Milwaukee
Green Bay
Marquette
Fort Myers
Miami
Lake/Island Reporting
Mountain Reporting
Swamp Reporting
Flight Service Station/Facility
Prepared by the Coast and Geodetic Survey at the direction of the FEDERAL AVIATION ADMINISTRATION
15 NOV 1968

Figure 159.

109° 104°
43° 43°

# ROCKY MOUNTAIN REPORTING SERVICE

This service is available to aircraft operating VFR (with or without flight plan) between Denver and Grand Junction via Corona Pass or Monarch Pass. This service cannot be offered on an area or random route basis due to the unreliability of communications caused by terrain extremes. The routes via Corona Pass and Monarch Pass have been flight checked and communications found reliable at no less than the altitudes published on the chart.

"Rocky Mountain Reporting Service" shall be available to all VFR aircraft upon request, whether or not operating on a regular flight plan.

REQUIREMENTS

1. Aircraft must be equipped with functioning 2-way radio capable of receiving VOR and transmitting 122.1 MHz
2. Pilot must fly prescribed route at an altitude no less than established MRA (Minimum Reception Altitude).
3. Pilot must arrange for accurate time of departure to be received by the FAA facility through which the service was requested.
4. Pilot must radio a position report every ten minutes as follows, whether eastbound or westbound:

   CORONA ROUTE

   To Denver Radio while operating between Denver and Corona Pass (vicinity of Moffat Tunnel, west of Tolland, Colorado).

   To Eagle Radio (through Kremmling VOR) between Corona Pass and vicinity of Glenwood Springs.

   To Grand Junction Radio between vicinity of Glenwood Springs and Grand Junction.

   MONARCH ROUTE

   To Denver Radio (through Denver or Kiowa VOR) between Denver and Lake George Intersection (Cheesman Reservoir). To Denver Radio (through Kiowa VOR only) between the Intersection and Elevenmile Reservoir. or west of Cripple Creek, Colorado.

   To Pueblo Radio between Elevenmile Reservoir or west of Cripple Creek and Salida, Colorado.

   To Grand Junction Radio (through Gunnison VOR) between Salida and Grand Junction.

   Pilots may file for all or any portion of either route, i.e. on a flight from Denver to Grand Junction, a pilot may request the service from Denver to Kremmling only.

   (If contact with the aircraft is lost for more than 15 minutes, Search and Rescue will be alerted.)
5. The service may be terminated at any time upon request of the pilot.
6. In case of aircraft radio failure, the pilot must land at nearest airfield and make telephone notification to either Denver, Eagle, Pueblo or Grand Junction Radio.

42° 42°

41° 41°

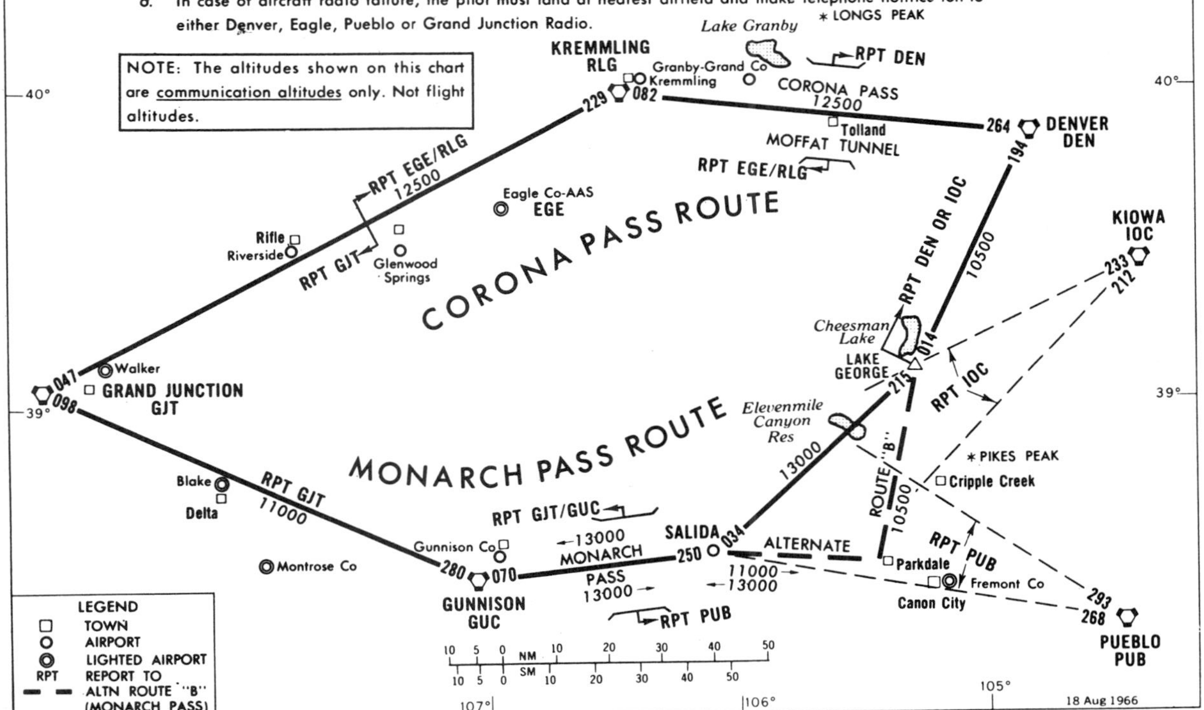


Prepared by the Coast and Geodetic Survey at the direction of the FEDERAL AVIATION ADMINISTRATION

Figure 160.

request assistance and navigation services. A vector from one point to another will be provided under the following conditions.

1. We request it.
2. The controller suggests it, and we agree.
3. When air safety requires it, in the opinion of the controller.

We must use caution, however, when flying on a radar vector. The radar controller cannot determine the weather at our altitude, and hence may give us a heading on which we cannot fly and maintain visual flight rules. It is the pilot's responsibility to conduct the flight safely, and, therefore, we must watch for other traffic and maintain outside visual references even when being vectored on radar.

*Radar Traffic Information Service*

Radar equipped air traffic control facilities may provide information concerning other traffic which may be in such proximity to our position as to warrant our attention. In other words, the radar operator will advise us if another aircraft appears on his radar scope which may create a collision hazard. However, this service *does not* relieve us of the responsibility of constant vigilance. No altitude information can be provided. The radar controller will simply advise us as to the position of the other aircraft and its distance from us.

Example: "Traffic at twelve o'clock, three miles, westbound."

We may request a vector to avoid the reported traffic.

The availability of this service depends on whether the controller feels he is in a position to provide it.

## Arrival

Upon arrival at an airport we should take advantage of the facilities available for radio communications.

*Unicom*

If the Airman's Information Manual designates that a Unicom facility exists at an airport with no Flight Service Station or control tower, we should tune to the Unicom frequency when about 15 miles from the airport, and listen for other aircraft transmissions which may indicate the whereabouts of other aircraft which are approaching the same airport. Then, when about five miles from the airport, we should contact the Unicom operator and furnish him with our position, altitude, and intentions. The Unicom operator will advise us as to the runway in use, an estimated wind, and any known traffic (at his discretion). Since final responsibility rests with the pilot in command, and since Unicom is *advisory* in nature, we may wish to use a runway other than that suggested. If we do so, however, we should inform the Unicom operator of our intentions.

*Flight Service Station Operational*

If there is not an operational control tower at an airport, but there is a Flight Service Station, we should contact it when about 15 miles from the destination. The Flight Service Station operator will provide information concerning the surface wind, the designated runway, and any *known* traffic. This information is advisory in nature, and responsibility for safe flight still rests with the pilot.

*Control Tower Operational*

When we operate into an airport at which there is an operating control tower, we must maintain two-way radio contact with the tower when within the airport traffic area, unless the tower authorizes other action. (The airport traffic area is the airspace within a five-mile radius from the center of the airport, extending from the surface up to, but not including, 2,000 feet above the surface.)

Our initial call to the tower should occur when we are about 15 miles from the airport. The tower operator will advise us as to the runway in use, the altimeter setting, surface wind, and will request another contact at a specified location. The tower issues clearances, which must be followed unless otherwise authorized. Final authority still rests with the pilot in command, however, so a clearance from the tower may be amended upon the request of the pilot.

*Radar Arrival Service*

At certain airports (listed in the Airman's Information Manual), radar service is provided via approach control. Visual traffic will be merged with instrument traffic in an orderly fashion. We should contact approach control instead of the preliminary tower contact. After radar contact is established, we will be directed to fly specific headings which will position our flight in the traffic pattern and in the proper landing sequence. Radar services will end when we are instructed to contact the control tower. Instructions issued by radar approach control are clearances, and must be followed unless amended. Final authority for safe flight, however, remains with the pilot in command.

*Speed Adjustment*

In order to avoid excessive vectoring in the establishment of arrival sequences, Air Traffic Control may request us to adjust our airspeed. All speed adjustments will be issued in knots and shall be predicated on indicated airspeed. We will be expected to comply with the speed adjustment request within plus or minus ten knots of the requested speed. Naturally, we can reject the speed adjustment if the minimum safe flight speed is less than that requested or if maximum safe flight speeds will be exceeded, but we must advise Air Traffic Control of the speed we will use.

## After Landing

After our landing is accomplished, and we have cleared the active runway, the control tower will advise us to contact Ground Control. The ground controller will inform us as to the taxi route to be used inbound, and will, if requested, close our flight plan.

## Automatic Terminal Information Service

In selected high-activity terminal areas, a continuous message will be broadcast from the VOR located on or near the airport. We may, prior to departure, while enroute, or prior to landing, receive the ceiling, visibility, surface wind, altimeter setting, and runways in use at a terminal airport by tuning to the appropriate VOR frequency (designated in the Airman's Information Manual).

We will be expected to listen to the Automatic Terminal Information Service broadcasts whenever they are in operation. The controller will assume that we are aware of the information contained in the message and, unless specifically requested, will not repeat the information.

## Summary

An entire text could be written concerning the proper use of communications equipment and facilities. The best way to learn a great deal about the subject is to use the radio often, visit the Federal Aviation Administration facilities, and ask questions of radio operators.

By all means, we must remember that communications specialists are employed to *help us*. If we do not know the proper terminology, we may use plain English. If we cannot find the correct frequency as listed in the Airman's Information Manual, we need only ask the nearest Flight Service Station.

Too many pilots become involved in situations which lead to an emergency without seeking the help of specialists on the ground. Many of these men and women of the Federal Aviation Administration are themselves pilots. They know the feeling of despair when aloft and in trouble. Ask for their help when needed, and ask *early*. There is no sense in requesting aid, for example, after the engine has failed from fuel starvation. We will discuss emergency communications in detail in Chapter 49.

FEDERAL AVIATION AGENCY

VFR PILOT EXAM-O-GRAM* NO. 14
RADIO COMMUNICATIONS FREQUENCIES

CAN YOU FIND THE FOLLOWING MOST COMMONLY USED FREQUENCIES?

| NOTE: Use data on page 3. (Assume you can transmit on any of the 10 frequencies listed on page 2.) | You Transmit On | You Receive On |
|---|---|---|
| 1. Tucumcari, N. M., FSS (Enroute Radio Contact) | | |
| 2. Tucumcari, N. M., FSS Arpt. Advisory Service | | |
| 3. Clovis, N. M., UNICOM | | |
| 4. Albuquerque, N. M., (Sunport) Tower | | |
| 5. Albuquerque, N. M., (Sunport) Ground Control | | |
| 6. Albuquerque, N. M., (Sunport) UNICOM | | |
| 7. Santa Fe, N. M., Tower | | |
| 8. Santa Fe, N. M., Ground Control | | |

TEST YOUR KNOWLEDGE
BY WRITING IN THE APPROPRIATE FREQUENCIES
CHECK YOUR ANSWERS WITH THOSE LISTED ON PAGE 4 -- THEN READ ON

Many beginning pilots get the idea that radio communications is a complex process involving an infinite number of frequencies and requiring a practiced art in operation. For this reason these same pilots shy away from normal usage of their radio equipment and consequently fail to realize the full benefit to be gained from it.

Air traffic is increasing steadily, and orderly movement of aircraft in flight and on airports is as imperative for aviation safety as highway traffic regulations are for automobile travel. For this reason the FAA has established certain requirements for two-way radio capability at airports having control towers operated by the FAA, and additionally, the use of radio, if so equipped, at airports served by other towers.

The FAA radio facilities are for all pilots' use. In the interest of safety for yourself and others, take advantage of these facilities on every flight. By keeping tuned to the appropriate frequencies along the route, you will receive much important information; and when you enter a congested area you will indeed find your radio a true friend.

Learning to use the aircraft radio is not difficult. Correct radiotelephone phraseologies and techniques should be used if possible; however, pilots should never hesitate to use the radio because of lack of familiarity with these procedures. A message in the pilot's own words is always acceptable. SO USE YOUR RADIO; LEARN IN ACTUAL PRACTICE.

Now that we've explained the importance of radio communications, and hope it has made an impression, let's examine the question of appropriate frequencies.

*Exam-O-Grams are non-directive in nature and are issued solely as an information service to individuals interested in Airman Written Examinations.

WHAT ARE THE MOST IMPORTANT FREQUENCIES? It is realized that many light airplanes are equipped with transmitters having a very limited number of channels. For this reason you must be very selective in choosing which frequencies would be most important. The following list of aircraft transmitting frequencies are listed in order of importance for most general aviation-type flying in most areas.

1. 121.5 mc - Emergency (world-wide)
2. 122.1 mc - Flight Service Station (FSS)
3. 122.5 mc - Most FAA towers
4. 122.8 mc - UNICOM (airports without towers) indicated on charts - U - and in Airmans' Information Manual - U-1
5. 122.7 mc - Some FAA towers
6. 122.6 mc - Some FAA towers and pilot to weather forecaster
7. 121.7 mc - Tower ground control
8. 121.9 mc - Tower ground control
9. 123.0 mc - UNICOM (airports with towers) indicated in Airmans' Information Manual - U-2
10. 122.4 mc - A few FAA towers

It would be advisable to choose the frequencies to be installed in order of importance, depending upon the capability of your transmitter. For example, if your set has only 4 channels, then use 121. 5, 122. 1, 122. 5, and 122. 8. (NOTE: For your particular area, a different frequency might be more important.) Should your set have a greater capability than 10 channels, we recommend this standard group plus additional frequencies as necessary to suit your particular needs. For more information on frequency utilization, see the AIRMANS' INFORMATION MANUAL. (NOTE: The reverse side of many Sectional Charts also contains a list of assigned frequencies.) Most aircraft receivers can be tuned to any frequency within the normal VHF band, and therefore present little or no problem from the standpoint of frequency restriction.

WHERE CAN YOU FIND FREQUENCIES FOR SPECIFIC TOWERS AND OTHER RADIO AIDS? The most convenient source is on the Sectional Chart. Frequencies which the facility can transmit on are enclosed in a box and placed adjacent to the facility. Should the facility have no voice transmitting capability, then the words "no voice" will appear on the bottom line of the box. A frequency followed by the letter "G" -- shown thus 122. 7G -- means that this facility listens or receives only (guards) on this frequency. Remember, most FAA towers can receive on 122. 5 and all Flight Service Stations (FSS) can receive on 122. 1. Although the map is a handy source for determining assigned frequencies, we wish to emphasize that the Airmans' Information Manual is the most current source. The Visual Navigation Chart Bulletin, the Airport-Airport/Facility Directory and the FSS Frequencies List will show the permanent changes. Remember to check the NOTAMS Section for any temporary or late changes.

## SECTIONAL CHART EXCERPTS

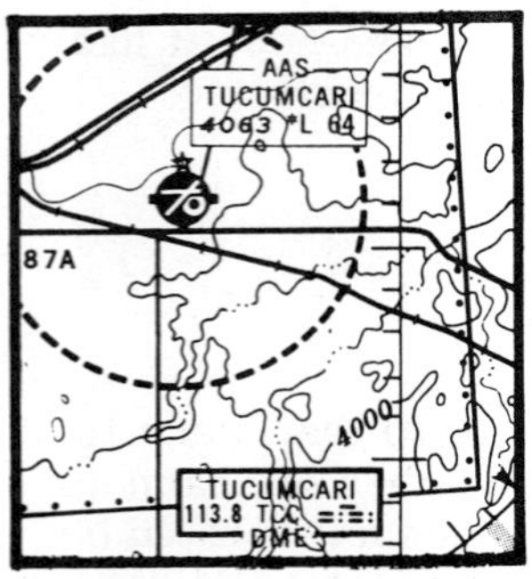


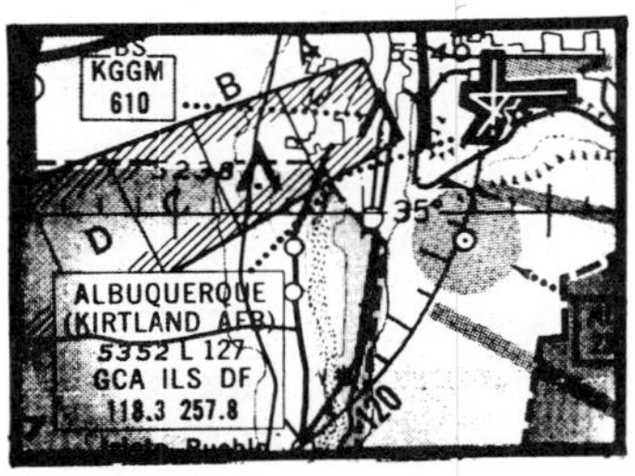


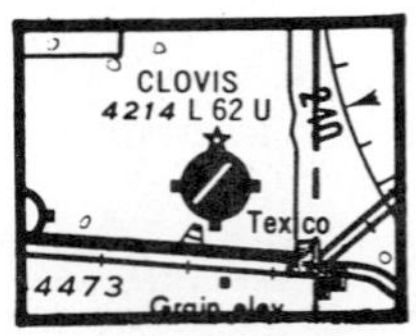

## EXCERPTS FROM THE AIRMAN'S INFORMATION MANUAL

### AIRPORT DIRECTORY

NEW MEXICO

o o o o o o o o o o o o o o o o o o o o o o o o o

**CLOVIS**

**CLOVIS MUNI** 6 E

4214 H62 (1) BL4 S5 F4 U-1 FSS: TUCUMCARI

**HILLCREST** 2E

4244 36 (4) S3 F2 FSS: TUCUMCARI

Remarks: P-lines E, SE, S & SW.

o o o o o o o o o o o o o o o o o o o o o o o o o

### FSS FREQUENCIES AND TELEPHONE NUMBERS

| Location and Frequencies | | Code Area | Telephone |
|---|---|---|---|
| NEW MEXICO | | | |
| Alburquerque 113.2 230 (ABQ) | FSS | (505) | CH 3-7832 |
| | WB | (505) | 242-2661* |
| | WB | (505) | CH 3-7832 |
| Carlsbad (AAS) 116.3 (CNM) | FSS | (505) | TU 5-2042 |
| Clayton | WB | (505) | 374-9511 |
| o o o o o o o o o o o o o o o o o o o o o o o o o | | | |
| Silver City | WB | (505) | 538-9023 (0530-2000 Mon-Fri) (0530-1330 Sat) |
| Truth of Consequences (AAS) 112.7 (TCS) | FSS | (505) | 894-3277 |
| Tucumcari (AAS) 113.6 (TCC) | FSS | (505) | 461-2900 |
| Zuni 113.4 (ZUN) | FSS | (505) | 782-3770 |

AIRPORT/FACILITY DIRECTORY

NEW MEXICO

**ALBUQUERQUE-SUNPORT/KIRTLAND AFB** *IFR* 3 SE (077° 9.6 NM from ABQ VOR)
5352 H127/8-26 (4) (S-100, T-155, TT-445) BL4, 6, 8, 10
S5 F5, JP1, 4 U2 **VASI:** Rnwy 8
**RVV:** Rnwy 35 **FSS:** ALBUQUERQUE on Fld
**REMARKS:** Tkofs rnwy 3 and 35 restr to DC-3 and smaller.
**ALBUQUERQUE TOWER** 118.3 126.2 122.5R 119.2
**CLRNC DEL** 119.2
**ALBUQUERQUE GND CON** 121.9
**RADAR SERVICES:** (BCN)
**ALBUQUERQUE APP CON** 124.3 134.1 126.2 123.9 122.5R 121.1 113.2T 110.3T
**ALBUQUERQUE DEP CON** 121.1
**TFC INFO** •123.9
**PAR** Rnwy 8[1] Ceil 300 Vsby 3/4 mi Min Alt 5652
**ASR** Rnwys 3[1], 8[1], 17, 35 Ceil 400 Vsby 1 mi Min Alt 5752
**ILS** 110.3 I-ABQ Apch Brg 350°
**REMARKS:** [1]500-1 1/2 rqrd for civil jet acft. IFF svc avbl.
**VOT:** 111.0

o o o o o o o o o o o o o o o o o o o o o o o o o

**SANTA FE COUNTY MUNI** *IFR* 9 SW (332° 4.2 NM from SAF VOR)
6344 H83/2-20 (3) (S-45, T-60, TT- ) BL4 S5 F4, JP1
**FSS:** SANTA FE on Fld
**REMARKS:** Rnwy 10-28 rest to air carrier acft.
**TOWER** 126.2 122.5R 119.5 **GND CON** 121.7

☆ ☆ ☆ ☆ ☆

**Student Pilots Radio Identification**

The FAA desires to help the student pilot in acquiring sufficient practical experience in the environment in which he will be required to operate. To receive additional assistance while operating in areas of concentrated air traffic, a student pilot need only identify himself as a student pilot during his initial call to an FAA radio facility. For instance, "Dayton Tower, this is Fleetwing 1234, Student Pilot, over." This special identification will alert FAA air traffic control personnel and enable them to provide the student pilot with such extra assistance or consideration as he may need. This procedure is not mandatory.

☆ ☆ ☆ ☆ ☆

IS LANDING AND TAKEOFF GUIDANCE AVAILABLE FROM FAA FLIGHT SERVICE STATIONS? Yes. Airport Advisory Service is provided by Flight Service Stations at many non-tower airports, and will be indicated thus on Sectional Charts.

This means that, when operating to or from the Tucumcari Airport, the FSS will furnish you with takeoff and landing information, favored runway, and current known traffic, but no airport traffic control. Normally this information is transmitted to you on 122.2; however, it may be transmitted on other frequencies if you so request.

WHAT PROBLEMS EXIST WHEN MANUALLY TUNING YOUR RECEIVER? (1) The receiver may not be properly calibrated -- that is, the final setting of the dial to obtain maximum reception does not agree with the published frequency; (2) the receiver may vibrate off the selected frequency because of shock produced by taxiing, landing, turbulence, etc.; and (3) not monitoring for other transmissions before transmitting.

WHAT CAN YOU DO TO HELP ELIMINATE THESE PROBLEMS? When tuning to a new frequency, adjust your set to the published frequency. If you receive nothing, move the tuner back and forth to either side of this setting to try to obtain reception. If it is a voice communications frequency only and you receive no transmissions after this procedure, transmit for initial contact with the ground station. Immediately start moving the tuner back and forth to either side until you receive the ground station; then tune for maximum reception. Note the final setting; if it is different than the published frequency, apply this to future settings and use the same technique. If you have two receivers with one adjusted to ground control frequency prior to landing, you may have to reset it after landing since the landing jar may vibrate it off frequency. ALWAYS MONITOR YOUR RECEIVER BEFORE TRANSMITTING AND ALWAYS IDENTIFY AURALLY ANY NAVIGATIONAL RADIO AID STATION TO WHICH YOU TUNE. "Whistle stop" tuning (an aural signal that indicates when transmitting and receiving frequencies are identical) eliminates some of the problems of manual tuning.

WHAT IS THE RECEPTION DISTANCE OF VHF RADIOS? VHF frequencies are described as having line-of-sight usability. While this is only theoretically true, best reception is normally obtained when a line of sight does exist. In actual practice the following figures can usually be depended upon.

| Aircraft Height Above Station * | Reception Distance |
|---|---|
| 1,000 feet | 45 statute miles |
| 5,000 feet | 100 statute miles |

* (Based on no intervening physical obstructions.)

DO YOUR ANSWERS AGREE WITH THESE?

| | | | | | |
|---|---|---|---|---|---|
| 1. | 122.1 | 113.6 | 5. | 121.9 | 121.9 |
| 2. | 122.1 | 122.2 | 6. | 123.0 | 123.0 |
| 3. | 122.8 | 122.8 | 7. | 122.5 | 126.2 (or 119.5) |
| 4. | 122.5 | 118.3 | 8. | 121.7 | 121.7 |

Department of Transportation
FEDERAL AVIATION ADMINISTRATION

## VFR PILOT EXAM-O-GRAM° NO. 35

### UNICOM FREQUENCIES AND USES

WHAT IS UNICOM? UNICOM is a private aeronautical radio station. It provides a communication channel for many airports without control towers (122.8 MHz) and a channel for many airports with control towers (123.0 MHz). Both the ground station and aircraft transmit and receive on the same frequency. Many of the FAA Written Tests contain test items concerning this subject.

WHAT USES MAY BE MADE OF UNICOM (122.8 MHz) AT THOSE AIRPORTS NOT SERVED BY A CONTROL TOWER?

1- It may be used for communications with private aircraft concerning runway and wind conditions, types of fuel available, weather, dispatching, availability of ground transportation, food, and lodging.

2- It may be very useful in an emergency. To illustrate this point, here is a sample transmission which might be used at a non-controlled airport: "Great Bend UNICOM, this is . . . . I am unable to receive a green light 'gear down' indication. Will you have your mechanic check my landing gear as I make a low pass over Runway 17?" In addition to observing the position of the landing gear during a fly by, the mechanic might also review the emergency gear lowering procedures with the pilot.

3- At certain airports, a pilot can turn on the runway lights by tuning his transmitter to the airport UNICOM frequency and then "pressing the microphone button" a predetermined number of times within a predetermined time interval. This is true at these locations as long as the UNICOM station receiver is turned on, even though it is unattended. Note the Airman's Information Manual (AIM) excerpt, above.

**OHIO**

§ **LIMA** 4 NW
827 H35 (1) BL4 S3 F4 U-1 FSS: FINDLAY
**Remarks:** P-line N, S. For rnwy lgts press mike button 4 times within 5 seconds on UNICOM freq; lgts will remain on for 15 min.

4- It may be useful in a wide variety of other ways, such as: (a) A student calls to advise his instructor that he is experiencing a rough engine and seeks advice, (b) A private pilot calls in and asks if a mechanic is available to work on his inoperative aircraft tachometer, (c) A doctor requests that an ambulance meet his airplane upon landing to pick up a hospital patient.

○○○

HOW CAN A PILOT DETERMINE IF UNICOM IS AVAILABLE AT AN AIRPORT WITHOUT A CONTROL TOWER? The letter "U" in the airport information on the Sectional Chart indicates UNICOM. ⇨

§ GREAT BEND MUNI 4 SW
1890 H80 (3) BL4 S5 F4 U-1 FSS: RUSSELL
Remarks: Ngt opern N/S rnwy only. 1006' (2873' MSL) TV twr 6.5 NM NNE.

The availability of UNICOM can also be determined by referring to the Airport Directory Section of AIM. Note the UNICOM symbol "U-1" is used at airports without a control tower.

CAN COMMUNICATIONS ALWAYS BE ESTABLISHED ON UNICOM? No. Most pilots who land regularly at UNICOM equipped airports, have on occasion been unable to get a reply from UNICOM stations. This situation is usually caused by a shortage of personnel at small airports. Sometimes this results in the UNICOM being "on" but unattended. In some instances, the volume control on the station receiver may have been turned down and then forgotten. Although these situations are unfortunate, pilots should realize they may occur.

○○○○○

DISTRIBUTION: ZC-307

HOW IS UNICOM USED AT AIRPORTS SERVED BY A CONTROL TOWER? Communications on 123.0 MHz are identical to those permitted on 122.8 MHz with the exception of information such as runway and wind conditions, weather, etc., which should be furnished by the tower.

HOW DOES A PILOT DETERMINE IF UNICOM IS AVAILABLE AT AN AIRPORT WITH A CONTROL TOWER? The Airport/Facility Directory excerpt to the left shows how UNICOM is listed by the symbol "U-2" in AIM.

NEW ORLEANS
§NEW ORLEANS-LAKEFRONT IFR 4 NE FSS: NEW ORLEANS on Fld
10 H57/17-35(4) (S60 T100 TT160) BL4 S5 F5, JP1, 5 U2
Remarks: Acft [illegible]

The letter "U" also appears in the airport data box on the Sectional Chart.

HOW ARE UNICOM STATIONS IDENTIFIED? Usually by the name of the airport, but sometimes by the name of the fixed-base operator, or even the town where the station is located. For example: In the New Orleans area there are two airports--Lakefront and International--located about 13 miles apart, and both have UNICOM stations on 123.0. A transient pilot who desires to use UNICOM and doesn't know the name of the operators in the area would normally address his call to "Lakefront UNICOM" or "New Orleans International UNICOM," rather than "New Orleans UNICOM."

HOW DO PILOTS ABUSE OR MISUSE UNICOM? Perhaps one of the most common abuses is the situation where several aircraft are flying at high altitudes and using one of the UNICOM frequencies for lengthy aircraft-to-aircraft radio chatter. At high altitudes their transmissions reach out in all directions and tend to block out many local airport UNICOM transmissions. Remember, use UNICOM like a party line telephone--be brief, transmit only essential messages.

FAA Aeronautical Center
Flight Standards Technical Division
Operations Branch
P. O. Box 25082
Oklahoma City, Oklahoma 73125



Note: UNICOM stations are crystal controlled to transmit and receive only on one frequency--either 122.8 or 123.0

UNICOM is never used for Air Traffic Control purposes.

UNICOM frequency 123.05 MHz has been added for use at Heliports.

☆☆☆☆☆

## CHECK YOUR READING

1. List the problems of speech associated with *articulation*.
2. Why is it particularly important to keep anger from showing in our voice when talking with a communicator?
3. Why should we refrain from broadcasting while smoking or eating?
4. How can we best improve our speech habits?
5. List examples of telescoping that you have commonly used in the past.
6. List the common errors associated with using the microphone.
7. When identifying the aircraft during the initial radio contact, we must use the full identification number. When may we shorten the identification number?
8. When initially contacting a surface facility, must we use the ground station identifier or the aircraft identifier first? Why?
9. Define the terms *Wilco*, *Roger*, *Over*, and *Out*.
10. Why should we refrain from calling a Flight Service Station at 15 minutes past the hour?
11. Convert the following local times to Greenwich Mean Time.
    0200 EST 0620 Pacific Daylight Time 1414 CDT
12. How would you broadcast 17,500 feet?
13. Use the phonetic alphabet to spell your full name.
14. Using the latest edition of the Airman's Information Manual, list the frequencies for transmitting and receiving the following facilities:
    Unicom; Flight Service Station at Dallas, Texas; Control Tower at New Bedford, Mass.; Approach Control at LaGuardia Airport, New York; Ground Control at Los Angeles International Airport, Los Angeles, Cal.
15. List the checks you would perform if you received no reply after making three calls to a ground station.
16. You are cleared to "taxi to" a certain runway. While enroute you must cross a runway that has been active. Can you cross this runway without an additional clearance?
17. You are cleared to "taxi to runway 32." Can you perform the run-up on runway 32?
18. While on approach, you receive a flashing red light signal from the tower. What does this mean?
19. After receipt of a clearance, who is the person solely responsible for safe flight?
20. If we do not reject a clearance, how are we expected to act?
21. If we cannot comply with a clearance, what action must we take?
22. After requesting Lake Reporting Service for a crossing of Lake Michigan, we fail to contact a Flight Service Station for 15 minutes. What action will be initiated?
23. We request radar vectoring from our departure airport to a specific VOR. While being vectored, we fly into an area of cloud. Who is responsible for avoiding the cloud and for maintaining visual flight rules conditions?

24. Who is responsible for avoiding other aircraft when a flight is being monitored by Radar Traffic Information Services?

25. What is Automatic Terminal Information Service?

# PART V

# THEORY AND HAZARDS OF INSTRUMENT FLIGHT

# 46

# THE THEORY AND HAZARDS OF INSTRUMENT FLYING

Instrument flight is the control of the aircraft attitude, speed, and thrust by reference to instruments in the cockpit rather than by outside references. Aircraft control without outside reference is difficult because the human body is not accustomed to defining and changing its environment with only one of its five senses. When flying in the clouds or in reduced visibility four of the five senses are useless and at times make us want to disbelieve the only reliable sense—vision.

## The Five Senses

The five senses are taste, smell, touch, sound, and sight. In flight, whether on instruments or not, taste is a useless sense, unless the dry, brassy taste that is common with nervousness aids in alerting us to impending danger. Smell, too, is a useless sense in flight. (Some pilots, however, testify that they know that they are home because of the smell of the city below.) Touch is a very useful sense when flying by outside reference. We can feel the pressure on the control column, the added weight on the seat of our pants during a steep turn, or the light-headedness which accompanies a too-rapid level-off. Sound also is an aid when flying with outside reference. We can hear the increase in the speed of the slipstream as we commence a dive, or the changing engine speed as we increase pitch. Sight is the primary sense when flying by outside reference. We can see the pitch change, the bank of a steep turn, and the pullout of a dive.

So, at best, during "visual flight" we have the use of three of our five senses: touch, hearing, and sight.

Flying on instruments, however, presents a different problem. The sense of touch becomes a demon, warning us of abnormal *g* forces, but not differentiating between the *g* force encountered in a steep turn or in a pullout of a steep dive. We can feel the back pressure on the control column, but cannot tell if that back pressure is causing an increase in pitch attitude or simply maintaining altitude in a steep turn. Neither can we tell if we are pulling out of a dive or raising the nose from a level flight attitude into a vertical climb.

The sense of hearing becomes a trap for the unsuspecting. Our ear warns us of increasing rpm, but does not tell us if that increase is due to a nose-down pitch attitude or an increased throttle setting. We hear the decreasing slipstream, but cannot tell if that decrease is caused by an increased pitch attitude or by the decreased airspeed in a turn.

Sight is the only remaining sense, and on sight we must rely. The aircraft instrument panel is equipped with instruments which will tell us about pitch attitude, bank, yaw, direction, speed, thrust, and trim. We only have to *look at them and believe them*. There is the trick. We must look at the instruments and interpret them. We must believe the interpretation and perform control actions in accordance with them.

## *Looking at the Instruments*

When we learn to fly on instruments, which we do from the very beginning of flight training, we learn to see in a new way. Seeing an instrument panel is quite different than seeing a beautiful landscape or a lovely girl. Cross-checking, or scanning the instruments, is a very important skill and must be practiced until mastered. When we begin cross-checking, we will probably look at each individual instrument without being particularly aware of what we are looking for. As we gain experience, we will learn what to look for, when to look for it, and what reaction must be made to it. Eventually, we will develop our own scanning rate and sequence to match the demand of the situation.

*Fixation.* A common error in the development of good scanning habits is to stare at one particular instrument. We may find ourselves staring at the turning directional gyro, wondering why it is turning. Or perhaps we will stare at the altimeter, trying to maintain altitude, while the nose of the aircraft wanders due to lack of attention.

*Omission.* Another likely error is to leave one instrument out of the cross-check. For example, when we roll out of a bank, we may be trying so hard to maintain the heading that we omit the altimeter from the cross-check.

*Emphasis on One Instrument.* It is understandable that we might emphasize the instrument which we best understand during our cross-check, even though it provides inadequate information. Probably the most common instrument to be overemphasized is the attitude indicator. However, we must not allow ourselves to concentrate on that one instrument for maintaining altitude, for example, because we cannot hold our altitude with precision without also using the altimeter.

*Instrument Interpretation*

After looking at an instrument, we must interpret the information it presents. If we understand the operating principle of each instrument, we will learn, with experience, to apply that knowledge to the particular aircraft in which we are flying and to the specific flight condition. As we learn new maneuvers, we will learn what indications to expect from the instruments and what instruments are critical to the performance of the maneuver.

*Aircraft Control*

As soon as we learn to substitute the instrument readings for outside references, we will realize that the control actions and thought processes are identical with those used for controlling the aircraft when flying with outside references.

*Pitch Control:* the control of the rotation of the aircraft about its lateral axis. After we interpret the pitch attitude from the appropriate instruments, we will exert control pressures to effect the desired change.

*Bank Control:* the control of the rotation of the aircraft about its longitudinal axis. We use the instruments to determine the bank attitude and exert control pressures to roll the aircraft to the desired bank attitude.

*Power Control:* used when the instruments indicate a need for a change in thrust.

*Trim Control:* When flying on instruments, proper trim control becomes even more important than when flying visually. An untrimmed aircraft adds to our tension because it requires our constant attention. Control pressures should be induced by the pilot, and not induced by an untrimmed aircraft. A properly trimmed aircraft will maintain an attitude by itself if we let it fly "hands off." If we allow the aircraft to go out of control (as sometimes occurs during the early stages of instrument training), we need only to remove our hands and feet from all of the controls and the aircraft will return to the attitude for which it was trimmed.

## Aircraft Performance

No effort is made in this text to explain the intricacies of every flight maneuver when flying on instruments. Our flight instructor will spend many hours with that chore. We must keep in mind that the aircraft performs exactly the same whether we fly it visually by outside references or by visual reference to the instruments only.

## The Hazards of Instrument Flight

The aircraft, its engine, and its instrument panel present no greater hazard during instrument flight than during visual flight. There are, however, physiological and psychological effects which may present hazards unless properly understood and controlled.

*Psychological Problems*

In past years, instrument flying was reserved for the end of the flight training sequence. Pilots flew for many hours before attempting flight by reference to instruments alone. Thanks to the modern thinking on the part of those responsible for FAA approved flight training curricula, instrument flight training now begins during the initial training period. This method has served to remove many of the mysteries surrounding flight "on the gauges." There still is, however, a feeling of apprehension in the minds of many pilots when they attempt their first actual instrument flight.

As mentioned, we have become so used to relying on three of our senses to determine and control our environment that it becomes psychologically difficult to adjust to sight alone when flying on instruments and to ignore the false sensations received from the other senses.

*Physiological Problems*

As we will learn in Chapter 47, the human body must adjust to many physical changes which occur in flight. It adjusts amazingly well, but we will require time and experience aloft to recognize the effects of physiological changes in vision, circulation, breathing, etc.

*Disbelief*

Most people find it hard to believe that four of their senses become useless during instrument flight, and that much instruction and practice is necessary before controlling the aircraft by sight alone. Too many pilots attempt flight without sufficient outside reference and end up flying on instruments inadvertently. It is safe to say that instrument flight without proper knowledge and experience will likely end in near or complete disaster.

Year after year, pilots find themselves trapped on top of an overcast, in heavy precipitation, or in reduced visibility conditions. They attempt flight by visual reference outside the aircraft, only to find that aircraft control is impossible. Instrument flight for them is useless, for their instruction has been sparse and their practice negligible. Their senses battle each other for attention, each giving the pilot a false indication. Panic begins as the heart beats rapidly. The aircraft is forced out of control by uncalled-for control action, and the beginning of the end is near.

Tape recordings of pilot broadcasts during these conditions are heartbreaking. Screams for help, breathless prayer, and unreadable gibberish precede the final and unending silence which signifies the ultimate crash into oblivion. Pilots read and hear, but do not always believe, the many documented case histories of those before them who did not believe. They read; they hear; but they do not believe. We must not fall into the trap of disbelief.

We must be convinced that instrument flight requires training, practice, and vigilance, for without these preparations our own body and its physiological and psychological reactions will be the cause of our despair. We must believe.

An old, but true quotation seems appropriate when discussing unqualified pilots who attempt instrument flight without preparation. "There are old pilots, and there are bold pilots. There are *no* old, bold pilots."

## CHECK YOUR READING

1. What are the five senses?
2. What is the only usable sense during instrument flight?
3. Describe the problems concerned in *looking at the instruments.*
4. Describe aircraft control response when flying on instruments as compared with visual flight.
5. Why should the aircraft be properly trimmed at all times?
6. Why must all pilots receive proper training and constantly practice prior to attempting flight on instruments?

# PART VI

# PHYSIOLOGY OF FLIGHT

# 47

# THE PHYSIOLOGY OF FLIGHT

Man was designed to live at the bottom of an ocean of air. As our technology advanced into the age of flight, we learned that man could adjust to an environment quite different from that in which we live. Space travel has increased our knowledge and adaptability to strange surroundings.

If we are to be successful in our attempt to learn to fly, we must learn what changes to expect in our bodies and what restrictions to a normal existence these changes present.

## Physiological Altitude

Our body must have oxygen to survive. If we deprive ourselves of oxygen, either by increasing our altitude to heights where there is less oxygen available, or by changing the chemistry of our blood so that it absorbs less oxygen we have, in effect, increased our physiological altitude.

Oxygen is transported to the tissues by combining with the hemoglobin in our blood. Oxygen enters the bloodstream through the walls of tiny air sacs within our lungs, called *alveoli*. The amount of oxygen which combines with the hemoglobin depends on two things; first, the pressure of the oxygen within the lungs, and second, the willingness of the hemoglobin to accept the available oxygen. Therefore, our physiological altitude can be increased in either of the two ways described here.

*Partial Pressure*

Since oxygen represents only about one-fifth of the total volume of the atmosphere, its pressure is only one-fifth of the total pressure of the atmosphere. At sea level the oxygen partial pressure is sufficient to nearly saturate the hemoglobin, but as we increase our altitude, the total pressure of the atmosphere decreases, causing the partial pressure of the oxygen to decrease. Therefore, less oxygen will be absorbed by the hemoglobin for delivery to the tissues, as altitude increases.

*Toxicity*

Even though the partial pressure of the oxygen is sufficient to allow it to pass through the alveoli and be absorbed into the hemoglobin, the chemistry of the hemoglobin may be changed by another substance or gas which is introduced into the bloodstream. When this happens, the hemoglobin is unable to transport as much oxygen, which results in an increased physiological altitude. We will discuss this phenomenon fully later, when we learn about specific toxics.

## Effects of Reduced Total Pressure

A change in atmospheric pressure results in disorders in various parts of the body. We will discuss the more common disorders so as to become familiar with their symptoms and effects.

*Aerotitus*

Expanding and contracting air within the middle ear may cause inflammation, pain, and temporary deafness if the air within the ear is not adjusted to outside air pressure. As shown in Fig. 161, the middle ear is separated from the outer ear by a thin membrane.

The cavity of the middle ear is surrounded by bones and filled with air. Normally, air pressure within the ear is equalized with outside pressure through the Eustachian tube, which is a duct leading from the middle ear to the throat.

During ascent into lower atmospheric pressure, the air within the middle

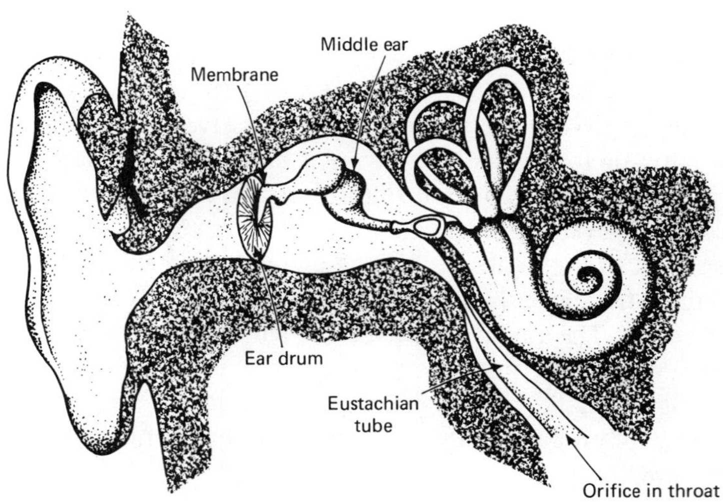


Fig. 161. The human ear.

ear escapes through the Eustachian tube quite easily. The construction of the tube makes it reluctant to allow outside air to return to the middle ear, however, which is necessary during a descent to higher atmospheric pressure. As the differential between outside air pressure and middle ear air pressure increases, the eardrum begins to bulge inward, causing severe pain. If the pressures are not equalized, the eardrum will rupture. Normally, we can help open the Eustachian tube to equalize air pressure during descent by frequently yawning or swallowing. In extreme cases, it may help to hold the nose and mouth closed and blow gently, forcing the Eustachian tube to open. If the pressure cannot be equalized, we may need to stop the descent or initiate a short climb to relieve the pain.

Naturally, any excess mucous in or around the end of the Eustachian tube will decrease the chances of opening the tube. Therefore, we must not attempt flight with a head cold or other mucous producing illness. (See page 401 where cold remedies and decongestants are discussed.)

### *Sinus Barotrama*

Our skull contains cavities, called sinuses, which are grouped around the nasal passages. Each sinus is connected to the nasal cavity by narrow passages in the bone structure. These openings are lined with mucous membranes. Inflammation of the membranes, caused by a cold, hay fever, or other illness, will make them swell and possibly close the passages which vent the sinuses. We have no control over these bony passages, and, if they become closed, we are apt to experience pain (severe at times) when we change the outside atmospheric pressure. This is particularly noticeable during a rapid descent.

The only remedy for sinus pain is to immediately stop the descent or initiate a climb. A slow descent may allow the pressure within the sinuses to equalize with outside atmospheric pressure.

Flight during periods of even mild head or respiratory congestion, should be avoided if we are to operate efficiently and avoid physical pain and/or physical damage.

## Hypoxia

Hypoxia is caused by an insufficient supply of oxygen to the body tissues. It is extremely dangerous, as it leads to total unconsciousness and eventual death, but its symptoms are insidious; i.e. the conditions gradually worsen while the patient actually feels better.

### *Effects of Hypoxia*

The eye is most susceptible to a deficiency in the amount of oxygen delivered to its tissues. As the quantity of oxygen in the bloodstream diminishes, either due to an increase in altitude or toxics in the blood, vision begins to deteriorate. Mental confusion soon begins, along with euphoria or a feeling of well-being. At this point, hypoxia is forming a bottomless pit into which we will fall, but we will be tickled about it. Tests prove that pilots feel great (and want to tell the world so) just before they reach the point where they can no longer concentrate, lose their judgement ability, suffer from slow reflexes, and ultimately lose consciousness.

*Hypoxic Altitude*

At altitudes up to ten thousand feet, the mild hypoxic effects encountered do not materially deteriorate our flying ability, if our bloodstream is not carrying any toxics. Night vision deteriorates rapidly with the lack of oxygen. Therefore, when flying at night, we must use supplemental oxygen above five thousand feet if we are to maintain normal vision.

## Toxics

Toxics in the bloodstream may increase our physiological altitude before we leave the ground. Then, when we increase altitude, our physiological altitude will be much higher than the actual altitude.

*Carbon Monoxide*

Think of the hemoglobin in the bloodstream as a tiny ship in a river. As the ship passes through the dock (the lung), it picks up passengers (oxygen molecules) for transport to the tissues throughout the body. Suppose, however, when the ship reaches the dock, other passengers are waiting to board, each one holding a baseball bat in his hand. The weapon gives the unwanted passenger a definite advantage, and for every oxygen passenger who gets on board, two hundred other passengers board. The ship can only hold a limited number of passengers, so the number of oxygen passengers is severely decreased.

This is exactly what happens when carbon monoxide enters the lungs. It is the unwanted passenger, but it is absorbed by the hemoglobin about two hundred times more readily than oxygen.

There are two primary sources of carbon monoxide in flight. The first is due to leaks in the engine exhaust system or cabin heater. The odorless, colorless, tasteless gas seeps into the cockpit where it will be drawn into our lungs with every breath.

The second source of carbon monoxide is from smoking. Three cigarettes smoked at sea level can raise our physiological altitude to eight thousand feet. If we then take off and climb to ten thousand feet, smoking another cigarette in the climb, our physiological altitude will be about eighteen thousand feet, which is approximately the altitude at which unconsciousness occurs within thirty minutes without supplemental oxygen.

*Alcohol*

The effects of alcohol in the body are twofold. Not only does alcohol prevent oxygen from reaching the tissues, but it interferes with the normal *use* of oxygen by the tissues.

Pilots do not fly when drunk. But statistics show that many fatal accidents are connected to alcohol use. One two-ounce drink at sea level does not seriously affect most people. However, at ten thouand feet, the two-ounce drink is equal to four ounces. Therefore, "one for the road" at the airport cocktail lounge becomes "two for the money" at ten thousand feet. If we also smoke a cigarette or two in the lounge and during the flight, we seriously affect our ability to fly safely.

Two commonly believed misconceptions produce trouble for pilots. The first is that alcohol in small quantities is a stimulant, and that a cup of coffee will burn up the alcohol, but leave the recipient in a stimulated condition. The fact is that the rate of oxidation is quite constant and the decline of alcohol in the blood is very slow. The second misconception is that the lack of severe symptoms from alcohol consumption on the ground indicates that no impairment of flying ability will appear in flight. We must remember that every drink on the ground may equal two or more aloft.

### *Drugs*

Too many pilots ignore the advice of aviation medics concerning the use of drugs prior to or during flight. The word "drugs" is not meant to infer hallucinatory, hard narcotic, or marijuana products, although their use is certainly prohibited. The drugs which cause most problems are the antihistamines, stimulants, muscle relaxants, barbiturates, tranquilizers, weight-control agents, and antismoking agents. Most pilots, for instance, will notice no ill effects after using a nasal decongestant spray when on the ground, so they naturally use it in flight. Some nasal decongestant sprays can cause drowsiness, dizziness, loss of visual acuity, and loss of equilibrium in flight.

Tranquilizers are becoming more popular in the high-pressure atmosphere of today's society, and yet, in flight, they may cause blurred vision, mental depression, and sleepiness. Other very common drugs which may be purchased without prescription are weight-control agents. Severe nervousness, judgement errors, and euphoria may result from using weight-control agents. Antismoking agents may cause nausea and distraction in flight.

In short, pilots must use extreme care in selecting *any* medicine. Most doctors can suggest a substitute medicine for one which may deteriorate our flight ability, or will specify the length of time required for possible side effects to dissipate. **Before using any medication whatsoever**, we must be positively certain that our flying ability will not be impaired.

## Prevention of Hypoxia

The method employed to prevent hypoxia depends on whether we wish to be protected from hypoxia induced by low oxygen partial pressure or from that induced by toxics.

### *Oxygen Systems*

As the partial pressure of oxygen in the atmosphere decreases, we can introduce pure oxygen to the lungs by breathing from oxygen tanks. Even light aircraft can be equipped with oxygen systems, and if we intend to do much night flying above 5,000 feet and daytime flying above 10,000 feet, we should use stored oxygen to supplement that found in the atmosphere. Oxygen systems and controls vary with aircraft models, but all are simple to operate and are very reliable.

*Precautions Against Toxicity*

If we are to avoid toxic hypoxia we must refrain from introducing toxics into the bloodstream. Heavy smokers should reduce the number of cigarettes consumed during the few hours prior to each flight, as well as during the flight. This is particularly important when planning a night flight, because hypoxia causes night vision to deteriorate more rapidly. Sufferers from hay fever, head colds, etc. must exercise extreme caution when using cold remedies and nasal sprays. In short, if in doubt about *any* medication, we must check with an aeromedic.

Alcohol, narcotics, and barbiturates are to be strictly avoided.

## Warning Signals of Hypoxia

Since vision is the first bodily function to deteriorate because of hypoxia, we should constantly be aware of our visual acuity while in flight. Many pilots pick a finely printed label or placard in the aircraft, that can barely be read from their position in the cockpit. During the flight they occasionally read the print. If it becomes impossible to read, they suspect the onslaught of hypoxia.

If our body is subjected to oxygen lack for any length of time, our lips and fingernails will take on a blue tint. Dizziness, tingling, and hot flashes may be experienced. By then, however, euphoria will have begun, and the fact that we are turning blue will have no alarming effect whatever. In fact, we may feel better than we have for a long time. Therefore, it is obvious that we must detect hypoxia early, before euphoria sets in and we sit and marvel at our well-being, giggling frequently to emphasize the point.

Unfortunately, the warning signals of hypoxia are not identical for every person, nor are they the same for any one person every day. This fact increases the necessity for remaining alert to the symptoms of this insidious illness.

## Recovery From Hypoxia

If hypoxia is caused by a lack of oxygen partial pressure, recovery is almost instantaneous after the introduction of sufficient oxygen. We can descend to a lower altitude, or breathe stored oxygen to effect an immediate recovery.

Toxic hypoxia is more difficult to cure. The oxidation rate (how fast the toxic is washed from the blood) of the toxic determines the time required for recovery.

## Hyperventilation

Carbon dioxide is a waste product of the body. Activity causes an increase in the amount of carbon dioxide produced and dispelled into the blood, which triggers an unconscious increase in the breathing rate in an attempt to bring more oxygen into the body. If we increase our rate of breathing for any reason other than increased activity, we will cause a carbon dioxide deficiency in the blood. The resulting symptoms are quite similar to those associated with hypoxia. The unwary pilot often mistakes hyperventilation for hypoxia, and in an attempt to acquire more oxygen, increases further his rate of breathing, thus intensifying the hyperventilation. In extreme cases muscle cramps and tingling in the fingers and toes precede total collapse.

*Recovery from Hyperventilation*

Recovery begins almost immediately after the carbon dioxide content of the blood is increased. We can slow our rate of breathing to allow this build-up, or rebreathe our exhaled air by breathing into a paper bag or other container for a few minutes.

*Precautions to Avoid Hyperventilation*

When we are scared, excited, or tense, our normal tendency is to increase the rate of breathing, or overbreathe. Stress and fatigue are constant flying companions, particularly during times when we experience new situations. The pilot who flirts with a thunderstorm, for example, is likely to begin hyperventilating. We must exercise discipline over our body to insure its healthy operation at all times.

## Vision

As we have learned, no other human sense is as vital to the safety of flight as is vision. Therefore, we must learn the effects of flight upon the eye, and what to expect from this marvelous camera-like organ.

*The Construction of the Eye*

The eye is almost an extension of our brain, connected to it by nerves. Images are impressed on the eye and sent to the brain in split-second time. The eye is extremely adaptable to varying light intensities, having the ability to record images in the brightest sunlight as well as dim moonlight. Two primary mechanisms are responsible for this adaptability. First, the dilating pupil acts like a camera aperture to adjust the amount of light which enters the eye. But of more importance, especially to pilots, is the network of rods and cones which record the image.

*The Rods and Cones.* The center of the eye is made up of cones which are used when bright light is available and for detecting color. Surrounding the cones are the rods, which "see" in dim light, but do not register color. During daylight or bright artificial light conditions, the cones record the bulk of the information entering the eye, and at night, or in dim light, the rods come into play, recording the bulk of the information, but only in black and white and intermediate tones of gray.

*Night Vision*

Night vision then, is dependent upon the rods in the eye. As we have learned, the eye is the first organ to suffer from oxygen deficiency. Within the eye, the rods will lose their efficiency prior to the cones when the oxygen content of the blood is reduced. Therefore, night vision decreases first with hypoxia, and it is for this reason that supplemental oxygen should be used during night flight when above 5,000 feet.

If we wish to receive a good picture from our eye during dimly lit conditions, we must not look directly at the object we wish to see, but rather to one side of it. In this way, we bring the rods into play, which are much more light sensitive than the cones in the center of the field of vision.

*Adaptation*

All of us have experienced night adaptation of the eye. We have entered a darkened movie theatre, unable to see the first few rows of seats. A few moments later, however, we not only can see the seats around us, but may read the movie program. Our eyes have become adapted to the dim light. Conversely, as we leave the theatre and enter a bright sunlit street, our eye automatically sends a signal to the brain to close the pupil in order to reduce the amount of light entering the eye.

Our eyes will adapt to very dim light, but will require time to do so. Once the rods have adapted, a bright light, even the flare of a match, will destroy their adaptation, and night vision will be severely decreased. Therefore, when planning a night flight, we should avoid bright lights for about thirty minutes prior to flying, and make every effort to avoid bright lights when taxiing or during flight. Some pilots shut one eye when they find that a momentary bright light is unavoidable (such as the taxi light of another aircraft) so that their night vision will be preserved in one eye at least.

*Care of the Eye*

Because our eyes are important to us, we must care for them. As we gain altitude, the intensity of the infrared and visual light rays increases. Normally these rays are harmless, but a direct look at the sun may seriously injure the eye, burning the retina. Pilots should wear high quality dark glasses to protect the light adaptability of their eyes and to absorb ultraviolet radiation.

Frequent eye examinations will insure longer eye life, as will proper sight habits (proper light for reading, avoiding glare, etc.).

*Illusions*

Our eyes can play tricks on us. Most often, we will be susceptible to illusions at night, and while flying on instruments.

*Nighttime Illusions.* When we stare for a long time at a stationary pinpoint of light, we will begin to see it move erratically. For instance, suppose we see a star through a hole in an overcast. As we look at it, it will appear to move around in the sky, or move slowly in one direction. This illusion can cause a hazard if, for instance, we see the light of another aircraft dead ahead. If it is on a collision course with us, its relative direction will not change. As we stare at it, trying to determine its direction of movement, we see that it begins to move to one side, indicating a course change. However, the movement of the light may only be an illusion, and we may remain on a collision course. We can eliminate much of this illusion by scanning back and forth across the horizon instead of staring at one object. We may also note the position of the light relative to the cockpit window, nose cowling, etc. and check its relative motion to that point.

Quite often it is difficult to tell if another aircraft is moving toward or away from us. Again, scanning instead of staring will aid in seeing the aircraft in its true perspective.

Lights reflecting on a canopy or windshield can give the illusion of an impending midair collision.

Lights on the horizon may suddenly appear at a much higher altitude.

*Illusions In Clouds.* Flight in and out of cloud banks may give the illusion of rapidly changing speed or altitude.

The rotating beacon (anticollision light) may produce false sensations of turning or unusual attitudes.

### *Hypnotic Effects*

Recent tests have shown that the rotating beacon or flashing wing lights may cause the viewer to be hypnotized.

## Movement

So far, our discussion of physiological reactions to flight has not been associated with high or changing velocities. Yet, flying certainly is a new environment of speed and rapidly changing direction. We should become familir with some of the more important effects of movement.

Movement affects our physical mechanisms, either routinely or disastrously. Movement provides enormous quantities of information and, unless we are careful, this information can easily be misinterpreted by the brain, resulting in an erroneous picture of our surroundings.

### *Velocity and Acceleration*

Man is not affected by continuous, constant speed in a straight line. That fact has certainly been proven by astronauts as they hurtle through space at speeds of 25,000 miles per hour or more. Without visual references, they would not realize any motion. But when velocity *changes*, either by acceleration, deceleration, or changing direction, man may be seriously affected.

### *The Measurement of Inflight Forces*

In our every day life on earth we are subjected to the force of gravity, which will produce an acceleration of a falling body of 32 feet per second, per second. That is, a falling body will increase its speed by 32 feet per second, every second, if it is not prevented from doing so by encountering solid objects. This acceleration is called "one *g*" force. The force of one normal gravity is used as a yardstick to which we compare other forces.

In flight, the force of gravity is replaced by a new force which is the resultant of all forces acting on the machine. Man is affected by the equal and opposite forces of inertia, which are generated by the tendency of a body to remain in the same place or to move at a uniform speed in a straight line. In other words, a body at rest tends to remain at rest; a body in motion in a straight line tends to remain in motion in a straight line. It is when we attempt to oppose inertia that we affect the body. If we apply a force which accelerates a body four times faster than does the force of gravity, for instance, it is called a "four *g* force."

That is why our apparent body weight changes in flight. Under a force produced by the motions of flight, man's weight may apparently increase or decrease. If the resultant force of inertia is four *g*'s, we will weigh four times more than normal. If the forces of flight add up to zero, we will be weightless.

## Classification of *g* Forces

We may classify the *g* forces in flight as *positive*, *negative*, or *transverse*. If the resultant forces act in the same direction as normal gravity, that is from our head toward our feet, they are called *positive g*. If they act opposite to normal gravity, from our feet toward our head, they are called *negative g*. Resultant forces acting across our body, or perpendicular to normal gravity, are called *transverse g*.

Further classification of *g* forces is derived from the time the force lasts. Normal forces in flight last for several seconds or more and are called forces of *long duration*. The time period may be very short, however, especially during an accident, lasting only fractions of a second. These are called forces of *short duration*.

Since, in light aircraft, resultant forces in flight are generally of long duration and positive, we will discuss them in more detail.

## Positive *g*—Long Duration

During normal flight maneuvers, centrifugal force acts from our head toward our feet, causing positive *g*. Many maneuvers, including the turn, pullout, loop, etc. produce positive *g*. The first sensation we will detect when experiencing positive *g* is heaviness in the seat. Our limbs will seem to weigh more than normal and our facial tissues will sag. We will have difficulty speaking as the *g* force increases and body movement will become difficult if not impossible. When the positive *g* force becomes two to four times normal gravity, we will notice a dimming of vision or a complete *black out*. This is caused by insufficient oxygen being delivered to the eye, because the blood is too heavy to be pumped to the head in sufficient quantities by our heart. The term *black out* refers only to a loss of vision. Unconsciousness will eventually occur if the *g* force is of very long duration.

It is difficult to pinpoint the exact amount of positive *g* which will cause a particular symptom, as the ability to withstand *g* forces varies with each individual. The time required to build up the *g* force also determines how much we can stand. We can survive a much higher *g* force if the acceleration which causes it is of long duration.

As with many abnormal situations, the body will attempt to bring corrective actions into play. The heart will increase its pumping action in an effort to maintain normal blood pressure throughout the body. Blood vessels will constrict in the lower extremities of the body to prevent minor pooling of the blood.

#### *Effects of Positive g*

Obviously, loss of vision and unconsciousness are dangerous for the pilot. However, even less dramatic symptoms may have an effect on our flying ability. The increased weight we will experience will cause slower body movements, and we will be less accurate with movements such as reach. A feeling of fatigue is often instilled in the pilot if excessive *g* forces are applied. A false sense of the vertical is also common, but we will discuss that in more detail later, when we learn about disorientation.

#### *Protection from Positive g*

In light aircraft, there are no facilities for pressure suits or couches on which to lie when experiencing high *g* forces, because the aircraft structure itself is not designed to withstand *g* forces in excess of 3.8 *g*'s (normal category). We can, however, increase our ability to function normally while influenced by unusual positive *g* forces by maintaining good health, exercising, and remaining within the suggested weight range for our physique.

### Negative *g*—Long Duration

Luckily, negative *g* of long duration is a rare occurence in flight, for it is a very uncomfortable condition indeed.

One negative *g* force balances out the positive *g* which normal gravity provides, leaving us weightless, which is not at all uncomfortable or dangerous. Two negative *g*'s, however, gives us the sensation of hanging upside down, with blood rushing to the head. At two or three negative *g*'s we would feel congestion of blood in the head in the form of throbbing pain. Mental confusion accompanies the feeling that the skull will burst. Loss of vision follows the "red out" which is caused by either ruptured veins in the eye or by the eyelid which is pulled over the eye. Sustained negative *g* may cause permanent brain damage. No practical method of protection from negative *g* forces now exists.

### Transverse *g*—Long Duration

Since transverse *g* forces must be in excess of ten *g* before producing bodily malfunctions, and light aircraft cannot sustain forces of that magnitude, we will not discuss transverse *g* forces here.

## Equilibrium

Although vision is the most important of all our senses in receiving accurate information concerning our whereabouts, we do receive information from other sources. Even though we are blindfolded, for instance, we can normally tell where our limbs are in relation to space and may be able to walk quite well. Our brain receives messages from the vestibular mechanism in the inner ear, nerve endings, muscle joints, tendons, skin, and organs. Unfortunately, when in flight, we may easily be fooled by any of the messages received from the body, except those perceived by the eye.

### *The Vestibular Mechanism*

Part of the inner ear is the vestibular mechanism which consists of three semicircular canals set at right angles to each other. The canals are constructed from membrane and contain fluid and hairlike structures. As our body accelerates (changes its *rate* of motion), the nerve endings are stimulated by the deflection of the hairlike structures. Pressure changes on the flow of liquid in the canals cause this deflection.

When we tip our heads from side to side or fore and aft, the semicircular canals telegraph the movement to our brain. However, the sensitivity of the canals is not high, and only the *beginning* of the motion is perceived, not continuous motion. Therefore, if we tip our heads very slowly to the left, for instance, and then reverse the motion, bringing our heads rapidly back to the vertical, we will only perceive the rapid motion. If our eyes are closed, we will receive the impression of tipping to the right.

When in flight, a slow steady turn may be interpreted as straight flight; and the roll-out from a turn may be interpreted as a turn in the opposite direction. The vestibular mechanism cannot distinguish between centrifugal force and gravity, which often join hands in flight. Without visual aids, we cannot properly interpret the messages received from the semicircular canals, so we must learn to disregard them when visual references are lost.

### *The Postural Sense*

The expansion and contraction of muscles and tendons, touch and pressure, and shifting abdominal muscles contribute to our postural sense. However, in flight the postural sense often misinterprets centrifugal force as a sensation of rising or falling. Without visual assistance, a steep turn may be interpreted as a steep climb, or a shallow descending turn to be straight and level flight.

## Spacial Disorientation

It is obvious that vision is a necessity to combat spacial disorientation. Instrument flight without proper training and experience is dangerous because we need to learn which senses to disregard and how to rely on vision, the only sense that remains true.

Vertigo, or spacial disorientation, may attack even the most experienced pilot. Inexperienced pilots often succumb to the dizziness and complete inability to control the aircraft which vertigo may induce.

## Fatigue and Boredom

Surprisingly, our scientific knowledge about fatigue is quite vague. We know that we can cause a man to become fatigued by requiring his muscles to contract. But we can also cause fatigue in the man by requiring him to perform boring duties. For instance, if we sit and write our name for a long period of time we will become tired, and with passing hours our writing will become illegible. However, at that moment when deterioration is at its peak, we can perform another task such as writing a column of figures quite perfectly. The inability to write our name properly was caused by fatigue, but it could not have been caused by changes in the muscle of the hand or arm. Science is unable to explain this phenomenon.

Classification usually falls into two broad groups. First, those feelings of tiredness and inadequacy, and second, a falloff in performance. The latter form of fatigue, when combined with an undesirable task, can lead to a neurosis of avoidance such as "battle fatigue" or "operational fatigue."

This type of fatigue is extremely important to us as pilots, for after a certain length of time in flight, a definite reduction in skill may be shown. Our reaction to stimulus will be slow; we may hold onto the controls more rigidly, and move more slowly than normal.

### *Prevention of Fatigue*

Aircraft design is an important factor in the prevention of fatigue. The cabin should be comfortable, controls easily manipulated, and for high altitude flight, cabin pressurization must be sufficient to maintain a safe and comfortable physiological altitude.

We must not allow ourselves to reach a state of boredom aloft. Long cross-country flights, particularly over familiar terrain can be boring. We must create interest in the landscape, the weather, and our navigation, to remain alert. One of the hazards of flight is the tendency to be lulled into a bored, almost trancelike state of unconcern. If an emergency arises while we are in this condition, we will be slow to recognize it and to react to it.

## Fear

The first and most important thing to learn about fear is that it is normal. Everyone who has ever flown has known anxiety at times, airline pilots and students alike.

We must learn to control fear. We should not be ashamed of it or think of it as a form of cowardice, for fear can make us think and act faster than normal. Fear, then, is useful if we understand it.

Fear produces an intensification of our bodily processes, which may be unpleasant. We have all experienced the pounding heart, rapid pulse, muscular tenseness, dry throat and mouth, sweating, frequent urination, and loose bowels which fear causes. Nervous energy is created by fear, and instinct releases it in forms of self-preservation and anger.

If we do not release this pent-up energy produced by fear, we will suffer from tiredness, loss of appetite, nausea, insomnia, irritability, moroseness, bad dreams, or obsession with thoughts of harrowing flights. These are all normal symptoms, and do not indicate the onslaught of organic diseases or a mental breakdown.

Fear is our body's way of physiologically marshalling its forces to meet danger. Since tranquilizing drugs inhibit this normal response of the body, *we must not fly* when using them.

If fear is a nearly constant companion, particularly before flight, we should discuss it with a flight surgeon or an experienced pilot friend. Competence and self-confidence are the best antidotes for fear.

We all have a panic button. We do not *have* to use it.

## Summary

We have learned that our bodily functions must adjust to the varying conditions of flight. The adaptability of man is truly amazing. Experience in flight will add to our knowledge of our own body and will help us to know what to expect from it during normal and emergency conditions.

### CHECK YOUR READING

1. Explain physiological altitude.
2. Why does an increase in altitude decrease the amount of oxygen entering the bloodstream?
3. Explain toxicity.
4. How is the middle ear vented to outside atmospheric pressure?
5. Explain briefly the cause and symptoms of sinus barotrama.
6. Which bodily function is first to suffer from hypoxia?
7. Describe euphoria.
8. What is the best method to determine the first onslaught of hypoxia?
9. What effect does smoking have on the physiological altitude of our body?
10. What precautions must all pilots take prior to using prescription drugs or non-prescription medicines?
11. At what altitude should supplementary oxygen be used at night? During daytime?
12. Why must pilots detect hypoxia early in its inception?
13. Describe the symptoms of hyperventilation.
14. What causes hyperventilation?
15. How can we prevent or combat hyperventilation?
16. Which part of the eye sees best at night?
17. Which part of the eye is most susceptible to hypoxia?
18. Describe the common nighttime illusions.

19. What is inertia?
20. How does inertia affect man in flight?
21. Define positive *g*.
22. Define negative *g*.
23. Describe the bodily symptoms experienced during periods of excess positive *g*.
24. Is the amount of positive *g* required to cause *black-out* the same for every pilot? Is it the same every day for each individual?
25. What is the only reliable sense on which we must depend when flying without outside references?
26. Why does the vestibular mechanism report false information concerning equilibrium?
27. How can we prevent fatigue and boredom in flight?
28. Describe the symptoms of fear.

# PART VII

# THE AIRMAN'S INFORMATION MANUAL

# 48

# THE AIRMAN'S INFORMATION MANUAL

The Airman's Information Manual (AIM) is published by the Federal Aviation Administration for the purpose of providing pilots with a wealth of information, most of which is not found in any other publication conveniently available.

No other Federal aeronautical publication contains so much information necessary for day-to-day flight operations. Every pilot, no matter if he is a student, hobby flier, or professional, should overlook the tremendous value of this publication. Every pilot should subscribe to the Airman's Information Manual and refer to it often. This section is *not* designed to replace the Airman's Information Manual, but will make us familiar with its contents.

Subscription orders for the Airman's Information Manual should be placed with the Superintendent of Documents, G.P.O., Washington, D.C. 20402.

## Contents

The contents of the Airman's Information Manual is divided into four parts. Each part is issued for a period of time which varies with the frequency of change of the contents of that specific part.

## Part I—Basic Flight Manual And ATC Procedures

Part I is issued quarterly. It contains basic fundamentals which are required knowledge when flying within the National Airspace System. The annual subscription price is $2.50. More important information contained in Part I is described here in order to familiarize us with the contents.

*Good Operating Practice*

This section contains information which all pilots should learn in order to improve the safety of all who fly. It describes, in general terms, ways to reduce the chance of a midair collision (including the use of airways and traffic patterns), radio frequency congestion, and communications procedures.

*Air Navigation and Radio Aids*

A clear and simplified description of low frequency, VHF Omni-Directional Range, and instrument landing system installations is presented in this section.

*Radar*

A thorough knowledge of this section will aid us when we begin to use controlled airports and fly into densely populated areas, because surveillance radar, radar controlled approaches, and the Air Traffic Control radar beacon system are all described.

*Airport and Air Navigation Lighting and Marking Aids*

This section contains information concerning various airport lighting installations, runway markings, and light beacons.

*Radiotelephone Phraseology and Techniques*

Very helpful information concerning the use of the two-way radio is contained in this section. Although no pilot should refrain from using the radio for communications because of unfamiliarity with the proper phraseology and technique, the communications network will certainly operate more efficiently if we all make an effort to learn and follow the suggestions presented in these few pages of the Airman's Information Manual.

*The NOTAM Code*

Information concerning the National Airspace system is disseminated in three ways. Aeronautical charts comprise the primary method. The second method is the Airman's Information Manual, and the third is the National Notice to Airmen system. Each system varies as to the frequency of issuance of information. The most up to date, indeed almost instantly revised information, is contained in the Notice to Airmen, or NOTAM, and Airmen Advisories.

The Airmen Advisory contains information which is "nice to know" but not time-critical. That is, the information may be given to the airman upon request or on an "as needed" basis. Examples of messages contained in the Airmen Advisory are: taxiway closed, grass cutting in progress along the runway edges, men and equipment crossing a runway, etc. Airmen Advisories are distributed nationally via voice communications, teletypewriter, telephone, etc.

The NOTAM contains critical information which is required for flight planning, and which may affect a pilot's decision to make a flight. Examples of NOTAM messages are: airport closed, radar out of service, braking action conditions due to snow or ice, runway lighting out of service, etc. NOTAM codes are disseminated via teletypewriter, voice communications, or telephone.

In order to reduce the telecommunications congestion and translation difficulties, a standard abbreviated code is used for the broadcast of NOTAMS. The code group contains five letters, the first letter always being Q. The second and third letters identify the facility, danger to aircraft, or service being reported. The fourth and fifth letters denote operating status of the facility, the danger to aircraft, or the service being reported. The NOTAM codes, as shown in the Airman's Information Manual are depicted in Fig. 162.

For example, the NOTAM code list shows the NOTAM "QEDES" to mean that the VOR/DME is out of service. If this NOTAM appeared on the teletypewriter circuit following the Kansas City aviation weather report, we would be immediately alerted to the status of the Kansas City VOR/DME.

Prior to flight, we should contact the nearest Flight Service Station to obtain current NOTAMS and Airmen Advisories for the intended route of flight.

### *Weather*

This section of Part I discusses weather broadcasts, inflight weather advisories, pilot reports, and clear air turbulence, as well as weather radar and thunderstorms.

### *Safety of Flight*

Information in this section serves as good review reading to refresh our memory on wake turbulence and the avoidance of vortices.

### *Medical Facts For Pilots*

These few pages of Part I provide an excellent capsule review of the discussion in this guide on the physiological changes to our body in flight (see Chapter 47), including the dangers of fatigue, alcohol and drugs, and carbon monoxide. Vision, ear discomfort, and panic are also discussed.

### *Bird Hazards*

Birds, due to their abundance, large size, or dense flocks, constitute a possible hazard to flight. Therefore, the migratory patterns of many birds are described in Part I. We should periodically review this section, noting any migratory patterns which may occur in the geographic areas in which we fly.

### *Preflight*

Many of the preflight planning activities are described in these pages of Part I, including the weather briefing and flight plan, providing an excellent review of the material presented in this text concerning cross-country flight planning (see Chapter 56).

# THE NOTAM CODE

## DECODE

### SECOND AND THIRD LETTERS

| 2nd & 3rd Letters | SIGNIFICATION | 2nd & 3rd Letters | SIGNIFICATION |
|---|---|---|---|
| | **RADIO AIDS** | | |
| **AA** | . . . [*specify TWR, APP, ACC or FIC*] air traffic services receiver . . . kHz (*or* . . . MHz) | **EB** | Broadcasting station (public) (VS) |
| **AB** | Inner marker, Instrument Landing System (IM) (for runway number . . .) | **EC** | CONSOL or CONSOLAN station |
| **AC** | . . . [*specify TWR, APP, ACC or FIC*] air traffic services transmitter . . . kHz (*or* . . . MHz) | **ED** | DECCA or DECTRA |
| **AD** | Middle marker, Instrument Landing System (MM) (for runway number . . .) | **EE** | Ground Controlled Approach System (GCA) |
| **AE** | Outer marker, Instrument Landing System (OM) (for runway number . . .) | **EF** | Terminal Area Surveillance Radar (TAR) |
| **AF** | Fan-type marker | **EG** | |
| **AG** | Glide path, Instrument Landing System (GP) (for runway number . . .) | **EH** | Elevation element of the Precision Approach Radar (PAR) (for runway number . . .) |
| **AH** | Non-directional radio beacon (NDB) | **EI** | Monitoring device associated with . . . (*specify*) radio aid |
| **AI** | Instrument Landing System (ILS) (for runway number . . .) | **EJ** | All air ground facilities (except . . .) |
| **AJ** | Radio range (other than VOR) and associated voice communications | **EK** | Precision Approach Radar (PAR) (for runway number . . .) |
| **AK** | Radio receiving facilities | **EL** | LORAN |
| **AL** | Localizer, Instrument Landing System (LLZ) (for runway number . . .) | **EM** | Azimuth element of the Precision Approach Radar (PAR) (for runway number . . .) |
| **AM** | Locator, middle, Instrument Landing System (LM) (for runway number . . .) | **EN** | DME (1000 MHz Distance-Measuring Equipment)* |
| **AN** | TACAN | **EO** | VOR/DME |
| **AO** | Locator, outer, Instrument Landing System (LO) (for runway number . . .) | **EP** | Radar responder beacon (RSP) |
| **AP** | VOR (VHF Omnidirectional Radio Range) and associated voice communications | **EQ** | Surveillance Radar Element (SRE) of GCA |
| **AQ** | VOR (VHF Omnidirectional Radio Range) | **ER** | Radio transmitting facilities |
| **AR** | Radio range (other than VOR) (RNG or VAR) | **ES** | A radio navigation facilities (except . . .) |
| **AS** | Radio range leg | **ET** | Teletypewriter transmitting facility(ies) |
| **AT** | Attention signal | **EU** | Radio direction-finding station . . . (*frequency or type*) |
| **AU** | Meteorological communications . . . kHz (*or* . . . MHz) | **EV** | VORTAC (the combination of VOR and TACAN) |
| **AV** | Voice communications . . . kHz (*or* . . . MHz) | **EW** | Ground interrogator, Secondary Surveillance Radar (SSR) (System |
| **AW** | VOLMET | **EX** | SELCAL |
| **AX** | Non-directional radio beacon (NDB) and voice facility | **EY** | Surface movement radar (SMR) |
| **AY** | 200 MHz Distance-Measuring Equipment | **EZ** | |
| **AZ** | Station location marker VHF (MKR) | | |
| **EA** | En-route Surveillance Radar (RSR) | | |

* Where more than one DME is installed at the same location, the type of associated facility (e.g., ILS, VOR' TACAN or VORTAC) should be indicated.

Figure 162(a).

## DECODE
### SECOND AND THIRD LETTERS (Continued)

| 2nd & 3rd Letters | SIGNIFICATION | 2nd & 3rd Letters | SIGNIFICATION |
|---|---|---|---|
| | **LIGHTING FACILITIES** | | |
| IA | Boundary lights | IP | Approach light system [*type* . . . (*specify ligth intensity*)] (for runway number . . .) |
| IB | Aerodrome beacon | IQ | Runway alignment indicator (for runway number . . .) |
| IC | Circling guidance lights (for runway number . . .) | IR | Runway edge lights [*type* . . . (*specify light intensity*)] (for runway number . . .) |
| ID | Runway end identification lighting (for runway number . . .) | IS | |
| IE | Marine light beacon | IT | Taxiing guidance system |
| IF | Floodlights | IU | Runway centre line light (for runway number . . .) |
| IG | Visual approach slope indicator system (*all types*) (for runway number . . .) | IV | Approach light beacon (for runway number . . .) |
| IH | Taxiway edge lights | IW | Stopway lights (for runway number . . .) |
| II | Hazard beacon | IX | Sequenced flashing lights (for runway number . . .) |
| IJ | Threshold lights (for runway number . . .) | IY | Taxiway centre line lights |
| IK | Flares (for runway number . . .) | IZ | Runway touchdown zone lights (for runway number . . .) |
| IL | All landing area lighting facilities | | |
| IM | Identification beacon | | |
| IN | Range lights | | |
| IO | Obstruction lights | | |
| | **AERODROMES—SEARCH AND RESCUE—DANGERS TO AIRCRAFT IN FLIGHT** | | |
| OA | Land aerodrome | UA | Alighting area |
| OB | Beaching facilities | UB | Mooring buoys |
| OC | Water aerodrome | UC | |
| OD | Meteorological forecast service | UD | Prohibited, restricted or danger area designated as . . . (*name or reference identification*) |
| OE | Meteorological observation service | UE | Aircraft |
| OF | Meteorological watch service | UF | Fixed balloons |
| OG | Runway arresting gear | UG | Bombing or aerial depth charge dropping |
| OH | Heliport or helicopter alighting area | UH | Air exercises (*or* flying displays) |
| OI | | UI | Gun or missile firing |
| OJ | | UJ | Glider flying |
| OK | | UK | Demolition of explosives |
| OL | | UL | Landing direction indicator |
| OM | All runway [except number(s) . . .] | UM | Mooring and docking facilities |
| ON | Stopway for runway number . . . | UN | Parachute jumping exercises |
| OO | Taxiway(s) | UO | Wind direction indicator |
| OP | Rescue vessel | UP | Free balloon |
| OQ | Ocean station vessel | UQ | Apron |
| OR | Refuelling [. . . type fuel(s) *or* . . . octane] | UR | Runway(s) number(s) . . . |
| OS | Search and rescue aircraft [*specify VLR, LRG, MRG, SRG or HEL*] | US | Strip . . . (*number or magnetic direction*) |
| OT | Rescue and fire fighting facilities | UT | Grass landing area |
| OU | | UU | |
| OV | . . . [*specify TWR, APP, ACC or FIC*] air traffic service | UV | Fog dispersal equipment |
| OW | | UW | |
| OX | | UX | |
| OY | | UY | |
| OZ | Warship | UZ | Runway threshold (number . . .) |

Figure 162(b).

## DECODE
### FOURTH AND FIFTH LETTERS

| 4th & 5th Letters | SIGNIFICATION | 4th & 5th Letters | SIGNIFICATION |
|---|---|---|---|
| | **HAZARD OR STATUS OF OPERATION OR CONDITION OF FACILITIES** | | |
| **AA** | | **EA** | |
| **AB** | Usable for length of . . . and width of . . . | **EB** | Location change to . . . effective . . . (*date/time*) |
| **AC** | Covered by dry snow to a depth of . . . *Note.—This snow is not compacted.* | **EC** | Characteristics *or* identification *or* radio call sign changed to . . . |
| **AD** | Cleared of soft snow, full length and width | | |
| **AE** | Totally free of snow and ice (cleared and dry) | **ED** | Operating frequency(ies) will be changed to . . . kHz (*or* . . . MHz) effective . . . (*date/time*) |
| **AF** | Covered by [ . . . (*type*)] ice to a depth of . . . | | |
| **AG** | Covered by dry snow (patches or partially cleared) | | |
| **AH** | Covered by rime or frost | **EE—EG** | |
| **AI** | Operating without tone modulation | **EH** | Not heard |
| **AJ** | Operating without coding or without flashing | **EI—EL** | |
| **AK** | Covered by compacted or rolled snow to a depth of . . . | **EM** | Military operations only |
| | | **EN** | Not available due to . . . (*specify reason*) 1) until further notice; *or* 2) until . . . (*date/time*); *or* 3) from . . . (*date/time*) to . . . (*date/time*) |
| **AL** | Operating on reduced power | | |
| **AM** | Snow clearance in progress [estimated time of completion is . . . (*date/time*)] | | |
| **AN** | Grass cutting in progress [estimated time of completion is . . . (*date/time*)] | **EO** | Obscured by snow |
| | | **EP** | Available on prior permission (of . . .) only |
| **AO** | Marked by . . . | **EQ** | Covered by wet snow or slush (patches or partially cleared) |
| **AP** | Work is in progress [estimated time of completion is . . . (*date/time*)] | | |
| **AQ** | Work completed | **ER** | Covered by frozen ruts or ridges |
| **AR** | Snow clearance completed | **ES** | Out of service 1) until further notice; *or* 2) until . . . (*date/time*); *or* 3) from . . . (*date/time*) to . . . (*date/time*) [due to the following condition(s) . . .] |
| **AS** | Grass cutting completed | | |
| **AT** | Sanding is in progress [estimated time of completion is . . . (*date/time*)] | | |
| **AU** | Appears unreliable | | |
| **AV** | Covered by ice (patches or partially cleared) | **ET** | Test operation only. NOT for operational use |
| **AW** | Height of snow-banks is . . . (*figures and units*) | **EU—EV** | |
| | | **EW** | Completely withdrawn |
| **AX*** | Braking action is . . . (1 = poor, 2 = medium/poor, 3 = medium, 4 = medium/good, 5 = good) | **EX** | |
| | | **EY** | Is outside the limits of its assigned ocean station |
| | | **EZ** | Is within the limits of its assigned ocean station |
| | | **IA—IB** | |
| **AY** | Are to avoid area, radius of danger being . . . (about the point . . .) | **IC** | Report of apparent unreliability or track displacement hereby is cancelled |
| **AZ** | Will take place 1) until further notice; *or* 2) until . . . (*date/time*); *or* 3) from . . . (*date/time*) to . . . (*date/time*) (on the days of . . . between the hours of . . . and . . .) at . . . (*location*) (within the sector of . . . and a radius of . . .) at . . . height above . . . (*datum*) | **ID** | Available on request to . . . |
| | | **IE** | |
| | | **IF** | Flight checked and found reliable |
| | | **IG—IJ** | |
| | | **IK** | Available on request (to . . .) immediately [*or* at . . . (*time period*) notice] |
| | | **IL** | Hours of service are now . . . |

Figure 162(c). * For bad or no braking action, use "NIL" for domestic use.

# DECODE

## FOURTH AND FIFTH LETTERS

### HAZARD OR STATUS OF OPERATION OR CONDITION OF FACILITIES

| 4th & 5th Letters | SIGNIFICATION |
|---|---|
| **IM** | |
| **IN** | Operative (*or* re-operative), activated (*or* re-activated)<br>1) until further notice; *or*<br>2) until . . . (*date/time*); *or*<br>3) from . . . (*date/time*) to . . . (*date/time*) |
| **IO** | Operating normally |
| **IP** | Track(s) reported to be displaced (. . . degrees) (. . . direction) of published bearing(s), other tracks probably have shifted |
| **IQ** | To bu used as radio beacon only |
| **IR** | Magnetic track(s) towards station is (are now . . . [will be . . . at . . . (*date/time*)] |
| **IS** | Operative (*or* re-operative) subject to conditions/limitations already published |
| **IT** | Aircraft restricted to runways and taxiways |
| **IU** | Unserviceable for aircraft heavier than . . . tons |
| **IV** | Unsafe<br>1) until further notice; *or*<br>2) until . . . (*date/time*); *or*<br>3) from . . . (*date/time*) to . . . (*date/time*) |
| **IW—IZ** | |
| **OA—OF** | |
| **OG** | Operative but ground checked only, awaiting flight check |
| **OH—OJ** | |
| **OK** | Resumed normal operation |
| **OL** | Track(s) ground checked, approved for instrument flying |
| **OM** | Shut down for maintenance<br>1) until further notice; *or*<br>2) until . . . (*date/time*); *or*<br>3) from . . . (*date/time*) to . . . (*date/time*);<br>disregard all signals |
| **ON** | |
| **OO** | On operational trials |
| **OP** | Approved for operational use |
| **OQ** | |
| **OR** | Previously promulgated shutdown has been cancelled |
| **OS** | |
| **OT** | New facility in operation |
| **OU** | Operating without interruption for voice transmissions<br>1) until further notice; *or*<br>2) until . . . (*date/time*); *or*<br>3) from . . . (*date/time*) to . . . (*date/time*) |
| **OV** | |
| **OW** | Wet (including standing water in pools) |
| **OX** | Exercising at . . . (*date/time, location and height above the specified datum*) |
| **OY** | Exercises completed |
| **OZ** | |
| **UA** | Closed to all operations<br>1) until further notice; *or*<br>2) until . . . (*date/time*); *or*<br>3) from . . . (*date/time*) to . . . (*date/time*) |
| **UB—UC** | |
| **UD** | Closed to all night operations<br>1) until further notice; *or*<br>2) until . . . (*date/time*); *or*<br>3) from . . . (*date/time*) to . . . (*date/time*) |
| **UE** | |
| **UF** | Closed for an unknown duration due to flood |
| **UG** | Closed for an unknown duration [*or* until . . . (*date/time*) due to ice or snow |
| **UH** | Closed for an unknown duration [*or* until . . . (*date/time*) due to thaw |
| **UI** | Closed<br>1) until further notice; *or*<br>2) until . . . (*date/time*); *or*<br>3) from . . . (*date/time*) to . . . (*date/time*)<br>for maintenance |
| **UJ—UL** | |
| **UM** | Operating in an unmonitored status |
| **UN—US** | |
| **UT** | Operative but caution advised due to following condition(s) . . . |
| **UU** | Suitable for . . . (*specify*) equipped aircraft only |
| **UV** | Covered by wet snow or slush to a depth of . . . |
| **UW** | Covered by water to a depth of . . . |
| **UX—UZ** | |

Figure 162(d).

*Departure* Communications and taxi procedures of concern to both the VFR and the IFR pilot are reviewed in this section.

*Enroute* The enroute section of Part I describes the Airways Route Systems used for air navigation and good operating practices while enroute. Although some of the information is of concern only to the instrument pilot, much of it is of interest to all pilots. Of particular importance is the pictorial display of visual cruising altitudes which must be used to comply with Federal Air Regulations 91.109 and 91.121, shown in Fig. 163.

*Arrival* Many procedures concerning arrival at controlled and uncontrolled airports are discussed, including Automatic Terminal Information Service, Radar Traffic Information Service, and Terminal Radar Service (see Chapter 45).

*Landing* In this section, tower-controlled and nontower airports are discussed, including details concerning traffic patterns and traffic pattern indicators.

*General* The section of Part I entitled *General* is divided into two subsections. The first contains information concerning United States Customs, Immigration and Naturalization, Public Health, and Agricultural Department requirements. Prior to attempting international flight we should review these procedures and requirements. A list of *Airports of Entry* (customs services available) is also provided.

The second subsection concerns the Security Control of Air Traffic. Special security instructions may be issued during air defense emergency conditions, in accordance with the Security Control of Air Traffic and Air Navigation Aids Plan (SCATANA). Air Traffic Control facilities will broadcast instructions received from the military services over all available Air Traffic Control frequencies. Flights may be directed to land at the nearest available airport. Excerpts from the Airman's Information Manual concerning the Security Control of Air Traffic are shown in Figs. 164 and 165.

In view of this information concerning the Security Control of Air Traffic, we should always guard an Air Traffic Control frequency or Flight Service Station frequency while in flight.

*Emergency Procedures* This section discusses emergency procedures to be used in any emergency phase (uncertainty, alert, or distress). We will discuss many of these procedures in detail in Chapter 49, Emergency Procedures. However, the Airman's Information Manual provides an excellent review of emergency situations including search and rescue, communications failure, lost aloft, overdue aircraft, etc.

# CRUISING ALTITUDES

## FAR PARTS 91.109,91.121

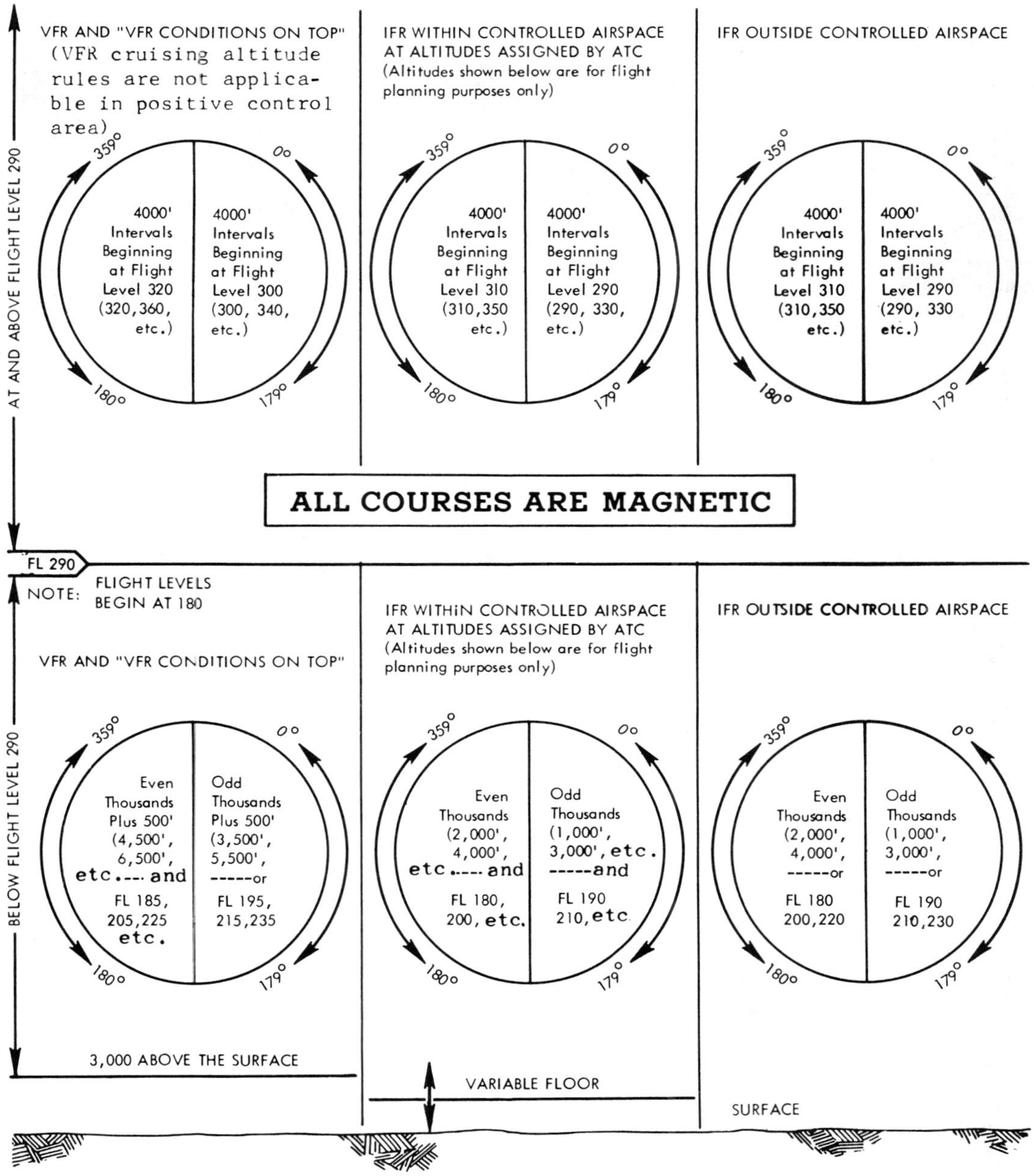


Figure 163.

# GENERAL

## FAR 99—SECURITY CONTROL OF AIR TRAFFIC

### Operational Requirements, Abbreviated Form

(Detailed procedures to be followed are contained in FAR 99)

| | *Domestic ADIZ* | *Coastal ADIZ* | *DEWIZ* |
|---|---|---|---|
| Flight Plan | Required. For flight in Panama C.Z. Domestic ADIZ, flight plan must be approved through ATC. | Required. | Required for takeoff except as permitted in FAR 99.13. |
| Two-way Radio | Required except as stated in FAR 99.1(c). | Required except as stated in FAR 99.1(c). | Required. |
| Position Reports | IFR-normal reports. DVFR-give ETP at least 15 minutes before penetration. | Same as domestic ADIZ; or Inbound foreign aircraft report at least one hour from U.S. | IFR-normal reports. DVFR-report prior to penetration. Correlation of ground filed data may be required. |
| Recommended Tolerances | Within 5 minutes of estimate and 10 miles of course centerline. | Within 5 minutes of estimate and 20 miles of course centerline. | Within 5 miles of estimate and 20 miles of course centerline. |

Except for Section 99.7, FAR 99 does not apply to aircraft operating:

- Southbound from U.S. through Southern Border Domestic ADIZ and not penetrating a Coastal ADIZ
- Within Continental U.S. and remaining within 10 miles of departure point
- In Coastal or Domestic ADIZ north of 28° N or west of 85° W at TAS of less than 180 knots
- In Alaskan DEWIZ at TAS of less than 180 knots while pilot maintains listening watch on appropriate frequency
- Over or within three miles of any island in Hawaiian Coastal ADIZ

Exemptions (except to Far 99.7) may also be granted by an ARTC Center, on a local basis, for certain operations (see FAR 99.1(d)).

Note:—An "Airfiled" flight plan makes the aircraft subject to interception for positive identification. Pilots are therefore urged to file required DVFR flight plans either in person or by telephone.

---

#### FAR 99.7 SPECIAL SECURITY INSTRUCTIONS

1. During defense emergency or air defense emergency conditions, additional special security instructions may be issued in accordance with the SCATANA (Security Control of Air Traffic and Air Navigation Aids) Plan.

2. Under the provisions of the SCATANA Plan, the military will direct the action to be taken—in regard to landing, grounding, diversion, or dispersal of aircraft and the control of air navigation aids—in the defense of the United States during emergency conditions.

3. At the time a portion or all of SCATANA is implemented, ATC facilities will broadcast appropriate instructions received from the military over available ATC frequencies. Depending on instructions received from the military, VFR flights may be directed to land at the nearest available airport, and IFR flights will be expected to proceed as directed by ATC.

4. Pilots on the ground may be required to file a flight plan and obtain an approval (through FAA) prior to conducting flight operation.

5. In view of the above, all pilots should guard an ATC or FSS frequency at all times while conducting flight operations.

See chart on page 1–97 which depicts the FAA and North American Air Defense Command (NORAD) Region boundaries.

Figure 164.

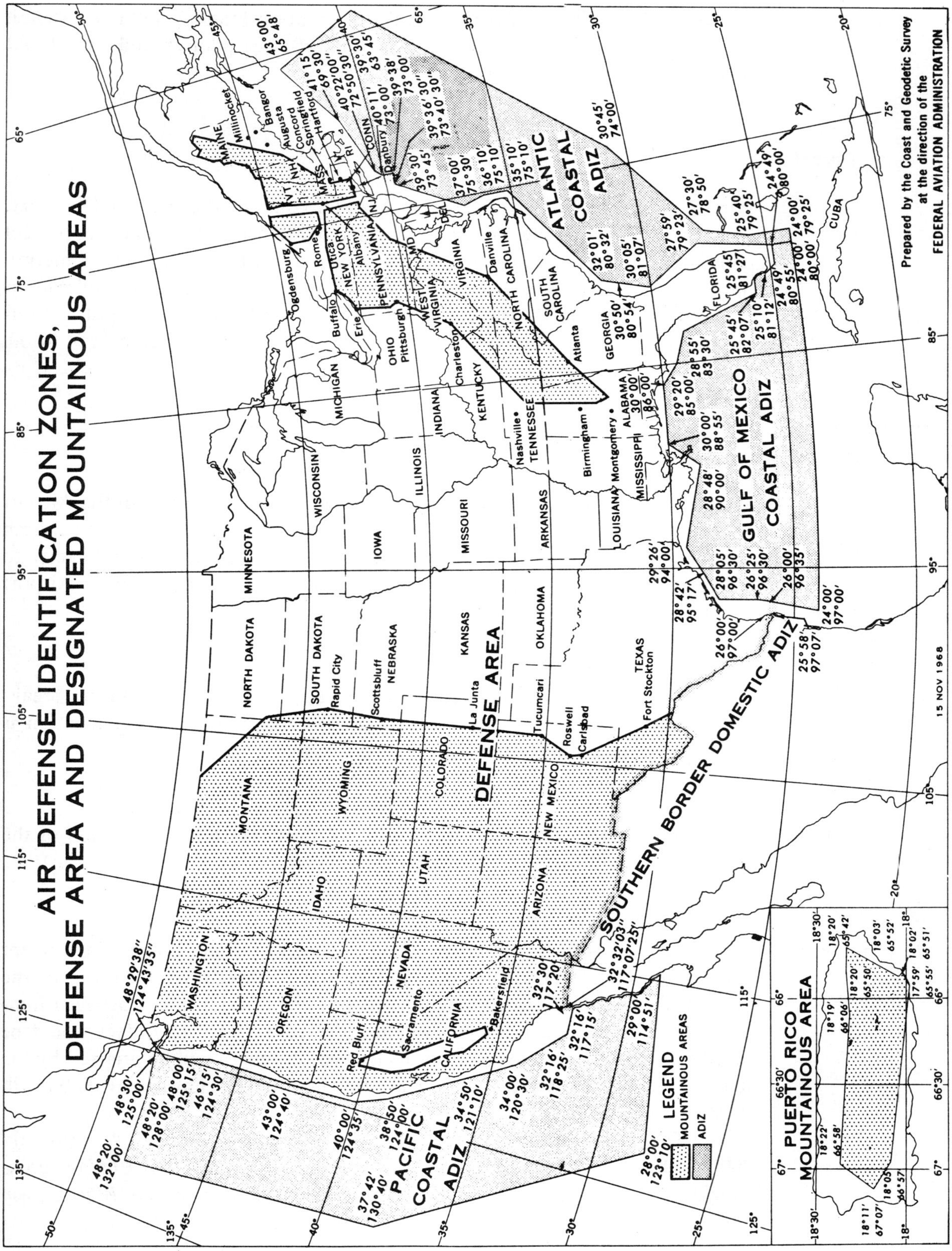
AIR DEFENSE IDENTIFICATION ZONES,
DEFENSE AREA AND DESIGNATED MOUNTAINOUS AREAS
ATLANTIC COASTAL ADIZ
GULF OF MEXICO COASTAL ADIZ
SOUTHERN BORDER DOMESTIC ADIZ
PACIFIC COASTAL ADIZ
DEFENSE AREA
PUERTO RICO MOUNTAINOUS AREA
LEGEND
MOUNTAINOUS AREAS
ADIZ
15 NOV 1968
Prepared by the Coast and Geodetic Survey
at the direction of the
FEDERAL AVIATION ADMINISTRATION

Figure 165.

## Summary—Part I

As we have seen, Part I of the Airman's Information Manual contains a wealth of information for pilots of all types of aircraft. No pilot should be without this government publication.

## Part II

Part II of the Airman's Information Manual contains the Airport Directory. It is issued semiannually, and contains a list of all facilities and services (except communications) at all airports, seaplane bases, and heliports in the conterminous United States, Puerto Rico, and the Virgin Islands which are available for transient use. Included also is a list of selected commercial broadcast stations operating at 100 watts power or more. The annual subscription cost is $4.00.

## The Airport Directory Legend

We will not discuss all of the codes and other material included in the Airport Directory legend, but only the more important ones. When we wish to research the facilities and/or services available at a specific airport, we should refer to the legend of codes which precedes the Airport Directory. A sample Airport Directory Legend is shown in Fig. 166.

*Location* The Airport Directory is arranged alphabetically by state, with a further breakdown within each state alphabetically by city name. The airport location is stated in nautical miles from the center of the referenced city.

*Elevation* The height, in feet above mean sea level, of the highest usable portion of the landing area will be published as the *field elevation.*

*Runways* The length, surface, and weight bearing capacity of the longest instrument runway, or the longest active landing portion of the runway is given to the nearest hundred feet. (Seventy feet is used as the division point between hundreds of feet. That is, 1,568 feet would be listed as "15." 1,572 feet would be listed as "16.") If the runway is hard-surfaced, its length will be prefixed by the letter *H* (concrete, asphalt, etc.). The total number of runways available will be shown in parenthesis.

The weight-bearing capacity of the longest runway is noted. We must add "000" to the digits which follow the code letters *S*, *T*, *TT*, and *MAX* for the gross weight capacity of the runway. The code letters stand for *Single Wheel*, *Twin Wheel*, and *Twin-Tandem* type landing gears.

# AIRMAN'S INFORMATION MANUAL

PART 2

## Airport Directory

### LEGEND

#### LOCATION

The airport location is given in nautical miles (to the nearest mile) and direction from center of referenced city.

#### ELEVATION

Elevation is given in feet above mean sea level and is based on highest usable portion of the landing area. When elevation is sea level, elevation will be indicated as "00." When elevation is below sea level, a minus sign (−) will precede the figure.

#### RUNWAYS

The runway surface length, and weight bearing capacity are listed for the longest instrument runway or sealane, or the longest active landing portion of the runway or strip, given to the nearest hundred feet, using 70 feet as the division point, i.e., 1468 feet would be shown as "14"; 1474 feet would be shown as "15". Runways lengths prefixed by the letter "H" indicates that runways are hard surfaced (concrete; asphalt; bitumen, or macadam with a seal coat). If the runway length is not prefixed, the surface is sod, clay, etc. The total number of runways available is shown in parenthesis. (However, only hard surfaced runways are counted at airfields with both hard surfaced and sod runways.)

#### RUNWAY WEIGHT BEARING CAPACITY

Add 000 to figure following S, T, TT and MAX for gross weight capacity, e.g., (S-000).

S–Runway weight bearing capacity for aircraft with single-wheel type landing gear. (DC-3), etc.

T–Runway weight bearing capacity for aircraft with twin-wheel type landing gear. (DC-6), etc.

TT–Runway weight bearing capacity for aircraft with twin-tandem type landing gear. (707), etc.

Quadricycle and twin-tandem are considered virtually equal for runway weight bearing considerations, as are single-tandem and twin-wheel.

A blank space following the letter designation is used to indicate the runway weight bearing capacity to sustain aircraft with the same type landing gear, although definite figures are not available, e.g., (T- ).

Omission of weight bearing capacity indicates information unknown. Footnote remarks are used to indicate a runway with a weight bearing greater than the longest runway.

**SEAPLANE BASE FACILITIES**

A number preceding the parenthetical designation, indicates the number (quantity) available.

Beaching gear, consisting of the quantity and type of beaching gear available.

The number (quantity) if available, of Mooring Buoys (MB) and Crash Boats (CB) available.

#### LIGHTING

**B: Rotating Light** (Rotating beacon). (Green and white, split-beam and other types.) Omission of **B** indicates rotating light is either not available or not operating standard hours (sunset-sunrise).

NOTE.—Code lights are not codified, and are carried in Remarks.

● **L: Field Lighting** (when code **L4–7** is indicated, lighting **4, 5, 6, 7** is available). An asterisk (*) preceding an element indicates that it operates on prior request only (by phone call, telegram or letter). Where the asterisk is not shown, the lights are in operation or available sunset to sunrise or by request (radio call). **L** by itself indicates temporary lighting, such as flares, smudge pots, lanterns.

**1**—Portable runway lights (electrical)
**2**—Airport Boundary
**3**—Runway Floods
**4**—Low Intensity Runway
**5**—Medium Intensity Runway
**6**—High Intensity Runway
**7**—Instrument Approach (neon)
**8A, B, or C**—High Intensity Instrument Approach

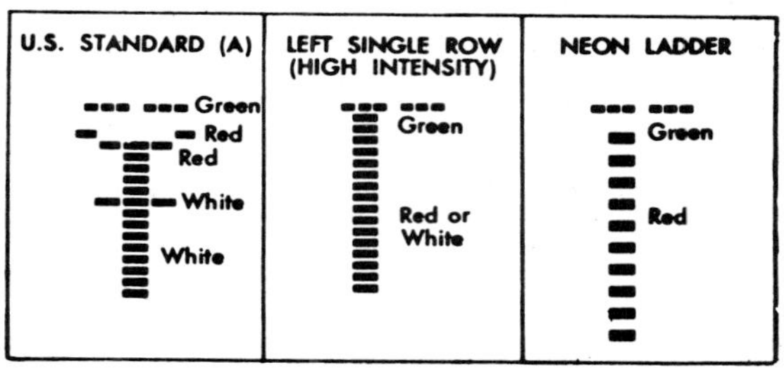


**9**—Sequence Flashing Lights (3,000′ out unless otherwise stated)
**10**—Visual Approach Slope Indicator (VASI)
**11**—Runway end identifier lights (threshold strobe (REIL)
**12**—Short approach light systems (SALS)

Figure 166(a).

**13**—Runway alignment lights (RAIL)

**14**—Runway centerline

**15**—Touchdown zone

Because the obstructions on virtually all lighted fields are lighted, obstruction lights have not been included in the codification.

## SERVICING

**S1**: Storage.

**S2**: Storage, minor airframe repairs.

**S3**: Storage, minor airframe and minor powerplant repairs.

**S4**: Storage, major airframe and minor powerplant repairs.

**S5**: Storage, major airframe and major powerplant repairs.

## FUEL

| Code | Grade |
|---|---|
| F12 | 80/87 |
| F15 | 91/98 |
| F18 | 100/130 |
| F22 | 115/145 |
| F30 | Kerosene, freeze point −40°F |
| F34 | Kerosene, freeze point −58°F |
| F40 | Wide-cut gasoline, freeze point −60°F |
| F45 | Wide-cut gasoline without icing inhibitor, freeze point −60°F |

## OTHER

§ Notam Service is provided. Applicable only to airports with established instrument procedures or high volume VFR activity.

**AOE**—Airport of Entry.

**VASI**—Visual Approch Slope Indicator, applicable runway provided.

**RVV**—Runway Visibility Value, applicable runway provided.

**RVR**—Runway Visual Range, applicable runway provided.

**TPA**—Traffic Pattern Altitude—This information is provided for only those airports without a 24-hour operating control tower or FSS. Directions of turns are indicated only when turns of the pattern(s) are to the right (non-standard). TPA data are related to the runway listed under the tabulated airport information. Generally, only one altitude is listed; however, separate altitudes may be shown for aircraft of different performance or size.

**FSS**—The name of the associated FSS and/or combined station/tower (CS/T) is shown in all instances. When the FSS is located on the named airport, "on fld" is shown following the FSS name. When the FSS can be called through the local telephone exchange, (Foreign Exchange) at the cost of a local call, it is indicated by "(LC)" (local call) with the phone number immediately following the name of the FSS. i.e., "FSS: WICHITA (LC481-5867)." When an Interphone line exists between the field and the FSS, it is indicated by "(DL)" (direct line) immediately following the name of the FSS, i.e., "FSS: OTTO (DL)."

**The availability of a VHF/DF at a FSS is indicated by the letters VHF/DF. For service, contact FSS on standard frequencies.**

## AIRPORT REMARKS

(£) Indicates that an air traffic control tower and/or an instrument landing system are associated with the airport. For specific details see the Airport/Facility Directory in Part 3 of the Airman's Information Manual.

**"Fee"** indicates landing charges for private or non-revenue producing aircraft. In addition, fees may be charged for planes that remain over a couple of hours and buy no services, or at major airline terminals for all aircraft.

**"Rgt tfc 13–31"** indicates right turns should be made on landings and takeoffs on runways 13 and 31.

**Remarks**—data is confined to operational items affecting the status and usability of the airport, traffic patterns and departure procedures

**Obstructions.**—Because of space limitations only the more dangerous obstructions are indicated. Natural obstructions, such as trees, clearly discernible for contact operations, are frequently omitted. On the other hand, all pole lines within at least 15:1 glide angle are indicated.

**UNICOM**

A private aeronautical advisory communications facility operated for purposes other than air traffic control, transmits and receives on one of the following frequencies:

**U1**—122.8 MHz for Landing Areas (except heliports) without an ATC Tower or FSS;

**U2**—123.0 MHz for Landing Areas (except heliports) with an ATC Tower or FSS;

**U3**—123.05 MHz for heliports.

**HOW TO ORDER THE COMPLETE AIRPORT DIRECTORY**

The following page contains excerpts from the Airport Directory section of Part 2 of the F.A.A. AIM.

Instructions for subscribing to the complete Airport Directory by the F.A.A. are given on the page following the Contents Page.

Figure 166(b).

*Lighting*

If a rotating beacon is available at the airport and operates standard hours (sunset to sunrise) the letter *B* will appear in the directory.

The code letter *L* indicates that field lighting is available. A number, from 1 through 15 inclusive, follows the letter *L* to indicate the type of lighting available. For example, 15 would indicate that medium intensity runway lighting is available.

*Servicing*

The code numbers S1 through S5 indicate the specific servicing available at the airport. For example, S3 would indicate that "storage, minor airframe, and minor powerplant repairs" are available.

*Fuel*

Code numbers preceded by the code letter *F* indicate the grade of fuel available. For example, F12 indicates that 80/87 octane fuel may be found on the airport.

Figure 167 depicts a typical Airport Directory. Using a current Airman's Information Manual we should practice decoding the information in the Airport Directory.

**Part III**

Part III of the Airman's Information Manual contains operational data and Notices to Airmen. It is issued every 28 days. The subscription rate is $18.00 per year (including Part III A). Abbreviations used in Part III and Part III A are shown in Fig. 168.

**Airport Facility Directory**

The Airport Facility Directory contains all of the information found in the Airport Directory of Part II, plus radio communications and radio navigational aids.

*Communications*

All frequencies available at each airport are listed and assumed to be for both transmitting and receiving unless otherwise specified by the letter *T* (transmit only), *R* (receive only), or *X* (on request). The primary frequency is listed as the first frequency in each grouping.

*Voice Call*

The voice call that we should use in order to contact the traffic control service is the airport name, followed by the call name of the particular service desired (for example, *Abilene Tower*).

*Services Available*

The particular communications services available will be listed, such as Tower, Automatic Terminal Information Service, Flight Service Station, etc.

# AIRPORT DIRECTORY

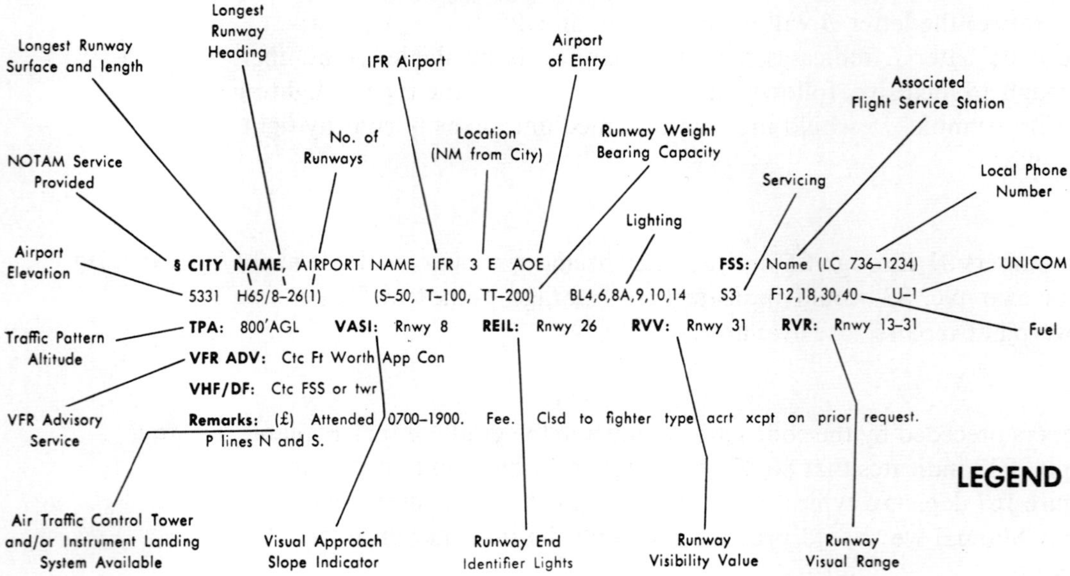


## EXCERPTS
(for study purposes only)

### GEORGIA

§ **ATLANTA,** DEKALB-PEACHTREE 8 NE IFR **FSS:** Atlanta
1002 H40/16-34 (3) (S-63, T-85, TT-135) BL4, 6, 8, S5
F12, 18, 34 U2
**Remarks:** (£) TV towers 1042' (2049' MSL) 7.5NM SSW; 1076' (2049 MSL) 8NM SSW; 1100' (2049' MSL) 5NM SSW. For additional info see Part 3.

§ **ATLANTA,** MUNI 7 SSW IFR **FSS:** Atlanta
1024 H100(9L-27R(4) (S-120, T-170, TT-310) BL5, 6, 8A, 9, 10, 11 S5 F22, 30 U2 **VASI:** Rnwy 27R, 27L
**RVR:** Rnwy 9L, 9R 33 **REIL:** Rnwy 27R.
**Remarks:** (£) TV towers 1042' (2049' MSL) 8.5 NM NE. 1076' (2049' MSL) 7.5 NM NE. 1100' (2049' MSL) 10 nmi NE. Commercial fee only. For additional info see Part 3.

### NORTH CAROLINA

**CHARLOTTE,** CARPENTER 9 SW **FSS:** Hickory (LC 332-8161)
650 40 (1) (S-4) S5 F12, 18 U-1
**Remarks:** P-line W. 2565' asphalt strip avbl.

§ **CHARLOTTE,** DOUGLAS MUNI 4 W IFR
**FSS:** Hickory (LC 332-8161)
748 H78/18-36 (2) (S-120, T-150, TT-240) BL6, 8A, 9 S5
F12, 18, 22, 34 U2 **RVR:** Rnwy 5
**Remarks:** (£) 1073' (1932' MSL) twr 10 NM ENE. 1071' (1871' MSL) twr 10.5 NM ENE. Twr 1404' (2049' MSL) 8 NM SE. Lndg fee for commercial acft. For additional info see Part 3.

**CHINA GROVE** 2 E **FSS:** Hickory
750 20 (1)
**Remarks:** P-line SE.

**CLINTON,** SAMPSON CO 3 SW **FSS:** Fayetteville
147 35 (1) S5 F12, 18 U-1

CONFEDERATE FLD **See RED SPRINGS**

Figure 167(a).

### SOUTH CAROLINA

§ **CHARLESTON,** AFB/MUNI 9 NW IFR **FSS:** Charleston on Fld
45 H90/15-33 (2) (S-100, T-140, TT-250) BL4,6,8A, 9,10 S5
F22, 45 U2 **VASI:** Rnwys 15-33 & 22 **RVR:** Rnwy 15
**VHF/DF:** Ctc FSS
**Remarks:** (£) U.S. Customs lndg rgts arpt. Fee for commercial acft only. For additional info see Part 3.

§ **CHARLESTON,** JOHN'S ISLAND IFR 6 SW **FSS:** Charleston
16 H50 (3) (S-30, T-50, TT-85) S5 F12, 18
**VFR ADV:** Ctc Charleston App Con
**Remarks:** 1044' (1049' MSL) twr 10 NM NE.

**CHERAW,** MUNI 4 NW **FSS:** Florence
240 H34 (1) (S-30) BL4 F12, 18[1] U-1
**Remarks:** Attended by call-phone 537-2619. [1]Avbl on req-phone 537-5311.

### TENNESSEE

**CENTERVILLE,** MUNI 3 N **FSS:** Nashville
760 H30 (1) (S-20) BL2, 4 S1 F12, 18
**Remarks:** Unattended.

**CHATTANOOGA,** DALLAS BAY SKY PARK 13 N
690 H31/5-23 (1) F18
**Remarks:** Attended wkends 0800–1800.

§ **CHATTANOOGA,** LOVELL FIELD 5E IFR **FSS:** Crossville (LC 267-5343)
682 H62 2-20 (2) (S-115, T-160, TT-240) BL4, 6, 8A, 9, 10 S5
F12, 15, 18, 22 U-2 **VASI:** Rnwy 2 **RVR:** Rnwy 20
**Remarks:** (£) For additional info see Part 3.

§ **CLARKSVILLE,** OUTLAW FLD IFR 6 NW **FSS:** Nashville
549 H50 16-34 (2) (S-16, T-30, T-50) BL4 S5 F12, 18, 22 U-1
**VFR ADV:** Ctc Campbell App Con 118.1

EXCERPTS
(for study purposes only)

**TEXAS**

§ **ABILENE,** MUNI 3 SE IFR **FSS:** Abilene on Fld
1778 H60/17-35(2) (S-70, T-89, TT-145) BL4,6,8A,9 S5
F18, 30 U2 **RVV:** Rnwy 35
**Remarks:** (£) For additional info see Part 3.

**ABILENE,** BUTTERFIELD TRAIL 5 NW **FSS:** Abilene
1670 H25(1) L4 S5[1] F12, 18 U-1
**Remarks:** Attended dalgt hrs. [1]Prior req.

**ABILENE,** ELMDALE ARPK 3 NE **FSS:** Abilene
1750 H22(1) S3[1] F12, 18
**Remarks:** Unattended. Rgt tfc rnwy 35. [1]Avbl on prior req.

§ **CORSICANA,** MUNI 5 SE **FSS:** Dallas
440 H34(1) BL4 S1 F12, 18 U-1
**Remarks:** P-line N. Rdo twr SE.

**CORSICANA,** CORSICANA "75" 2 N **FSS:** Dallas
480 20(1) F12
**Remarks:** Unattended. P-line N. Rdo twr SE.

**GRAFORD,** HARMAR MARINA 12 WSW
1100 H25(1)

§ **GRAHAM,** MUNI IFR 2 E **FSS:** Mineral Wells
1123 H33(1) (S-80) BL4 S3 F12, 18
**Remarks:** P-line W. Tower N. 300' twr 3/4 mi SW.

**GRANBURY** 2 W **FSS:** Mineral Wells
718 34(1)

§ **SHERMAN,** MUNI 1 SE IFR **FSS:** Dallas
745 H40(1) (S-20, T-32) BL4 S5 F12, 15, 18[1] U-1 **TPA-800**
**VFR ADV:** Ctc Perrin App Con 123.8, 126.8
**Remarks:** Rgt tfc rnwy 34. [1]On call at ngt. [2]Oper 1200-0600Z. **ruf**

§ **STEPHENVILLE,** CLARK MUNI 1 E **FSS:** Mineral Wells
1318 H42(1) BL4 S5 F12, 18 U-1

## HELIPORT DIRECTORY

***** FLORIDA *****

**CAPE CANAVERAL,** CAPE KENNEDY HELIPORT 6 N of Coca Beach
10 H100 x 100(1) S3 F12, 18
**Remarks:** Attended 0900-1700 lcl time.

**CLEARWATER,** MUNICIPAL HELIPORT 3 NE **FSS:** Tampa
73 F12, 18

***** IDAHO *****

**CHALLIS,** HELIPORT 1 NE **FSS:** Idaho Falls
5085 F12, 18

**LEWISTON,** LEWISTON NEZ PERCE CO HELIPORT 2 S
1438 BL4 F12, 18 **FSS:** Walla Walla

## SEAPLANE BASE DIRECTORY

**NEW YORK**—*Continued*

**ROCHESTER,** FLOATAIR 1 NE **FSS:** Buffalo
246 UNL (1) F12 S5
**Remarks:** Attended days. U.S. Customs 0800-1700 Mon-Sat.

**ROCHESTER,** GLEN HAVEN SKYPORT Adj NE **FSS:** Buffalo
246 211 S3 F12
**Remarks:** Land rises to 150' above water along E & W shores.

***** SOUTH CAROLINA *****

**CHARLESTON,** MUNI MARINA Adj SW **FSS:** Charleston
00 60 F12
**Remarks:** Land E and W Wapoo Creek. Bridge at W end. Ashley River E end. Dockage. Dual bridge N.

## COMMERCIAL BROADCAST STATIONS

The following listing contains only stations of 100 watts or more power. No station listing will be found for any locality for which we do not have any record of an airport. Also, because of space limitations, not all stations in every city are listed. We have attempted to show the most powerful station in a locality, and/or that station which maintains the longest hours of operation.

**LEGEND**

D................Days D–N.......Days-Nights N..............Nights

| NAME/LOCATION (Mileage & direction from center of city) | IDENT | FREQ. kHz | POWER (kws.) day | TIME OF OPERATION |
|---|---|---|---|---|
| **ALABAMA** | | | | |
| Albertville 1.5 SE | WAVU | 630 | 1.00 | D |
| Alexander City 2 W | WRFS | 1050 | 1.00 | D |
| **ARIZONA (Con't.)** | | | | |
| Douglas In City | KAWT | 1450 | 0.25 | D–N |
| Flagstaff 1.5 E | KEOS | 690 | 0.50 | D–N |

Figure 167(b).

# AIRMAN'S INFORMATION MANUAL

## PARTS 3 and 3A

**Operational Data and Notices to Airmen**

Selected material and excerpts from PARTS 3 and 3A of the F.A.A. AIM.

## ABBREVIATIONS

**Note:** "s" may be added for plural, or as appropriate.

| | |
|---|---|
| AAS | Airport Advisory Service at FSS-Located airport-shown in Location Column. |
| AB | Continuous automatic transcribed broadcast service. |
| AC | Approach Control Tower, FAA (transmits on voice channel or associated ranges and ILS in addition to frequencies listed). |
| acft | aircraft |
| ADF | Automatic Direction Finder |
| adjt | adjacent |
| admin | administration |
| AGL | above ground level |
| AID | Airport Information Desk |
| AIM | Airman's Information Manual. |
| AL | Approach and Landing Chart |
| ALS | Approach light System |
| alt | altitude |
| amdt | amendment |
| ANRA | air navigation radio aids |
| ant | antenna |
| apch | approach |
| apchg | approaching |
| APP CON | Approach control |
| aprxly | approximately |
| arpt | airport |
| arr | arrival/arrive |
| ARSR | Air Route Surveillance Radar. |
| ARTCC | Air Route Traffic Control Center |
| ASDE | airport surface detection equipment |
| ASR | Arpt Surveillance Radar. |
| ATC | air traffic control |
| ATCT | air traffic control tower |
| ATIS | Automatic Terminal Information Service |
| auto | automatic |
| aux | auxiliary |
| avbl | available |
| awy | airway |
| B | Scheduled Broadcasts weather at 15 and 45 minutes after the hour; Air Force Broadcasts, generally, 29 minutes). |
| BC | back course |
| bcn | beacon |
| bcst | broadcast |
| bldg | building |
| bndry | boundary |
| brg | bearing |
| btn | between |
| clsd | closed |
| clrnc | clearance |
| cntr | center |
| Co | county |
| comm | communication |
| comsnd | commissioned |
| comsng | commissioning |
| config | configuration |
| constr | construction |
| constrd | constructed |
| cont | continuous/continuously |
| crs | course |
| CS/T | combined station/tower |
| ctc | contact |
| ctl | control |
| ctld | controlled |
| dalgt | daylight |
| decmsnd | decommissioned |
| demol | demolition |
| dep | departure |
| DEP CON | Departure control |
| DF | direction finder |
| discntd | discontinued |
| DL | interphone (direct line) |
| DME | UHF standard (TACAN compatible) distance measuring equipment |
| dptg | departing |
| DVFR | Defense Visual Flight Rule. |
| efctv | effective |
| elev | elevation |
| emgcy | emergency |
| eng | engine |
| eqpmt | equipment |
| excp | except |
| extsn | extension |
| extsv | extensive |
| facil | facility |
| FL | Flight Level. |
| flashg | flashing |
| fld | field |
| flt | flight |
| flwg | following |
| FM | fan marker |
| fone | telephone |
| freq | frequency |
| FSF | Flight Service Facility. |
| FSS | Flight Service Station |
| GCA | groundcontrolled approach |
| gnd | ground |
| G/S | Glide slope |
| GWT | gross weight |
| H | Non-directional radio beacon (homing), power 50 watts to less than 2000 watts |
| HH | Non-directional radio beacon (homing), power 2000 watts or more |
| hi- | high- |
| hr | hour |
| H-SAB | Non-directional rbn providing auto trans wea service. |
| hvy | heavy |
| Hz | Cycles per second |
| hwy | highway |
| ident | identification |
| IFF | Identification Friend or Foe |
| IFR | Instrument Flight Rules |
| IFSS | International Flight Service Station |
| ILS | instrument landing system |
| imdtly | immediately |
| inbnd | inbound |
| indef | indefinitely |
| info | information |
| inop | inoperative |
| inst | instrument |
| int | intersection |
| intl | international |
| intsv | intensive |
| J-bar | jet runway barrier |
| kHz | kilo Hertz |
| lat | latitude |
| LC | foreign exchange (local call) |
| lcl | local |
| lclzr | localizer |
| lctd | located |
| lctn | location |
| LFR | Low/Medium frequency radio range |
| lgt | light |
| lgtd | lighted |
| lgtg | lighting |
| LMM | compass locator at middle marker ILS |
| lnd | land |
| lndg | landing |
| LOM | compass locator at outer marker ILS |
| long | longitude |
| LRCO | Limited Remote Communication Outlet. |
| M | magnetic (after a bearing) |
| MA | Range (adcock, vertical radiators), power less than 50 watts. |
| MAA | maximum authorized altitude |
| mag | magnetic |
| maint | maintain, maintenance |
| maj | major |
| max | maximum |
| MCA | minimum crossing altitude |
| MEA | minimum en-route IFR altitude |
| meml | memorial |
| MH | Non-directional radio beacon (homing), power less than 50 watts |
| MHz | mega Hertz |
| mi | mile |
| min | minimum or minute |
| ML | Range (loop radiators), power less than 50 watts. |

Figure 168(a).

## ABBREVIATIONS

MM------ middle marker ILS
MOCA---- minimum obstruction clearance altitude
mod------ modernization/modification
MRA----- minimum reception altitude
mrkd----- marked
mrkg----- marking
MRL----- Range (loop radiators), power 50 watts or more, but less than 150 watts.
MSL

PAR------ Precision Apch Radar.
patn------ pattern
permly---- permanently
p-line----- pole line
pwr------ power

quad----- quadrant

RA------- Range (adcock, vertical radiators), power 150 watts or more.
rad------- radial
RAE------ Royal Aircraft Establishment visual glide slope indicator
RAPCON--radar approach control (USAF)
RATCC--- radar air traffic control center (NAVY)
rbn------- radio beacon
RCO------ Remote Communications Outlet.
rcv------- receive
rcvg------ receiving
rcvr------ receiver
rdo------- radio
reconstr--- reconstruction
REIL----- runway end identifier lights
relctd----- relocated
Req------ Operates on request.
rgt------- right
RL----- Range (loop radiators), power 150 watts or more.
rng------- range
rnwy----- runway
ruf------- rough
RVR------ runway visual range
RVV------ runway visibility values

S-------- Simultaneous range, homing signal and/or voice.
SABH---- Non-directional rbn having limited navigational use. Provides auto wea bcst
SALS----- Short approach light system
SFL------ Sequence flashing lights
SIAP----- Standard Instrument Approach Procedures.
SID------ Standard Instrument Departure.
SIF------- Selective Identification Feature (of the basic Mark X radar beacon system)
sked------ schedule
SLMM-- Simultaneous middle marker compass locator and twr voice transmission.
SLOM-- Simultaneous outer marker compass locator and twr voice transmission.
smi------- statute mile/s
SR------- sunrise
SS------- sunset
stn------- station
STOL--- Short take-off and landing rnwy

sys------- system
T-------- true (after a bearing)
TACAN--- UHF navigational facility—omnidirectional course and distance information
tet------- tetrahedron
tfc------- traffic
til-------- Until
tkof------ take-off
tmprly---- temporarily
tmpry---- temporary
TRACON-- Terminal Radar Apch Ctl.
trans----- transcribed
trml------ terminal
trng------ training
TRSA----- Terminal Radar Service Area
tsmt------ transmit
tsmtg----- transmitting
tsmtr----- transmitter
TV------- television
twr------- tower
TWEB---- trans. wea bcst
txwy------ taxiway
UFN------ until further notice
UHF------ Ultra hi freq
unavbl---- unavailable
uncltd---- uncontrolled
unlgtd---- unlighted
unmrkd--- unmarked
VASI----- visual approach slope indicator

vcnty----- vicinity
VFR------ visual flight rules
VHF------ Very high frequency
VOR------ VHF navigational facility—omnidirectional, course only
VOR-DME- collocated VOR facility and UHF standard distance measuring equipment
VORTAC-- collocated VOR and TACAN navigational facilities
VOT------ a VOR Receiver Testing Facility
vsby------ visibility
W------ Without voice facilities on range frequency.
wea------ weather
WB------ Weather Bureau
WIP------ work in progress
wt-------- weight
Z------- VHF station location marker at a LF range station.
Z------- Greenwich mean time.

Figure 168(b).

## LOCATION

The airport location is given in nautical miles (to the nearest mile) and direction from center of referenced city.

## ELEVATION

Elevation is given in feet above mean sea level and is based on highest usable portion of the landing area. When elevation is sea level, elevation will be indicated as "00." When elevation is below sea level, a minus sign (−) will precede the figure.

## RUNWAYS

The runway surface length, and weight bearing capacity are listed for the longest instrument runway or sealane, or the longest active landing portion of the runway or strip, given to the nearest hundred feet, using 70 feet as the division point, i.e., 1468 feet would be shown as "14"; 1474 feet would be shown as "15". Runway lengths prefixed by the letter "H" indicates that runways are hard surfaced (concrete; asphalt; bitumen, or macadam with a seal coat). If the runway length is not prefixed, the surface is sod, clay, etc. The total number of runways available is shown in parenthesis. (However, only hard surfaced runways are counted at airfields with both hard surfaced and sod runways.)

## RUNWAY WEIGHT BEARING CAPACITY

Add 000 to figure following S, T, TT and MAX for gross weight capacity, e.g., (S-000).

S–Runway weight bearing capacity for aircraft with single-wheel type landing gear. (DC-3), etc.

T–Runway weight bearing capacity for aircraft with twin-wheel type landing gear. (DC-6), etc.

TT–Runway weight bearing capacity for aircraft with twin-tandem type landing gear. (707), etc.

Quadricycle and twin-tandem are considered virtually equal for runway weight bearing considerations, as are single-tandem and twin-wheel.

A blank space following the letter designation is used to indicate the runway weight bearing capacity to sustain aircraft with the same type landing gear, although definite figures are not available, e.g., (T-   ).

Omission of weight bearing capacity indicates information unknown. Footnote remarks are used to indicate a runway with a weight bearing greater than the longest runway.

Figure 169(a).

## LIGHTING

**B: Rotating Light** (Rotating beacon). (Green and white, split-beam and other types.) Omission of **B** indicates rotating light is either not available or not operating standard hours (sunset-sunrise).

NOTE.—Code lights are not codified, and are carried in Remarks.

**L: Field Lighting.** An asterisk (*) preceding an element indicates that it operates on prior request only (by phone call, telegram or letter). Where the asterisk is not shown, the lights are in operation or available sunset to sunrise or by request (radio call). **L** by itself indicates temporary lighting, such as flares, smudge pots, lanterns.

1—Portable runway lights (electrical)

2—Airport Boundary

3—Runway Floods

4—Low Intensity Runway

5—Medium Intensity Runway

6—High Intensity Runway

7—Instrument Approach (neon)

8A, B, or C—High Intensity Instrument Approach (ALS)

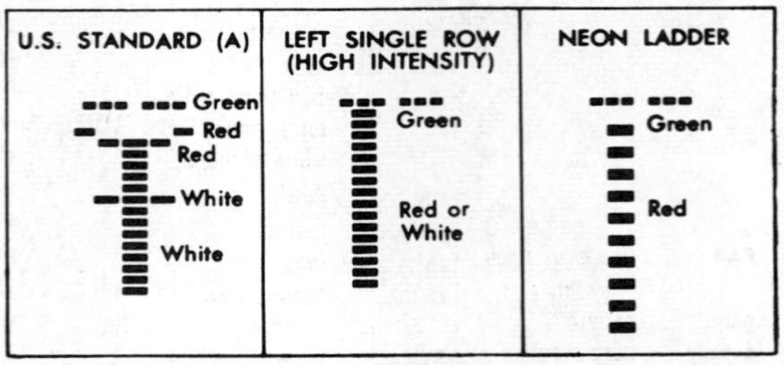


9—Sequence Flashing Lights (SFL) (3,000′ out unless otherwise stated)

10—Visual Approach Slope Indicator (VASI)

11—Runway end identifier lights (threshold strobe) (REIL)

12—Short approach light systems (SALS)

13—Runway alignment lights (RAIL)

14—Runway centerline

15—Touchdown zone

Because the obstructions on virtually all lighted fields are lighted, obstruction lights have not been included in the codification.

## SERVICING

**S1:** Storage.

**S2:** Storage, minor airframe repairs.

**S3:** Storage, minor airframe and minor powerplant repairs.

**S4:** Storage, major airframe and minor powerplant repairs.

**S5:** Storage, major airframe and major powerplant repairs.

## FUEL

| Code | Grade |
|---|---|
| F12 | 80/87 |

: −40°F
: −58°F
·eeze point −60°F
hout icing inhibitor, freeze

:R

ated FSS is shown in all
is located on the named air-
following the FSS name.
ılled through the local tele-
n Exchange) at the cost of
a local call, it is indicated by "(LC)" (local call) with the phone number immediately following the name of the FSS, i.e., "FSS: WICHITA (LC481–5867)." When an Interphone line exists between the field and the FSS, it is indicated by "(DL)" (direct line) immediately following the name of the FSS, i.e., "FSS: OTTO (DL)."

**RVV**—Runway Visibility Values, applicable runway provided.

**RVR**—Runway Visual Range, applicable runway provided.

**VASI**—Visual Approach Slope Indicator, applicable runway provided.

## AIRPORT REMARKS

**"FEE"** indicates landing charges for private or non-revenue producing aircraft. In addition, fees may be charged for planes that remain over a couple of hours and buy no services, or at major airline terminals for all aircraft.

**"Rgt tfc 13–31"** indicates right turns should be made on landings and takeoffs on runways 13 and 31.

Remarks data is confined to operational items affecting the status and usability of the airport, traffic patterns and departure procedures.

**Obstructions.**—Because of space limitations only the more dangerous obstructions are indicated. Natural obstructions, such as trees, clearly discernible for contact operations, are frequently omitted. On the other hand, all pole lines within at least 15:1 glide angle are indicated.

## FLIGHT SERVICE STATIONS

Flight Service Stations are listed alphabetically by state in the Airport/Facility Directory RCO's and LRCO's where available shown at the facility site following the three letter identifier. If located at other than a facility site, they are listed alphabetically.

Flight Service Stations (FSS) and Combined Station/Tower (CS/T) provide information on airport conditions, radio aids and other facilities, and process flight plans. CS/T personnel are not certificated pilot weather briefers; however, they provide factual data from weather reports and forecasts. Airport Advisory Service is provided at the pilot's request on 123.6 by FSSs located at airports where there are no control towers in operation. (See Part 1 ARRIVALS.)

In addition, they provide an aviation weather briefing service. Flight and weather briefing services are also provided by calling the telephone numbers listed in the Chapter entitled "FSS-CS/T Information and Weather Bureau Telephone Numbers".

**Civil communication frequencies used in the flight service station air/ground system are now operated simplex on 122.0, 122.2, 122.3, 122.6, 123.6 and emergency 121.5 plus 122.1 and 123.6 receive only as follows:**

**a. 122.0 is assigned at selected FSSs as a weather channel for both general aviation and air carriers.**

**b. 123.6 is designated as an airport advisory channel at all FSSs which provide this service at nontower locations. 123.6 is still in commission at some FSSs collocated with towers and may be used for en route communications at those locations.**

**c. Some FSS's use 123.65 or certain 50 KHz channels in the 122–123 MHz band (such as 122.05). Pilots using the FSS A/G system should refer to this directory or appropriate charts to determine frequencies available at the FSS or remoted facility through which they wish to communicate.**

Part time FSS hours of operation are shown in remarks under facility name.

## COMMUNICATIONS

Clearance is required prior to taxiing on a runway, taking off, or landing at a tower controlled airport.

When operating at an airport where the control tower is operated by the U.S. Government, two-way radio communication is required unless otherwise authorized by the tower. (When the tower is operated by someone other than the U.S. Government, two-way radio communication is required if the aircraft has the necessary equipment.)

Frequencies transmit and receive unless specified as: T—Transmit only, R—Receive only, X—On request. Primary frequencies are listed first in each frequency grouping, i.e., VHF, LF. Emergency frequency 121.5 is available at all TOWER, APPROACH CONTROL and RADAR facilities, unless indicated as not available in remarks.

Radar available is listed under "RADAR SERVICES" Radar beacons are indicated by "(BCN)" after "RADAR SERVICES", when available.

Figure 169(b).

## COMMUNICATIONS REMARKS

Remarks data are confined to operational items affecting the status and usability of navigational aids, such as: ILS component restrictions, part time tower hours of operation, frequency sectorization, VOT frequencies, proposed changes to navigational aids, etc.

## VOICE CALL

The voice call for contact with the traffic control services listed at each airport is the airport name followed by the call of the particular service desired, i.e., "LAGUARDIA TOWER." In these instances, only the name of the service is listed. When the voice call of the facility is not the same as the airport name, the complete voice call is listed.

## SERVICES AVAILABLE

**TOWER**

Pre-Taxi Clearance Procedure
Clearance Delivery (CLRNC DEL).
Approach Control (APP CON) Radar and Non-Radar.
Departure Control (DEP CON) Radar and Non-Radar.
VFR Advisory Service (VFR ADV) Non-Radar.
Traffic Information Service (TFC INFO) Radar.
Surveillance Radar Approach (ASR).
Precision Radar Approach (PAR).
Ground Control (GND CON).
VHF Direction Finding (VHF/DF).

**RADIO NAVIGATION AIDS**

Included in this section is a tabulation of all Air Navigation Radio Aids in the National Airspace System and those upon which the FAA has approved an instrument approach. Private or military Navigation Radio Aids not in the National Airspace System are not tabulated.

**AUTOMATIC TERMINAL INFORMATION SERVICE (ATIS)**

ATIS is continuous broadcast of recorded non-control information in selected areas of high activity. See Part 1.

**RADAR APPROACH PROCEDURE MINIMA**

Weather minima for precision and surveillance radar approaches (PAR/ASR) specify only the lowest straight-in authorized for the approach.

**FLIGHT SERVICE STATION (FSS)**

Airport Advisory Service (AAS).
Island, Mountain and Lake Reporting Service.
VHF Direction Finding (DF).

**UNICOM**

A private aeronautical advisory communications facility operated for purposes other than air traffic control, transmits and receives on one of the following frequencies:

**U-1**—122.8 MHz for Landing Areas (except heliports) without an ATC Tower or FSS;
**U-2**—123.0 MHz for Landing Areas (except heliports with an ATC Tower or FSS;
**U-3**—123.05 MHz for heliports with or without ATC Tower or FSS;
**U-4**—122.85 MHz for landing areas not open to the public;
**U-5**—122.95 MHz for landing areas not open to the public.

**NOTE.—UNICOM used for communications must be licensed by the Federal Communication Commission in order to be listed in this publication.**

Figure 169(c).

## RADIO CLASS DESIGNATIONS

**Identification of VOR/VORTAC/TACAN Stations by Class (Operational Limitations):**

**Normal Usable Altitudes and Radius Distances**

| Class | Altitudes | Distance (miles) |
|---|---|---|
| T | 12,000′ and below | 25 |
| L | Below 18,000′ | 40 |
| H | Below 18,000′ | 40 |
| H | 14,500′ — 17,999′ | 100* |
| H | 18,000′ — FL 450 | 130 |
| H | Above FL 450 | 100 |

***Applicable only within the contiguous 48 States.**

H=High L=Low T=Terminal

NOTE: An H facility is capable of providing L and T service volume and an L facility additionally provides T service volume.

The term VOR is, operationally, a general term covering the VHF omnidirectional bearing type of facility without regard to the fact that the power, the frequency-protected service volume, the equipment configuration, and operational requirements may vary between facilities at different locations.

B ___________ Scheduled Broadcast Station (broadcasts weather at 15 and 45 minutes after the hour.

DME _______ UHF standard (TACAN compatible) distance measuring equipment.

H __________ Non-directional radio beacon (homing), power 50 watts to less than 2,000 watts.

HH _________ Non-directional radio beacon (homing), power 2,000 watts or more.

H-SAB _____ Non-directional radio beacons providing automatic transcribed weather service.

ILS _________ Instrument Landing System (voice, where available, on localizer channel).

LMM _______ Compass locator station when installed at middle marker site.

LOM _______ Compass locator station when installed at outer marker site.

MA _________ Range (adcock, vertical raditors), power less than 50 watts.

MH _________ Non-directional radio beacon (homing) power less than 50 watts.

ML _________ Range (loop radiators), power less than 50 watts.

MRA _______ Range (adcock, vertical radiators), power 50 watts or more but less than 150 watts.

MRL _______ Range (loop radiators), power 50 watts or more, but less than 150 watts.

RA _________ Range (adcock, vertical radiators), power 150 watts or more.

RL _________ Range (loop radiators), power 150 watts or more.

S ___________ Simultaneous range, homing signal and/or voice.

# AIRPORT/FACILITY DIRECTORY

SABH ------ Non-directional radio beacon having limited navigational use. Provides automatic weather broadcasts.

TACAN ----- UHF navigational facility—omnidirectional course and distance information.

VOR -------- VHF navigational facility—omnidirectional, course only.

VOR/DME -- Collocated VOR navigational facility and UHF standard distance measuring equipment.

VORTAC --- Collocated VOR and TACAN navigational facilities.

W ---------- Without voice facilities on range frequency.

Z ----------- VHF station location marker at a LF range station.

**NOTES**

1. All FAA MH facilities operate continuously unless otherwise cited.
2. All FAA ranges operate continuously. Those which are not manned continuously are cited in the remarks with hours of operation in parentheses, e.g., (0600–2400).
3. LMF and VHF ranges listed at the same location are controlled by the same FSS.
4. Military navigational facilities which are not part of the common system are not listed in this publication.

**SAMPLE**

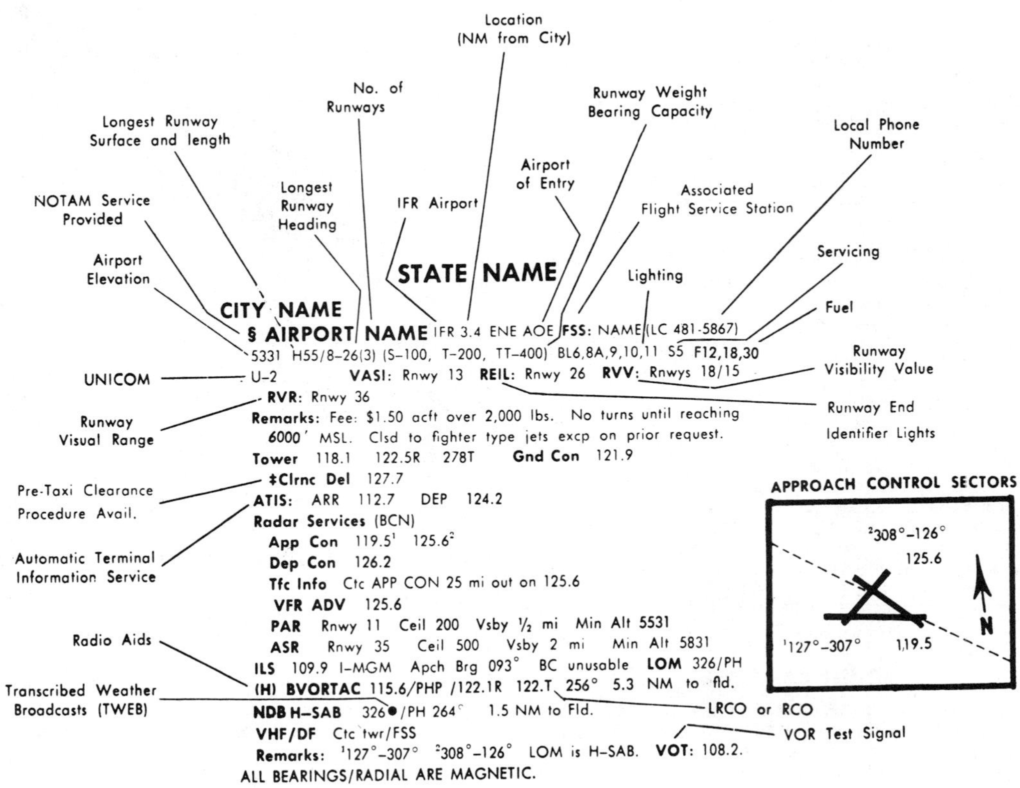


***PLEASE NOTE:***

*The following pages contain excerpts from PARTS 3 & 3A of the FAA A.I.M. Instructions for subscribing to the complete periodical by the FAA are given on the page following the Contents Page.*

Figure 169(d).

# AIRPORT/FACILITY DIRECTORY

## EXCERPTS
(for study purposes only)

## GEORGIA

**§ATLANTA** ***IFR*** 7SSW **FSS:** ATLANTA
1024 H100(9L-27R)(4) (S-100, T-170, TT-340) BL5,6,8A,9,10,11 S5 F22,30 U2 **VASI:** Rnwy 27R, 27L. **REIL:** 27R.
**RVR:** Rnwy 9L, 9R[1], 27L, 33.
**Remarks:** TV towers 1049 (2049' MSL) 9NM NNE. 1076' (2049' MSL) 7.5 NM NE. 1100' (2049' MSL) 10 nmi NE. Commercial Fee only. [1]Avail for apch and roll out.
**Tower** 119.1 119.5 122.7R **Gnd Con** 121.9 121.75
**‡Clrnc Del** 121.6
**ATIS:** 111.8 111.0
**Radar Services:** (BCN)
**App Con** 126.9 (270–089°) 118.1 (090–269°) 120.5[1]
**Dep Con** 125.7 (270–089°) 125.0 (090–269°) 123.8[2]
**Term Radar Service:** 119.3 (S of V18) 119.8 (N of V18)
**ASR**
**ILS** 109.9 I-ATL Apch Brg 089° **LOM:** 375/AT
109.3 **I-AZA** Apch Brg 329° BC unusable **LOM: 344/AZ**
110.5[4] I-ALR Apch Brg 089° **LOM:** 254/AL
**(H) BVORTAC** 115.6/ATL 016° 8.1NM to fld.
**NDB H-SAB** 266●/ATL 270° 4.7NM to rnwy 27R.
**Remarks:** [1]A/C only for Fulton County and DeKalb-Peachtree. [2]Radar monitoring of instrument apchs only at pilot's req. [3]D/C only for Fulton County and DeKalk-Peachtree. [4]Lclzr unusable beyond 60° either side of cntrline, BC unusable beyond 50° either side of cntrline. **VOT:** 111.0

**§MACON, LEWIS B. WILSON** ***IFR*** 9S **FSS:** MACON on Fld
354 H50/5-23(2) (S-30, T-50, TT-85) BL5,6[1],8A,9 S3
F22,30 **RVR:** Rnwy 5
**Remarks:** [1]HI rnwy lgts not avbl 2300–0700 lcl time. All other lgts avbl thru Macon FSS 2300–0700 lcl time. Rgt tfc rnwys 13 and 23.
**Tower**[1] 118.3 122.5R **Gnd Con** 121.7
**Radar Services:** (BCN)
**App Con** 119.6 134.1 124.2 122.7R
**Dep Con** 124.8
**Tfc Info** Acft lndg ctc app con on 119.6, acft overflying ctc 124.8.
**ASR** Rnwys 5, 13, 23 Ceil 500 Vsby 1 mi Min Alt 854.
**ILS** 109.5 I-MCN Apch Brg 047° BC unusable **LOM:** 290/MC
**(H) BVORTAC** 114.2/MCN at fld.
**Remarks:** [1]Opers 0700–2300 lcl time. RVR minimums not valid; or avbl when tower is clsd.

## SOUTH CAROLINA

**§CHARLESTON AFB/MUNI** ***IFR*** 9 NW **FSS:** CHARLESTON on Fld
45 H90/15-33(2) (S-100, T-140, TT-250) BL4,6,8A,9,10 S5 F22,45 U-2 **VASI:** Rnwys 15, 21, & 33 **RVV:** Rnwy 33 **RVR:** Rnwy 15
**Remarks:** U.S. Customs lndg rgts arpt. J-bar rnwys 15–33.
**Tower** 121.0 122.5R 109.7T **Gnd Con** 121.9
**‡Clrnc Del** 121.7
**ATIS** 113.5
**Radar Services:**
**App Con** 123.9[2] 119.3[3] 118.5 109.7T
**Dep Con** 123.9[2] 119.3[3]
**ASR** Rnwys 3, 15, 21, 33 Ceil 400 Vsby 1 mi Min Alt 445.
**Tfc Info** Ctc App Con. Expanded radar svc avbl.
**PAR** Rnwy 3 Ceil 300 Vsby 3/4 mi Min Alt 445
Rnwy 15 Ceil 200 Vsby 1/2 mi Min Alt 245
Rnwy 33 Ceil 300 Vsby 3/4 mi Min Alt 445

## SOUTH CAROLINA—*Continued*

**ILS**[1] 109.7 I-CHS Apch Brg 149° BC unusable **LOM:** 329/CH
**(H) VORTAC** 113.5/CHS on fld.
**NDB H-SAB** 329●/CH
**Remarks** LOM is CH NDB. Acft not 2-way eqptd use 122.5T and 119.3R. [1]Front crs unusable above 5000' MSL. [2]151–330°.

## TENNESSEE

**BROOKS NDB MHW** 287/ME **FSS:** MEMPHIS

**CHATTANOOGA**
**§LOVELL FIELD** ***IFR*** 5E **FSS:** CROSSVILLE (LC 267-5343)
682 H62/2-20(2) (S-115, T-160, TT-240) BL4,6,8A,9,10 S5 F12,15,18,22, U-2
**VASI:** Rnwy 2 **RVR:** Rnwy 20
**Remarks:** Ngt fueling callout with surcharge less than 100 gals 2300–0600 lcl time.
**Chattanooga Tower** 118.3 119.2 122.5R **Gnd Con** 121.9
**Radar Services:** (BCN)
**Chattanooga App Con** 119.2[1] 125.1[2]
**Chattanooga Dep Con** 119.2[1] 125.1[2]
**Tfc Info** Ctc App Con 25 NM out
**ASR** Rnwy 2 Ceil 400 Vsby 1 mi Min Alt 1082
Rnwy 20 Ceil 500 Vsby 1 mi Min Alt 1182
Rnwy 32 Ceil 600 Vsby 1 mi Min Alt 1282
**ILS** 109.5 I-CHA Apch Brg 196° **LMM:** 201/HA
**Chattanooga (H) BVORTAC** 115.8/CHA/121.5 122.1R 122.6 123.6 332° 4.9NM to fld
**Daisy NDB HW** 341/CQN 196° 7.7NM to fld.
**Remarks:** [1]016–196° [2]197–015° Glide slope unusable below 882' MSL.

**CLARKSVILLE (T) BVOR** 110.6/CKV/122.1R 123.6R **FSS:** NASHVILLE

**CROSSVILLE FSS** 121.5 122.1R 122.2T 122.6 123.6 **DF**

**CROSSVILLE (L) BVORTAC** 117.6/CSV **FSS:** CROSSVILLE
**NDB MHW**[1] 375/CMA
**Remarks:** [1]Operd by City. Unusable beyond 15 NM.

**DYERSBURG FSS** 121.5 122.1R 122.3 122.6 123.6
**Remarks:** Opers 0600–2200 other hrs ctc Jackson FSS.

**DYERSBURG (L) BVORTAC** 116.8/DYR **FSS:** DYERSBURG

**GRAHAM (L) BVOR** 111.6/GHM/122.1R **FSS:** NASHVILLE

**HOLSTON MOUNTAIN (L) BVORTAC** 114.6/HMV **FSS:** TRI CITY

## TEXAS

**§ABILENE MUNI** ***IFR*** 3SE **FSS:** ABILENE on Fld
1778 H60/17-35(2) (S-70, T-89, TT-145) **BL4,6,8A,9 S5** F18,30 U2 **RVV:** Rnwy 35
**Remarks:** Rgt tfc rnwys 4, 35. Rnwy 12–30 is txwy used for lgt acft dalgt hrs only.
**Tower**[1] 120.1 122.7R **Gnd Con 121.9**
**Radar Services:** (BCN)
**App Con** 124.1 126.5 121.3 134.1 113.7T
**Dep Con** 125.0
**Tfc Info** VFR arvg acft ctc ABI App Con 20 mi out, tsmt 124.1, 122.5; rcv 124.1. Dptg & enrout tsmt 121.3, 122.5; rcv 121.3.
**ILS** 110.3 I-ABI Apch Brg 350° **LOM:** 353/AB
**(H) BVORTAC** 113.7/ABI 102° 9.5 NM to fld.
**Remarks:** [1]Opers 0600–2300 lcl time.

**ALICE FSS** 121.5 122.1R 122.3 122.6 123.6

**ALICE (L) BVOR** 114.5/ALI **FSS:** ALICE

Figure 169(e).

Figure 169 shows excerpts from a typical Airport Facility Directory. We should use the Airport Directory code legend to decode the Airport Facility Directory.

*Flight Service Station and Combined Station/Tower Information and Weather Bureau Telephone Numbers*

Flight Service Stations and combined station/tower telephone numbers are listed in Part III, as shown in Fig. 170 for easy reference when we wish to receive a preflight briefing or file a flight plan.

**FSS–CS/T INFORMATION AND WEATHER BUREAU TELEPHONE NUMBERS**

Flight Service Stations (FSS) and Combined Station/Tower (CS/T) provide information on airport conditions, radio aids and other facilities, and process flight plans. CS/T personnel are not certificated pilot weather briefers; however, they provide factual data from weather reports and forecasts. Aiport Advisory Service is provided at the pilot's request on 123.6 by FSSs located at airports where there are not control towers in operation. (See Part 1 ARRIVALS.)

The telephone area code number is shown in parentheses. Each number given is the preferred telephone number to obtain flight weather information. Automatic answering devices are sometimes used on listed lines to given general local weather information during peak workloads. To avoid getting the recorded general weather announcement, use the selected telephone number listed.

★ Indicates Pilot's Automatic Telephone Weather Answering Service (PATWAS) or telephone connected to the Transcribed Weather Broadcast (TWEB) providing transcribed aviation weather information.

◆ Indicates a restricted number, use for aviation weather information

▬ Call FSS for "one call" FSS/WBAS briefing service.

EXCERPT

| Location and Identifier | | Area Code | Telephone |
|---|---|---|---|
| **TENNESSEE** | | | |
| Memphis MEM | FSS | (901) | 398-9268▬ |
| | FSS | (901) | 398-2347★ |
| Nashville BNA (Metro) | FSS | (615) | 254-6552/3 |
| | FSS | (615) | 255-8732★ |
| | WB | (615) | 256-4468◆ |

| Location and Identifier | | Area Code | Telephone |
|---|---|---|---|
| **TEXAS** | | | |
| Tyler TYR (Pounds) | CS/T | (214) | LY 4-6301 |
| Victoria VCT | WB | (512) | 575-1276◆ |
| Waco ACT | CS/T | (817) | PL 2-4811 |
| | WB | (817) | PL 3-6484 |
| Wichita Falls SPS (Sheppard AFB/Wichita Falls Air Trml) | FSS | (817) | 322-1302▬ |

Figure 170.

*Restrictions to Enroute Navigation Aids*

Specific restrictions to particular enroute navigation aids are listed alphabetically by state. A sample excerpt is shown in Fig. 171.

*Sectional Chart Bulletin*

This section of Part III provides us with a tabulation of the major changes in aeronautical information that have occurred since the last publication date of each sectional chart. Only those changes in controlled airspace and special-use airspace that present a hazardous condition or impose a restriction on the pilot are included.

Figure 171 shows a sample excerpt from the sectional chart bulletin.

*VOR Receiver Check Points*

A list of airborne and ground check points is provided in Part III. Each entry will be given in the following manner: facility name, bearing in degrees magnetic from the VOR, location of the check point (distances in nautical miles), and altitude in feet above mean sea level. A sample excerpt from the list of VOR receiver check points is shown in Fig. 172.

## RESTRICTIONS TO ENROUTE NAVIGATION AIDS

**Radio Facility Restrictions are cited until cancelled by the Associated Station.**
**Restricted areas are defined in degrees from magnetic North.**

EXCERPT

**SOUTH CAROLINA**

ALLENDALE VOR: Unusable on 171° radial beyond 20 mi below 4,500′ MSL.
FLORENCE VORTAC: VOR portion unusable 360–010° beyond 20 mi below 5000′ MSL.
MYRTLE BEACH VORTAC: Unusable beyond 25 nmi below 2500′ MSL.

**TENNESSEE**

KNOXVILLE VORTAC: VOR portion unusable 140–175° all alts, 175–186° above 17,000′ MSL.

**TEXAS**

AUSTIN VORTAC: VOR portion unusable 260–320° beyond 35 mi below 3,500′ MSL.
BROWNSVILLE VORTAC: DME portion unusable 330–350° beyond 15 nmi below 4000′ MSL.
EL PASO VORTAC: DME portion unusable 265–300° beyond 30 mi below 9,200′ MSL.
HOUSTON VORTAC: DME portion unusable 085–210° beyond 35 mi below 2,500′ MSL.
SALT FLAT VORTAC: Unusable 031–049° beyond 30 NM below 15,000′ MSL.

## SECTIONAL CHART BULLETIN

The purpose of this Bulletin is to provide a tabulation of the major changes in aeronautical information that have occurred since the last publication date of each Sectional Aeronautical Chart. The general policy is to include only those changes to controlled airspace and special use airspace that present a hazardous condition or impose a restriction on the pilot; major changes to airports and radio navigational facilities, thereby providing the VFR pilot with the essential data necessary to update and maintain his chart current. When the Sectional Aeronautical Chart is republished, the corrective tabulation will be removed from this Bulletin.

**NOTE: New data which have been added to this issue are shown below the rule line under the appropriate chart.**

EXCERPT

**•NEW ORLEANS**
**62nd Edition, September 14, 1967**

Delete Youngsville Arpt 30°05′50′′N, 91°58′50′′W. Add UNICOM Welsh Muni Arpt 30°14′30′′N, 92°49′45′′W. Change New Orleans Int ctl twr freq 119.1 to 119.9: 29°59′30′′N, 90°15′37′′W. Relocate VOR Beaumont freq 114.9 ident BPT from 29°54′59.9′′N, 94°06′13.2′′W to 29°56′45′′N, 94°00′57.5′′W and realign arwys to new

**•PHOENIX**
**58th Edition, January 4, 1968**

Add obstn & elev 2190′ MSL 33°35′50′′N, 112°05′15′W.

---

Delete Pumpkin Center Arpt 33°51′N, 111°18′W. Delete Kansas Settlement Arpt 32°05′N, 109°46′W. Change scheduled publication date of next edition from July 18, 1968 to July 25, 1968.

## AIR ROUTE TRAFFIC CONTROL CENTER COMMUNICATIONS FREQUENCIES

**SEE PART 1, ENROUTE, DIRECT COMMUNICATIONS-CONTROLLERS/PILOTS**

**(This information supplements the RCAG data shown on Enroute Low Altitude Charts)**

***Center Sector Discrete Frequencies for the Low and High Altitude Airway Systems***

EXCERPT

| | LOW | HIGH | | LOW | HIGH |
|---|---|---|---|---|---|
| **ALBUQUERQUE** | 125.2 | 135.05 | **CHICAGO** | | |
| Amarillo | 127.2 132.3 | 134.75 | Aurora | 120.6 127.6 | 133.5 |
| Animas | 124.6 | 133.0 | Bradford | 126.8 | 133.95 |

Figure 171.

*Special Notices* The section of Part III called *Special Notices* contains information of a general nature or universal application and information pertinent to a specific geographic area. One of the most important notices which appears in the special notice section is the *Special Operations—Oil Burner Routes.*

# VOR RECEIVER CHECK POINTS

The list of VOR airborne check points and ground check points is given on the following pages. Use of these Check Points is explained in Part 1.

NOTE: The information is provided in the following order: Facility name (plus airport name, if needed); bearing in degrees magnetic from the VOR; location of the check point (distances in nautical miles); and altitude (in feet MSL, if any).

EXCERPT

## AIRBORNE

●**Aberdeen, S.D.**: 301°; over apch end rnwy 30; 1700′.

**Abilene, Tex.** (Municipal): 047°; 10.1 nmi silos in center of Ft. Phantom alternate for lgt acft 099° 9 nmi ctl twr and bcn.

**Lufkin, Tex.**: 328°; 4.6 mi, rotating bcn on Angelina Co. Arpt; 1300′.

**Macon, Ga.** (Muni): 050°; 6.3 NM, rdo twr; 1800′.

## GROUND

**Albany, Ga.**: 149.5°; center of tie down area on N edge or ramp.

**Alexandria, La.** (Esler Fld): 151°; 3.5 nmi in front Admin Bldg.

**Dalhart, Tex.**: 170°; SE corner of main ramp.

**Dallas, Tex.** (Addison): 159°; intersection of center and parallel taxiways.

# SPECIAL NOTICES

Special Notices of a general nature or universal application, and other than for a specific geographical location, are grouped together under General Notices. Special Notices pertinent to a specific geographic area are grouped together under Area Notices by state, then city, airport, or location within the state. The month and year the notice is initially inserted into the manual is provided at the conclusion of each Special Notice. A solid dot ● prefixes new or revised Special Notices.

# SPECIAL NOTICES—GENERAL

## HEAVY TRAFFIC AROUND MILITARY FIELDS

**Pilots are advised to exercise vigilance when in close proximity to most military airports. These airports may have jet aircraft traffic patterns extending up to 2500 feet above the surface. In addition, they may have an unusually heavy concentration of jet aircraft operating within a 25 nautical mile radius and from the surface to all altitudes. This precautionary note also applies to the larger civil airports.**

**June 1967**

## GREAT LAKES WATER POLLUTION PROGRAM

The U.S. Coast Guard is endeavoring to identify and gather detailed data on sources of water pollution in the Great Lakes. Of crucial importance in this program is the appearance of a film of oil or die-offs of large schools of dead fish floating on the surface of the water.

The FAA is cooperating in this program by requesting pilots to report any such sightings in the Great Lakes area to the nearest Flight Service Station.

## SPECIAL OPERATIONS

### OIL BURNER ROUTES

**MILITARY LOW-LEVEL OPERATION "OIL BURNER"**: The USAF and the U.S. Navy conduct low-level navigational/bombing training flights in jet aircraft in both VFR and IFR weather conditions along the routes shown in the following pages. Operations are contained within 9 nautical miles on either side of the center-line except that on the VFR segments they will be contained within 4 nautical miles of the centerline. When route widths shown on the chart are less than those indicated above, the operation will be confined to the route width shown. Aircraft will operate at the altitudes specified on the route charts. Near the end of the low-level navigation portion of the route, aircraft will conduct simulated bomb release maneuvers within a "Bomb Run Corridor" at and between the altitudes shown. "Short Look" and "Lay Down" bomb runs are generally conducted from 4,000 to 6,000 feet MSL "Long Look" bomb runs are generally conducted from 18,000 to 21,000 feet MSL. On some route segments when the weather conditions are 3,000 feet ceiling and 5 miles' visibility or better, aircraft may descend to an altitude of not less than 500 feet above terrain in daylight hours or 800 feet above terrain during hours of darkness. Flights are conducted only during the times specified for each route. The current operational status of a particular route is included in scheduled weather broadcasts or may be obtained by calling a Flight Service Station near the route. Revised information is underlined.

## NEW YORK/VERMONT

# WATERTOWN OB–4

### Revised Effective July 30, 1967

Aircraft shall cross 43°57′N, 76°04′W (Watertown, N.Y. VOR) at FL 250 or as assigned by ARTCC: descend to cross 44°09′N, 75°4 at FL 230; continue descent to cross 44°24′N at 17,000 feet MSL;

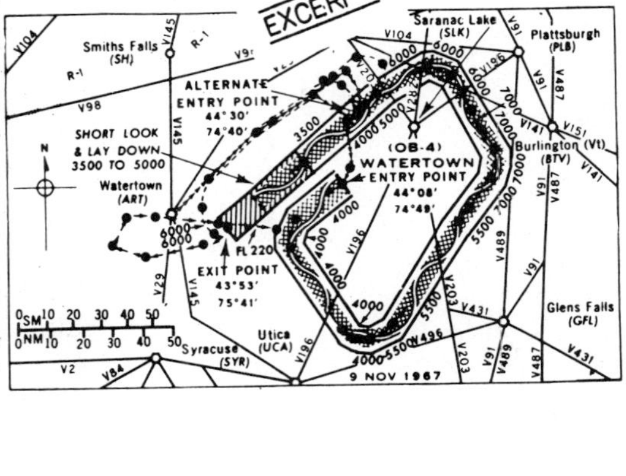


Figure 172.

*Oil Burner Routes. Oil Burner* is the name given to low-level military operations. Both Air Force and Navy jet aircraft conduct navigation and bombing training in VFR and IFR weather conditions. Their routes of flight may be as low as 500 feet above the surface and may extend upward to 21,000 feet. The time when Oil Burner routes are active will be specified.

Due to the nature of Oil Burner flights, that is, high-speed aircraft at low altitudes, they may constitute a hazard to other aircraft. We should check the Airman's Information Manual for Oil Burner routes in our area, and contact the nearest Flight Service Station for information concerning the operational status of the route.

A sample excerpt of the Oil Burner route segment of the Special Notices section is shown in Fig. 172.

## Part III A

Part III A is issued every 14 days and contains Notices to Airmen which are considered essential to the safety of flight plus new or revised Oil Burner routes. The information in Part III A is arranged alphabetically by state and within each state by city or locality. New or revised data are indicated by underlining the first line of the item to call attention to the new item.

A sample excerpt from Part III A is shown in Fig. 172.

## Summary

It cannot be overemphasized that the Airman's Information Manual is part of the equipment needed for safe flight. Every pilot should subscribe to and make use of this publication.

### CHECK YOUR READING

1. Where can the Airman's Information Manual be purchased?
2. In which Part of the AIM would the following information be found?
   A. Oil Burner routes
   B. Communications procedures during departure
   C. Medical facts for pilots
   D. Notices to Airmen
   E. NOTAM codes
   F. Airport lighting
   G. VOR test stations and frequencies
   H. VOR test procedure
   I. Fuel and services available at a specific airport
3. Using the excerpt from the Airport Directory found in this chapter, decode in full the information concerning the Atlanta Municipal Airport, Atlanta, Georgia.

4. Using the excerpt from the Airport Facility Directory found in this chapter, find the following information concerning Lovell Field, Chattanooga, Tennessee.
   A. Tower frequencies
   B. Ground control frequencies
   C. Field elevation
   D. Is VFR traffic information available? On what frequency?

# PART VIII

# EMERGENCY PROCEDURES

# 49

# EMERGENCY PROCEDURES

Most emergency situations which occur in flight could be avoided. Those that are unavoidable can be minimized by following the tried and proven suggestions outlined here. Naturally, every possible emergency cannot be covered in detail in this or any text. Many specific remedies to emergency situations are dependent on the particular configuration of the aircraft involved. However, by learning and using a few basic rules concerning emergencies, we will diminish the number of situations which lead to serious emergencies, and lessen the seriousness of those that do occur.

## Emergency Communications

Unfortunately, few pilots in trouble seek aid early enough in the sequence of events which leads to serious difficulties. If they had requested help early, the emergency probably would not have developed. The most effective tool we have in the aircraft to counteract emergencies is the two-way radio. A virtual army of specialists on the ground can be brought to our defense, if we simply contact them.

### *The Four C's*

In the event of any inflight difficulty, we can simplify communications by memorizing and using the *four C's*.

*Confess*. It is not easy for any man to confess that he is embroiled in a situation which has become greater than his capabilities. We must learn, first of all, to recognize our own capabilities and those of our aircraft. When any situation develops so as to approach the limit of those capabilities, we must confess first to ourselves, and then to a surface radio facility. No man goes through life without erring. Wise men confess their errors to themselves and to those who may help them.

*Communicate*. Tell someone! Pick up the microphone, tune to the emergency frequency (121.5 MHz) and call for help. All we have to say is "Anyone hearing (aircraft no.) please answer on 121.5MHz."

Although 121.5 MHz is the standard world-wide emergency frequency, an emergency message can be given to Flight Service Stations, towers, radar stations, indeed anyone, on their respective frequencies.

If time permits, and we can remember the standard phraseology and procedures, the ground communicator will appreciate the following sequence of statements.

1. Alert call: Tune to 121.5 MHz and say "Mayday" three times. Mayday is the internationally known distress call. This will alert all receiving stations in our vicinity that a call of an emergency nature is about to be broadcast.
2. Aircraft identification: Repeat three times, distinctly.
3. Type and color of aircraft.
4. Position: If our position is definitely known, we should state it. If it is estimated, we should say that it is an estimated position. If it is unknown, we should confess that fact. When giving our position, we must be as accurate as possible and try to pinpoint our position. Large landmarks (cities, lakes, mountains, etc.) make good reference points.
5. True or magnetic heading: Although either the true or the magnetic heading may be given, the magnetic heading is more readily available. We should state which heading we are giving.
6. True or estimated airspeed: Either the true or estimated airspeed may be given. We must state which one we are using and whether in knots or miles per hour.
7. Altitude: We must use caution not to misread the altimeter in a moment of stress and to report that altitude exactly.
8. Fuel remaining: The ground communicator needs to know the amount of fuel remaining, in hours and minutes, so that he will know how much time he has in which to work.
9. Nature of distress: An honest evaluation of our predicament will hasten a proper remedy. We must not hedge. Frank, concise language in simple terms will best convey our message.
10. Pilot's intentions: If we know what we are going to do for the emergency, we should inform the communicator.
11. Assistance desired: We should ask for any assistance required,

whether it be an escort, a position fix, a steer to an airport, the opinion of a mechanic (or other specialist), etc.

12. Two 10-second dashes with the microphone button open, followed by the aircraft identification number will signify the end of the emergency message.

Naturally depending on the nature of the emergency, we may wish to eliminate or abbreviate some of this information.

*Climb.* For almost every emergency (except fire in flight), we should immediately initiate a climb if weather permits. The higher we get, the better will be the radio reception on the surface, the more radar facilities may spot us, and the more distance we will have to the ground. Remember, except for fire in flight, altitude is insurance. The aircraft will perform reasonably well in most emergencies if we can keep it flying, and we need altitude to do that.

*Comply.* Once we establish communications and request aid, we must do what we are told, if at all possible. Our instructions will be issued in a clear, calm tone. If we cannot (or do not know how to) comply with any instruction, we must advise the communicator. We must *not* comply with any instruction which makes us feel uncomfortable. Remember, we are human and under stress. The communicator is not going to turn in a grade for our performance. His only desire is to get us on the ground safely.

*Mental attitude.* One of the most important factors concerning the overcoming of an emergency is our mental attitude. Fear is a helpful human trait if we control it. Panic is not helpful. We must not allow our mind to wander to the possibility of a dreadful ending to our emergency. Thinking positively will help us to keep our wits, and determination to succeed will help in bringing us to a happy end. Hundreds of specialists on the ground will depend on us to keep a cool head and handle the situation to the best of our ability. No man can be criticized for doing his best, no matter how minimal that is.

## Direction Finding Assistance

Many FAA control towers are equipped with facilities which may provide direction finding assistance to pilots. Any Flight Service Station will advise us as to the availability of a DF facility.

### *The DF Steer*

The DF steer is a procedure which allows the surface facility to provide the pilot with his direction and distance from the facility, and an estimated heading to the facility or to any requested facility.

We must simply request a DF steer from the DF facility (or any Flight Service Station). The DF station will ask us for a short transmission. We should key the microphone and count slowly to ten. The DF facility communicator will furnish us with homing information. We should acknowledge his information before leaving the frequency.

## Radio Communications Failure

If we should become lost or experience any other emergency in conjunction with radio communications failure, we may choose to follow a standard procedure in an attempt to alert radar stations in the vicinity of our distress. However, because of the high density of air traffic in many areas, radar operators may overlook these procedures. Therefore, they should be used with discretion and as a last resort in seeking help.

### *Receiver Only Operating*

If our radio receiver is operating, but we cannot transmit, we may fly a triangular pattern to the *right* in order to alert a radar facility of our position. This pattern is shown in Fig. 173.

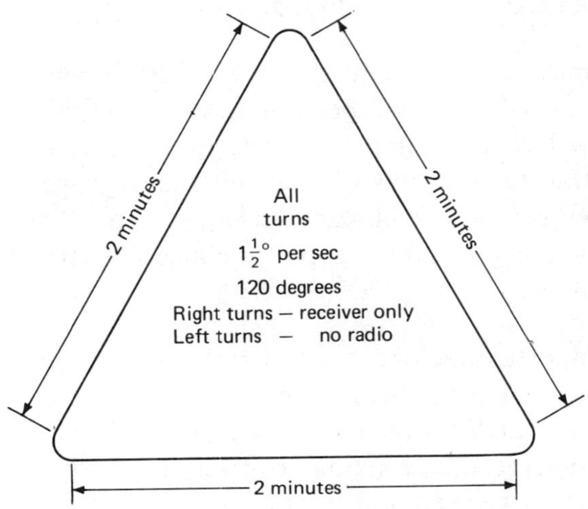


Fig. 173. Emergency flight pattern.

We must complete a minimum of two patterns before resuming our original course and after about 20 minutes we should perform another two patterns. We must tune our radio receiver to the emergency frequency (121.5 MHz) and listen for instructions.

### *Transmitter and Receiver not Operative*

If we experience complete radio failure, we should fly the same pattern (Fig. 173), except all turns are made to the *left*. If the patterns are observed by radar operators, an escort aircraft will be dispatched if possible.

Because of the time consumed and the unreliability of positive observation of the emergency flight patterns, we must use careful judgement before relying on them for a solution to any emergency.

## Engine Failure

Today's aircraft engines are extremely reliable if properly maintained and carefully operated. It is a rare occurrence indeed for an engine to fail without advance warning. We must develop a habit pattern of systematically checking the engine operating instruments. The oil temperature and pressure gauges, tachometer and/or manifold pressure gauges, and the cylinder head temperature gauge will normally indicate advance warnings of impending engine malfunction.

Because aircraft engines fail so infrequently, we face the danger of being lulled into a sense of security if we do not consciously remember that the engine is a mechanical device and is subject to failure. The occasional practice of simulated forced landings ***with a flight instructor*** will provide the insurance on which we may rely in the event of engine malfunction.

### *Engine Failure on Takeoff*

During the takeoff segment of the flight, the engine is normally developing maximum power, and is, therefore, more subject to mechanical failure than at any other time during the flight. Because of our proximity to the surface, we will have little time to plan an emergency landing. The most important thing to remember in case of engine failure is that the aircraft *must* remain in control.

Most pilots agree that a *controlled* landing on any terrain is better than a crash onto the best terrain. If an engine failure occurs at low altitude, the best glide angle must be *immediately* established to maintain safe flying speed. Pilots have always argued about the advisability of making a turn at low altitude to afford a better choice of landing area. In general, our safest course of action if the engine fails on takeoff is to establish the best glide angle and plan a landing in a direction straight ahead (or nearly so). Too many pilots have lost control of the aircraft while trying to negotiate a turn at low altitude to support that theory. If time permits, fuel, magneto, and electrical switches may be turned off prior to landing.

### *Engine Failure at Altitude*

As mentioned, an engine failure without advance warning is very rare. We will cut our chances of being caught unawares by constant vigilance. However, if the engine does fail at altitude, we should *first* establish the best glide angle. A suitable landing area must be found quickly. It is better to pick a poor landing area that we know is within gliding distance than to pick a perfect area which is at a questionable distance. Our landing must be into the surface wind if at all possible. Surface smoke, ripples on bodies of water, or runway indicators should be checked to determine wind direction.

While gliding toward the landing area, the engine controls and fuel selectors should be checked. Many pilots have executed a forced landing due to fuel starvation only to realize later that the fuel selector valve was on an empty tank and could have been switched to another tank to tap an unused reservoir of fuel. The mixture control should be placed in the full rich position, with the carburetor heat on *hot*. If time permits, the magneto switch may be checked.

As we approach the landing area, we should, if at all possible, plan on a normal approach; a base leg will aid us in determining the proper final approach altitude and path. We should accomplish a thorough search of the approach and landing area for obstructions or hazards. Remember, we will be much better off if we *land the aircraft* while it is under our control, instead of crashing uncontrolled.

## Fire in Flight

Although a rare occurrence, inflight fire is one of the most hazardous of all emergencies. Careless smoking habits or shoddy maintenance of the aircraft and power plant may lead to fire in flight.

### *Engine Fire*

The usual cause of fire in the engine compartment is a flooded carburetor. During a prolonged power-off descent the carburetor may become flooded, and the escaping fuel may be ignited. We may normally extinguish this type of fire by simply adding power.

A thorough preflight inspection of the engine compartment must be made to detect fuel and oil leaks around the engine. A *wet* engine is a fire hazard.

If the fire is not *immediately* extinguished, the fuel must be shut off and a forced landing must be made with all dispatch. After landing, we should evacuate the aircraft as quickly as possible.

### *Electrical Fire*

A fire in the electrical system can be detected by its peculiar odor. Wiring insulation is not easily ignited, but as its temperature increases, it emits the odor of smoldering cloth. At the first indication of an electrical wiring fire, all electrical equipment must be shut off, including the master switch. This action will not affect the engine operation, as the electrical equipment is totally independent of the ignition system.

If it is imperative that some electrical equipment be used for safe flight (such as position lights at night), we may turn on the master switch *only* for a few moments. If no odor prevails, we may turn on the desired equipment, one item at a time, checking each time for odor. Using this procedure, we will isolate the defective circuit.

### *Generator (Alternator) Failure*

The generator warning light or ammeter will indicate generator failure. Since the aircraft battery will *not* support all the electrical equipment on board for any length of time, all unnecessary equipment must be shut off. Only the most important electrical equipment should be used, and only for the shortest possible time. Depending on the specific flight condition, a landing at the nearest suitable airport may be advisable.

### *Engine Overheat*

An overheated engine may be the result of insufficient oil quantity or oil pressure, prolonged use of high power with low airspeed, or preignition and detonation.

At the first indication of high operating temperature (high oil temperature or cylinder head temperature), we should reduce the power and increase the airspeed. This will allow for more effective engine cooling. If the aircraft is equipped with adjustable cowl flaps, they should be opened fully. The mixture control should be in the full rich position.

A landing should be planned at the nearest suitable airport if the engine temperature cannot be reduced.

*Low Oil Pressure*

At the first indication of reduced oil pressure, we should reduce the power to that required to maintain maneuvering speed and plan a landing as soon as possible. Power setting *changes* should be at a minimum.

*Low Fuel Quantity*

The emergency condition caused by low fuel quantity is one which is easily prevented. Careful flight planning, with special attention to fuel quantity and consumption will preclude the necessity of coping with this emergency situation.

However, if low fuel quantity threatens the safety of our flight, we *must* plan a landing *prior to* fuel starvation. It is far better to accomplish an emergency landing before running out of fuel than to be forced to land with no power.

*Use of Military Facilities*

Civilian pilots may use military airports in emergency conditions. If possible, the military control tower should be contacted on the emergency frequency (121.5 MHz) prior to landing. We must use extreme caution due to the presence of high-speed military aircraft. After landing, we must *follow instructions explicitly*.

**Summary**

The best attack against emergency conditions is to prevent them from happening. Careless preflight planning, slipshod habits, and inflight daydreaming invite disaster. We should constantly review emergency procedures outlined in the Aircraft Operations Manual and mentally perform those procedures while flying in every conceivable configuration.

FEDERAL AVIATION AGENCY

VFR EXAM-O-GRAM* NO. 18
LOST PROCEDURES - PILOTAGE

It seems that all who fly cross-country are destined to lose their way or become "temporarily misplaced" at one time or another. Therefore, we should give some forethought to procedures and practices that may be used to lead wandering birdmen out of the wilderness. Confining our problems to the typical VFR dilemma, we can start with the general and proceed to more specific rules.

1. GENERAL:

Don't fight the problem -- try to solve it! Stay loose -- don't hit the panic button, thus virtually assuring that all the thinking gears will grind to a halt.

Analyze and evaluate as to: Fuel available and consumption rate. In other words, how much longer can you fly insofar as fuel is concerned? Be conservative -- not hopelessly optimistic.

Weather -- is it good, bad, indifferent, improving, or deteriorating?

Equipment -- is everything functioning? Do you have lights (cockpit, landing, navigation, etc.) or survival gear of any description?

Terrain -- is it open, flat country, mountains, marshes, semi-desert, sparsely or thickly populated?

Daylight -- hour remaining (if any). Have you had night or instrument flying?

Once you have a reasoned assessment of the situation, you are better prepared to make vital decisions. One of the first is to decide if help is available, or are you all by your lonesome?

KNOW and ABIDE BY YOUR LIMITATIONS AND THOSE OF YOUR EQUIPMENT

URONUROWN
(with no radio)

2. SPECIFIC:

Condition One:

(a) Low on fuel
(b) Weather deteriorating
(c) Inadequate experience and darkness imminent
(d) Engine or equipment malfunctioning

While (d) is not necessarily associated with being lost, the solution which follows would certainly apply if the situation were serious enough.

SOLUTION:

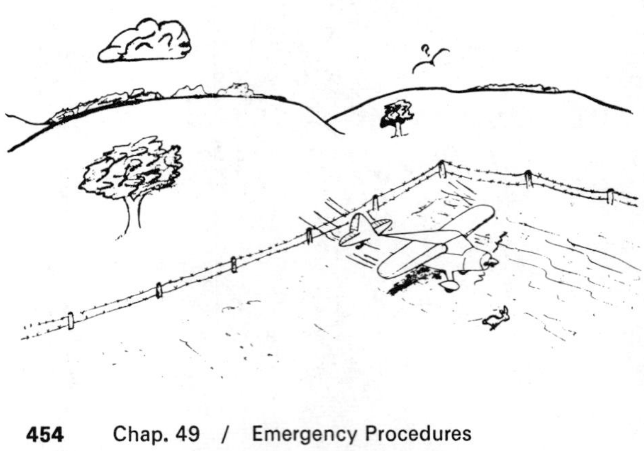

GET IT ON THE GROUND! Most accidents are the product of mistakes which have multiplied over a period of time. Getting lost is no exception. Don't push your luck. It may well be that in doing so you have added the final mistake which will add another figure to the statistics on accidents. How much better to be on the ground than in it. If terrain or other conditions make it impossible at the moment, don't waste time, for it is of the essence. Don't search for a field comparable to Idlewild. Anything usable will do. Remember, most people on the ground know where they are. You know you do not. This state of ignorance may well become permanent because the triple "whammy" of getting caught while "dangling, dark, and dreary" will rob you of virtually all control of the situation. If there is any alternative whatsoever . . .

NEVER fly until the petrol peters out. There are few things so nerve shattering as the rustle of the wind when an engine has coughed its last.

NEVER fly until the sun slowly sinks in the golden west. It may be a beautiful sight but the goblins will get you if you don't watch out.

NEVER fly until the biggest, meanest goblin of them all, Ole Bad Weather, falls flat on his face. He will do his best to take you with him.

Remember the NEVERs, lest in the blink of an eye, they become FOREVERs.

## KNOW and ABIDE BY YOUR LIMITATIONS AND THOSE OF YOUR EQUIPMENT

Condition Two:

(a) Plenty of fuel (b) Plenty of daylight
(c) Plenty of good weather

SOLUTION:

Establish an "Error Semi-Circle".

1. Straighten up and fly right -- straight, that is. Establishing a course by hunch or because you "got a feeling" is for the birds. Don't wander aimlessly.

2. Use knowledge of last known position, elapsed time, approximate wind and ground speed, (airspeed is better than nothing), to establish how far you may have traveled since your last check point.

3. Use this distance as a radius and draw a semi-circle ahead of last known position on the chart. For example, you estimate your ground speed at 120 mph. If you have been flying 20 minutes since your last check point, then the no-wind radius of your semi-circle is 40 miles projected along the direction of your estimated track. If you believe your wind is from the right, then you are most probably in the left quadrant of your semi-circle. Of course, unless you were sure about the wind, you could not ignore the right quadrant. The use of a simple computer can materially reduce the effort required in solving problems of speed, time, distance, and fuel consumption.

4. If you have been flying a steady compass heading and keeping a reasonably accurate navigation log, it's not likely you will have too much difficulty. If you've been operating "fat, dumb, and happy" too long, your search is going to be more difficult.

5. In either case, loosen up the eyeballs and start some first-class pilotage. DON'T OVERLOOK THE POSSIBILITY OF BEING LOST, YET RIGHT ON COURSE OR VERY NEARLY SO. First, look for something big. Don't concern yourself with the minute or trivial at this point unless nothing better is available. Often there will be linear features such as rivers, mountain ranges, or prominent highways and railroads easy to spot and identify. By turning either to the right or the left, you CAN'T miss them. If preferable, you can use them simply as references for orientation purposes. In other words, you can use them as landfalls or backstops, and thus find them of great value in fixing your approximate position.

## KNOW and ABIDE BY YOUR LIMITATIONS AND THOSE OF YOUR EQUIPMENT

Once you have utilized such features to the fullest extent possible, or if there are none available, you can use anything that might help -- don't pass up a thing. Double check all landmarks. Compare and analyze -- analyze and compare railroads and highways, topographical features (man-made and natural). Check for available air markers with names on 'em. But don't go down on the deck and stay there. As a general rule, it is both safer and easier at higher altitudes.

Remember this point. Be sure you have up-to-date charts, including those adjacent to the one in use. Everything which appears on the chart will usually be on the ground, but no standard chart is so detailed that everything you can see on the ground can also be found on the chart. If you either habitually or occasionally fly without suitable pilotage charts, you deserve to incur the full wrath of every goblin that ever hounded airmen. Every VFR flight is involved with pilotage!

One final word -- an ounce of preflight planning is worth far more than a pound of in-flight desperation. Sound, adequate preflight work will always pay off. It can prevent getting lost in the first place or at least simplify the task of fixing a position if you are "temporarily misplaced".

It can mean the difference between an enjoyable, satisfying experience and a palm-sweating ordeal.

Let's face it, intrepid airman, you may some day find yourself behind an airborne "eight-ball". Now, verily, this is unfortunate, but for the informed, assistance may be at hand. Though many may suffer from rude exposure to the rigors of navigation up "Trouble Creek", few need labor without a paddle. The "paddles", or aids, available are many and varied, especially for those with radio capability and knowledge of all the facilities available, to summon the willing, able help that is ready. Before we examine the "aids" available to the pilot in any emergency phase, let's define the latter term. Emergency phase simply means a situation involving distress, the need to resolve uncertainty, or to alert those able to help with a pressing problem.

Department of Transportation

FEDERAL AVIATION ADMINISTRATION

## VFR PILOT EXAM-O-GRAM° NO. 19

## EMERGENCY OR LOST PROCEDURES (RADIO)

### STATISTICS ON RADIO AIDS FOR PILOTS IN TROUBLE
(AS OF SEPT. 1962)

| | |
|---|---|
| 18,000 | FAA Controllers and Flight Service Specialists |
| 961 | Air/Ground Communication Channels |
| 700 | VORs and VORTACs |
| 340 | Flight Service Stations |
| 260 | Towers |
| 200 | Approach Control Facilities |
| 67 | Military VHF/DF Stations |
| 65 | FAA VHF/DF Stations |
| 10 | FCC HF/DF Stations |
| 68 | Airport Surveillance Radars |
| 49 | Long Range Radars |
| 28 | Precision Approach Radars |
| 36 | Air Route Traffic Control Centers |
| 20,784 | . . . Individuals and aids ready and willing to help!! |

NOTE: These radio aids should first be used to keep you out of trouble.

## THE MANY AIDS AVAILABLE TO YOU AS A PILOT

I. THE "U" AID. This is Old Faithful. If you panic or have not mastered this one, you are likely to find the other aids both difficult to apply and limited in utility. With this, you develop good habits of planning and performance. It is the only thing reliable if you find yourself in some of the situations outlined in this and the previous Exam-O-Gram. Only through the use of this aid will you become acquainted with those that follow.

II. THE RADIO COMMUNICATION AID. This will vary from the most sophisticated and expensive equipment to that which is barely adequate. Therefore, its utility will also vary. Regardless of this, a pilot should know the

A. MEANS OF DECLARING AN EMERGENCY.

1. Transmission of a Radio-Telephone Message.
2. Flying an Appropriate Triangular Pattern.

B. THREE ELECTRONIC MEANS OF OBTAINING ASSISTANCE.

1. Receipt of Radio-Telephone Message.
2. Direction Finding (D/F) Bearings.
3. Radar Detection of Triangular Patterns.

Radio Communication will generally prove to be the most useful aid. You may contact many FAA, military, or FCC Stations or facilities on the emergency frequency (121.5 MHz), or other assigned frequencies within the frequency and range capability of your equipment. When you do, utilize

### The Four "C" Procedure

CONFESS -- The doctors aren't likely to treat your troubles if U won't admit you've got 'em! DON'T WAIT TOO LONG!

COMMUNICATE -- if in doubt, shout"! But be prepared to do more than simply "holler" for help. You may need to give any or all of the following: identification, type of aircraft, estimated position, heading, estimated speed, altitude, fuel, nature of problem, and assistance desired. Clearly understand that the facility with which you have established contact may not be able to supply the type of assistance you need or seek, but if you communicate pertinent information, facility personnel should be able to alert those who can. Personnel at FSS stations are

trained to do so, and by voice directions alone, can get you back on course or to an airport. Often they can tell you how to use your equipment in order to get the assistance you need.
Even if a forced landing is imminent, try to communicate with somebody. The reasons for doing so should be obvious. If you cannot establish contact with a specific facility, you may communicate IN THE BLIND. Precede your message (3 times) with the word MAYDAY (distress), and simply transmit on your emergency frequency or available air-to-ground frequencies for reception by any facility or even other aircraft.

CLIMB if you can. Altitude improves VHF capability and increases likelihood of Radar and D/F identification.

COMPLY with the instructions of your ground contact. He is a trained specialist, anxious to help. If you cooperate, he usually can.

## III. THE DIRECTION FINDING (D/F) AID.

Types include both HF and VHF fixes and/or steers. In other words, they locate your position and give magnetic direction to fly to reach a desired destination.

Use is limited by: Radio equipment, your position in relation to facilities able to provide help, and the extent you are able to follow the four C's.
D/F stations and network include not only FAA, but FCC and military facilities. Those civilian airports capable of providing such aids are listed in the Airport/Facility Directory of the Airman's Information Manual, but when necessary, you may make the request of any station or tower you are capable of contacting. If possible, they will relay the request to the appropriate facility or station in the D/F net system. Check the Airman's Information Manual for information on stations, frequencies, and procedures.

## IV. THE RADAR AID.

Types of radar are classified as to function, e.g., Air Route Surveillance Radar, Airport Surveillance Radar, Precision Approach Radar, and Radar Beacons. There are over 140 FAA-operated units plus many military installations.
Recommended emergency procedures are covered in Part 1 (ATC Procedures) of the Airman's Information Manual. Assuming the four "C" factors are favorable, radar may provide assistance when an EMERGENCY SITUATION EXISTS in a general sense, or specifically when LOST with:

A. Radio transmitter and receiver both inoperative.

B. Radio transmitter alone functioning.

C. Radio receiver alone functioning.

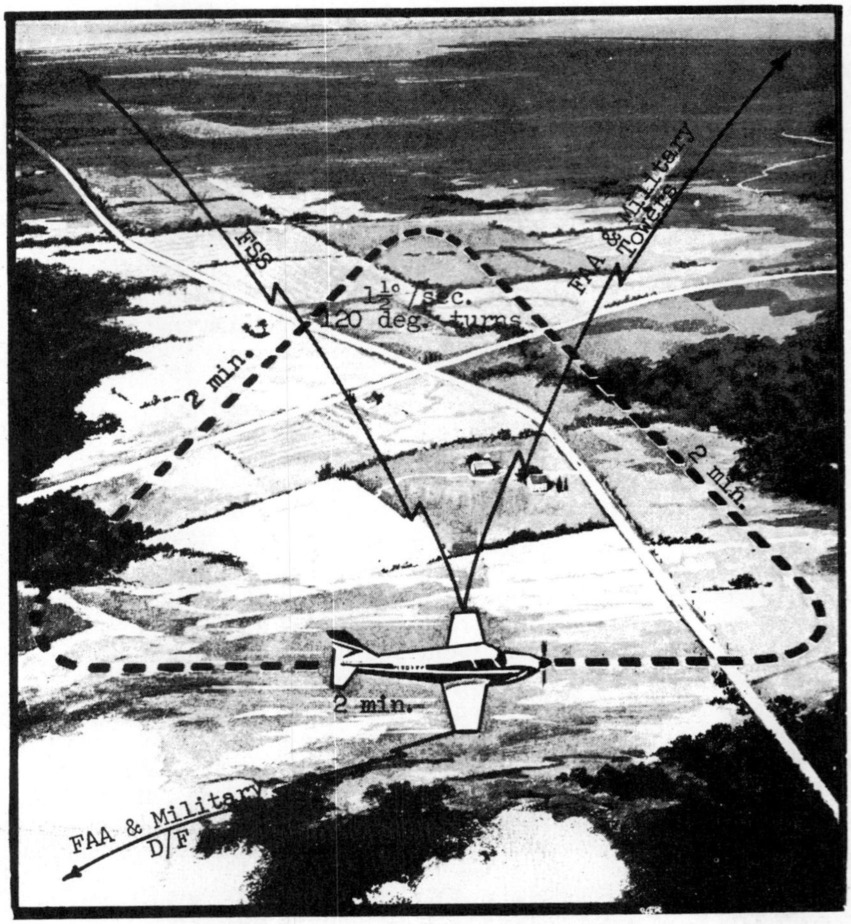


For situations A and B on page 3, fly the triangular pattern depicted below. For situation C (Radio Receiver Alone Functioning), follow the same procedure except make all turns to the right.

Fly a minimum of two patterns and repeat every twenty minutes until instructions are received on the emergency frequency (121.5 MHz), or interception and guidance by Air Rescue Service aircraft is accomplished. Note that for situations B (Radio Transmitter Alone Functioning) or C (Radio Receiver Alone Functioning), you might get help both by flying the radar triangle and by using radio . . . . Thus you may have three aids: Radio, Radar, and "U".

Space will not permit a detailed analysis of all procedures to follow in using these "aids". This is where the "U Aid" comes into the picture. You should get a current Airman's Information Manual. You should carefully study the section dealing with D/F, emergency, radio failure, and search and rescue procedures.

FEDERAL AVIATION AGENCY
VFR EXAM-O-GRAM *NO. 7

## GETTING CAUGHT ON TOP OF AN OVERCAST

1. DO YOU HAVE ALL OF THE FOLLOWING REQUISITES FOR INSTRUMENT FLIGHT? (1) An Instrument Rating, (2) an aircraft fully equipped for instrument flight, and (3) recent instrument experience. If not, you should heed the following bit of advice concerning flight over an extensive overcast cloud condition -- DON'T!

2. SHOULD YOU AVOID FLYING VFR OVER CLOUDS ENTIRELY? No. Many times it is both practical and desirable to select a cross-country cruising altitude above a scattered cloud condition to take advantage of smoother air, improved visibility, more favorable winds, or provide for more terrain and obstacle clearance, provided (1) you have legal cloud separation for climb, cruise, and destination descent, (2) weather conditions are stable or improving, and (3) you stay alert and take immediate action if the clouds beneath you increase and the "sucker holes" start to shrink. Don't wait too long to descend or make a 180° turn (one of aviation's oldest safety devices) if the situation warrants it.

3. WHAT OTHER PRECAUTIONS SHOULD YOU TAKE TO AVOID AN "ON-TOP-OF-AN OVERCAST" TRAP? (1) Prior to your cross-country flight, visit or telephone the local Weather Bureau Airport Station or your nearest FAA Flight Service Station or other flight service facility for a thorough weather briefing (see EXAM-O-GRAM No. 5). Select an altitude that will be compatible with terrain and cloud separation requirements. (2) Use reasonable restraint in exercising the prerogative of VFR flight when conditions are close to minimums. Remember that, with the right conditions, a low overcast can form under you in a matter of minutes. Consider the weather, the terrain you are flying over, and allow yourself a margin of safety commensurate with your experience level. (3) When you file your VFR flight plan with an FSS, request Flight Following Service in order to take advantage of special weather briefings from your en route designated Flight Watch Station(s) (see EXAM-O-GRAM No. 6). Also, monitor appropriate frequencies for scheduled weather broadcasts at 15 and 45 minutes past each hour.

4. WHAT SHOULD YOU DO IF YOU SHOULD GET CAUGHT ABOVE AN OVERCAST? You are admittedly "in a jam." Loss of orientation, a very probable sequel to loss of ground references, will further complicate your problem. However, you can improve your chances of avoiding disaster by following a few logical procedures. (See Airman's Information Manual - "Emergency Procedures".) For example, you should (1) establish communications with an FSS or other ground stations and confess your predicament. The personnel in these stations are well trained in assisting airmen in distress; give them a chance to help you before it's too late. If necessary, they can alert available VHF Direction Finding and Radar Stations (including military stations) to stand by for possible assistance. (2) If you have trouble establishing contact with a ground station, climbing will increase the range of your VHF radio equipment and improve the chances of ground radar detection. (3) Conserve your fuel by using an economical or maximum endurance power setting. (4) If you really need help, comply with instructions received from your ground station.

5. Prevention is a much better approach to this problem than the cure. If you are a VFR pilot, AVOID GETTING CAUGHT ON TOP OF AN OVERCAST!

* Exam-O-Grams are non-directive in nature and are issued solely as an information service to individuals interested in Airman Written Examinations.

## CHECK YOUR READING

1. What is the most effective tool in the aircraft for use in combatting emergencies?
2. What are the *four C's*?
3. What is the standard worldwide emergency frequency?
4. What is the standard worldwide distress signal for use over airborne radio?
5. When stating fuel quantity on board during an emergency, what units of measure should be used?
6. Why should we climb during emergency conditions (except for fire in flight)?
7. What should we do if unable to comply with emergency instructions?
8. Describe the procedure to be used for a DF steer.
9. What is the *first* item to be accomplished if the engine fails on takeoff?
10. If the engine fails at altitude, what should we do while gliding to the landing area?
11. If we detect an electrical fire while in flight and shut off all electrical equipment and the master switch, what effect will this action have on the engine operation?
12. Why must we shut off all unnecessary electrical equipment in the event of a generator failure?
13. What procedures must be accomplished in the event of engine overheating?
14. What procedures must be accomplished in the event of low oil pressure?
15. May military airports be used by civilian pilots in an emergency?

# PART IX

# FEDERAL AIR REGULATIONS

# 50

# FEDERAL AIR REGULATIONS

## Introduction

The Federal Air Regulations are our "rules of the road." They are, of course, written in legal language, which makes them difficult to understand. However, if we are to operate our aircraft safely we must be fully knowledgeable of all of the rules which pertain to our operation.

This text does not attempt to quote all of the Federal Air Regulations. To do so would require a very ponderous book. What is attempted is to quote many of the pertinent Federal Air Regulations, and, where the legal language is difficult to understand, to give a brief explanation of each regulation.

In some instances a portion of the regulation concerns only helicopter, glider, or seaplane pilots. In these cases, and where all of the regulation is not pertinent to student and private pilots, that portion will be omitted.

The Federal Air Regulation is printed first, followed by any comments deemed necessary. Italics are used by the author to emphasize a word or words and are not part of the regulation.

Deleted words are indicated by a series of dots (. . . ).

The author's comments which follow some of the regulations are preceded by the word "Note."

As student or private pilots, we will particularly need to know Federal Air Regulation Parts 1, 61, 71, 73, 91, 93, 99, and 43, and Department of Transportation Regulation 430.

*CAUTION:* Federal Air Regulations are frequently revised, deleted, and conceived. Due to their importance, we should order current Federal Air Regulations from the Government Printing Office, Washington, D.C.

## Federal Air Regulations Part I—Definitions & Abbreviations

Part I contains legal definitions of commonly used technical words found in the Federal Air Regulations.

***Administrator:*** the Administrator of the Federal Aviation Administration or any person to whom he has delegated his authority in the matter concerned.

***Aircraft:*** a device that is used or intended to be used for flight in the air.

Note: Because the definition of *aircraft* includes a device which is "*intended* to be used for flight," any regulation which mentions *aircraft* is pertinent to the machine that has *not yet flown*, such as a home-built or experimental machine.

***Aircraft Engine:*** an engine that is used or intended to be used in propelling aircraft, including engine appurtenances and accessories necessary for its functioning, but not including the propellers.

***Airplane:*** an engine-driven fixed-wing aircraft heavier than air, that is supported in flight by the dynamic reaction of the air against its wings.

Note: The *airplane* differs from the *aircraft* in that it is engine-driven, fixed-wing, and supported by the dynamic reaction of air against its wings. Therefore, a balloon for instance, is an *aircraft* but not an *airplane*.

***Airport:*** an area of land or water that is used or intended to be used for the landing and takeoff of aircraft, and includes its buildings and facilities, if any.

***Airport Traffic Area:*** unless otherwise specifically designated in Part 93, that airspace within a horizontal radius of 5 statute miles from the geographical center of any airport at which a control tower *is operating*, extending from the surface, up to, but not *including* 2000 feet above the surface.

Note: The airport traffic area exists only at an airport where there is a control tower *which is in operation*. Therefore, if the tower is closed during the nighttime hours, the Airport Traffic Area does not exist.

***Air Traffic:*** aircraft operating in the air or on the surface, exclusive of loading ramps and parking areas.

***Air Traffic Clearance:*** an authorization by Air Traffic Control for the purpose of *preventing collisions* between *known* aircraft, for an aircraft to proceed under specified traffic conditions within controlled airspace.

Note: An Air Traffic Clearance is *not* an instruction of how to fly. An air traffic controller has the responsibility of preventing collision of *known aircraft*. He is *not* responsible for *unknown* aircraft.

***Air Traffic Control:*** a service operated by an appropriate authority to promote the safe, orderly, and expeditious flow of air traffic.

Note: Again, no mention is made of issuing instructions on how to fly as part of the duties of Air Traffic Control.

***Approved:*** unless used with reference to another person, means approved by the administrator.

***Category:***

(1) as used with respect to the certification, ratings, privileges, and limitations of *airmen*, means a broad classification of aircraft. Examples include: airplane, watercraft, glider, and lighter-than-air.

(2) as used with respect to the operation of *aircraft*, means a grouping of aircraft based upon intended use or operating limitations. Examples include: transport, normal, utility, acrobatic, limited, restricted, and provisional. Note: *Category* has two meanings, which may be confusing if not completely understood. When used in connection with airman certification, ratings, privileges, and limitations, it means a broad classification of aircraft (not airplanes, for an airplane is *one kind of aircraft*). When used in connection with certification of aircraft, it refers to one grouping of *aircraft* based upon its use.

***Ceiling:*** the height *above the earth's surface* of the *lowest* layer of clouds or obscuring phenomena that is reported as "broken," "overcast," or "obscuration," and *not* classified as "thin" or "partial."

Note: Scattered clouds, thin cloud layers, or partial obscurations do *not* constitute a ceiling. Cloud heights are always reported in *height above the surface.*

***Class:***

(1) as used with respect to the certification, ratings, privileges, and limitations of *airmen*, means a classification of aircraft within a category having similar operating characteristics. Examples include: single engine, multiengine, land, water, gyroplane, helicopter, airship, and free balloon.

(2) as used with respect to the certification of *aircraft*, means a broad grouping of aircraft having similar characteristics of propulsion, flight, or landing. Examples include: airplane, rotorcraft, glider, balloon, landplane, and seaplane.

Note: *Class* also has two meanings. One refers to airmen and the other to aircraft.

***Flight Plan:*** specified information relating to the intended flight of an aircraft, that is filed orally or in writing with Air Traffic Control.

Note: The great advantage to be gained by filing a visual flight plan is that of alerting search and rescue if need be.

***Flight Time:*** the time from the moment the aircraft first moves under its own power for the purpose of flight until the moment it comes to rest at the next point of landing (*block to block* time).

Note: The definition specifically says that flight time begins when the aircraft begins to move for the purpose of flight, under its *own* power, and ends the moment the aircraft comes to rest. Therefore, taxiing from the ramp to the hangar is not considered flight time.

***Flight Visibility:*** the *average* forward horizontal distance, from the cockpit of an aircraft in flight, at which prominent unlighted objects may be seen and identified by day, and prominent lighted objects may be seen and identified by night.

***Large Aircraft:*** aircraft of more than 12,500 pounds maximum certificated takeoff weight.

***Medical Certificate:*** acceptable evidence of physical fitness on a form prescribed by the administrator.

***Night:*** the time between the end of evening civil twilight and the beginning of morning civil twilight, as published in the American Air Almanac, converted to local time.

***Pilot in Command:*** the pilot responsible for the operation and safety of an aircraft during flight.

Note: There is *no* exception to this rule. The pilot in command always carries full responsibility for the operation and safety of the aircraft.
***Prohibited Area:*** designated airspace within which the flight of an aircraft is prohibited.
***Restricted Area:*** airspace designated under Part 73 of this chapter within which the flight of an aircraft, while not wholly prohibited, is subject to restriction.
Note: In other words, flight in a restricted area is possible, but only within the specific restrictions. Restricted areas are shown on navigation charts and listed in the Airman's Information Manual.
***Small Aircraft:*** aircraft of 12,500 pounds or less, maximum certificated take-off weight.
***Traffic Pattern:*** the traffic flow that is prescribed for aircraft landing at, taxiing on, or taking off from, an airport.
Note: Specified taxi routes are part of the traffic pattern.
***Type:***

(1) as used with respect to the certification, ratings, privileges, and limitations of *airmen*, means a specific make and basic model of aircraft, including modifications thereto that do not change its handling or flight characteristics. Examples include: DC-9, B-707, DC-8.

(2) as used in respect to the certification of *aircraft*, means those aircraft which are similar in design. Example includes: DC-7 and DC-7C; F-27 and F27F; DC-8 and DC-8A.

Note: As was the case with category and class, the definition of *type* differs when applied to *airmen* certification and *aircraft* certification.

## CHECK YOUR READING

1. What is the difference between an *airplane* and an *aircraft* as far as Federal Air Regulations are concerned?
2. Describe an *airport traffic area*. When is it in force?
3. What is the function of Air Traffic Control?
4. When used in respect to *airman* certification, what does *category* mean?
5. When used in respect to an *airman* what does *class* mean?
6. When used in respect to an *airman* what does *type* mean?
7. Are cloud heights given in height above sea level or height above the surface?
8. What constitutes a *ceiling*?
9. Define *flight time*.
10. What is the difference between *prohibited* and *restricted* areas?

# 51

# FEDERAL AIR REGULATIONS

## Part 61

Part 61 of the Federal Air Regulations concerns the certification of pilots. We will discuss the more important regulations contained in subparts A, B, and C only, which concern student and private pilots.

### Subpart A—General

*FAR 61.3 Certificates and Ratings Required*

(a) Pilot Certificate: No person may act as a pilot in command or in any other capacity as a required pilot flight crewmember of a civil aircraft of U.S. registry unless he has in his *personal possession* a current pilot certificate issued to him under this part. However, when the aircraft is operated within a foreign country . . .

(b) Pilot Certificate; foreign aircraft . . .

(c) Medical Certificate: Except for glider pilots piloting gliders, no person may act as pilot in command or in any other capacity as a required pilot flight crewmember of an aircraft under a certificate issued to him under this part, unless he has in his *personal possession* an appropriate current medical certificate issued under Part 67 of this chapter. However, in the case of a pilot certificate issued under Part 61.33 . . .

(d) Flight Instructor Certificate . . .

(e) Inspection of Certificate: Each person who holds a pilot certificate or medical certificate shall present either or both for inspection upon the request of the administrator or an authorized representative of the National Transportation Safety Board, or of any Federal, state, or local law enforcement officer.

(f) Instrument Rating . . .

(g) Category II Pilot Authorization . . .

Note: Both the Pilot Certificate *and* the Medical Certificate must be in our personal possession at all times, available for inspection.

*FAR 61.9 Duration of Certificates*

(a) Student Pilot Certificate: A student pilot certificate expires at the end of the 24th month after the month it was issued.

(b) Flight Instructor Certificates . . .

(c) . . .

(d) Special Purpose Certificate . . .

(e) Other Pilot Certificates: Any pilot certificate issued under this Part and not covered by paragraphs (a) through (d) of this section is issued *without an expiration date.*

(f) Surrender, Suspension, Revocation: Any pilot certificate issued under this Part ceases to be effective if it is surrendered, suspended, or revoked.

(g) Return of Certificate: The holder of any certificate issued under this Part that is suspended or revoked shall, upon the administrator's request, return it to the administrator.

Note: Although the student pilot certificate expires on the last day of the 24th month after the month in which it was issued, the private pilot certificate is good until suspended or revoked.

*FAR 61.15 Aircraft Ratings*

(a) The *category* ratings to be placed on private, commercial, and airline transport pilot certificates are:

(1) Airplane
(2) Rotorcraft
(3) Glider, and
(4) Lighter-than-air

(b) When applicable, the airplane *class* ratings to be placed on private, commercial, and airline transport pilot certificates are:

(1) Single engine Land
(2) Multiengine Land
(3) Single engine Sea
(4) Multiengine Sea

Subparts (c) and (1) not applicable to this text.

*FAR 61.16 General Limitations*

(a) Type ratings required: No person may act as pilot in command of any of the following aircraft unless he holds a type rating for that aircraft:

(1) A large aircraft (except lighter-than-air)
(2) A helicopter . . .
(3) After March 31, 1966, a turbojet aircraft . . .

(b) Small aircraft; carrying another person or operating for compensation or hire. Unless he holds a *category* and *class* rating for that aircraft, a person may not act as pilot in command of a small aircraft that is carrying another person or is operated for compensation or hire. In addition, he may not act as pilot in command of that aircraft for compensation or hire.

(c) Small aircraft: Soloing not for compensation or hire. No person may act as pilot in command of a small aircraft in operations conducted other than under paragraph (b) of this section, unless he meets at least one of the following conditions:

(1) he holds a category and class rating appropriate to that aircraft.
(2) he has soloed and logged that flight time in that category and class of aircraft before December 16, 1965.
(3) He has made and logged at least three takeoffs and landings to a full stop in that category and class of aircraft, as the sole manipulator of the controls, while accompanied by a pilot who is entitled to carry passengers in that aircraft.
(4) He has made and logged at least three takeoffs and landings to a full stop while operating under an authorization issued by a Flight Standards District Office.

However, the holder of a pilot certificate with an airplane category rating may solo gliders without complying with this paragraph.

(d) Exception: This section does not require a class rating for gliders. In addition, the rating limitations of this section do not apply to:

(1) *The holder of a student pilot certificate;*
(2) The holder of a pilot certificate when operating an aircraft under authority of an experimental provisional type certificate;
(3) The holder of a pilot certificate when taking a flight test given by the Administrator; or
(4) . . .

Note: This Section of Part 61 spells out in detail the category, class, and type requirements of pilots. Notice that it does *not* apply to student pilots. Subpart B, FAR 61.61–61.73 will describe in detail the privileges of a student pilot.

A private pilot may *not* act as pilot in command (either solo or with passengers) unless he holds a category *and* class rating for the aircraft. A private pilot may *not* solo an aircraft unless he holds an appropriate category and class rating or has soloed and logged that flight time in that category and class before December 16, 1965, or has made three takeoffs and landings to a full stop in that category and class as the sole manipulator of the controls, when accompanied by another qualified pilot, or has made three takeoffs and landings to a full stop under an authorization by the district office. In other words, this regulation requires a private pilot to have a "check out" in the aircraft before soloing or carrying passengers in it.

*FAR 61.20 Written Tests: Cheating or Other Unauthorized Conduct*

(a) Except as authorized by the Administrator, no person may:

(1) copy or intentionally remove a written test under this part;
(2) give to another, or receive from another, any part or copy of that test;

(3) give help on that test to, or receive help on that test from any person during the period that test is being given;
(4) take any part of that test in behalf of another person;
(5) use any material or aid during the period that the test is being given; or
(6) intentionally cause, assist, or participate in any act prohibited by this paragraph.

(b) No person who commits an act prohibited by paragraph (a) of this section is eligible for any airman or ground instructor certificate or rating under this chapter for a period of one year after the date of that act. In addition, the commission of that act is a basis for suspending or revoking any airman or ground instructor certificate or rating held by that person.

Note: In short, this rule says "Don't Cheat." How foolish it would be to cheat, as we *must* know what we are doing in flight; and the tests we must take really are for our own protection. Notice that the law also punishes one who *receives* information from a test.

*FAR 61.39 Pilot Logbooks; Except Airline Transport Pilots*

(a) General: The flight time used to meet the experience requirements for any pilot certificate or rating, or to meet the recent flight experience requirements of 61.47, must be shown by a reliable record. The logging of other flight time is not required. Each pilot who keeps a logbook under this section shall enter into it the following information as to each flight:

(1) General
  (a) Date
  (b) Duration
  (c) Points of departure and arrival; and
  (d) Identification mark of the aircraft
(2) Type of Piloting Time
  (a) Pilot in command or solo
  (b) Second in command
  (c) Synthetic trainer; or
  (d) Flight instruction from an appropriately rated flight instructor, in the case of airplanes, gliders, or rotorcraft
(3) Conditions of flight
  (a) Day or night
  (b) Actual instrument flight, or
  (c) Simulated instrument flight

(b) Solo flight time: A student, private, or commercial pilot may log as solo only that flight time during which he is the *only* occupant of the aircraft.

(c) Pilot-in-command time: A private or commercial pilot may log as pilot in command only the flight time during which he is the only manipulator of the controls of an aircraft for which he is rated or the flight time during which he is the only occupant of the aircraft. A flight instructor . . .

(d) Second-in-command time: A private or commercial pilot may log as second in command that flight time during which he is performing the duties of a second in command. He may be credited with not more than 50 per cent of that kind of flight time toward the total flight time required for a higher certificate or rating. However, a private pilot may not be credited with more than 50 hours of flight time under this paragraph.

(e) Instrument time. . . .

(f) Inspection of pilot logbooks: A pilot who keeps a logbook under this section shall present it for inspection upon the request of, and after reasonable notice by, the administrator, an authorized representative of the National Transportation Safety Board, or any state or local law enforcement officer.

Note: In recent years the Federal Aviation Administration has become more particular concerning pilot logbooks. It would be wise for us to review the specific information required in the log, and make sure that the logbook is properly filled out. Although the law only requires the logging of pilot time which is used for recent experience requirements or experience to qualify for a pilot certificate or rating, it is wise to log all flight time so that the logbook becomes an accurate diary of flight.

*FAR 61.43 Medical Certificates: Duration*

(a) A first-class medical certificate expires at the end of the last day of

(1) the sixth month after the month in which it is issued, for operations requiring an airline transport pilot certificate;
(2) the 12th month after the month in which it is issued, for operations requiring only a commercial pilot certificate; and
(3) the 24th month after the month in which it is issued, for operations requiring only a private or student certificate or . . .

(b) A second-class medical certificate expires at the end of the last day of

(1) the 12th month after the month in which it is issued, for operations requiring a commercial pilot certificate; and
(2) the 24th month after the month in which it is issued, for operations requiring only a private or student pilot certificate or . . .

(c) A third-class medical certificate expires at the end of the last day of the 24th month after the month in which it is issued, for operations requiring a private or student pilot certificate or . . .

Note: The operation in which the holder of a medical certificate is involved determines the duration of the certificate. In other words, a first-class medical certificate is good until the *last day* of the 24th month after the month it was issued if used *only* for operations requiring a private or student pilot certificate. If the holder is involved in an operation requiring an airline transport certificate, the medical certificate expires on the last day of the 6th month after the month it was issued.

*FAR 61.45 Operation During Physical Deficiency*

No person may act as pilot in command or in any other capacity as a required pilot flight crewmember while he has a known physical deficiency, or increase of a known physical deficiency, that would make him unable to meet the physical requirements for his current medical certificate.

Note: This regulation makes it *unlawful* to fly with any known physical deficiency which would make us unqualified to pass our medical certificate physical. The section in this text concerning physiology discusses some of the effects of physical disabilities.

*FAR 61.47 Recent Flight Experience*

(a) General. No person may act as pilot in command of an aircraft carrying passengers unless, within the preceding 90 days, he has made at least five takeoffs and five landings to a full stop in an aircraft of the same category, class, and type. This section does not apply to operations requiring an airline transport pilot certificate, or to operations conducted under Part 135.

(b) Night experience. No person may act as pilot in command of an aircraft carrying passengers during the period beginning one hour after sunset and ending one hour before sunrise (as published in the American Air Almanac) unless, within the preceding 90 days, he has made at least five takeoffs and five landings to a full stop during that period of the day . . .

(c) . . .
(d) . . .
(e) . . .
(f) . . .
(g) . . .

Note: We must have made, within the preceding 90 days, at least five takeoffs and landings to a full stop in an aircraft of the same category, class, and type prior to carrying passengers (except airline transport and Part 135 pilots).

Notice that the nighttime experience paragraph specifies one hour after sunset and one hour before sunrise, even though *night* is officially from sunset to sunrise.

*FAR 61.51 Change of Address*

Within 30 days after any change in the permanent mailing address, the holder of a pilot or instructor certificate shall notify the FAA Airman Certification Branch, Oklahoma City, Oklahoma, in writing, of his new address.

## Subpart B—Student Pilots

*FAR 61.61 Eligibility Requirements: General (Effective May, 1967)*

(a) To be eligible for a student pilot certificate, a person must
- (1) be at least 14 years of age for a certificate authorizing him to pilot a glider, or at least 16 years of age for a certificate authorizing him to pilot any other aircraft;
- (2) be able to read, speak, and understand the English language, or have such operating limitations on his student pilot certificate as are necessary for the safe operation of aircraft, to be removed when he shows that he can read, speak, and understand the English language; and
- (3) hold at least a third-class medical certificate issued under Part 67 of this chapter within the preceding 24 months, or, in the case of gliders . . .

(b) Student pilot certificates are issued by the following:
   (1) FAA inspectors.
   (2) Designated pilot examiners.
   (3) Aviation medical examiners, when requested by persons being examined by them for medical certificates under Part 67 of this chapter. However, an aviation medical examiner may not issue a student pilot certificate if operating limitations are required under paragraph (a) (2) of this section.

*FAR 61.63 Requirements for Solo Flight*

(A) A student pilot may not operate an aircraft in solo flight until
   (1) He is familiar with the general and visual flight rules of Part 91 of this chapter;
   (2) He has had
      (a) In the case of an airplane category, flight instruction from a holder of a flight instructor certificate with an airplane rating in preparatory and flight procedures such as preflight inspection, starting, warming up, and operating and stopping the engine; in taxiing, takeoff, landing, and parking; in traffic pattern procedures; in level flight, turns, climbs, and glides; and in stalls and emergency landings.
      (b) In the case of a rotorcraft category . . .
      (c) In the case of a glider category, . . .
   (3) An appropriately rated flight instructor finds that the student has complied with this section and is otherwise able to make solo flights, and has endorsed his student certificate to that effect.
(B) In the case of a single-place gyroscope . . .
(C) This section does not apply to student lighter-than-air pilots.

*FAR 61.65 Airplane Operations: Flight Area Limitations*

A student pilot may not operate an airplane in solo flight outside of a local area designated by his flight instructor until

(a) he is familiar with flight planning elements such as plotting courses, estimating time enroute and fuel required, and obtaining and evaluating weather reports;

(b) he has received flight instruction (from a holder of a flight instructor certificate with an airplane rating) in
   (1) crosswind and simulated soft field takeoffs and landings,
   (2) climbing and gliding turns at minimum safe speeds,
   (3) cross country navigation by reference to aeronautical charts,
   (4) safe operating procedures in simulated emergencies such as engine failures, loss of flying speed, marginal visibility, deteriorating weather, getting lost, and similar critical situations,
   (5) conforming with air traffic control instructions by radio and lights,
   (6) the proper use of two-way radio communications and VFR navigational procedures and techniques,
   (7) level flight, turns, climbs, and glides by referring solely to flight instruments,

(8) the use of the magnetic compass; and

(c) his student pilot certificate is endorsed by an appropriately rated flight instructor who finds that the student has complied with this section and is otherwise able to make solo cross-country flights. If ground electronic equipment and navigation aids are not available within 100 miles of the base of operations, a synthetic trainer may be used for training in air traffic procedures, phraseology, and radio navigation. The instrument flight instruction required by paragraph (b) (7) of this section shall be given in an airplane equipped with at least a sensitive altimeter, turn and bank indicator, and a means of simulating instrument flight.

*FAR 61.73 General Limitations*

(a) Except as provided in paragraph (b) of this section, a student pilot may not act as pilot in command of an aircraft

(1) that is carrying a passenger;
(2) that is on an international flight;
(3) for compensation or hire;
(4) in furtherance of a business; or
(5) other than the make and model endorsed on his certificate by his certificated flight instructor.

(b) Notwithstanding paragraph (a) (2) of this section, a student pilot may make international flights for the purpose of solo cross-country training from the Haines, Gustavus, or Juneau Airports, Alaska, to White Horse, Yukon Territory, over the Province of British Columbia, and return.

(c) A student pilot who has not acted as pilot in command of a powered aircraft within the preceding 90 days, may not pilot such an aircraft in solo flight until he has passed a flight check given by a certified flight instructor who so endorses his student pilot logbook.

(d) A student pilot may not operate an airplane or rotorcraft in solo flight unless within the preceding 90 days

(1) he has received flight instruction in that category of aircraft from a certificated flight instructor with an appropriate rating on his flight instructor certificate;
(2) he has demonstrated to that flight instructor that he is competent to solo that category of aircraft; and
(3) that flight instructor has endorsed in the student pilot's logbook that he has given that flight instruction and found the student competent for solo flight. However, this paragraph does not apply if the student pilot meets the requirements of paragraph (c) of this section, has acquired the aeronautical experience required for a private pilot certificate, and obtains an endorsement by a flight instructor that the student pilot is considered competent to make solo flights without mandatory periodic flight checks.

(e) A student pilot may not operate an airplane or rotorcraft in solo cross-country flight until a certificated flight instructor with an appropriate rating on his flight instructor certificate has reviewed the student's preflight preparation and planning, determined that the student is competent to make the flight, and has so endorsed the student's pilot logbook. The student must carry that logbook on each solo cross-country flight. However, a student pilot may perform repeated solo cross-country flights over a specified course of not more than 50 miles in length, without an endorsement for each flight, if a certificated flight instructor with an appropriate rating on his flight instructor certificate has

(1) given him flight instruction over the course in both directions, and in takeoffs and landings at both landing areas involved; and
(2) found that the student is competent to make flights over the course without an authorization for each flight and has so endorsed the student's pilot logbook.

However, this paragraph does not apply if the student pilot has acquired the aeronautical experience required for a private pilot certificate and obtains an endorsement by a certificated flight instructor that the student pilot is considered competent to exercise solo cross-country privileges without a flight instructor's supervision.

Note: Student pilots may *not* carry passengers. A student pilot cannot make a cross-country flight unless a flight instructor has endorsed his logbook which must be carried on the flight [except as stated in paragraph (e) (2)].

## Subpart C—Private Pilots

*FAR 61.81 Eligibility Requirements: General*

To be eligible for a private pilot certificate, a person must

(a) be at least 16 years of age for a glider rating, . . . , and at least 17 years of age for any other rating; (airplane or rotorcraft)

(b) be able to read, speak, and understand the English language, or have such operating limitations placed on his pilot certificate as are necessary for the safe operation of aircraft, to be removed when he shows that he can read, speak, and understand the English language.

(c) hold at least a third-class medical certificate issued under Part 67 of this chapter within the preceding 24 months, or, in the case of a glider rating, . . .

(d) comply with the sections of this Part that apply to the rating he seeks.

*FAR 61.83 Aeronautical Knowledge*

An applicant for a private pilot certificate must pass a written test on

(a) the regulations of this chapter governing private pilot privileges and limitations, the general operating and air traffic rules, and the rules of the National Transportation Safety Board governing accident reporting;

(b) practical aspects of cross-country flying including, for an airplane or rotorcraft rating, flight planning, map reading, pilotage, the use of radio aids to VFR navigation, radio communication procedures, and emergency procedures;

(c) recognizing dangerous weather conditions and evaluating weather reports; and

(d) general safety practices in operating airplanes, rotorcraft, or gliders, as the case may be.

*FAR 61.85 Airplane Rating: Aeronautical Experience*

(a) An applicant for a private pilot certificate (airplane) must have had

(1) a total of at least 40 hours of flight instruction (from an appropriately rated flight instructor) and solo flight time;
(2) at least 20 hours of solo flight time of which at least 15 were in an airplane;
(3) except as provided in paragraph (b) of this section, at least 10 hours of solo cross-country flight time, during which each flight included a landing at a place more than 25 miles from the place of departure, and during which at least one flight included a landing at a place more than 100 miles from the place of departure [Effective: April 11, 1968], and;
(4) at least 3 hours of flight instruction, from a holder of a flight instructor certificate with an airplane rating, after his first solo cross-country flight, including a review of procedures and maneuvers previously learned, additional instruction in procedures and maneuvers required for the private pilot flight test, and flight instruction in primary flight maneuvers in controlling the airplane solely by reference to flight instruments.

The instrument instruction required by subparagraph (4) of this paragraph shall be given in an airplane equipped with at least a sensitive altimeter, turn and bank indicator, and a means of simulating instrument flight.

(b) An applicant who shows that he is located on an island from which the required flights cannot be accomplished without flying over water more than 10 miles . . .

(c) The pilot certificate issued to a person under paragraph (b) of this section . . .

(d) If an applicant for a private pilot certificate under paragraph (b) of this section . . .

(e) The holder of a private pilot certificate with the limitation or endorsement prescribed in paragraph (c) and (d) of this section or in paragraph (b) . . .

*FAR 61.87 Airplane Rating: Aeronautical Skill*

(A) General: An applicant for a private pilot certificate (airplane) must pass a practical test on the procedures and maneuvers listed in paragraph (b) of this section. The test is given in three phases, an oral operation test; a basic piloting techniques test; and a cross-country test.

(B) Procedures and maneuvers. The applicant must perform the following procedures and maneuvers competently.

(1) Phase I—Oral Operational Test.
  (a) Airplane registration, airworthiness, and equipment documents.

(b) Airplane log books and airworthiness inspection reports.
(c) Airplane performance, range, and operation.
(d) Airplane loading, including fuel, oil and baggage capacities.
(e) Airplane line check.
(f) Use of radio for voice communication.

(2) Phase II—Basic Piloting Technique Test.
(a) Preflight operations.
(b) Taxiing or sailing and docking.
(c) Normal and crosswind takeoffs and landings.
(d) Climbs, level flight, and descents at normal speeds and at minimum controllable speeds.
(e) Stalls from all normally anticipated flight attitudes, with and without power.
(f) 720° steep turns about a point.
(g) Wheel landings in tailwheel-type airplanes, and full stall landings in tricycle-type airplanes.
(h) Short field takeoff and power approach and landing.
(i) Soft field takeoff and landing.
(j) Slips and a slip to a landing (if a three-control airplane is used).
(k) Emergency operation of airplane equipment.
(l) Engine-out emergencies, if a multiengine aircraft is used, including
(i) maneuvering with one engine out (feathered if possible);
(ii) engine-out minimum control speed demonstration;
(iii) use of engine-out best rate-of-climb speed;
(iv) effect on engine-out performance of failing to feather, extension of gear and flaps, and combination of them; and
(v) approach and landing with an engine set for zero thrust, or the drag of a feathered propeller, as applicable.

(3) Phase III—Cross-country Test:
(a) Cross-country flight planning,
(b) Cross-country flying.
(c) Cross-country emergencies (lost, weather, overheating engine, power failure, etc.)
(d) Use of radio aids to VFR navigation.

(C) Instrument flight. In addition, during Phase III of the practical test, the applicant must demonstrate, in simulated instrument flight, his ability to safely control an aircraft manually solely by reference to instruments including the following:

(1) Recovery from the start of a power-on spiral.
(2) Recovery from the approach to a climbing stall.
(3) Normal turns of at least 180° left and right to within ±20° of a preselected heading.
(4) Shallow climbing turns to a predetermined altitude.

(5) Shallow descending turns at reduced power to a predetermined altitude.
(6) Straight and level flight.

*FAR 61.101 General Privileges and Limitations*

(A) Compensation or hire. Except as provided in subparagraphs (1) through (5) of this paragraph, a private pilot may not act as pilot in command of an aircraft that is carrying passengers or property for compensation or hire; nor may he, for compensation or hire, act as pilot in command of an aircraft.

(1) A private pilot may, for compensation or hire, act as pilot in command of an aircraft in connection with any business or employment if the flight is only incidental to that business or employment and the aircraft does not carry passengers or property for compensation or hire.
(2) A private pilot may share the operating expenses of a flight with his passengers.
(3) A private pilot who is an aircraft salesman and who has at least 200 hours of logged flight time may demonstrate an aircraft in flight to a prospective buyer.
(4) A private pilot may act as pilot in command of an aircraft used in a passenger-carrying airlift sponsored by a charitable organization, and for which the passengers make a donation to the organization, if
  (a) the sponsor of the airlift notifies the FAA General Aviation District Office having jurisdiction over the area concerned, at least 7 days before the flight, and furnishes any essential information that the office requests;
  (b) the flight is conducted from a public airport adequate for the aircraft used, or from any other airport that has been approved for the operation by an FAA inspector;
  (c) he has logged at least 200 hours of flight time;
  (d) no acrobatic or formation flights are conducted;
  (e) each aircraft used is certified in the standard category and complies with the 100-hour inspection requirement of FAR 91.169 of this chapter; and
  (f) the flight is made under VFR during the day.

For the purpose of this subparagraph (4), a "charitable organization" means an organization listed in Publication No. 78 of the Department of the Treasury called the "Cumulative List, Organization Described in Section 170(c) of the Internal Revenue Code of 1954," as amended from time to time by published supplemental lists.

(5) A private pilot (lighter-than-air). . . .
  (b) Lighter-than-air . . .

Note: As stated in this regulation, a private pilot may not act as pilot in command of an aircraft for compensation. Certain exemptions are granted in subparagraph (1) through (5).

## CHECK YOUR READING

1. When operating as pilot in command, must we have a medical certificate in our personal possession, or should we keep it as part of a permanent file in a safe place?
2. To whom must we show our pilot certificate?
3. What is the duration of a student pilot certificate?
4. If a student pilot certificate is issued on April 1, 1970, on what date will it expire?
5. What is the duration of a private pilot certificate?
6. List the category ratings to be placed on a private pilot certificate.
7. List the class ratings to be placed on a private pilot certificate.
8. What ratings, if any, must a private pilot hold in order to carry passengers as pilot in command of a small aircraft?
9. May a private pilot act as pilot in command for compensation?
10. Why is it important to keep an accurate logbook?
11. May a private pilot log second-in-command time? How much of that flight time may be credited to a private pilot?
12. What date will a First Class medical certificate expire, if held by a private pilot who received it on June 20, 1970?
13. What is the duration of a Third Class medical certificate?
14. A private pilot suffers from vertigo or dizziness due to medication. May he legally act as pilot in command?
15. List the requirements which must be fulfilled prior to carrying passengers as pilot in command at night.
16. What constitutes *night* regarding pilot experience?
17. What is the minimum age for a student pilot?
18. After one endorsement by a flight instructor, may a student pilot operate an airplane in solo flight outside the "local area"?
19. May a student pilot carry passengers? If so, list the occurrences when he may carry passengers.
20. What is the minimum age for a private pilot?
21. What medical certificate may a private pilot hold?
22. May a private pilot share the operating expense of an airplane with his passengers?
23. May a private pilot demonstrate an aircraft in flight to a prospective buyer? If so, list the requirements prior to doing so.
24. May a private pilot fly on actual instruments as pilot in command?
25. May a private pilot fly for compensation as pilot in command in connection with a business? If so, list the limitations.

# 52

# FEDERAL AIR REGULATIONS

**Parts 71 & 73**

Part 71 of the Federal Air Regulations contains information concerning the designation of Federal Airways, Controlled Airspace, and Reporting Points.

Part 73 defines Special Use Airspace.

*FAR 71.5 Extent of Federal Airways*

(a) Each Federal airway is based on a centerline that extends from one navigational aid or intersection to another navigational aid (or through several navigational aids or intersections) specified for that airway.

(b) Unless otherwise specified in Subpart B or C of this part

(1) each Federal airway includes the airspace within parallel boundary lines 4 miles each side of the centerline. Where an airway changes direction, it includes that airspace enclosed by extending the boundary lines of the airway segments until they meet.

(2) Where the changeover point for an airway segment is more than 51 miles from either of the navigational aids defining that segment, and . . .

(c) Unless otherwise specified in Subpart B or C of this part

(1) each Federal airway includes that airspace extending upward from 700 feet (until designated from 1,200 feet or more) above the surface of the earth to, but not including, 18,000 feet MSL, except that Federal airways for Hawaii have no upper limits. Variations of the lower limits of an airway are expressed in digits

representing hundreds of feet above the surface (AGL) or mean sea level (MSL) and, unless otherwise specified, apply to the segment of an airway between adjoining navigational aids or intersections; and

(2) the airspace of a Federal airway within the lateral limits of a transition area has a floor coincident with the floor of the transition area.

(d) One or more alternate airways may be designated between specified navigational aids or intersections along each VOR Federal airway described in Subpart C of this part. Unless otherwise specified, the centerline of an alternate VOR Federal airway and the centerline of the corresponding segment of the main VOR Federal airway are separated by 15 degrees.

(e) A Federal airway does not include the airspace of a prohibited area.

Note: A Federal airway is 8 nautical miles wide; 4 nautical miles either side of the centerline. Unless otherwise designated, the airway extends upward from 700 feet *above the surface*, up to but not including 18,000 feet MSL.

*FAR 71.7 Control Areas*

Control areas consist of the airspace designated in Subparts B, C, and E of this part, but do not include the continental control area. Unless otherwise designated, control areas include the airspace between a segment of a main VOR Federal airway and its associated alternate segments with the vertical extent of the area corresponding to the vertical extent of the related segment of the main airway.

*FAR 71.9 Continental Control Area*

The continental control area consists of the airspace of the 48 contiguous states, the District of Columbia, and Alaska south of latitude 68°00′00″ N., excluding the Alaska peninsula west of longitude 160°00′00″ W., at and above 14,500 feet MSL, but does not include

(a) the airspace less than 1,500 feet above the surface of the earth; or

(b) prohibited and restricted areas, other than restricted area military climb corridors and the restricted areas listed in subpart D of this Part.

*FAR 71.11 Control Zones*

The control zones listed in subpart F of this part consist of controlled airspace extending upward from the surface of the earth. A control zone may include one or more airports and is normally a circular area with a radius of 5 miles and any extensions necessary to include instrument approach and departure paths.

Note: The control zone begins at the surface and extends upward to infinity. It includes one or more airports and is normally a circular area with a radius of 5 miles, with extensions for instrument approaches and departures. The control zones are depicted on the sectional chart by outlining with black dashed lines, often looking like a keyhole.

*FAR 71.13 Transition Areas*

The transition areas listed in subpart G of this Part consist of controlled airspace extending upward from 700 feet or more above the surface of the earth when designated in conjunction with an airport for which an approved instrument approach procedure has been prescribed; or from 1,200 feet or more above the surface of the earth when designated in conjunction with airway route structures or segments. Unless otherwise specified, transition areas terminate at the base of the overlying controlled airspace.

*FAR 71.15 Positive Control Areas*

The positive control areas listed in Subpart H of this part consist of controlled airspace within which there is positive control of aircraft.

Note: Figure 174 depicts a cross-section view of the control zone, airport traffic area, positive control area and airways.

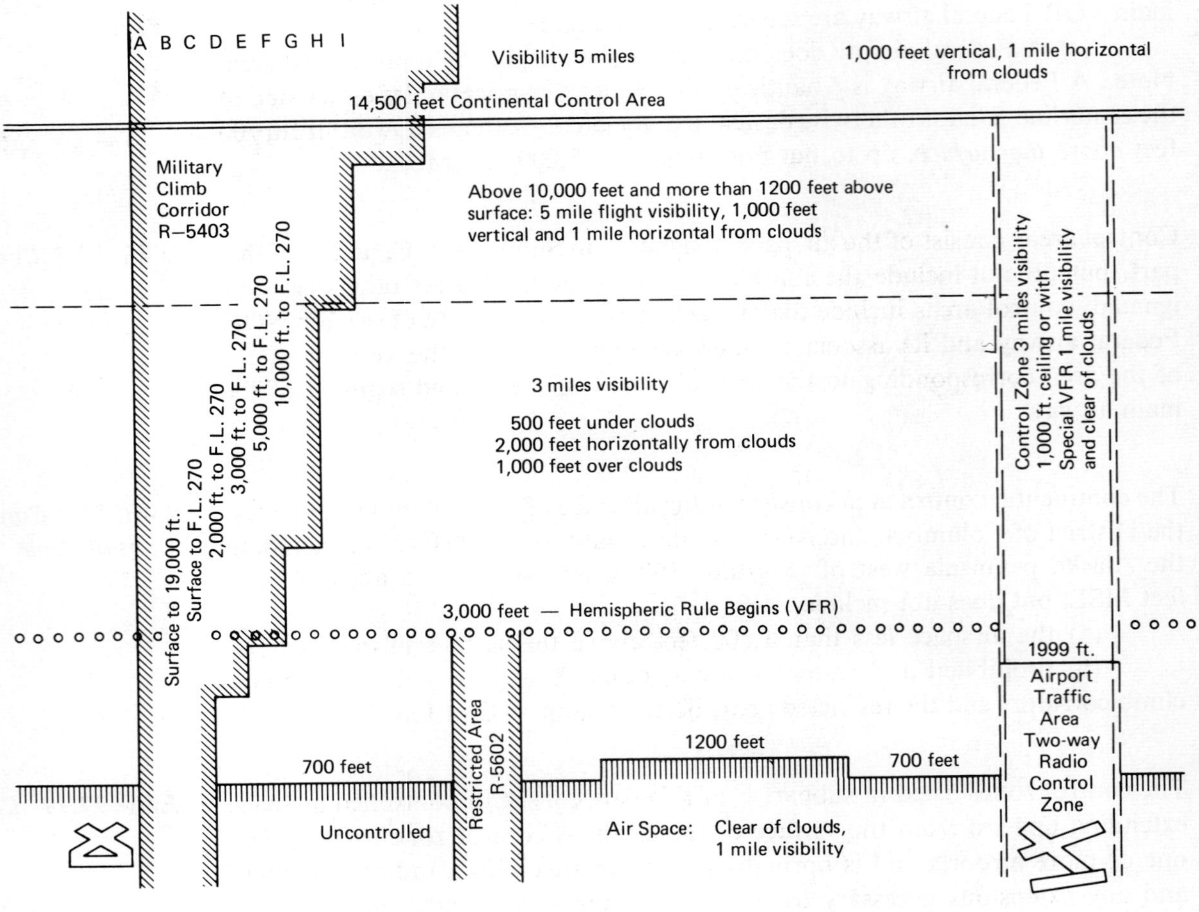


Fig. 174. Cross section of United States airspace.

*FAR 73.1 Applicability*

The airspace that is described in Subpart B and Subpart C of this part is designated as special use airspace. This part describes the requirements for the use of that airspace.

*FAR 73.3 Special Use Airspace*

(a) Special use airspace consists of airspace of defined dimensions identified by an area on the surface of the earth wherein activities must be confined because of this nature, or wherein limitations are imposed upon aircraft operations that are not a part of those activities, or both.

(b) The vertical limits of special use airspace are measured by designated altitude floors and ceilings expressed as flight levels or feet above mean sea level.

(c) The horizontal limits of special use airspace are measured by boundaries described by geographic coordinates or other appropriate references that clearly define their perimeter.

(d) The period of time during which a designation of special use airspace is in effect is stated in the designation.

Note: Many areas which are designated as "special use airspace" are used for gunnery practice, missile testing, or other very hazardous activities. We must use *extreme* caution to avoid these areas when they are active. The Airman's Information Manual lists the areas of special use airspace and defines their use and active time.

*FAR 73.5 Bearings; Radials; Miles*

(a) All bearings and radials in this part are *true* from point of origin.

(b) Unless otherwise specified, all mileages in this part are stated as *statute miles.*

## Subpart B—Restricted Areas

*FAR 73.11 Applicability*

This subpart designates restricted areas and prescribes limitations on the operation of aircraft within them.

*FAR 73.13 Restrictions*

No person may operate an aircraft within a restricted area between the designated altitudes and during the time of designation, unless he has the advance permission of

(a) the using agency described in 73.15; or

(b) the controlling agency described in 73.17.

*FAR 73.15 Using Agency*

(a) For the purposes of this subpart, the following are using agencies:

(1) The agency, organization, or military command whose activity within a restricted area necessitated the area being so designated.

(2) In the case of a Restricted Area/Military Climb Corridor that does not have a designated controlling agency, the Military Air Traffic Control facility may be contacted for permission for transit through the climb corridor.

(b) Upon the request of the FAA, the using agency shall execute a letter establishing procedures for joint use of a restricted area by the using agency and the controlling agency, under which the using agency would notify the controlling agency whenever the controlling agency may grant permission for transit through the restricted area in accordance with the terms of the letter.

(c) The using agency shall

(1) schedule activities within the restricted area;

(2) authorize transit through, or flight within, the restricted area as feasible; and
(3) contain within the restricted area all activities conducted therein in accordance with the purpose for which it was designated.

*FAR 73.17 Controlling Agency*

For the purposes of this part, the controlling agency is the FAA facility that may authorize transit through or flight within a restricted area in accordance with a joint-use letter issued under 73.15.

**Subpart C—Prohibited Areas**

*FAR 73.81 Applicability*

This subpart designates prohibited areas and prescribes limitations on the operation of aircraft therein.

*FAR 73.83 Restrictions*

No person may operate an aircraft within a prohibited area unless authorization has been granted by the using agency.

*FAR 73.85 Using Agency*

For the purpose of this subpart, the using agency is the agency, organization, or military command that established the requirement for the prohibited area.

Department of Transportation
FEDERAL AVIATION ADMINISTRATION
**VFR PILOT EXAM-O-GRAM® NO. 41**

CONTROLLED AIRSPACE (SERIES 1)

Incorrect answers selected by many applicants in FAA Airman Written Examinations indicate that knowledge of controlled airspace and the application of pertinent Federal Aviation Regulations is a very weak area in pilot education. This weakness is further confirmed in investigations of accidents and violations. Just as there are traffic laws and restrictions regulating the operation of automobiles in certain areas or zones, there are regulations governing the operation of aircraft within designated airspace. These regulated areas are established only in the interest of safety. To apply the regulations pertinent to a particular airspace, the pilot must first identify and determine whether he is within that airspace. However, unlike streets and highways where automobile operation is restricted, airspace with special flight restrictions has no posted signs or markers. Consequently, in locating controlled airspace, you must visualize it by correlating the chart symbols with ground reference points. Where the vertical or horizontal limits are not shown on the chart (for example, the limits of an airport traffic area), a knowledge of the boundaries is essential.

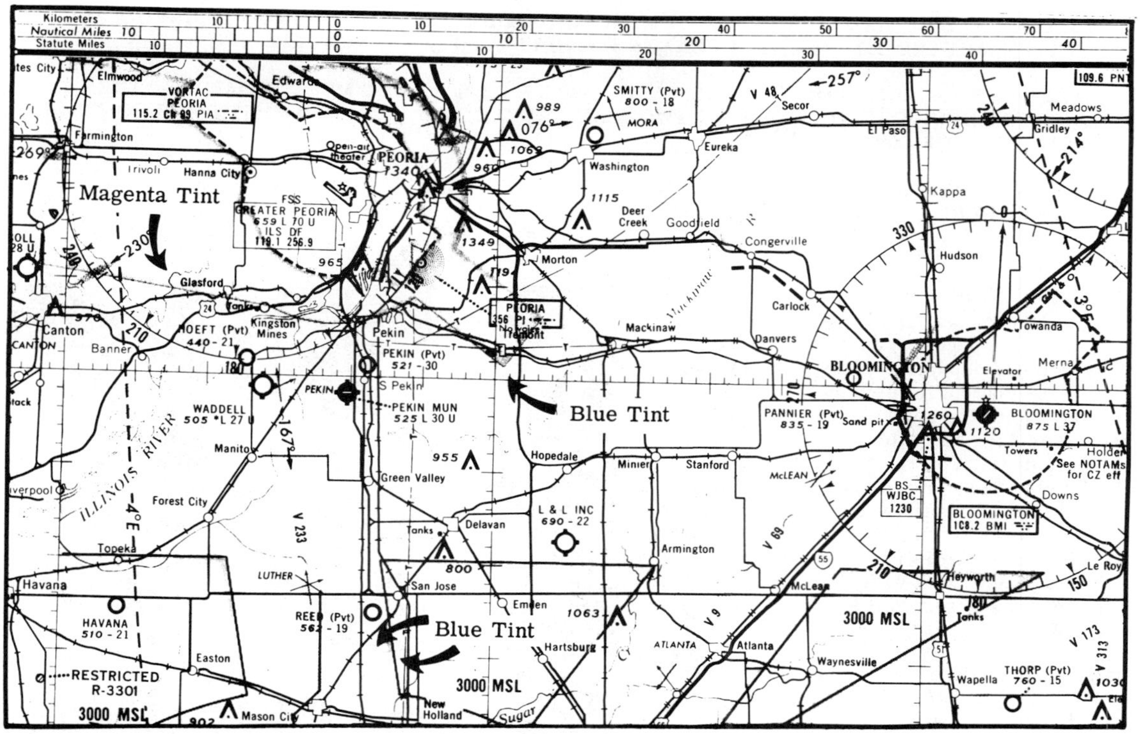


To assist in locating and identifying various controlled airspace, this Exam-O-Gram attempts to show in three-dimensional perspective, only that airspace closely related to VFR flights. While the dimensions of the airspace are indicated, the illustrations are not drawn in true proportions. An accurate and complete description of each airspace will be found in FAR Part 1, Definitions and Abbreviations, or Part 71, Designation of Federal Airways, Controlled Airspace and Reporting Points.

As you read this Exam-O-Gram relate each illustration to the Sectional Chart excerpt above; then consult the applicable regulation referenced in the accompanying text. Although each airspace has its individual boundaries and rules, you should remember that frequently one overlaps or is collocated with another, as in the case of an airport having an airport traffic area within a control zone. Appearing with each illustration are review questions dealing with VFR flights on the above chart segment. Check your answers against the answers and explanations given in the next Exam-O-Gram (No. 42).

AIRPORT TRAFFIC AREA - airspace within 5 statute miles of an airport at which a control tower is in operation. This airspace extends from the ground up to but not including 2,000 feet above ground level (AGL). The boundaries are not depicted on aeronautical charts but the presence of an airport traffic area is indicated when tower radio frequencies are shown in the airport data box. (Letters CT appear on new series Sectional Charts.) While this area is not designated as controlled airspace, the radio communications and operating requirements affect all flights taking off and landing at those airports.

Reference: FAR 1, 91.85, 91.87.

(Use chart excerpt on first page)

1. IS THERE AN AIRPORT TRAFFIC AREA AT GREATER PEORIA AND BLOOMINGTON AIRPORTS?

2. IF YOU HAVE NO RADIO, AND THE CEILING AND VISIBILITY ARE UNLIMITED (CAVU), ARE YOU PERMITTED TO LAND AT GREATER PEORIA? AT BLOOMINGTON?

3. WHEN OVERFLYING GREATER PEORIA AIRPORT AT 3,000 FEET MSL IN CAVU WEATHER, ARE YOU REQUIRED TO MAINTAIN TWO-WAY RADIO COMMUNICATIONS WITH THE CONTROL TOWER? AT 2,500 FEET MSL?

4. WHAT IS THE MAXIMUM AUTHORIZED SPEED FOR RECIPROCATING ENGINE AIRCRAFT WHEN APPROACHING GREATER PEORIA FOR LANDING?

<u>CONTROL ZONE</u> - airspace that is normally a circular area with a 5 mile radius and extensions as necessary for instrument approach and departure paths at certain airports. (A control tower may or may not be located within the control zone.) It starts at the surface, and extends up to the base of the Continental Control Area. The lateral limits are shown on aeronautical charts by a dashed line. Specific weather minimums are required for VFR flight within this zone, and an ATC clearance is required when the weather is less than VFR minimums. Certain other flight restrictions are also in effect within this airspace.

Reference: FAR 1, 71.11, 91.71, 91.105, 91.107.

(Use chart excerpt on first page)

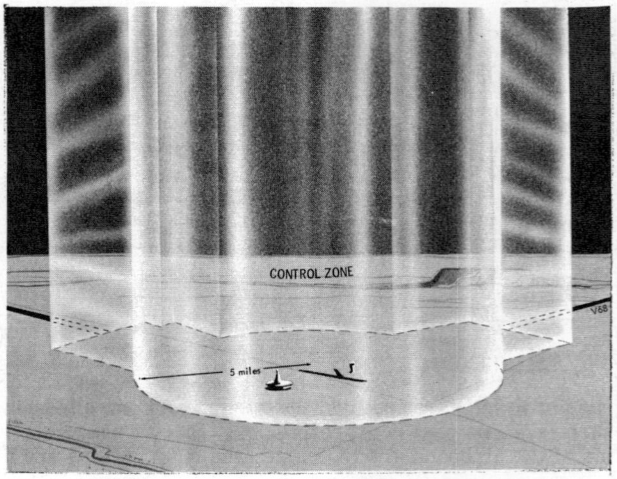


1. WHAT WEATHER CONDITIONS MUST EXIST TO LAND OR TAKE OFF AT BLOOMINGTON WITHOUT OBTAINING A SPECIAL VFR CLEARANCE? AT GREATER PEORIA?

2. WITH WHOM WOULD YOU COMMUNICATE TO RECEIVE A SPECIAL VFR CLEARANCE AT GREATER PEORIA? AT BLOOMINGTON?

3. AT WHAT ALTITUDE ARE YOU PERMITTED TO PRACTICE STALLS, SPINS, CR ACROBATICS OVER BLOOMINGTON AIRPORT?

VFR – No. 41

<u>CONTROL AREA</u> - designated airspace starting at 700 feet above ground level (unless a higher altitude is specified) and extending upward to the base of the overlying Continental Control Area (14,500 feet MSL). The lateral limits, normally identical to Federal Airways but frequently including far-reaching extensions, are shown on Sectional Charts with colored boundaries. Those areas bounded by magenta (purplish-red) tint start at 700 feet AGL, while those bounded by blue tint start at 1,200 feet AGL (unless a higher altitude is specified). The dark edges of tinted borders indicate the outer boundary of the airspace. VFR flights within control areas require certain weather minimums.

Reference: FAR 1, 71.7, 71.163, 91.105.

(Use chart excerpt on first page)

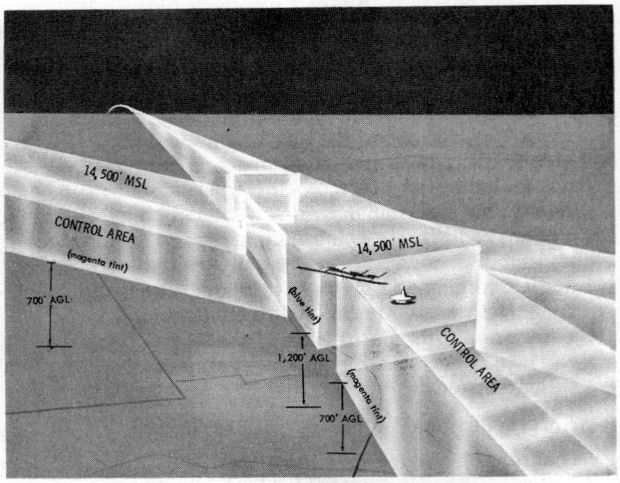


1. WHEN OVER THE TOWN OF MACKINAW (CENTER OF CHART) AT 3,000 FEET ABOVE THE GROUND, ARE YOU WITHIN CONTROLLED AIRSPACE?

2. WHEN FLYING IN THE IMMEDIATE VICINITY OF MACKINAW AT 15, 500 FEET MSL WOULD YOU BE IN THE CONTROL AREA?

3. WHAT ARE THE VISIBILITY AND CLOUD CLEARANCE MINIMUMS WHEN FLYING VFR AT 9, 500 FEET MSL IN THE IMMEDIATE VICINITY OF MACKINAW?

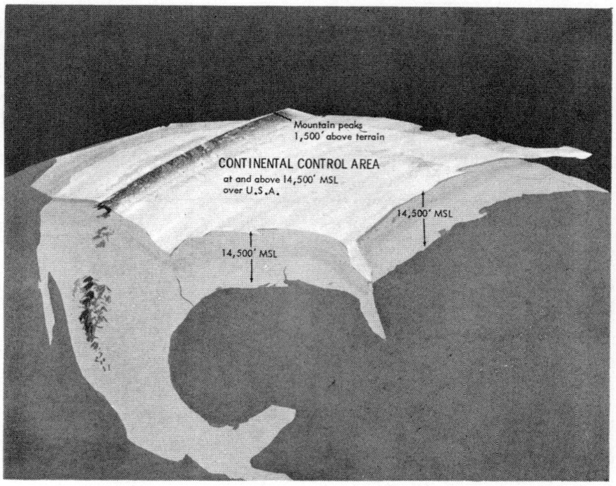


CONTINENTAL CONTROL AREA - Consists of the airspace of the 48 contiguous states, the District of Columbia, and Alaska south of latitude 68 degrees North, excluding the Alaska peninsula west of Longitude 160° West, at and above 14, 500 MSL, but does not include (a) the airspace less than 1500 feet above the surface; and (b) certain prohibited or restricted areas.

Reference: FAR 1, 71. 9, 91. 105.

(Use chart segment on first page)

1. WHAT VISIBILITY IS REQUIRED TO FLY VFR AT 16, 500 FEET MSL ON V9 V69 AIRWAY?

2. IF THERE ARE CLOUDS AT 17, 000 FEET MSL, HOW FAR BELOW THE CLOUDS SHOULD YOU FLY ON V9 V69 AIRWAY VFR?

3. IF YOU ARE OUTSIDE THE LATERAL LIMITS OF AN AIRWAY AT 15, 500 FEET MSL, WHAT ARE THE MINIMUM WEATHER CONDITIONS FOR VFR FLIGHT?

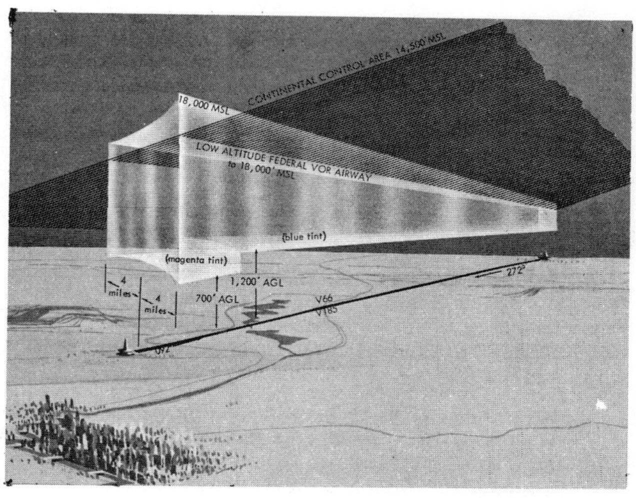


FEDERAL AIRWAY - each airway is based on a centerline that extends from one navigational aid or intersection to another navigational aid specified for that airway. The centerline is shown on aeronautical charts, and the magnetic course and the airway identity are indicated. Each airway includes airspace within parallel boundary lines normally 4 nautical miles each side of the centerline, and extends from 1, 200 feet above the surface (unless a higher altitude is indicated) upward to, but not including 18, 000 feet MSL. Airways in Hawaii have no upper limits. Inasmuch as a Federal Airway is controlled airspace divided at 10, 000 feet MSL, VFR flight within the airway requires distinct weather minimums above and below 10, 000 feet. Certain flight maneuvers are restricted in this airspace.

Reference: FAR 1, 71. 5, 91. 71, 91. 105.

(Use chart excerpt on first page)

1. HOW FAR EAST OR WEST OF WADDELL AIRPORT (SOUTHWEST OF PEORIA) SHOULD YOU GO TO PRACTICE STALLS, SPINS, OR ACROBATICS?

2. IF YOU ARE FLYING VFR AT 15, 500 FEET MSL ON V9 V69 AIRWAY (SOUTHWEST OF BLOOMINGTON), WHAT ARE THE MINIMUM VISIBILITY AND CLOUD CLEARANCE REQUIREMENTS? AT 3, 000 FEET MSL?

3. WHAT MINIMUM VISIBILITY IS REQUIRED TO FLY VFR FROM WADDELL AIRPORT TO PEKIN MUNICIPAL AIRPORT AT 1, 100 FEET MSL?

ANSWERS WITH EXPLANATIONS ARE PRESENTED IN EXAM-O-GRAM NO. 42.


FAA Aeronautical Center
Flight Standards Technical Division
Operations Branch
P. O. Box 25082
Oklahoma City, Oklahoma 73125

Department of Transportation

FEDERAL AVIATION ADMINISTRATION

# VFR PILOT EXAM-O-GRAM® NO. 42

## CONTROLLED AIRSPACE (SERIES 2)

Illustrations and brief descriptions of airspace in which certain controls are imposed on VFR flights, along with references to pertinent Regulations, are presented in Exam-O-Gram No. 41. In addition, questions relating to a segment of a selected aeronautical chart are posed with the answers omitted, to inspire complete study of controlled airspace. Those questions are restated and the correct answers are briefly discussed in this Exam-O-Gram as an additional service to the reader.

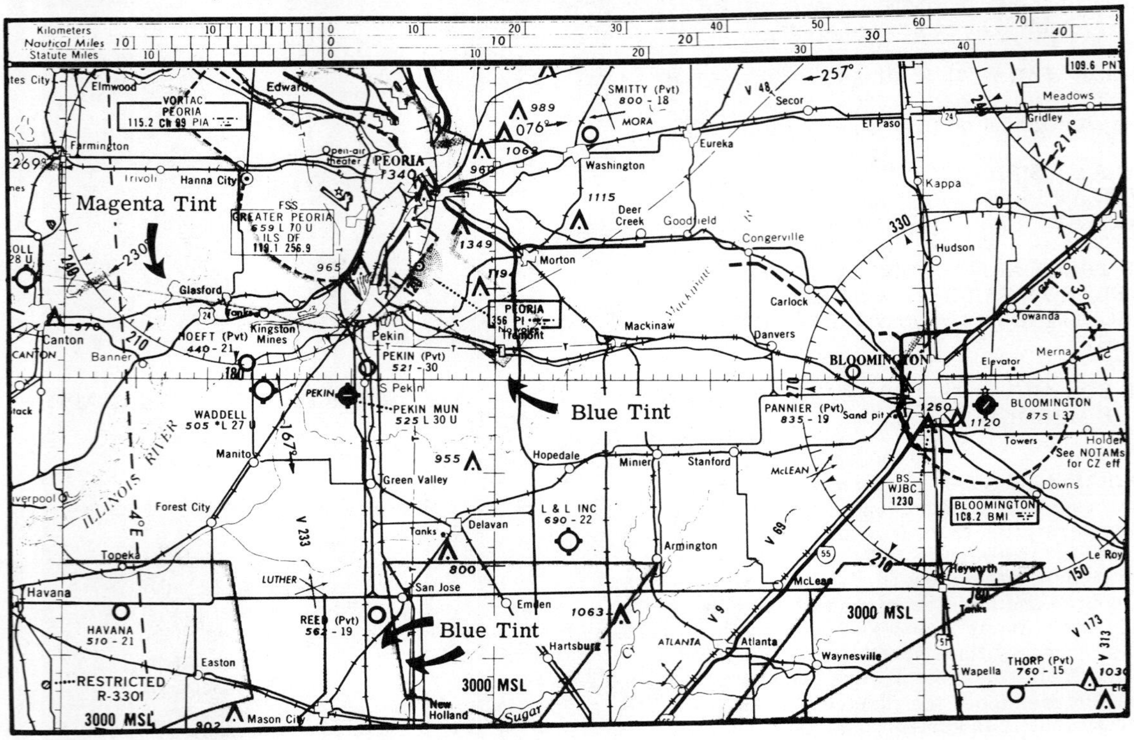


## AIRPORT TRAFFIC AREA

1. IS THERE AN AIRPORT TRAFFIC AREA AT GREATER PEORIA AND BLOOMINGTON AIRPORTS?

   There is an Airport Traffic Area at Greater Peoria, but not at Bloomington. Remember that Airport Traffic Areas exist only at airports where control towers are in operation. In checking the airport data box for Greater Peoria, you will note radio frequencies, indicating the existence of a control tower and therefore an Airport Traffic Area when the control tower is in operation. The lack of frequencies at Bloomington Airport indicates there is no tower, so there is no Airport Traffic Area at that airport. Two-way radio communications are normally required for takeoff and landing at Greater Peoria, but not at Bloomington during VFR weather.

2. IF YOU HAVE NO RADIO, AND THE CEILING AND VISIBILITY ARE UNLIMITED (CAVU), ARE YOU PERMITTED TO LAND AT GREATER PEORIA? AT BLOOMINGTON?

   Yes. Because Greater Peoria has an Airport Traffic Area, radio communications are normally required for takeoff and landing. However, Regulations provide that if the radio becomes inoperative during flight, a landing may be made if the ceiling and visibility are at least 1,000 feet and 3 miles, after obtaining clearance to land by light signal (green light) from the tower.

Similarly, an aircraft not equipped with radio may land if prior approval (by telephone) is obtained from the control tower. Inasmuch as Bloomington has no control tower, (therefore no Airport Traffic Area) there are no requirements for radio communications or prior approval for landing when the weather meets basic VFR minimums.

3. WHEN OVERFLYING GREATER PEORIA AIRPORT AT 3, 000 FEET MSL IN CAVU WEATHER, ARE YOU REQUIRED TO MAINTAIN TWO-WAY RADIO COMMUNICATIONS WITH THE CONTROL TOWER? AT 2, 500 FEET MSL?

   At 3, 000 feet MSL - no; at 2, 500 feet MSL - yes. Airport Traffic Areas extend from the surface up to but not including 2, 000 feet above the ground. Since the airport elevation is 659 feet, you would be above the Airport Traffic Area when at 3, 000 feet MSL and would not be required to communicate with the tower. (If the flight visibility is less than 3 miles, however, communications would be required since this is also a Control Zone. ) At 2, 500 feet MSL, you would be in the Airport Traffic Area, since this altitude is within 2, 000 feet of the ground, and you would be required to maintain two-way communications with the tower.

4. WHAT IS THE MAXIMUM AUTHORIZED SPEED FOR RECIPROCATING ENGINE AIRCRAFT WHEN APPROACHING GREATER PEORIA FOR LANDING?

   Any time you are flying below 10, 000 feet MSL, your indicated airspeed should be no more than 250 knots (288 mph). When within the Airport Traffic Area, your airspeed should be no more than 156 knots (180 mph).

## CONTROL ZONE

1. WHAT WEATHER CONDITIONS MUST EXIST TO LAND OR TAKE OFF AT BLOOMINGTON WITHOUT OBTAINING A SPECIAL VFR CLEARANCE? AT GREATER PEORIA?

   At least a 1, 000 foot ceiling and 3 miles visibility at either of these airports. Note on the chart that both Bloomington and Greater Peoria airports lie within a Control Zone. A special VFR clearance is not required to operate within a Control Zone if the ceiling is at least 1, 000 feet and the visibility is at least 3 miles. When the weather is below these minimums, but not less than one mile visibility, a special VFR clearance from ATC (Air Traffic Control) is required prior to operating in the Control Zone. (Visibility less than one mile requires an instrument clearance. )

2. WITH WHOM WOULD YOU COMMUNICATE TO RECEIVE A SPECIAL VFR CLEARANCE AT GREATER PEORIA? AT BLOOMINGTON?

   With Peoria Tower at Greater Peoria, and Bloomington Radio at Bloomington Airport. To separate and coordinate IFR traffic and special VFR traffic, appropriate clearances are issued by the ATC (Air Traffic Control) facility serving each airport. At Greater Peoria these clearances are normally obtained through the control tower or approach control. Since a control tower is not located at Bloomington, clearances should be obtained by communicating with Bloomington Radio (controlled by Peoria FSS) who would obtain and relay the clearances from ATC.

3. AT WHAT ALTITUDE ARE YOU PERMITTED TO PRACTICE STALLS, SPINS, OR ACROBATICS OVER BLOOMINGTON AIRPORT?

   Above 14, 500 feet MSL only. Regulations prohibit the performance of those maneuvers within a Control Zone (as well as an airway). Bloomington Airport lies within a Control Zone and since this zone extends upward to the base of the Continental Control Area, those maneuvers would not be permitted below 14, 500 feet over Bloomington Airport.

## CONTROL AREA

1. WHEN OVER THE TOWN OF MACKINAW (center of the chart) AT 3, 000FEET ABOVE THE GROUND, ARE YOU WITHIN CONTROLLED AIRSPACE?

   Yes. At first it may appear on the chart that Mackinaw does not lie within controlled airspace. However, remember that the outer limit of controlled airspace is shown by the darker and more definite edges of the tinted boundaries and the vanishing, or feathered edges show the direction of the controlled airspace. Since the darker edges of the nearby controlled airspace boundaries face away from Mackinaw, and the vanishing edges face toward Mackinaw, it is apparent that this town lies in a Control Area. This particular Control Area starts at 1, 200 feet above the ground (boundaries in blue tint); therefore at 3, 000 feet above the ground you would be within the Control Area.

2. WHEN FLYING IN THE IMMEDIATE VICINITY OF MACKINAW AT 15, 500 FEET MSL, WOULD YOU BE IN THE CONTROL AREA?

No. Control Areas extend from 700 feet or 1, 200 feet (as designated) above the ground, upward to the base of the Continental Control Area. Since the Continental Control Area starts at 14, 500 feet MSL, you would be in this overlying area (instead of the lower Control Area) when at 15, 500 feet MSL.

3. WHAT ARE THE VISIBILITY AND CLOUD CLEARANCE MINIMUMS WHEN FLYING AT 9, 500 FEET MSL IN THE IMMEDIATE VICINITY OF MACKINAW?

3 miles visibility and 1, 000 feet above or 500 feet below and 2, 000 feet horizontally from any cloud formation. We have established that the airspace surrounding Mackinaw between 1, 200 feet above the ground and 14, 500 feet MSL is a Control Area. Therefore, on a VFR flight at 9, 500 feet MSL, you must have at least 3 miles visibility and remain at least 1, 000 feet above or 500 feet below, and 2, 000 feet horizontally from clouds.

## CONTINENTAL CONTROL AREA

1. WHAT VISIBILITY IS REQUIRED TO FLY VFR AT 16, 500 FEET MSL ON V 9 V 69 AIRWAY?

At least 5 miles. Remember that when operating at or above 14, 500 feet MSL you are in the Continental Control Area, where visibility requirements are greater than those for a Control Area. Therefore, on a VFR flight at 16, 500 feet MSL on V 9 V 69, visibility must be at least 5 miles.

2. IF THERE ARE CLOUDS AT 17, 000 FEET MSL, HOW FAR BELOW THE CLOUDS SHOULD YOU FLY ON V 9 V 69 AIRWAY VFR?

At least 1, 000 feet. When on an airway below 14, 500 feet MSL you are in a Control Area where only 500 feet clearance beneath the clouds is required. However, if flying above 14, 500 feet MSL you are in Continental Control Area and are required to have at least 1, 000 feet separation beneath the clouds.

3. IF YOU ARE <u>OUTSIDE THE LATERAL LIMITS</u> OF AN AIRWAY AT 15, 500 FEET MSL, WHAT ARE THE <u>MINIMUM WEATHER CONDITIONS</u> FOR VFR FLIGHT?

At least 5 miles flight visibility, since a flight at or above 14, 500 feet MSL, on or off an airway, is within the Continental Control Area. Additionally, you cannot fly VFR at less than 1, 000 feet above or 1, 000 feet below and 1 mile horizontally from any cloud formation.

## FEDERAL AIRWAYS

1. HOW FAR EAST OR WEST OF WADDELL AIRPORT (SOUTHWEST OF PEORIA) SHOULD YOU GO TO PRACTICE STALLS, SPINS, OR ACROBATICS?

At least 5 miles east or 3 miles west. Waddell Airport lies within the 8-mile width of V 233 airway and Regulations prohibit the performance of those maneuvers within a Federal Airway. To be outside the limits of the airway you must fly 5 miles east or 3 miles west of Waddell.

2. IF YOU ARE FLYING VFR AT 15, 500 FEET MSL ON V 9 V 69 AIRWAY (SOUTHWEST OF BLOOMINGTON) WHAT ARE THE MINIMUM VISIBILITY AND CLOUD CLEARANCE REQUIREMENTS? AT 3, 000 FEET MSL?

Flight at 15, 500 feet MSL, on or off airways, is within the Continental Control Area, where you must have at least 5 miles visibility and remain at least 1, 000 feet above or 1, 000 feet below and 1 mile horizontally from any cloud formation. When flying at 3, 000 feet MSL on the airway you are in a Control Area and VFR flight requires that you have at least 3 miles visibility and remain at least 1, 000 feet above or 500 feet below and 2, 000 feet horizontally from any cloud formation.

3. WHAT MINIMUM VISIBILITY IS REQUIRED TO FLY FROM WADDELL AIRPORT TO PEKIN MUNICIPAL AIRPORT AT 1, 100 FEET MSL?

One mile visibility. Although a Federal airway lies between Waddell Airport and Pekin Municipal, and both airports are within the horizontal boundaries of a Control Area, the base of an airway or control area is never less than 700 feet above the surface. The elevation of the two airports and the terrain between them is approximately 500 feet, so at 1, 100 feet MSL, you would be beneath the controlled airspace. Therefore, this flight can be made at 1, 100 feet MSL under VFR with as little as 1 mile visibility.

○○○○○

** NOTE: Since the original printing of this Exam-O-Gram, the airspace at and above 18, 000 feet MSL in the area involved in this particular chart segment, has been designated Positive Control Area. Flights under Visual Flight Rules are prohibited above 18, 000 feet MSL in this area. (Northeast and north central U. S. A.)

VFR – No. 42

## CHECK YOUR READING

1. What is the width of a Federal airway?
2. Unless otherwise specified, at what height does the airway begin?
3. How high does an airway extend?
4. May a Federal airway include the airspace of a prohibited area?
5. At what height does a control zone begin?
6. Where may we find specific information about a special use airspace?
7. May a pilot fly in a restricted area? If so, list the requirements prior to doing so.
8. When may a pilot fly in a prohibited area?
9. At what height does the continental control area begin?
10. At what altitude does a control zone end?

# 53

# FEDERAL AIR REGULATIONS

**Part 91—General Operating & Flight Rules**
**Part 93—Special Air Traffic Rules and Airport Traffic Patterns**

Part 91 describes the General Operating and Flight Rules. They are the "rules of the road" for aviation.

## Subpart A—General

*FAR 91.1 Applicability*

(a) Except as provided in paragraph (b) of this section, this part prescribes rules governing the operation of aircraft (other than moored balloons, kites, unmanned rockets, and unmanned free balloons) within the United States.

(b) Each person operating a civil aircraft of U.S. registry outside of the United States shall

(1) when over the high seas, comply with Annex 2 (Rules of the Air) to the Convention on International Civil Aviation;
(2) when within a foreign country, comply with the regulations relating to the flight and maneuver of aircraft there in force; and
(3) except for 91.15 (b), 91.17, 91.38, and 91.43, comply with Subparts A and C of this part so far as they are not inconsistent with

applicable regulations of the foreign country where the aircraft is operated or Annex 2 to the Convention on International Civil Aviation.

Note: *All* aircraft (except moored balloons, kites, unmanned rockets, and unmanned free balloons) within the U.S. must operate within Part 91 regulations.

*FAR 91.3 Responsibility and Authority of the Pilot in Command*

(a) The pilot in command of an aircraft is directly responsible for, and is the final authority as to, the operation of that aircraft.

(b) In an emergency requiring immediate action, the pilot in command may deviate from any rule of this subpart or of Subpart B to the extent required to meet that emergency.

(c) Each pilot in command who deviates from a rule under paragraph (b) of this section shall, upon the request of the Administrator, send a written report of that deviation to the Administrator.

Note: As stated in this rule, the pilot in command is directly responsible for, and is final authority as to, the operation of that aircraft. Any deviation from Subpart A or B may require a letter to the Administrator upon his request.

*FAR 91.5 Preflight Action*

Each pilot in command shall, before beginning a flight, familiarize himself with all available information concerning that flight. This information must include, for a flight under IFR or a flight not in the vicinity of an airport, available weather reports and forecasts, fuel requirements, alternatives available if the planned flight cannot be completed, and any known traffic delays of which he has been advised by ATC.

Note: We learned that preflight action is good common sense. This rule makes it unlawful to delete preflight actions.

*FAR 91.9 Careless or Reckless Operation*

No person may operate an aircraft in a careless or reckless manner so as to endanger the life or property of another.

*FAR 91.10 Careless or Reckless Operation Other Than for the Purpose of Air Navigation*

No person may operate an aircraft other than for the purpose of air navigation, on any part of the surface of an airport used by aircraft for air commerce (including areas used by those aircraft for receiving or discharging persons or cargo), in a careless or reckless manner so as to endanger the life or property of another.

*FAR 91.11 Liquor and Drugs*

(a) No person may act as a crewmember of a civil aircraft while

(1) under the influence of intoxicating liquor; or

(2) using any drug that affects his faculties in any way contrary to safety.

(b) Except in an emergency, no pilot of a civil aircraft may allow a person who is obviously under the influence of intoxicating liquors or drugs (except a medical patient under proper care) to be carried in that aircraft.

*FAR 91.13 Dropping Objects*

No pilot in command of a civil aircraft may allow any object to be dropped from that aircraft in flight that creates a hazard to persons or property. However, this section does not prohibit the dropping of any object if reasonable precautions are taken to avoid injury or damage to persons or property.

*FAR 91.27 Civil Aircraft Certificates Required*

(a) No person may operate a civil aircraft of U.S. registry unless it has within it

(1) an appropriate and current airworthiness certificate (including a special flight permit or an authorization under 91.45); and
(2) a registration certificate issued to its owner.

(b) No person may operate a civil aircraft unless the airworthiness certificate required by paragraph (a) of this section is displayed at the cabin or cockpit entrance so that it is legible to passengers or crew.

Note: This regulation specifies two papers which must be in the aircraft (airworthiness certificate and registration certificate). Regulation 91.31 specifies that the Operations Limitations must also be in the aircraft. Local or state regulations may specify other papers which must be on board.

*FAR 91.29 Civil Aircraft Airworthiness*

(a) No person may operate a civil aircraft unless it is in an airworthy condition.

(b) The pilot in command of a civil aircraft is responsible for determining whether that aircraft is in condition for safe flight. He shall discontinue the flight when unairworthy, mechanical, or structural conditions occur.

*FAR 91.31 Civil Aircraft Operating Limitations and Marking Requirements*

(a) Each person operating a civil aircraft shall comply with the operating limitations for that aircraft prescribed by the certificating authority of the country of registry.

(b) No person may operate a U.S. registered civil aircraft unless there is available in the aircraft a current FAA approved Aircraft Flight Manual for that aircraft, placards, listings, instrument markings, or any combination thereof, containing each operating limitation prescribed for that aircraft by the administrator, including the following:

(1) Power plant (e.g., rpm, manifold pressure, gas temperature, etc.)
(2) Airspeeds (e.g., normal operating speed, flaps extended speed, etc.)
(3) Aircraft weight, center of gravity, and weight distribution, including the composition of the useful load in those combinations and ranges intended to insure that the weight and center of gravity position will remain within approved limits (e.g., combinations and ranges of crew, oil, fuel, and baggage).
(4) Minimum flight crew.
(5) Kinds of operation.
(6) Maximum operating altitude.

(7) Maneuvering flight load factors.
(8) Rotor speed (for rotorcraft).
(9) Limiting height-speed envelope (for rotorcraft).

(c) No person may operate a U.S. registered civil aircraft unless that aircraft is identified in accordance with Part 45.

Note: Placards, instrument markings, etc. may contain operations limitations.

*FAR 91.33 Powered Civil Aircraft with Standard Category U. S. Airworthiness Certificate; Instrument and Equipment Requirement*

(a) General. Except as provided in paragraphs (c)(3) and (e) of this section, no person may operate a powered civil aircraft with a standard category U.S. airworthiness certificate in any operation described in paragraphs (b) through (f) of this section unless that aircraft contains the instruments and equipment specified in those paragraphs (or FAA-approved equivalents) for that type of operation, and those instruments and items of equipment are in operable condition.

(b) Visual flight rules (day). For VFR flight during the day the following instruments and equipment are required:

(1) Airspeed indicator.
(2) Altimeter.
(3) Magnetic direction indicator.
(4) Tachometer for each engine.
(5) Oil pressure gauge for each engine using pressure system.
(6) Temperature gauge for each liquid-cooled engine.
(7) Oil temperature gauge for each air-cooled engine.
(8) Manifold pressure gauge for each altitude engine.
(9) Fuel gauge indicating the quantity of fuel in each tank.
(10) Landing gear position indicator, if the aircraft has a retractable landing gear.
(11) If the aircraft is operated for hire over water and beyond power-off gliding distance from shore, a Very Pistol, and approved flotation gear readily available to each occupant.
(12) Approved safety belts for all occupants. The rated strength of each safety belt shall not be less than that corresponding with the ultimate load factors specified in the current applicable aircraft airworthiness requirements, considering the dimensional characteristics of the safety belt installation for the specific seat or berth arrangement. The webbing of each safety belt shall be replaced as required by the administrator.

(c) Visual flight rules (night). For VFR flight at night the following instruments and equipment are required:

(1) Instruments and equipment specified in paragraph (b) of this section.
(2) Approved position lights.
(3) On large aircraft or when required by the aircraft's airworthiness certificate, an approved anticollision light system. In the event of failure of any light of the anticollision light system, operations with the aircraft may be continued to a stop where repairs or replacement can be made without undue delay.

(4) If the aircraft is operated for hire, one electric landing light.

(5) An adequate source of electrical energy for all installed electrical and radio equipment.

(6) One spare set of fuses, or three spare fuses of each kind required.

(d) Instrument Flight Rules. . .

(e) Flight at or above 24,000 MSL. . .

(f) Category II Operation. . .

Note: This regulation specifies that an approved safety belt is required for *all* occupants. This includes small children. An electric landing light is required *only* on aircraft operated for hire at night.

**Subpart B—Flight Rules**

*FAR 91.61 Applicability*

This subpart prescribes flight rules governing the operation of aircraft within the United States.

*FAR 91.63 Waivers*

(a) The administrator may issue a certificate of waiver authorizing the operation of aircraft in deviation of any rule of this subpart if he finds that the proposed operation can be safely conducted under the terms of that certificate of waiver.

(b) An application for a certificate of waiver under this section is made on a form and in a manner prescribed by the Administrator and may be submitted to any FAA office.

(c) A certificate of waiver is effective as specified in that certificate.

*FAR 91.65 Operating Near Other Aircraft*

(a) No person may operate an aircraft so close to another as to create a collision hazard.

(b) No person may operate an aircraft in formation flight except by arrangement with the pilot in command of each aircraft in the formation.

(c) No person may operate an aircraft carrying passengers for hire, in formation flight.

(d) Unless otherwise authorized by ATC, no person operating an aircraft may operate his aircraft in accordance with any clearance or instruction that has been issued to the pilot of another aircraft for radar Air Traffic Control purposes.

*FAR 91.67 Right of Way Rules; Except Water Operations*

(a) General. Except when, because of restriction to visibility beyond the pilot's control, another aircraft cannot be seen, each person operating an aircraft shall comply with this section. When a rule of this section gives another aircraft the right of way, he shall give way to that aircraft and may not pass over, under, or ahead of it, unless well clear.

(b) In distress. An aircraft in distress has the right of way over all other traffic.

(c) Converging. When aircraft of the same category are converging at approximately the same altitude (except head-on or nearly so) the aircraft to the other's right has the right of way. If the aircraft are of different categories

(1) a balloon has the right of way over any other category of aircraft;

(2) a glider has the right of way over an airship, airplane or rotorcraft; and

(3) an airship has the right of way over an airplane or rotorcraft.

However, an aircraft towing or refueling other aircraft has the right of way over all other engine-driven aircraft.

(d) Approaching head-on. When aircraft are approaching each other head-on, or nearly so, each pilot of each aircraft shall alter course to the right.

(e) Overtaking. Each aircraft that is being overtaken has the right of way and each pilot of an overtaking aircraft shall alter course to the right to pass well clear.

(f) Landing. Aircraft, while on final approach to land, or while landing, have the right of way over other aircraft in flight or operating on the surface. When two or more aircraft are approaching an airport for the purpose of landing, the aircraft at the lower altitude has the right of way; but it shall not take advantage of this rule to cut in front of another which is on final approach to land, or to overtake that aircraft.

(g) Inapplicability. This section does not apply to the operation of an aircraft on water.

Note: Since the right-of-way rules are important to us, both during everyday operations and on written examinations, we should review them often.

*FAR 91.70 Aircraft Speed (Effective: Dec. 15, 1967)*

(a) No person may operate an aircraft below 10,000 feet MSL at an indicated airspeed of more than 250 knots (288 mph).

(b) Unless otherwise authorized or required by ATC, no person may operate an aircraft within an airport traffic area at an indicated airspeed of more than

(1) in the case of a reciprocating engine aircraft, 156 knots (180 mph); or

(2) in the case of a turbine-powered aircraft, 200 knots (230 mph).

However, if the minimum safe airspeed for any particular operation is greater than the maximum speed prescribed in this section, the aircraft may be operated at that minimum speed.

Note: Paragraph (b) refers to an "airport traffic area." Remember that this area extends from the surface up to but not including 2,000 feet above the surface, and is only in operation when the control tower is in operation.

*FAR 91.71 Acrobatic Flight*

(A) No person may operate an aircraft in acrobatic flight

(1) over any congested area of a city, town, or settlement;

(2) over an open-air assembly of persons;

(3) within a control zone or Federal airway

(4) below an altitude of 1,500 feet above the surface, or

(5) when flight visibility is less than three miles.

For the purposes of this paragraph, acrobatic flight means an intentional maneuver involving an abrupt change in acceleration, not necessary for normal flight.

(B) Unless each occupant of the aircraft is wearing an approved parachute, no pilot of a civil aircraft, carrying any person (other than a crewmember) may execute any intentional maneuver that exceeds

(1) a bank of 60 degrees relative to the horizon; or

(2) a nose-up or nose-down attitude of 30 degrees relative to the horizon.

(C) Paragraph (b) of this section does not apply to

(1) flight tests for pilot certification or rating; or

(2) spins and other flight maneuvers required by the regulations for any certificate or rating when given by

(a) a certificated flight instructor; or

(b) an airline transport pilot instructing in accordance with FAR 61.163.

*FAR 91.73 Aircraft Lights*

No person may, during the period from sunset to sunrise (or, in Alaska, during the period a prominent unlighted object cannot be seen from a distance of three statute miles, or the sun is more than six degrees below the horizon)

(a) operate an aircraft unless it has lighted position lights;

(b) park or move an aircraft in, or in dangerous proximity to, a night flight operations area of an airport unless the aircraft

(1) is clearly illuminated;

(2) has lighted position lights; or

(3) is in an area which is marked by obstruction lights; or

(c) anchor an aircraft unless the aircraft

(1) has lighted anchor light; or

(2) is in an area where anchor lights are not required on vessels.

Note: Although night-flight time begins one hour after sunset and ends one hour before sunrise, nighttime for aircraft lighting begins at sunset and ends at sunrise.

*FAR 91.75 Compliance with ATC Clearances and Instructions*

(a) When an ATC clearance has been obtained, no pilot in command may deviate from that clearance, except in an emergency, unless he obtains an amended clearance. However, except in positive controlled airspace, this paragraph does not prohibit him from cancelling an IFR flight plan if he is operating in VFR weather conditions.

(b) Except in an emergency, no person may, in an area in which air traffic control is exercised, operate an aircraft contrary to an ATC instruction.

(c) Each pilot in command who deviates, in an emergency, from an ATC clearance or instruction shall notify ATC of that deviation as soon as possible.

(d) Each pilot in command who (though not deviating from a rule of this subpart) is given priority by ATC in an emergency, shall, if requested by ATC, submit a detailed report of that emergency within 48 hours to the chief of that ATC facility.

*FAR 91.79 Minimum Safe Altitudes: General*

Except when necessary for takeoff or landing, no person may operate an aircraft below the following altitudes:

(a) Anywhere. An altitude allowing, if a power unit fails, an emergency landing without undue hazard to persons or property on the surface.

(b) Over congested area. Over any congested area of a city, town, or settlement, or over any open air assembly of persons, an altitude of 1,000 feet above the highest obstacle with a horizontal radius of 2,000 feet of the aircaft.

(c) Over other than congested areas. An altitude of 500 feet above the surface, except over open water or sparsely populated areas. In that case, the aircraft may not be operated closer than 500 feet to any person, vessel, vehicle, or structure.

(d) Helicopters. . .

Note: "Buzzing" or low flying is against the law. It is also one of the most dangerous of all flight maneuvers, particularly for the inexperienced pilot. Altitude is flying insurance.

*FAR 91.81 Altimeter Settings*

(A) Each person operating an aircraft shall maintain the cruising altitude or flight level of that aircraft, as the case may be, by reference to an altimeter that is set, when operating

- (1) below 18,000 feet MSL, to
  - (a) the current reported altimeter setting of a station along the route and within 100 nautical miles of the aircraft;
  - (b) if there is no station within the area prescribed in subdivision (a) of this subparagraph, the current reported altimeter setting of an appropriate available station; or
  - (c) in the case of an aircraft not equipped with a radio, the elevation of the departure airport or an appropriate altimeter setting available before departure; or
- (2) at or above 18,000 feet MSL. . .

*FAR 91.83 Flight Plan; Information Required*

(a) Unless otherwise authorized by ATC, each person filing an IFR or VFR flight plan shall include in it the following information:

- (1) The aircraft identification number, and if necessary, its radio call sign.
- (2) The type of the aircraft or, in the case of a formation flight, the type of each aircraft and the number of aircraft in the formation.
- (3) The full name and address of the pilot in command or, in the case of a formation flight, the formation commander.
- (4) The point and proposed time of departure.
- (5) The proposed route, cruising altitude (or flight level), and true airspeed at that altitude.
- (6) The point of first intended landing and the estimated elapsed time until over that point.
- (7) The radio frequencies to be used.
- (8) In the case of an IFR flight plan, an alternate airport, except as provided in paragraph (b) of this section.

(9) In the case of an international flight, the number of persons in the aircraft.

(10) Any other information the pilot in command or ATC believes is necessary for ATC purposes.

When a flight plan has been filed, the pilot in command, upon cancelling or completing the flight under the flight plan, shall notify the nearest FAA Flight Service Station or ATC facility

(b) Paragraph (a) (9) of this section. . .

(c) IFR. . .

Note: Although a flight plan is not required for VFR flight, it is good procedure to file one. In the case of an emergency, search and rescue procedures will be instituted quicker and more efficiently if we file a flight plan. *We must either cancel or close the flight plan to prevent unnecessary search and rescue.*

*FAR 91.85 Operating on or in the Vicinity of an Airport; General Rules*

(a) Unless otherwise required by Part 93 of this chapter, each person operating an aircraft on or in the vicinity of an airport shall comply with the requirements of this section and of 91.87 and 91.89.

(b) Unless otherwise authorized or required by ATC, no person may operate an aircraft within an airport traffic area except for the purpose of landing at, or taking off from, an airport within that area. ATC authorizations may be given as individual approval of specific operations or may be contained in written agreement between airport users and the tower concerned.

*FAR 91.87 Operation at Airports with Operating Control Towers*

(a) General. Unless otherwise authorized or required by ATC, each person operating an aircraft to, from, or on an airport with an operating control tower shall comply with the applicable provisions of this section.

(b) Communications with control towers operated by the United States. No person may, within an airport traffic area, operate an aircraft to, from, or on an airport having a control tower operated by the United States unless two-way radio communications are maintained between that aircraft and the control tower. However, if the aircraft radio fails in flight, he may operate that aircraft and land if weather conditions are at or above basic VFR weather minimums, he maintains visual contact with the tower, and he receives a clearance to land. If the aircraft radio fails while in flight under IFR, he must comply with 91.127.

(c) Communications with other control towers. No person may, within an airport traffic area, operate an aircraft to, from, or on an airport having a control tower that is operated by any person other than the United States unless,

(1) if that aircraft's radio equipment so allows, two-way radio communications are maintained between the aircraft and the tower; or

(2) if that aircraft's radio equipment allows only reception from the tower, the pilot has the tower's frequency monitored.

(d) Minimum altitudes. When operating to an airport with an operating control tower (Effective: December 4, 1967), each pilot of

(1) a turbine-powered airplane or a large airplane shall, unless other-

wise required by the applicable distance from cloud criteria, enter the airport traffic area at an altitude of at least 1,500 feet above the surface of the airport and maintain at least 1,500 feet within the airport traffic area, including the traffic pattern, until further descent is required for a safe landing;

(2) a turbine-powered airplane or a large airplane approaching to land on a runway being served by an ILS, shall, if the airplane is ILS equipped, fly that airplane at an altitude at or above the glide slope between the outer marker (or the point of interception with the glide slope, if compliance with the applicable distance from clouds criteria requires interception closer in) and the middle marker; and,

(3) an airplane approaching to land on a runway served by a visual approach slope indicator, shall maintain an altitude at or above the glide slope until a lower altitude is necessary for a safe landing.

However, subparagraphs (2) and (3) of this paragraph do not prohibit normal bracketing maneuvers above or below the glide slope that are conducted for the purpose of remaining on the glide slope.

(e) Approaches. When approaching to land at an airport with an operating control tower, each pilot of

(1) an airplane shall circle the airport to the left; and

(2) a helicopter shall avoid the flow of fixed-wing aircraft.

(f) Departures. No person may operate an aircraft taking off from an airport with an operating control tower except in compliance with the following:

(1) Each pilot shall comply with any departure procedures established for that airport by the FAA.

(2) Unless otherwise required by the departure procedure or the applicable distance from clouds criteria, each pilot of a turbine-powered airplane and each pilot of a large airplane shall climb to an altitude of 1,500 feet above the surface as rapidly as practicable. [(f)(2) Effective: December 4, 1967]

(g) Noise abatement runway system. When landing or taking off from an airport with an operating control tower and for which a formal runway-use program has been established by the FAA, each pilot of a turbine-powered airplane and each pilot of a large airplane, assigned a noise abatement runway by ATC, shall use that runway. However, each pilot has final authority and responsibility for the safe operation of his airplane and, if he determines in the interest of safety that another runway should be used, ATC will assign that runway (air traffic and other conditions permitting).

(h) Clearances required. No pilot may, at an airport with an operating control tower, taxi an aircraft on a runway, or takeoff or land an aircraft, unless he has received an appropriate clearance from ATC. A clearance to "taxi to" the runway is a clearance to cross all intersecting runways but is not a clearance to "taxi on" the assigned runway.

Note: A clearance to "taxi to" a runway is a clearance to cross all runways enroute, unless "hold short of" runway is included in the clearance.

*FAR 91.89 Operations at Airports without Control Towers*

(a) Each person operating an aircraft to or from an airport without an operating control tower shall,

(1) in the case of an airplane approaching to land, make all turns of that airplane to the left unless the airport displays approved light signals or usual markings indicating that turns should be made to the right, in which case the pilot shall make all turns to the right;

(2) in the case of a helicopter...

(3) in the case of an aircraft departing the airport, comply with any FAA traffic pattern for that airport.

*FAR 91.95 Restricted and Prohibited Areas*

(a) No person may operate an aircraft within a restricted area (designated in Part 73) contrary to the restrictions imposed, or within a prohibited area, unless he has the permission of the using or controlling agency, as appropriate.

(b) Each person conducting, within a restricted area, an aircraft operation (approved by the using agency) that creates the same hazards as the operations for which the restricted area was designated, may deviate from the rules of this subpart that are not compatible with his operation of the aircraft.

*FAR 91.97 Positive Control Areas and Route Segments*

(a) Except as provided in paragraph (b) of this section, no person may operate an aircraft within a positive control area, or positive control route segment designated in Part 71 of this chapter, unless that aircraft is

(1) operated under IFR at a specific flight level assigned by ATC;

(2) equipped with instruments and equipment required for IFR operations;

(3) flown by a pilot rated for instrument flight; and

(4) equipped, when in a positive control area, with

(i) a coded radar beacon transponder...

(ii) a radio providing direct pilot/controller communication...

## Visual Flight Rules

*FAR 91.105 Basic VFR Weather Minimums (Effective: March 16, 1968)*

(a) Except as provided in 91.107, no person may operate an aircraft under VFR when the flight visibility is less, or at a distance from clouds that is less, than that prescribed for the corresponding altitude in the table on the following page.

(b) When the visibility is less than one mile, a helicopter may be operated outside controlled airspace at 1,200 feet or less above the surface if operated at a speed that allows the pilot adequate opportunity to see any air traffic or other obstruction in time to avoid collision.

(c) Except as provided in 91.107, no person may operate an aircraft, under VFR, within a control zone beneath the ceiling when the ceiling is less than 1,000 feet.

(d) Except as provided in 91.107, no person may take off or land an

| ALTITUDE | FLIGHT VISIBILITY | DISTANCE FROM CLOUDS |
|---|---|---|
| 1,200 feet or less above the surface (regardless of MSL altitude): | | |
| Within controlled airspace. .......... | 3 statute miles | 500 feet below.<br>1,000 feet above.<br>2,000 feet horizontal. |
| Oustide controlled airspace. .......... | 1 statute mile except as provided in 91.105(b) | Clear of clouds. |
| More than 1,200 feet above the surface but less than 10,000 feet MSL: | | |
| Within controlled airspace. .......... | 3 statute miles | 500 feet below.<br>1,000 feet above.<br>2,000 feet horizontal. |
| Outside controlled airspace. .......... | 1 statute mile | 500 feet below.<br>1,000 feet above.<br>2,000 feet horizontal. |
| More than 1,200 feet above the surface and at or above 10,000 feet MSL | 5 statute miles | 1,000 feet below.<br>1,000 feet above.<br>1 mile horizontal. |

aircraft, or enter the traffic pattern of an airport under VFR, within a control zone

(1) unless ground visibility at that airport is at least 3 statute miles; or

(2) if ground visibility is not reported at that airport, unless flight visibility during landing or takeoff, or while operating in the traffic pattern, is at least 3 statute miles.

(e) For the purposes of this section, an aircraft operating at the base altitude of a transition area or control area is considered to be within the airspace directly below that area.

Note: The visibility and cloud clearance requirements for VFR flight are *minimum*. In this day of high-speed aircraft, we should not plan on flying with VFR minimums. Aircraft are difficult to see, even in atmospheric conditions much better than minimum VFR, and converging times are very short in high-speed aircraft.

*FAR 91.107 Special VFR Weather Minimums (Effective: April 30, 1968)*

(a) Except as provided in 93.113 of this chapter, when a person has received an appropriate ATC clearance, the special weather minimums of this section instead of those contained in 91.105 apply to the operation of an aircraft by that person in a control zone under VFR.

(b) No person may operate an aircraft in a control zone under VFR except clear of clouds.

(c) No person may operate an aircraft (other than a helicopter) in a control zone under VFR unless flight visibility is at least one statute mile.

(d) No person may takeoff or land an aircraft (other than a helicopter) at any airport in a control zone under VFR

(1) unless ground visibility at that airport is at least 1 statute mile;

(2) if ground visibility is not reported at that airport, unless flight visibility during landing or takeoff is at least 1 statute mile.

Note: At certain airports, we may be allowed to takeoff or land when the visibility is one mile or more and the operation remains clear of clouds. This regulation may be revised or omitted in the future. Part 93 lists the airports at which Special VFR does not apply. Additional airports may be designated in the future.

Special VFR operations will be permitted but IFR traffic has priority over Special VFR.

*FAR 91.109 VFR Cruising Altitude or Flight Level*

Except while holding in a holding pattern of two minutes or less, or while turning, each person operating an aircraft under VFR in level cruising flight, at or above 3,000 feet *above the surface*, shall maintain the appropriate altitude prescribed below:

(a) When operating below 18,00 feet MSL and

(1) on a *magnetic course* of zero degrees through 179 degrees, any odd thousand foot MSL altitude plus 500 feet (such as 3,500, 5,500, or 7,500); or

(2) on a *magnetic course* of 180 degrees through 359 degrees, any even thousand foot MSL altitude plus 500 feet (such as 4,500, 6,500, or 8,500).

(b) When operating above 18,000 MSL. . .

(c) When operating above flight level 290. . .

Note: This altitude rule applies only when VFR *at or above* 3,000 feet *above the surface*. The *magnetic course* of the aircraft determines which semi-circle the aircraft is flying in. Figure 175 depicts this rule.

## Subpart C—Maintenance

*FAR 91.161 Applicability*

(a) This subpart prescribes rules governing the maintenance, preventive maintenance, and alteration of U.S. registered civil aircraft operating within or without the United States.

(b) Sections 91.165, 91.169, 91.171, and 91.173 of this subpart do not apply to an aircraft maintained in accordance with a continuous airworthiness program under Part 121 or 127 of this chapter.

*FAR 91.163 General*

(a) The owner or operator of an aircraft is primarily responsible for maintaining that aircraft in an airworthy condition, including compliance with Part 39 of this chapter.

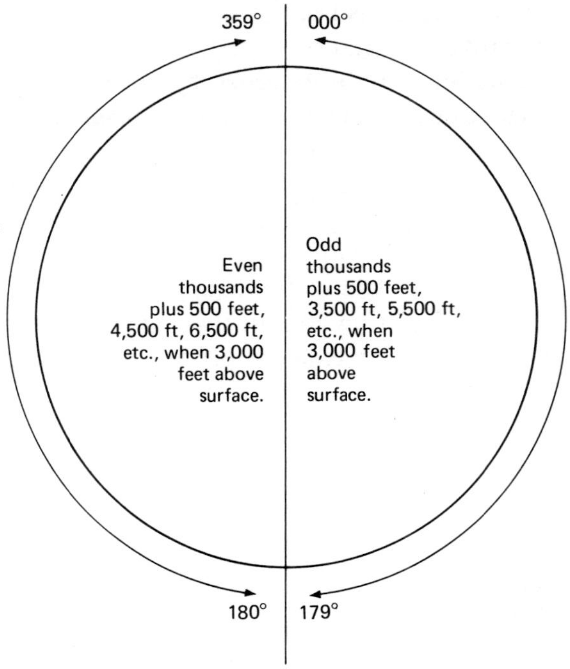


Fig. 175. Quadrantial altitude.

(b) No person may maintain an aircraft other than as prescribed in this subpart and other applicable regulations, including Part 43 of this chapter.

*FAR 91.165 Maintenance Required*

Each owner or operator of an aircraft shall have that aircraft inspected as prescribed in 91.169 and 91.170 of this chapter and shall, between required inspections, have defects repaired as prescribed in Part 43 of this chapter. In addition, he shall ensure that maintenance personnel make appropriate entries in the aircraft and maintenance records indicating the aircraft has been released to service.

Note: The *owner* or *operator* is primarily responsible for maintaining the aircraft in an airworthy condition.

*FAR 91.169 Inspections*

(a) Except as provided in paragraph (c) of this section, no person may operate an aircraft unless, within the preceding 12 calendar months, it has had

(1) an annual inspection in accordance with Part 43 of this chapter and has been approved for return to service by a person authorized by 43.7; or

(2) an inspection for the issue of an airworthiness certificate.

No inspection performed under paragraph (b) of this section may be substituted for any inspection required by this paragraph unless it is performed by a person authorized to perform annual inspections, and is entered as an "annual" inspection in the required maintenance records.

(b) Except as provided in paragraph (c) of this section, no person may operate an aircraft carrying any person (other than a crewmember) for hire or to give flight instruction for hire unless, within the preceding 100 hours of time in service, it has been inspected in accordance with Part 43 of this chapter and approved for return to service by a person authorized by 43.7. The 100-hour limitation may be exceeded by not more than 10 hours if necessary to reach a place at which the inspection can be done. The excess time, however, is included in computing the next 100 hours of time in service.

(c) Paragraphs (a) and (b) of this section do not apply to

(1) any aircraft for which its registered owner or operator complies with the progressive inspection requirements of 91.171 and part 43 of this chapter; or

(2) an aircraft that carries a special flight permit or a current experimental or provisional certificate.

## Part 93—Special Air Traffic Rules and Airport Traffic Patterns

Part 93 lists the locations at which special VFR minimums do not apply. This list is subject to change.

### Subpart A—General

*FAR 93.1 Applicability*

(a) This part prescribes special airport traffic patterns and airport traffic areas. It also prescribes special air traffic rules for operating aircraft in those traffic patterns and traffic areas and in the vicinity of airports described in this part.

(b) Unless otherwise authorized by ATC (with the exception of 93.113), each person operating an aircraft shall do so in accordance with the special air traffic rules in this part in addition to other applicable rules in Part 91 of this chapter.

(c) . . .

(d) Subpart I of this part prescribes the locations at which the special VFR weather minimums do not apply to fixed-wing aircraft.

### Subpart I—Locations at which Special VFR Weather Minimums do not Apply

*FAR 93.111 Applicability*

This subpart specifies the control zones in which special VFR minimums prescribed in 91.107 of this chapter do not apply, except for inflight emergencies.

*FAR 93.113 Control Zones within which Special VFR Weather Minimums Are not authorized.*

No person may operate a fixed-wing aircraft under the special VFR weather minimums prescribed in 91.107 of this chapter within the following control zones:

1. Atlanta, Ga. (Atlanta Airport).

2. Baltimore, Md. (Friendship International Airport).
3. Boston, Mass. (Logan International Airport).
4. Buffalo, N.Y. (Greater Buffalo International Airport).
5. Chicago, Ill. (O'Hare International Airport).
6. Cleveland, Ohio (Cleveland-Hopkins International Airport).
7. Columbus, Ohio (Columbus Municipal Airport).
8. Covington, Ky. (Greater Cincinnati Airport).
9. Dallas, Tex. (Love Field).
10. Denver, Colo. (Stapleton Municipal Airport).
11. Detroit, Mich. (Metropolitan Wayne County Airport).
12. Honolulu, Hawaii (Honolulu International Airport).
13. Houston, Tex. (William P. Hobby Airport).
14. Indianapolis, Ind. (Weir-Cook Municipal Airport).
15. Kansas City, Mo. (Kansas City Municipal Airport).
16. Los Angeles, Calif. (Los Angeles International Airport).
17. Louisville, Ky. (Staniford Field).
18. Memphis, Tenn. (Memphis Metropolitan Airport).
19. Miami, Fla. (Miami International Airport).
20. Minneapolis, Minn. (Minneapolis-St. Paul International Airport).
21. Newark, N.J. (Newark Airport).
22. New York, N.Y. (John F. Kennedy International Airport).
23. New York, N.Y. (LaGuardia Airport).
24. New Orleans, La. (New Orleans International Airport-Moisant Field).
25. Oakland, Calif. (Metropolitan Oakland International Airport).
26. Philadelphia, Pa. (Philadelphia International Airport).
27. Pittsburgh, Pa. (Greater Pittsburgh Airport).
28. Portland, Ore. (Portland International Airport).
29. San Francisco, Calif. (San Francisco International Airport).
30. Seattle, Washington (Seattle-Tacoma International Airport).
31. St. Louis, Mo. (Lambert-St. Louis Municipal Airport).
32. Tampa, Fla. (Tampa International Airport).
33. Washington, D.C. (Washington National Airport).

Note: The above list is subject to change.

## FEDERAL AVIATION AGENCY
## VFR PILOT EXAM-O-GRAM NO. 1

### CONTROL ZONE VFR WEATHER MINIMUMS

A pilot plans a VFR cross-country flight with his destination airport located in a control zone. The terminal forecast indicates that the ceiling and visibility will be decreasing but remain above VFR minimums until his estimated time of arrival. Upon arrival, he enters the control zone, contacts the tower, and indicates that he desires to land. He is cleared to land by the tower.

We shall assume that one or more of the following conditions actually existed at the time he entered the control zone:

1) Flight or ground visibility was less than three miles but not less than one mile; and

2) Ceiling was less than 1,000 feet.

### Analysis

1- WAS THE PILOT LEGAL? No! The fact that the control tower operator cleared him to land does not mean that he is legal. The tower controller is concerned with the safe, orderly, and expeditious movement of air traffic. He will refuse landing only on the basis of other traffic.

2- WHY WAS THE PILOT NOT LEGAL? FAR (Part 91.105) states in part, that no person may operate an aircraft under VFR within a control zone beneath the ceiling when the ceiling is less than 1,000 feet, or unless the flight visibility is at least three statute miles. It further states that no person may take off or land an aircraft, or enter the traffic pattern of an airport, under VFR, within a control zone unless ground visibility is at least three statute miles.

3- WHAT ACTION SHOULD THE PILOT HAVE TAKEN TO BE LEGAL? He should have remained clear of the control zone, called the control tower, and requested an air traffic control clearance to land. He should remember that such a clearance does not constitute authority for him to deviate from the minimum safe altitudes as given in FAR, Part 91.79.

4- WHAT ACTION IS DICTATED BY GOOD OPERATING PRACTICES? He should have used reasonable restraint in exercising the prerogative of VFR flight, especially in terminal areas. The weather minimums and distances from clouds are minimums. Giving oneself a greater margin in specific instances is just good judgment. Conducting a VFR operation in a control zone at weather minimums is not prohibited, but good judgment would dictate that pilots flying VFR keep out of the approach area.

Department of Transportation

FEDERAL AVIATION ADMINISTRATION

## VFR PILOT EXAM-O-GRAM° NO. 2

### VFR CRUISING ALTITUDES

Assume that you plan to make a VFR cross-country flight over terrain which has a constant elevation of 2,900 feet. After charting the course you determine that the true course is 188° and the magnetic variation is 12°E. According to the latest aviation weather reports, there is a broken layer of clouds at 7,000 feet all along the route, and the visibility is unlimited along your intended route. The winds aloft forecast indicates that the higher the altitude the more favorable the wind direction and speed.

If you intend to take advantage of the most favorable wind and still comply with Federal Aviation Regulations, you should decide upon a cruising altitude of:

a. 5, 500 feet MSL
b. 6, 500 feet MSL
c. 7, 500 feet MSL
d. 9, 500 feet MSL

Analysis

1- You wish to fly as high as legally possible to take advantage of the most favorable wind.

2- The base of the broken clouds is reported in height above the surface. Therefore, the base of the clouds is approximately 2, 900 feet plus 7, 000 feet, or 9, 900 feet above sea level.

3- Cruising altitude is a level above mean sea level (MSL), but the rules pertaining to the selection of a cruising altitude appropriate to the flight's magnetic course are applicable only when flying at or above 3, 000 feet above the ground.

4- This flight will be made at an altitude of 3, 000 feet or more above the surface in order to take advantage of the more favorable winds at higher altitudes. Since you will be flying at 3, 000 feet or more above the surface, you must, according to Federal Aviation Regulations, fly at a cruising altitude appropriate to the magnetic course. In this instance the magnetic course is 176° (true course 188° - 12° E magnetic variation = 176°.

5- A magnetic course of 176° in this case requires that you fly at an altitude (above sea level) of an odd thousand plus 500 feet.

6- In this example, you must maintain a vertical distance under the base of any cloud formation of at least 500 feet. This rules out a cruising altitude of 9, 500 feet. You do not choose 5, 500 feet since you want to take advantage of better tail winds at higher altitudes. You eliminate 6, 500 feet because you must be at an odd thousand altitude plus 500 feet. Therefore, you select a cruising altitude of 7, 500 feet, which meets legal requirements and gives you the advantage of more favorable winds.

Department of Transportation
FEDERAL AVIATION ADMINISTRATION
VFR PILOT EXAM-O-GRAM° NO. 22

## POTENTIAL MID-AIR COLLISIONS

Analyses of answers to Pilot and Ground Instructor Written Examinations indicate that many applicants do not fully understand several areas in Regulations and procedures that were devised as safety measures for VFR flying. Two of the areas will be covered in this Exam-O-Gram that seem to give applicants the most difficulty. They concern the Hemispherical Rule for VFR cruising altitudes and Airport Advisory Service at uncontrolled airports.

A pilot who does not keep abreast of and comply with the latest Regulations and procedures could be a source of danger to himself and to others in his vicinity. A recent Federal Aviation Agency report indicated that 549 "near mid-air" collisions were reported within the United States during calendar year 1962. This compared with 516 reports for 1961 and 470 reports during 1960. It would be reasonable to assume that other "near mid-air" collisions occurred that were not reported.

Failure to comply with Regulations and Procedures increases the degree of potential mid-air collision hazards!

Could any pilot with considerable flying experience truthfully say that he has never been involved in a "near miss" with other aircraft - or - that he is not seriously concerned about mid-air collisions? It is often so easy to fly for a long period of time with our head in the cockpit while we study charts or change radio frequencies. Finally, something tells us that we should start looking around, and then we suddenly realize how foolish we were to expose ourselves to the potential hazards of a mid-air collision while we were preoccupied.

Most pilots know very well the danger of not properly guarding the airplane from other aircraft while their attention is divided between things inside and outside the cockpit -- yet is there a pilot flying today who will not some day break this rule of common sense?

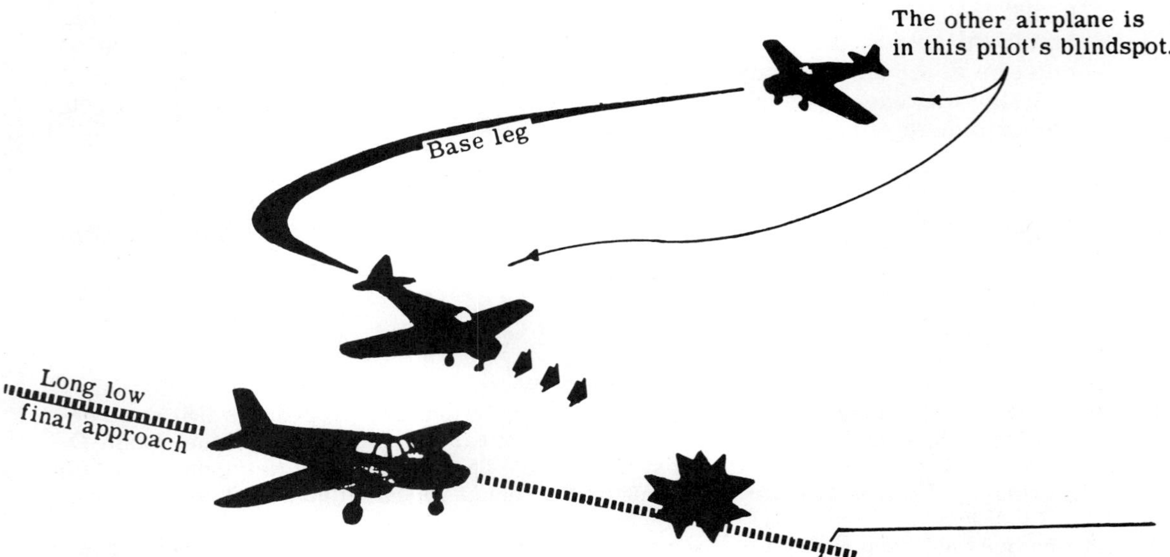


The collision that is about to happen as illustrated above can happen at any airport. A number of such accidents have already occurred - LOOK AROUND - DO NOT LET IT HAPPEN TO YOU.

IS TRAFFIC INFORMATION AVAILABLE AT CERTAIN NONCONTROLLED AIRPORTS? Yes, at certain noncontrolled airports (no control tower) where FAA Flight Service Stations are operating, there is available to you Airport Advisory Service. Use of this radio service will aid you in avoiding mid-air collisions.

WHAT SERVICE DOES THE AIRPORT ADVISORY SERVICE PROVIDE? The Flight Service Station (FSS) at uncontrolled airports provides airport advisory service to aircraft operating to or from the airport on which the station is located. The airport advisory service provides the following information to aircraft which are in communication with the station: Wind Direction and Velocity; Favored Runway; Altimeter Setting; Pertinent Known Traffic; Pertinent Known Field Conditions; Airport Taxi Routes and Traffic Patterns, etc.

NOTICE! There may be other aircraft in the vicinity of the airport not in communication with and thus not known by the FSS.

HOW DOES THE PILOT KNOW WHERE TO FIND AIRPORT ADVISORY SERVICE LOCATIONS? The locations are appropriately depicted on the Sectional Charts in this manner:

* * * * *

WHAT IS A SAFE WAY TO CLIMB OR DESCEND ON VICTOR AIRWAYS? The AIM Good Operating Practices section states: "During climb or descent, pilots are encouraged to fly to the right side of the center line of the radial forming the airway in order to avoid IFR and VFR cruising traffic operating along the center line of the airway."

TO AVOID OR REDUCE THE HAZARD OF TOO MUCH "EYES-INSIDE-THE-COCKPIT" FLYING, WHAT ACTION SHOULD A PILOT TAKE IN VFR CROSS-COUNTRY PREFLIGHT PLANNING?

(a) He should obtain from proper charts all the information pertinent to his route of flight. Information such as: headings, distances, checkpoints, altitudes, etc., should be placed in a flight log format. On the reverse side of the FLIGHT PLAN (FAA Form [redacted]) a flight log is provided for pilots.

(b) All necessary charts should be folded in proper sequence and conveniently located in the cockpit.

(c) The current issue of Airman's Information Manual (AIM) should be referred to with particular attention to NOTAMS and Airport/Facility Directory sections. All radio frequencies to be used on the flight should be written on the flight log for ready reference during the flight.

(d) The AIM Airport Directory or Airport/Facility Directory sections should be consulted to obtain airport data and to review VFR procedures for approaches to busy air terminals. For example: The Airport/Facility Directory section for Nashville Metropolitan Airport states: "Traffic Information - Contact Approach Control on 120.6 . . 25 miles out."

(e) The Airman's Information Manual should be reviewed for additional information under such headings as: Good Operating Practices, Air Navigation Radio Aids, Airport-Air Navigation Lighting and Marking Aids, Weather, Preflight, Departure, Radar Assistance to VFR Aircraft, VFR Cruising Altitudes, Arrival, and Emergency Procedures.

(f) A careful study of the Sectional or World Aeronautical Charts should be made to determine if your route of flight will traverse a Prohibited, Restricted, Alert, or Warning Area.

HOW DOES THE TOWER ASSIST IN PREVENTING MID-AIR COLLISIONS AT A CONTROLLED AIRPORT? Although it is always the direct responsibility of the pilot, when flying in VFR weather conditions, to avoid collision with other aircraft, the information and clearances issued by the controller in the tower are intended to aid pilots to the fullest extent in avoiding collisions. The controller in the tower issues clearances that can be safely followed without collision hazard if reasonable caution is exercised by the pilot. By advising the tower of your position well in advance of entering the control zone (normally a minimum of 15 miles out), you will be able to receive information on other aircraft which might be in your vicinity as well as being assured of a safe and orderly entry into the traffic pattern under the direction of the control tower.

> Note to Student Pilots: To receive additional assistance while operating in areas of concentrated air traffic, a student pilot should identify himself as a student pilot during his initial call to an FAA radio facility (Control Tower, FSS, Approach Control, etc.). For example: "Dayton Tower, this is Fleetwing 1234, Student Pilot, over."

At some busy airports (examples are: Memphis, El Paso, and Seattle) an expansion of the normal tower service is made possible through the use of radar. Here the tower will probably request that you listen on the approach control frequency which he will supply. You will then receive essentially the same information and direction as though the tower had you in visual contact, but at a much greater range. The number of such airports with this service is increasing rapidly and is another step toward reducing the possibility of mid-air collisions.

WHY SHOULD A PILOT CHECK THE <u>SPECIAL NOTICES - AREA</u> SECTION OF THE AIRMAN'S INFORMATION MANUAL? Before departing on an extensive cross-country flight in unfamiliar country, the pilot should check the <u>Special Notices - Area</u> section for such notices as "Terminal Radar Service Areas" and "Terminal Area Notices." For example: Special Air Traffic Rules apply to VFR flights in the Valparaiso, Florida, Terminal Area because of the high speed special activities conducted in the vicinity of Eglin AFB.

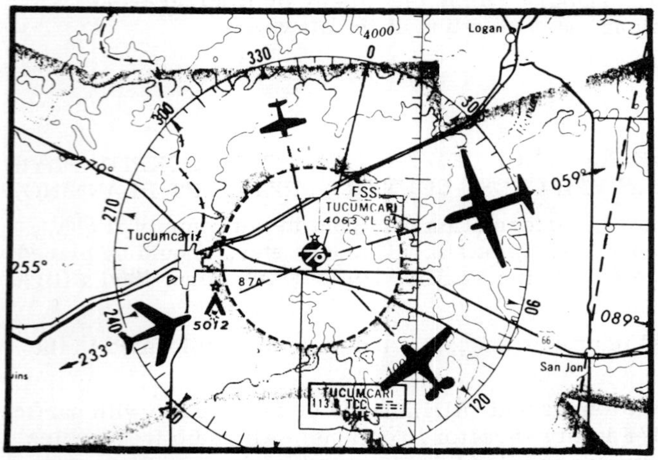


Possible Widely Scattered Aluminum

**DESTINATION:**

## Over The VOR Station!

The FAA Near Mid-Air Collision Report for Calendar Year 1962 indicates that 446 (81.24%) of the incidents occurred in clear skies and unrestricted visibility conditions. Of the 549 incidents reported <u>256 (46.6%) occurred over a VOR facility</u>, and the aircraft were utilizing VOR as the navigational aid in 89% of the en route incidents. <u>BE ALERT AT ALL TIMES</u>: Unlimited visibility appears to encourage a sense of security which is not at all justified.

* * * * *

DOES THE HEMISPHERICAL RULE OF CRUISING ALTITUDES PLAY AN IMPORTANT ROLE IN THE AVOIDANCE OF MID-AIR COLLISIONS? Yes, the rule is specifically designed to provide altitude separation, and applies to <u>local</u> as well as <u>cross-country flights.</u>

Many Airman Written Examination applicants are incorrectly answering questions pertaining to the Hemispherical Rule for VFR cruising altitudes. When an aircraft is operated in VFR level cruising flight at <u>3000 feet or more above</u> the surface and below 29,000 feet, the cruising altitudes (shown in the illustration to the right) shall be observed in accordance with the magnetic course being flown. (NOTE: See Airman's Information Manual.)

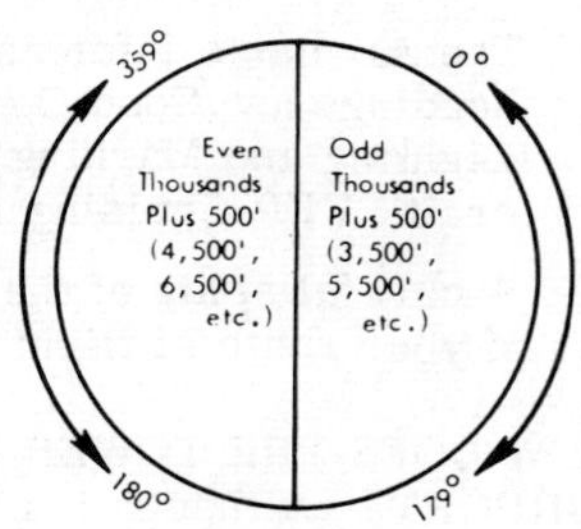


WHEN 3,000' ABOVE THE SURFACE

DO THE VFR CRUISING ALTITUDES APPLY BELOW 3000 FEET? No, only when you are flying at <u>3000 feet or more</u> above the surface.

Assume that in the diagram below your flight traverses terrain with the approximate elevations as depicted. You desire to select a <u>constant</u> cruising altitude which will conform to VFR cruising altitude requirements and also have sufficient altitude above mountain peaks to avoid downdrafts or extreme turbulence. Altitudes above the surface in mountainous areas should be based on the lowest general terrain (excluding deep crevices or canyons)

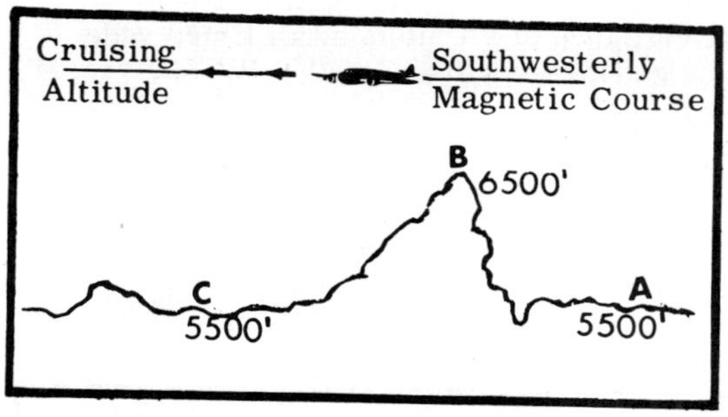


For example: + 5500' general terrain elev.<br>
3000' above terrain<br>
8500' effective altitude

10, 500' correct (even +500') at points A, B, C.<br>
10, 000' incorrect at points A, B, C.<br>
9, 500' incorrect at points A, B, C.<br>
9, 000' correct at point B-incorrect at points A, C.<br>
8, 500' correct (even +500') at points A, B, C.<br>
8, 000' correct at A, B, C.<br>
7, 500' correct at A, B, C.<br>
7, 000' correct at A, B, C.<br>
} less than 3000' above surface but inadequate safety above peaks.

Department of Transportation

FEDERAL AVIATION ADMINISTRATION

## VFR PILOT EXAM-O-GRAM° NO. 29

## POTENTIAL MID-AIR COLLISIONS (Series 2)

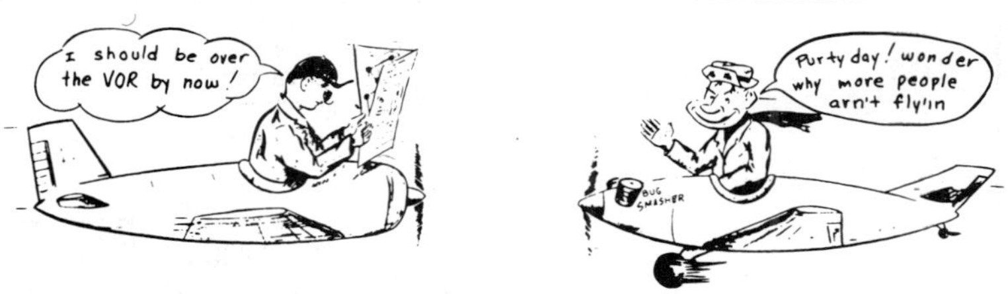


During 1963 there were 626 "near mid-air" collision incidents reported -- an increase of 77 over the preceding year. Due to the seriousness of this problem, most of the general aviation written examinations contain one or more test items that deal with FAA "right-of-way" rules. (NOTE: See VFR Exam-O-Gram No. 22 which also deals with Potential Mid-Air Collisions.)

This Exam-O-Gram will be concerned primarily with the right-of-way rule for "converging" aircraft. The "approaching head-on" rule is one that is simple and easily understood, a rule that has been in existence for many years and one that has an added safety factor in that each pilot of each aircraft shall alter course to the right.

CAN YOU ANSWER THE FOLLOWING SAMPLE TEST ITEM?

Assume that during a flight in a 4-place single-engine airplane you observe a light twin-engine airplane at your altitude. The light twin is approaching from your right on an apparent collision course. Although each pilot must do his best to avert a collision, which airplane, according to regulations, should give way and why should it give way?

1- You should give way since you are flying a small single-engine airplane.
2- You should give way since the light twin is on your right.
3- Both airplanes should alter their courses to the right since each pilot must do his best to avert a collision.
4- The light twin should give way since your airplane is to the light twin's left.

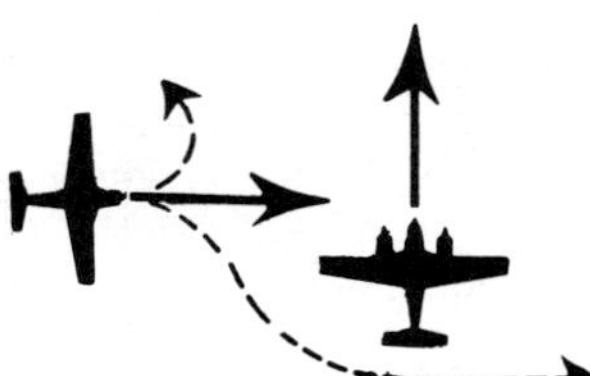

Figure 1.

The regulation concerning "converging" aircraft states in part: "When aircraft of the same category are converging at approximately the same altitude (except head-on, or nearly so), the aircraft to the other's right has the right-of-way." Therefore, response number 2 is correct since the other aircraft is on your right and both aircraft are of the same category - both are airplanes. See Figures 1 and 2.

☆ ☆ ☆ ☆ ☆

On the back page of this Exam-O-Gram, you will find an excerpt from a U. S. Naval Aviation Safety Bulletin. It is an excellent portrayal of the rapid rate of closure in a head-on view of a T-33 Jet, with a wing span of 42' 5", which closely approximates the span of several present day light twin-engine aircraft.

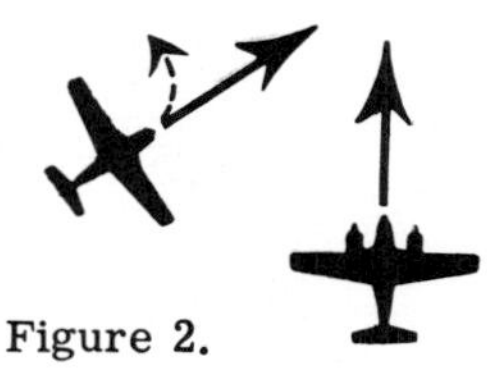

Figure 2.

Collision avoidance requires the necessity for constant vigilance of all pilots at all times under all circumstances. When another aircraft appears to be getting too close -- GIVE WAY! There is no equivalent to "fender bending" in aviation.

• Exam-O-Grams are non-directive in nature and are issued solely as an information service to individuals interested in Airman Written Examinations.

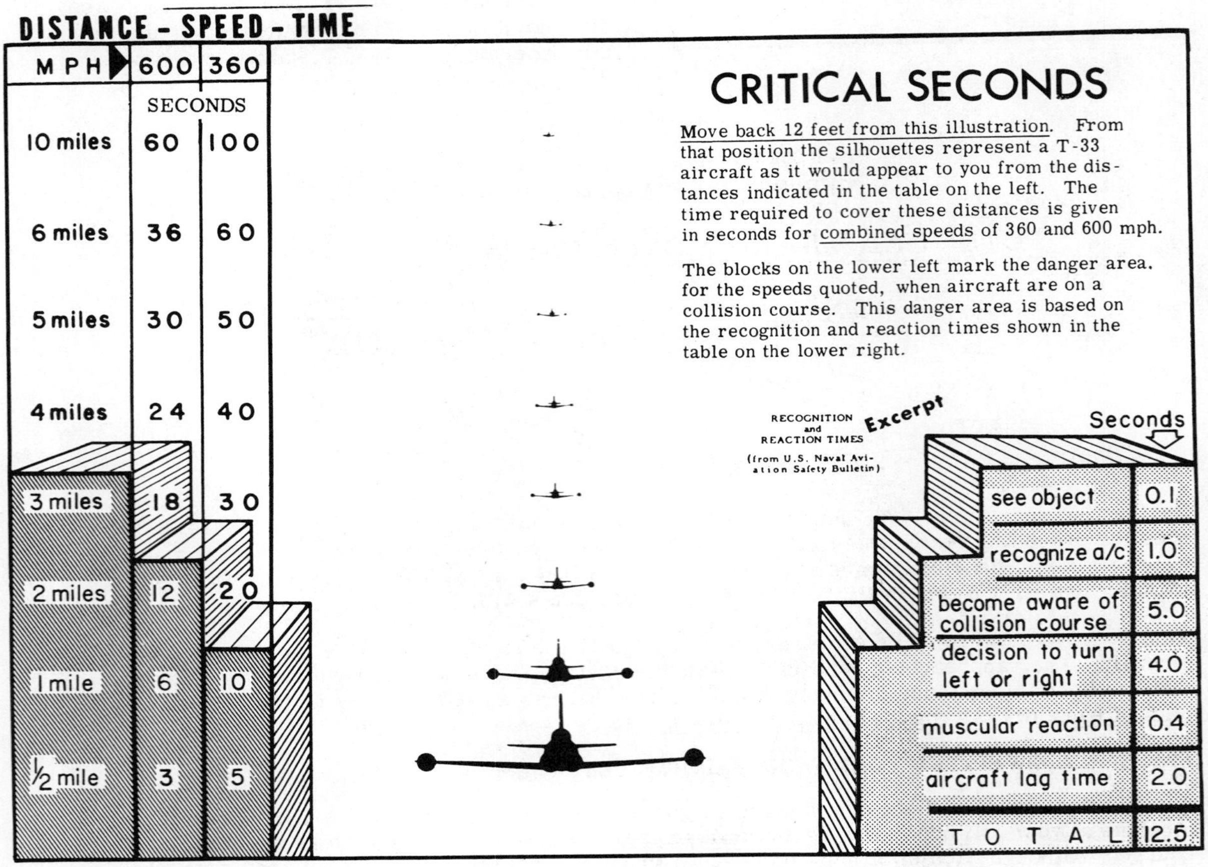

## CHECK YOUR READING

1. Who is responsible for the operation of an aircraft?
2. May the pilot in command deviate from the rules in Part 91 of the FAR?
3. May a pilot transport an obviously intoxicated person? If so, list the circumstances.
4. What certificates, manual, or other documents must be carried in an aircraft according to *Federal* law?
5. Who is responsible for determining if an aircraft is airworthy?
6. Many light aircraft have airspeed indicators on which the various speed ranges are shown in colors. Are these designations necessary on these aircraft?
7. We plan to carry three adults and one three-year-old child on a sightseeing flight. Must the child have an individual seat belt?
8. We wish to conduct formation flight. What requirements must be fulfilled prior to doing so?
9. Two *airplanes* are converging at a 90-degree angle. Which has the right of way?
10. Two *airplanes* are converging approximately head-on. Describe the procedure used to avoid a collision.
11. We intend to pass another aircraft which is ahead and to our right, on approximately the same heading. Describe the procedures we must use.
12. When in the traffic pattern, an airplane on final approach has the right of way over certain other aircraft. List them.
13. We are flying an aircraft which cruises at 280 knots (indicated airspeed). What precaution must we take when below 10,000 MSL?
14. We are flying a reciprocating-engine aircraft which cruises at 170 knots. We wish to enter an airport traffic area. What precaution must we take?
15. During what time period must the aircraft lights be operated?
16. What is the minimum safe altitude *anywhere*?
17. What is the minimum safe altitude over a congested area?
18. What is the minimum safe altitude while passing over a ship on the high sea?
19. Must we file a VFR flight plan?
20. May we fly through an airport traffic area if not intending to land (assume no ATC authorization)?
21. At what height may we fly (VFR) over an airport with an operating control tower, without contacting the tower?
22. If an airport is equipped with a Visual Approach Slope Indicator (VASI) on the runway in use, must we maintain an altitude at or above the visual glide slope?
23. When departing an airport, the control tower issues a clearance to "taxi to" a runway. While enroute to the runway, we must cross another runway. Do we need further clearance to cross?
24. May a private pilot fly VFR in a positive control area?
25. What are the VFR minimums for flight within controlled airspace, 1,200 feet or less above the surface?

26. What are the VFR minimums outside controlled airspace between 1,200 feet above the surface and 10,000 feet MSL?
27. What are the VFR minimums outside controlled airspace when 1,200 feet or less above the surface?
28. What altitude, between 4,000 and 7,000 feet, must we use when above 3,000 feet above the surface, and on a magnetic course of 050 degrees?
29. Assume the magnetic course is 005 degrees. Variation is 15 degrees west. We are flying over terrain which is at or near sea level. What altitude must we maintain between 4,000 and 6,000 feet?
30. Who is primarily responsible for maintaining the aircraft?
31. What is the duration of an airworthiness certificate?

# 54

# PART 99—SECURITY CONTROL OF AIR TRAFFIC

This part specifies the rules regarding the security control of air traffic. Special precautions have been taken to identify aircraft operating across the borders of the U.S. The Airman's Information Manual depicts each area requiring security control flight planning.

*FAR 99.1 Applicability*

(a) This subpart prescribes rules for operating civil aircraft in a defense area, or into, within, or out of the United States through an air defense identification zone (ADIZ), designated in Subpart B.

(b) Except for 99.7, this subpart does not apply to the operation of an aircraft

(1) in a coastal or domestic ADIZ of 28 degrees north latitude or west of 85 degrees west longitude at a true airspeed of less than 180 knots;

(2) in the Alaskan DEWIZ at a true airspeed of less than 180 knots while the pilot maintains a continuous listening watch on the appropriate frequency;

(3) from any point in the 48 contiguous United States on an outbound track through the Southern Border ADIZ that does not penetrate a coastal ADIZ;

(4) within the 48 contiguous United States and the District of Columbia, or within the State of Alaska, which remains within 10 nautical miles of the point of departure; or

(5) over any island, or within three nautical miles of the coastline of any island, in the Hawaiian ADIZ.

(c) Except as provided in 99.7, the radio and position reporting requirements of this subpart do not apply to the operation of an aircraft within the 48 contiguous United States and the District of Columbia, or within the State of Alaska, if that aircraft does not have two-way radio and is operated in accordance with a filed DVFR flight plan containing the time and point of domestic or coastal ADIZ penetration, and that aircraft departs within five minutes of the estimated departure time contained in the flight plan.

(d) An FAA, ATC center may exempt the following operations from this subpart (except 99.7), on a local basis only, with the concurrence of the military commanders concerned:

(1) Aircraft operations that are conducted wholly within the boundaries of an ADIZ and are not currently significant to the air defense system.

(2) Aircraft operations conducted in accordance with special procedures prescribed by the military authorities concerned.

*FAR 99.3 General*

(a) Air defense identification zones (ADIZ's) are areas of airspace over land or water in which the ready identification, location, and control of civil aircraft is required in the interest of national security. They are classified as

(1) coastal air defense identification zones (coastal ADIZ's);

(2) domestic air defense identification zones (domestic ADIZ's); and

(3) distant early warning identification zones (DEWIZ's).

(b) Unless designated as an ADIZ, a defense area is any airspace of the United States in which the control of an aircraft is required for reasons of national security.

(c) For the purposes of this part, a Defense Visual Flight Rules (DVFR) flight is a flight within an ADIZ conducted under the visual flight rules in Part 91.

*FAR 99.5 Emergency Situations*

In an emergency that requires immediate decision and action for the safety of the flight, the pilot in command of an aircraft may deviate from the rules in this part to the extent required by that emergency. He shall report the reasons for the deviation to the communications facility where flight plans or position reports are normally filed (referred to in this part as *an appropriate aeronautical facility*) as soon as possibile.

*FAR 99.7 Special Security Instructions*

Each person operating an aircraft in an ADIZ or defense area shall, in addition to the applicable operating rules of this part, comply with special security instructions issued by the Administrator in the interest of national security and that are consistent with appropriate agreements between the FAA and the Department of Defense.

No person may operate an aircraft in an ADIZ unless the aircraft has a functioning two-way radio.

*FAR 99.9 Radio Requirements*

(a) No person may operate an aircraft in or penetrating a coastal or domestic ADIZ unless he has filed a flight plan with an appropriate aeronautical facility

(b) Unless ATC authorizes an abbreviated flight plan,

(1) a flight plan for IFR. . .

(2) A flight plan for VFR flight must contain the information specified in 91.88 (a)(1) through (7).

(c) The pilot shall designate a flight plan for VFR flight as a DVFR flight plan.

*FAR 99.11 Flight Plan Requirements; Coastal or Domestic ADIZ*

(a) No person may operate an aircraft in or penetrating a DEWIZ unless he has filed a flight plan before takeoff with an appropriate aeronautical facility. If there is no facility for filing a DVFR flight plan, the pilot must comply with 99.25 (a)(2) and proceed according to the instructions issued by the appropriate aeronautical facility. These instructions normally require the flight to proceed to a specified area for visual identification or to land at a stated location.

(b) Unless ATC authorizes an abbreviated flight plan,

(1) a flight plan for IFR. . .

(2) a flight plan for VFR flight must contain the information in 91.83 (a)(1) through (7) and the estimated time and point of DEWIZ penetration (ETDP).

(c) The pilot shall designate a flight plan for VFR flight as a DVFR flight plan.

*FAR 99.13 Flight Plan Requirements DEWIZ*

The pilot in command of an aircraft for which a flight plan has been filed shall file an arrival or completion notice with an appropriate aeronautical facility unless the flight plan states that no notice will be filed.

*FAR 99.15 Arrival or Completion Notice*

No pilot may penetrate a domestic ADIZ under DVFR unless

(a) he reports to an appropriate aeronautical facility before penetration: the time, position, and altitude at which the aircraft passed the last reporting point before penetration and the estimated time of arrival over the next appropriate reporting point along the flight route;

(b) if there is no appropriate reporting point along the flight route, he reports at least 15 minutes before penetration: the estimated time, position, and altitude at which he will penetrate; or

(c) if the airport of departure is so close to the domestic ADIZ boundary that it prevents his complying with paragraphs (a) or (b) of this section, he has reported immediately after taking off: the time of departure, altitude, and estimated time of arrival over the first reporting point along the flight route.

*FAR 99.19 Position Reports; Aircraft Operating in or Penetrating a Domestic ADIZ; DVFR*

*FAR 99.21 Position Reports; Aircraft Entering the United States Through a Coastal ADIZ; United Stated Aircraft*

The pilot of an aircraft entering the United States through a Coastal ADIZ shall make the reports required in 99.17 or 99.19 to an appropriate aeronautical facility.

*FAR 99.25 Position Reports; Aircraft Entering the United States Through a DEWIZ*

(a) The pilot of an aircraft entering the United States through a DEWIZ,

(1) if under IFR. . .

(2) if under DVFR, shall report when within radio range of an appropriate aeronautical facility but before penetration: the time, altitude, and position at which he passed the last reporting point and the estimated time, altitude, and point of penetration.

(b) If requested, the pilot of an aircraft entering the United States through a DEWIZ shall advise an appropriate aeronautical facility of the difference between the actual time and point of penetration and the same data recorded in the original ground-filed flight plan.

*FAR 99.27 Deviation from Flight Plans and ATC Clearance and Instructions*

(a) No pilot may deviate from the provisions of an ATC clearance or ATC instruction except in accordance with 91.75 of this chapter.

(b) No pilot may deviate from his filed IFR flight plan. . .

(c) No pilot may deviate from his filed DVFR flight plan unless he notifies an appropriate aeronautical facility before deviating.

*FAR 99.29 Radio Failure; DVFR*

If the pilot operating an aircraft under DVFR in an ADIZ cannot maintain two-way radio communications, he may proceed in accordance with his original DVFR flight plan or land as soon as practicable. The pilot shall report the radio failure to an appropriate aeronautical facility as soon as possible.

*FAR 99.33 Flight Plans: Panama Canal Zone Domestic ADIZ*

Civil aircraft may operate within the Panama Canal Zone Domestic ADIZ only under a flight plan that has been approved by appropriate military authority acting through an FAA air traffic control facility.

## Subpart B—Designated Air Defense Identification Zones

*FAR 99.41 General*

The airspace above the areas described in this subpart is established as a domestic ADIZ, coastal ADIZ, DEWIZ, or defense area. The lines between points described in this subpart are great circles except that the lines joining adjacent points on the same parallel of latitude are rhumb lines.

*FAR 99.43, FAR 99.45, FAR 99.47.*

The descriptions of FAR 99.43 Domestic ADIZ's, FAR 99.45 Coastal ADIZ's, and FAR 99.47 Alaskan DEWIZ's contained in these sections are depicted on the aeronautical charts.

All airspace of the United States is designated as defense area except that airspace already designated as air defense identification zones.

*FAR 99.49 Defense Area*

Note: Because the regulations pertaining to the security control of air traffic are complicated, we should review them periodically, particularly before entering or departing an ADIZ or DEWIZ.

When in doubt about ADIZ or DEWIZ operations, any pilot may check with the nearest Flight Service Station.

## CHECK YOUR READING

1. What is the purpose of ADIZ and DEWIZ?
2. What radio requirements exist for flight in an ADIZ?
3. What information must be included in a report prior to penetrating a domestic ADIZ?
4. What procedure must be followed if radio failure occurs while operating under DVFR?
5. What FAA facility will aid the pilot concerning questions of ADIZ or DEWIZ operations?

# 55

# DOT PART 430—
# RULES PERTAINING TO AIRCRAFT ACCIDENTS, INCIDENTS, OVERDUE AIRCRAFT, AND SAFETY INVESTIGATIONS: FAR PART 43

## Subpart A—General

*DOT 430.1 Applicability*

This part contains rules pertaining to:

(a) Giving notice of and reporting aircraft accidents, and incidents, and certain other occurrences in the operation of aircraft when they involve civil aircraft of the United States wherever they occur, or foreign civil aircraft when such events occur in the United States, its territories, or possessions.

(b) Preservation, access to, and release of aircraft wreckage, mail, cargo, and records involving all civil aircraft in the United States, its territories or possessions.

(c) Investigation of aircraft accidents, certain incidents, and overdue aircraft, and special studies and investigations conducted by the Board pertaining to safety in air navigation and the prevention of accidents.

*DOT 430.2 Definitions*

As used in this part the following words or phrases are defined as follows:

***Aircraft Accident*** means an occurrence associated with the operation of an aircraft which takes place between the time any person boards the aircraft with the intention of flight, until such time as all such persons have disembarked, in which any person suffers death or serious injury as a result of being in or upon the aircraft or by direct contact with the aircraft or any thing attached thereto, or the aircraft receives substantial damage.

***Fatal Injury*** means any injury which results in death within seven days.

***Serious Injury*** means any injury which

(1) requires hospitalization for more than 48 hours, commencing within seven days from the date the injury was received;

(2) results in a fracture of any bone (except simple fractures of fingers, toes, or nose);

(3) involves lacerations which cause severe hemorrhages, nerve, muscle, or tendon damage;

(4) involves injury to any internal organ; or

(5) involves second or third degree burns, or any burns affecting more than five per cent of the body surface.

***Substantial Damage:***

(1) Except as provided in subparagraph (2) of this paragraph, substantial damage means damage or structural failure which adversely affects the structural strength, performance, or flight characteristics of the aircraft, and which would normally require major repair or replacement of the affected component.

(2) Engine failure, damage limited to an engine, bent fairings or cowling, dented skin, small punctured holes in the skin or fabric, ground damage to rotor or propeller blades, damage to landing gear, wheels, tires, flaps, engine accessories, brakes, or wing tips are not considered *substantial damage* for the purpose of this part.

Note: Reread the very specific definition of an aircraft accident, a fatal injury, a serious injury, and substantial damage. We must be familiar with these terms as the regulations use them often.

## Subpart B Initial Notification of Aircraft Accidents, Incidents, and Overdue Aircraft

*DOT 430.5 Immediate Notification*

The operator of an aircraft shall immediately, and by the most expeditious means available, notify the nearest National Transportation Safety Board, Bureau of Aviation Safety Field Office when:

(a) An aircraft accident or any of the following listed incidents occur:

(1) Flight control system malfunction or failure;

(2) Inability of any required flight crewmember to perform his normal flight duties as a result of injury or illness;

(3) Turbine engine rotor failures excluding compressor blades and turbine buckets;

(4) Inflight fire;

(5) Aircraft collide in flight.

(b) An aircraft is overdue and is believed to have been involved in an accident.

Note: Obviously, reporting of items (a)(1) through (5) will benefit others by preventing a possible accident. Item (b) provides for reporting overdue aircraft so as to begin search and rescue procedures as quickly as possible.

*DOT 430.6 Information to be Given in Notification*

The notification required in 430.5 shall contain the following information, if available:

(a) Type, nationality, and registration marks of the aircraft;

(b) Name of owner and operator of the aircraft;

(c) Name of the pilot in command;

(d) Date and time of the accident;

(e) Last point of departure and point of intended landing of the aircraft;

(f) Position of the aircraft with reference to some easily defined geographical point;

(g) Number of persons aboard, number killed, and number seriously injured;

(h) Nature of the accident including weather and the extent of damage to the aircraft so far as is known;

(i) A description of any explosives, radioactive materials, or other dangerous articles carried.

## Subpart D Reporting of Aircraft Accidents, Incidents, and Overdue Aircraft

*DOT 430.15 Reports and Statements to be Filed*

(a) Reports. The operator of an aircraft shall file a report as provided in paragraph (c) of this section on National Transportation Safety Board Form 6120.1 or 6120.2 (which forms are attached hereto and incorporated herein as part of this section)

(1) within ten (10) days after an accident for which notification is required by DOT 430.5 (a) or when after seven (7) days, an overdue aircraft is still missing.

(2) A report on an incident for which notification is required by 430.5 (a) shall be filed only as requested by an authorized representative of the National Transportation Safety Board.

(b) Crewmember statement. Each crewmember, if physically able at the time the report is submitted, shall attach thereto a statement setting forth the facts, conditions, and circumstances relating to the accident or incident as they appear to him to the best of his knowledge and belief. If the crewmember is incapacitated, he shall submit the statement as soon as he is physically able.

(c) Where to file the reports. The operator of an aircraft shall file, with the field office of the National Transportation Safety Board nearest the accident or incident, any report required by this section.

Note: The National Transportation Safety Board field offices are listed under U.S. Government in the telephone directories in the following cities: Anchorage, Alaska; Chicago, Ill.; Denver, Colo.; Fort Worth, Texas; Kansas City, Mo,; Los Angeles, Calif.; Miami, Fla.; New York, New York; Oakland, Calif.; Seattle, Wash.; Washington, D.C.

## Part 43—Maintenance, Preventive Maintenance, Rebuilding, and Alteration

Even though we, as pilots, will not be directly involved with aircraft maintenance, we may be responsible for it. Therefore, we should be familiar with some of the FAR's concerning maintenance.

*FAR 43.1 Applicability*

(a) Except as provided in paragraph (b) of this section, this part prescribes rules governing the maintenance, preventive maintenance, rebuilding, and alteration of any

(1) aircraft having a U.S. airworthiness certificate; and
(2) airframe, aircraft engine, propeller, or appliance of such an aircraft.

(b) This part does not apply to any aircraft for which an experimental airworthiness certificate has been issued unless a different kind of airworthiness certificate had previously been issued for that aircraft.

*FAR 43.3 Persons Authorized to Perform Maintenance, Preventive Maintenance, Rebuilding, and Alterations*

(a) Except as provided in this section, no person may maintain, rebuild, alter, or perform preventive maintenance on an aircraft airframe, aircraft engine, propeller, or appliance to which this part applies. . .

(b) The holder of a mechanic certificate may perform maintenance, preventive maintenance, and alteration as provided in Part 65 of this chapter.

(c) The holder of a repairman certificate may perform maintenance and preventive maintenance as provided in Part 65 of this chapter.

(d) A person working under the supervision of a holder of a mechanic or repairman certificate may perform the maintenance, preventive maintenance, and alterations that his supervisor is authorized to perform, if the supervisor personally observes the work being done to the extent necessary to ensure that it is being done properly and if the supervisor is readily available, in person, for consultation. However, this paragraph does not authorize the performance of 100-hour or annual inspections, nor inspections performed after a major repair or alteration.

(e) The holder of a repair station certificate may perform maintenance, preventive maintenance, and alterations as provided in Part 145 of this chapter.

(f) The holder of an air carrier certificate. . .

(g) The holder of a commercial operator certificate. . .

(h) The holder of a pilot certificate issued under Part 61 of this chapter may perform preventive maintenance on any aircraft owned or operated by him that is not used in air carrier service.

(i) A manufacturer may

(1) rebuild or alter any aircraft, aircraft engine, propeller, or appliance manufactured by him under a type or production certificate;
(2) rebuild or alter any appliance or part of an aircraft, aircraft engine, propeller, or appliance manufactured by him under a Technical Standard Order Authorization, an FAA-Parts Manu-

facturer Approval, or Product and Process Specification issued by the Administrator; and

(3) Perform 100-hour, annual, and progressive inspections on aircraft manufactured by him, while operating under a production certificate or under an approved production inspection system for such aircraft.

Note: A pilot may perform preventive maintenance on an aircraft operated by *him* if it is not used in air carrier service. Most maintenance must be performed by or supervised by a licensed repairman.

*FAR 43.7 Persons Authorized to Approve Aircraft Airframes, Aircraft Engines, Propellers, and Applicances for Return to Service after Maintenance, Preventive Maintenance, Rebuilding, or Alteration*

(a) Except as provided in this section, no person, other than the administrator, may approve any aircraft airframe, aircraft engine, propeller, or appliance for return to service after it has undergone maintenance, preventive maintenance, rebuilding, or alteration.

(b) The holder of a mechanic certificate or an inspection authorization may approve an aircraft airframe, aircraft engine, propeller, or appliance for return to service as provided in Part 65 of this chapter.

(c) The holder of a repair station certificate may approve an aircraft airframe, aircraft engine, propeller, or appliance for return to service as provided in Part 145 of this chapter.

(d) A manufacturer may approve for return to service any aircraft airframe, aircraft engine, propeller, or appliance that he has worked on under 43.3(i). However, except for minor alterations, the work must have been done in accordance with technical data approved by the Administrator.

(e) The holder of an air carrier certificate. . .

(f) The holder of a commercial operator certificate. . .

*FAR 43.9 Content, Form and Disposition of Maintenance, Rebuilding and Alteration Records (Except 100-hour, Annual and Progressive Inspections)*

(a) Maintenance record entries. Except as provided in paragraphs (b) and (c) of this section, each person who maintains, rebuilds, or alteιs an aircraft airframe, aircraft engine, propeller, or appliance shall make an entry in the permanent maintenance record of that equipment containing the following information:

(1) A description (or reference to data acceptable to the Administrator) of the work performed.

(2) The date or completion of the work performed.

(3) The name of the person performing the work.

(4) If the aircraft airframe, aircraft engine, propeller, or appliance is approved for return to service, the signature (and if a certified mechanic, the certificate number) of the person who approved it.

In addition to the entry required by this paragraph, major repairs and major alterations shall be entered on a form, and the form disposed of, in the manner prescribed in Appendix B, by the person performing the work.

(b) . . .

(c) . . .

Note: Common sense would dictate that most aircraft maintenance should be done by a certificated repairman. Too many pilots fancy themselves as mechanics, however, and become involved in maintenance. Without prior training and experience, we should not perform maintenance on an aircraft, even though it may be legal.

## Summary

We must remember that no attempt has been made here to present all FAR's. Only those considered most important have been included.

## CHECK YOUR READING

1. What is the definition of an *aircraft accident*?
2. What constitutes a *fatal injury*?
3. What constitutes a *serious injury*?
4. List the occurrences which must be immediately reported to the National Transportation Safety Board.
5. What maintenance may the pilot perform?

# PART X

# PREFLIGHT PLANNING

# 56

# PREFLIGHT PLANNING

Preflight planning should begin well in advance of the flight. Thorough preparation, with attention to all of the details will insure a safer, less hectic flight.

Our diet, rest habits, and mental attitude should be such that we will arrive at the airport in top condition. While driving to the airport, we may detect clues concerning the weather, such as patches of ground fog, frost, instability clouds, etc.

After arriving at the airport, the following sequence of events should generally be adhered to. As we gain experience in flight, we may add or delete items.

## Dispatch

We must check in with the dispatcher or operations manager, so that he will know that we are at the airport, and will be using the airplane as scheduled. We may request that the airplane be removed from the hangar, fueled to our specifications (if the weight and balance has been completed), and made ready for our preflight inspection.

The dispatcher may also have knowledge of the weather at or near our destination, or along our intended route of flight, which he gleaned from other pilots who have checked in after their flights.

## Weather Briefing

A detailed weather briefing is not only required by law, but certainly makes good sense. At some airports we may visit the United States Weather Bureau for a face-to-face briefing. At other fields, we may talk to the weatherman by telephone or direct land-line. In either case, we will assist the weatherman by supplying quite specific information to him. We should include at least the following information:

1. State the type of pilot we are. That is, student, private, commercial, etc.

2. State the type of aircraft we are planning to fly. That is, light single-engine, light twin engine, high-performance multiengine, etc.

3. State the name of the destination. If the airport is a little-known one, include the geographical area. Be careful to include the name of the state if flying to a city which has a namesake in two or more states (such as Kansas City, Missouri and Kansas City, Kansas).

4. State the estimated time of departure and an estimate of the time enroute.

5. State clearly whether the flight must be made VFR or if the airplane and pilot are qualified for IFR flight.

The weather briefer will use all of the charts, aviation weather reports, and pilot reports at his command to supply you with the weather information listed here.

1. General weather synopsis (the position of lows, fronts, pressure ridges, etc.).
2. Current weather conditions at the departure point, along the route, and at the destination.
3. Forecast weather conditions for the departure point, the route of flight, and the destination.
4. Alternate routes, if necessary.
5. Hazardous weather, if existent.
6. Forecast winds aloft.

We must not be shy in our request for any other information we desire. Remember, the weatherman is there to help in any way possible.

After forming a good mental picture of the weather we may encounter, we should begin to formulate an alternate course of action. During our navigation planning, we will firm up our alternate plans.

## Navigational Planning

Some pilots prefer to do as much navigational planning as possible on the day before their flight. Others do all of the planning at the airport after receiving the weather briefing. In either case, the following general sequence will be followed.

1. The destination airport should be checked out in detail by consulting the Airman's Information Manual. There is no sense in planning a flight to an airport which does not fill our needs. Runway length, surface, and direction are of particular importance. Primary communications frequencies should be jotted down. Obstructions on or near the airport should be noted.
2. The true course must be drawn distinctly on the appropriate charts. The date of publication on the chart must show that it is current. If the course takes us close to the edge of the chart, we should take along the chart of the surrounding area. Prominent check points along the route may be circled. (Caution: red pencil lines may completely disappear under the rose-colored cockpit lights at night. Therefore, red marking pens or pencils should not be used on charts or maps.)

   Obstructions to flight along the course and for 10 miles or so on either side of the course should be noted and marked. Airports which may be used as alternates should be noted. Radio facilities to be used and their respective frequencies should be noted on our log form. A sample log form is shown in Fig. 176.
3. Distances between check points (either visual or radio) must be carefully measured. (Caution: We must take care to check the scale on the chart, so as to avoid errors in distance measuring.) The distances are entered on the log form.
4. Wind computation on the computer will result in a true heading and groundspeed, which is entered in the log form.
5. Variation is applied to the true heading to find the magnetic heading. When we enter the airplane, we will check the compass correction card for deviation.
6. The groundspeed is applied to the distance between each check point and to the total length of the course. These items are entered on the log.
7. Fuel requirements are dependent on the time enroute, and fuel for an alternate course of action. It is foolhardy indeed to plan a flight which does not allow for at least one hour of extra fuel.

## Weight and Balance

After the fuel requirements have been established, the weight and balance may be computed. (If we are the sole occupant of the aircraft, and the aircraft operations manual does not forbid it, we may use full fuel capacity and may have computed the weight and balance well before the day of the flight.)

There is no excuse, legal or otherwise, which condones flight at a gross weight greater than the maximum allowable.

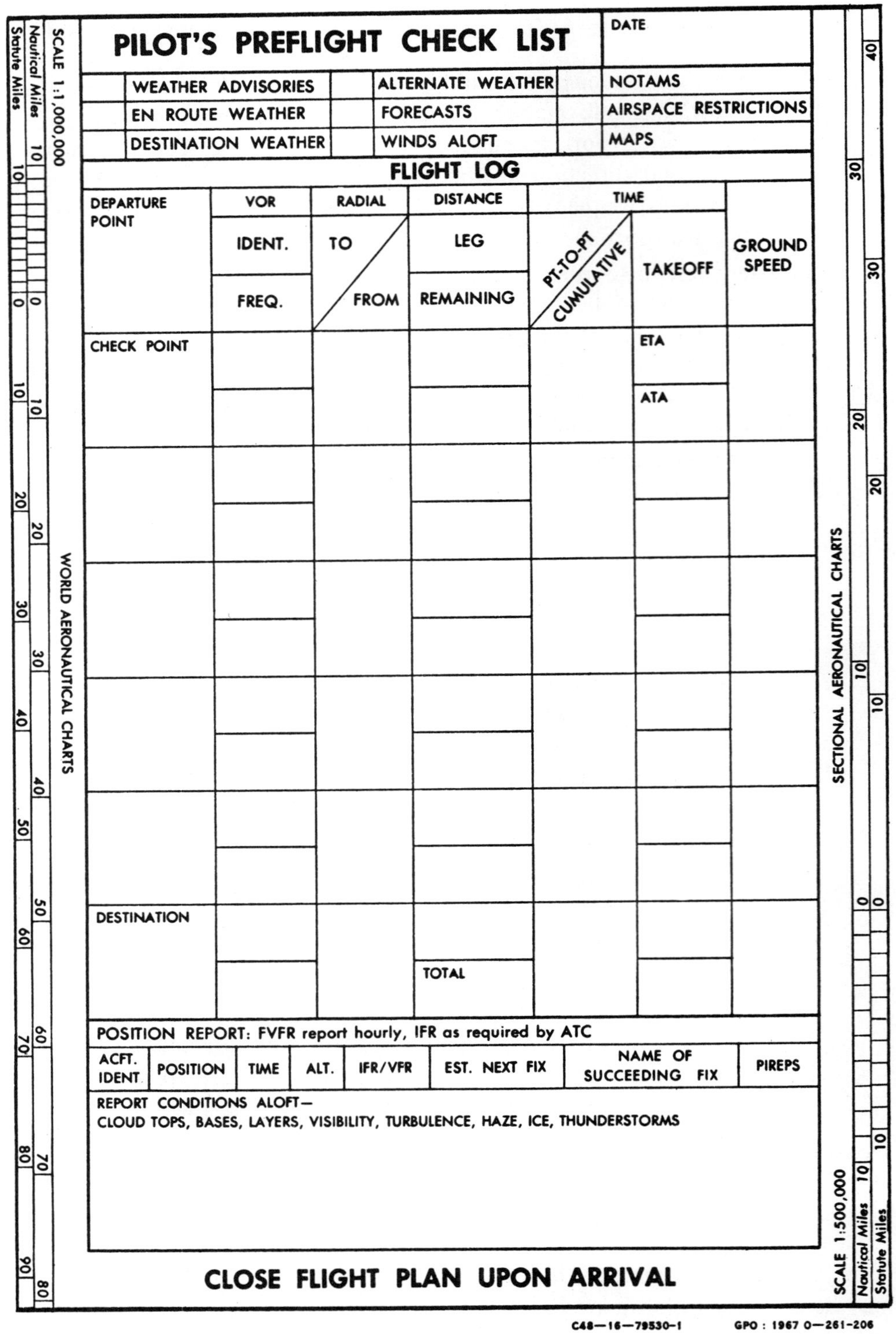

PILOT'S PREFLIGHT CHECK LIST

| | | | |
|---|---|---|---|
| WEATHER ADVISORIES | ALTERNATE WEATHER | NOTAMS | DATE |
| EN ROUTE WEATHER | FORECASTS | AIRSPACE RESTRICTIONS | |
| DESTINATION WEATHER | WINDS ALOFT | MAPS | |

FLIGHT LOG

| DEPARTURE POINT | VOR IDENT. / FREQ. | RADIAL TO / FROM | DISTANCE LEG / REMAINING | TIME PT-TO-PT / CUMULATIVE | TIME TAKEOFF | GROUND SPEED |
|---|---|---|---|---|---|---|
| CHECK POINT | | | | | ETA / ATA | |
| | | | | | | |
| | | | | | | |
| | | | | | | |
| | | | | | | |
| DESTINATION | | | TOTAL | | | |

POSITION REPORT: FVFR report hourly, IFR as required by ATC

| ACFT. IDENT | POSITION | TIME | ALT. | IFR/VFR | EST. NEXT FIX | NAME OF SUCCEEDING FIX | PIREPS |
|---|---|---|---|---|---|---|---|

REPORT CONDITIONS ALOFT—
CLOUD TOPS, BASES, LAYERS, VISIBILITY, TURBULENCE, HAZE, ICE, THUNDERSTORMS

CLOSE FLIGHT PLAN UPON ARRIVAL



C48—16—79530-1 GPO : 1967 O—261-206

Fig. 176. Sample log form.

## The Flight Plan

Although a flight plan is not required for VFR flight, common sense dictates that one be filed. Search and rescue procedures depend on flight plans for the speedy recovery of pilots in trouble.

A standard flight plan form is depicted in Fig. 177.

The flight plan form is self-explanatory. The following points should be particularly noted.

A. Block number 4 specifies *true* airspeed. We may compute true airspeed on the navigation computer, using the pressure altitude, temperature, and calibrated airspeed

B. Block number 6 specifies departure time in Greenwich Mean Time (Z). We will provide the proposed departure time. The actual departure time will be filled in by the ground station communicator after we depart. We must simply contact the nearest Flight Service Station and advise them to activate our flight plan.

C. Block number 7 requests our initial cruising altitude. We must remember to use an altitude in accordance with the quadrantial altitude rules for VFR flight.

D. Block number 8 requests our route of flight. We must be as specific as possible, listing either prominent check points or radio facilities between which we intend to fly. In the event of an emergency, search and rescue personnel will need to know our intended route. If, during the flight, we change our route, the nearest Flight Service Station should be informed of our new route.

E. Block number 10 is used for remarks, which may include points of intended landing, or any information which might aid search and rescue personnel should the need arise.

F. Blocks number 11 and 12 specify time enroute and fuel on board in *hours* and *minutes*.

The flight plan may be filed by telephone, by two-way radio, or in person at any Flight Service Station or other FAA communications facility. When we contact the Flight Service Station to file the flight plan, we should request any NOTAMS which pertain to our route of flight and any appropriate pilot reports.

Caution: The flight plan ***must be closed*** when the flight is completed, either by telephone, two-way radio, or in person. If the flight plan is not closed, search and rescue operations will be initiated within thirty minutes of our estimated arrival time. This is costly and may drain search and rescue personnel from being used on a legitimate emergency.

If, while enroute, we find that we will arrive more than 15 minutes early or late, we should notify the nearest Flight Service Station by radio.

Position reports radioed every hour to a Flight Service Station will increase the efficiency of search and rescue personnel should they be needed. The information to be included in the hourly position report is listed on the bottom of the sample log shown in Fig. 176.

## Preflight Inspection

A thorough preflight inspection should be accomplished with particular attention to fuel quantity.

Caution: We have completed a detailed weather and navigation log form and flight plan. We must be sure to take this information along on the flight. A reliable clock or watch is a must for cross-country flight. This should be part of the preflight inspection.

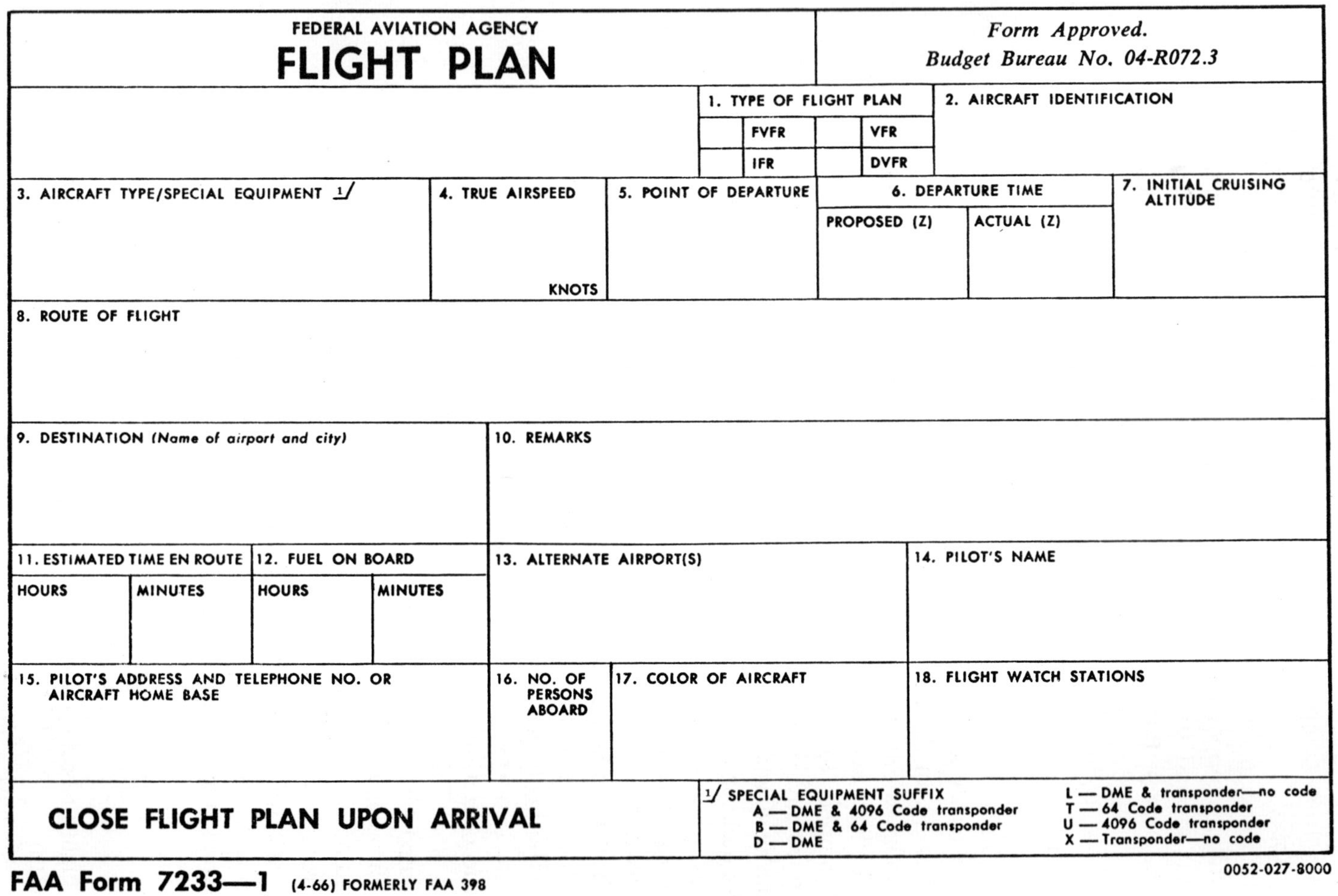
FEDERAL AVIATION AGENCY
FLIGHT PLAN

*Form Approved.*
*Budget Bureau No. 04-R072.3*

1. TYPE OF FLIGHT PLAN
FVFR
VFR
IFR
DVFR
2. AIRCRAFT IDENTIFICATION
3. AIRCRAFT TYPE/SPECIAL EQUIPMENT 1/
4. TRUE AIRSPEED
KNOTS
5. POINT OF DEPARTURE
6. DEPARTURE TIME
PROPOSED (Z)
ACTUAL (Z)
7. INITIAL CRUISING ALTITUDE
8. ROUTE OF FLIGHT
9. DESTINATION *(Name of airport and city)*
10. REMARKS
11. ESTIMATED TIME EN ROUTE
HOURS
MINUTES
12. FUEL ON BOARD
HOURS
MINUTES
13. ALTERNATE AIRPORT(S)
14. PILOT'S NAME
15. PILOT'S ADDRESS AND TELEPHONE NO. OR AIRCRAFT HOME BASE
16. NO. OF PERSONS ABOARD
17. COLOR OF AIRCRAFT
18. FLIGHT WATCH STATIONS

CLOSE FLIGHT PLAN UPON ARRIVAL

1/ SPECIAL EQUIPMENT SUFFIX
A — DME & 4096 Code transponder
B — DME & 64 Code transponder
D — DME
L — DME & transponder—no code
T — 64 Code transponder
U — 4096 Code transponder
X — Transponder—no code

FAA Form 7233—1 (4-66) FORMERLY FAA 398
0052-027-8000

Fig. 177. Sample flight plan.

Now that the formal preflight planning is completed, and the aircraft is prepared for flight, we should stop and relax for a moment or two.

We should review again, in our mind, the entire flight. Have we forgotten anything? Do we have time to complete the flight before darkness? Have we allowed time for a meal or coffee-break along the way? Are our passengers prepared for flight? (A last minute visit to the restroom might be suggested.)

The point to remember is "**don't rush**". If we think about the flight first, we can take off with a feeling of confidence.

FEDERAL AVIATION AGENCY
VFR EXAM-O-GRAM *NO. 4

PREFLIGHT PLANNING FOR A VFR CROSS-COUNTRY FLIGHT
(Series 1)

1. WHAT IS REQUIRED OF THE PILOT PRIOR TO THE FLIGHT? FAR, Part 91.5, states that "Each pilot in command shall, before beginning a flight, familiarize himself with all available information concerning that flight. This information must include, for . . . . a flight not in the vicinity of an airport, available weather reports and forecasts, fuel requirements, alternatives available if the planned flight cannot be completed, and any known traffic delays of which he has been advised by ATC."

2. WHY IS THIS REQUIRED? Careful preflight planning, in addition to satisfying FAR, enables the pilot to make his flight with greater confidence, ease, and safety. A review of fatal accident statistics for one year shows that as a "cause factor", inadequate flight planning was second only to "failure to maintain airspeed resulting in a stall."

3. WHAT ARE SOME SUGGESTED STEPS TO BE USED IN FLIGHT PLANNING?

(A) Assemble materials which will be needed on the flight such as current sectional charts, and other charts, for the route to be flown; the latest Airman's Information Manual (AIM), and plotter, computer, etc. Take along charts which adjoin those for the route of flight. Thus you are prepared in case it becomes necessary to circumnavigate bad weather, or in case you inadvertently fly off the chart on which your course is drawn.

(B) On the sectional chart, draw course to be flown; study terrain; select appropriate check points; consider caution, restricted, and prohibited areas and Air Defense Identification Zones; study airport information, including enroute airports that can be used in case of emergency; choose refueling stops; list frequencies of towers and navigational aids to be used and also Flight Service Stations reporting the weather.

(C) Review weather maps and forecasts, current weather reports, winds aloft forecasts, pilot weather reports, SIGMETS, advisories to light aircraft, Notices to Airmen (NOTAMS), and other information. Although you can get weather information by telephone, it is strongly recommended that a personal visit be made to the nearest Weather Bureau, Flight Service Station, or other flight service facility.

A chapter on Flight Planning is contained in the FAA publication, Private Pilot's Handbook of Aeronautical Knowledge, including a summary of flight assistance services available.

4. WHAT FURTHER ACTION IS DICTATED BY GOOD OPERATING PRACTICES? File a Flight Plan! This is not required by FAR but is dictated by good operating practice. It is extremely unlikely that air traffic rules can ever be written so as to eliminate the need for GOOD JUDGMENT in the planning and conduct of every flight. The pilot must make the final decision as to whether or not to make a flight. Use reasonable restraint in exercising this prerogative when preflight planning indicates the existence of marginal conditions of any kind.

BE SAFER -- FILE A FLIGHT PLAN

Department of Transportation

FEDERAL AVIATION ADMINISTRATION

## VFR PILOT EXAM-O-GRAM* NO. 5

### PREFLIGHT PLANNING FOR A VFR CROSS-COUNTRY FLIGHT (Series 2)

HOW MAY YOU OBTAIN WEATHER INFORMATION FOR PREFLIGHT PLANNING? Visit your local Weather Bureau (WB) airport station or your nearest FAA Flight Service Station (FSS), or other flight service facility for a thorough weather briefing. The latest weather maps, area forecasts, terminal forecasts, winds-aloft reports, winds aloft forecasts, advisories, hourly sequence reports, and pilot reports will be available. If a visit is impractical, telephone calls are welcomed. When telephoning, identify yourself as a pilot; state your intended route, destination, intended time of takeoff, and approximate time en route; and, advise if you intend to fly only VFR.

"FSS and Weather Bureau Telephone Numbers" section of the Airman's Information Manual (AIM) contains the location and telephone numbers of WB offices and FSS along with other pertinent information. Note the "restricted" telephone number listed for some WB Stations on which only aviation weather information is given. Some WB Stations have the Pilots' Automatic Telephone Weather Answering Service (PATWAS) which is a transcribed weather information service. For availability of weather information at various airports, check the Airport Directory, Airport/Facility Directory, or "FSS and Weather Bureau Telephone Numbers" section of AIM.

WHAT IMPROVEMENTS HAVE BEEN MADE TO PROVIDE MORE AND BETTER WEATHER INFORMATION FOR PILOTS? Equipment is provided at selected Flight Service Stations by which weather and Notice to Airman data will be recorded on tapes and broadcast continuously over the low-frequency (200-400 KHz) navigational aid facility, or VORs and VORTACs.

WHAT FURTHER PREFLIGHT WEATHER PLANNING SHOULD BE DONE TO OBTAIN IN-FLIGHT WEATHER INFORMATION? From your charts and the appropriate section of the Airman's Information Manual (FSS and Weather Bureau Telephone Numbers), make a list of the Flight Service Stations along your route that broadcast the weather information. In addition to the scheduled broadcast, you may also contact them at any time for further information.

WHAT IS RECOMMENDED BY GOOD OPERATING PRACTICES? If the preflight weather briefing reveals questionable or marginal weather, use reasonable restraint in flying VFR. File a flight plan. Maintain a close check on the weather through your Flight Service Stations. Be sure to close your flight plan upon arrival.

BE SAFER WITH A FLIGHT PLAN

FAA Aeronautical Center
Flight Standards Technical Division
Operations Branch
P. O. Box 25082
Oklahoma City, Oklahoma 73125



* Exam-O-Grams are non-directive in nature and are issued solely as an information service to individuals interested in Airman Written Examinations.

# FEDERAL AVIATION AGENCY
# VFR PILOT EXAM-O-GRAM * NO. 30
## FLIGHT PLANS (Series 1)

In normal day-to-day activities, people usually give some notification of their intentions, by one means or another. But, is sufficient information provided for timely action to be taken in case of failure to appear or return within a reasonable period? A flight plan as applicable to cross-country flying, is simply a brief description of the flight filed with the FAA, providing information of value to search and rescue activities in the event of an emergency while en route. The statement "NO FLIGHT PLAN FILED," which appears in many accident reports and accounts of extensive search operations for missing aircraft, indicates a degree of thoughtlessness on the part of the pilot for the safety of his flight.

> "Businessman-pilot flying daughter home from college crash lands in sparsely populated area resulting in injury to father and daughter. Rescue effected within three hours of ETA near flight plan course filed with FAA.
>
> "Family of five forced down by snow storm while on flight to winter resort. No flight plan filed with FAA and 5 days later family reported missing by relatives. Search parties after covering extensive area, find bodies with no injuries--expiration attributed to exposure."

Certain misconceptions concerning the filing of flight plans were brought to light by analyses of pertinent test items in airman written examinations and are explained in this Exam-O-Gram and Exam-O-Gram No. 31.

* * * * * * * * * * * * * * * *

DO FEDERAL AVIATION REGULATIONS (FAR) REQUIRE THAT FLIGHT PLANS BE FILED? With the exception of aircraft of foreign registry operating within the United States, Regulations do not require the filing of flight plans for VFR flights unless the flight is to penetrate an ADIZ (Air Defense Identification Zone). Filing is entirely at the pilot's discretion, but is recommended as good operating practice.

WHY SHOULD A FLIGHT PLAN BE FILED? A flight plan not only assures prompt search and rescue in the event you become overdue or missing, but it also permits en route stations and the destination station to render better service by having prior knowledge of your flight. It costs nothing except a few minutes of time to file a flight plan and may be the best "insurance" investment you ever made. On the other hand, failure to file could contribute to unpleasant moments during flight, or long periods of waiting for help when an emergency arises.

* Exam-O-Grams are non-directive in nature and are issued solely as an information service to individuals interested in Airman Written Examinations.

WHAT IS FLIGHT FOLLOWING SERVICE? It is a personalized lookout service to provide assistance specifically to you during your flight. In order to utilize this service, the aircraft must be equipped for two-way radio communications for reporting positions to FSS's and a flight plan requesting Flight Following Service (FVFR) must be filed. When Flight Following Service is requested, the departure station sends a flight notification to designated flight watch stations which are approximately one hour's flying time apart. Specialists at these watch stations continuously scan weather and aeronautical information pertinent to your flight. This information is given to you when you make your required position report to each watch station en route, or if it is considered urgent, attempts will be made to send it to you immediately through the nearest stations. An aircraft on a FVFR flight plan is considered overdue or unreported, and search action will be initiated when communication cannot be established after it fails to arrive 30 minutes after its ETA, or it misses contact with two successive flight watch points and 30 minutes have elapsed since the estimated time over (ETOV) the second of the two flight watch points. This type of "red carpet" service is yours just for the asking.

BY WHAT MEANS CAN A FLIGHT PLAN BE FILED? Whenever possible a personal visit should be made to the FSS to file the flight plan. This allows for a flight briefing by the FSS specialist and for the examining of weather information. If this is not practical, the flight plan may be filed by telephoning the FSS. In the event no other means are available, it may be filed by radio to the nearest FSS after becoming airborne.

MAY A FLIGHT PLAN BE FILED IF THE AIRCRAFT IS NOT RADIO EQUIPPED? YES! (However, Flight Following Service cannot be utilized.) In fact, it is of particular importance for pilots of non-radio equipped aircraft to file a flight plan since they have no means of obtaining in-flight assistance if the need should arise. Filing will at least alert the FAA to the route of flight and the expected time of arrival. However, this is not authorization to operate without radio to or from an airport at which there is an FAA control tower. The filing and closing of flight plans for these aircraft can be accomplished using normal ground communications.

SHOULD RADIO COMMUNICATIONS BE MAINTAINED WITH THE FSS WHEN OPERATING TO OR FROM THE AIRPORT ON WHICH IT IS LOCATED? Yes, although after May 17, 1965, it will no longer be a requirement for pilots of radio equipped aircraft to maintain communications with the FSS when they are within a five-mile radius of an uncontrolled airport where an FSS is located. Airport advisory service will continue to be available upon request and the procedure for obtaining this information is unchanged. Nonetheless, when a flight plan has been filed, pilots should report their actual takeoff time and arrival time to the FSS as soon as practicable.

WHAT DISPOSITION IS MADE OF FLIGHT PLANS? When the flight plan is filed before takeoff the FSS takes no action on it until informed of your actual departure time. This is normally done by radio. (In the case of non-radio equipped aircraft, the FSS will assume that you departed at a time upon which you agree.) The flight plan is then activated by the FSS transmitting through teletype circuits to the destination, the information pertinent to your estimated arrival.

MUST POSITION REPORTS BE MADE IF A FLIGHT PLAN HAS BEEN FILED? Reporting positions while en route on a VFR flight is required only if Flight Following Service has been requested (FVFR flight plan), or if the flight involves an ADIZ (DVFR flight plan). However, the making of periodic position reports for all VFR cross-country flights is strongly recommended whether or not a flight plan has been filed. It not only keeps the FSS's informed of your progress, but it provides you the opportunity to receive current information relative to the remainder of your flight such as changing weather conditions, airport advisories, and up-to-the-minute traffic information.

WHAT ACTION IS REQUIRED IF THE FLIGHT IS NOT CONDUCTED AS FILED? There are no objections to changing the route to other than that for which you filed, or to changing the destination while en route, provided the nearest FSS is notified as early as possible. So that you will not be considered overdue at your destination, the nearest FSS should also be informed if you do not expect to make good your planned ETA within 30 minutes. Remember, the rapidity of rescue, if it becomes necessary, will depend on how accurately your position may be determined. If the flight plan, as filed or revised, has been followed and your position is on course, rescue should be prompt.

HOW MAY THE ROUTE OR DESTINATION BE CHANGED EN ROUTE? Any significant change in the flight should be reported to the nearest FSS as soon as possible by radio. In addition to outlining the new route and/or destination, you will be required to provide the FSS with the revised estimated time en route and the remaining fuel on board. In the case of non-radio equipped aircraft, an alternate route or destination may be filed with the original flight plan, or a landing should be made to close and refile a flight plan.

WHAT MUST BE DONE AT THE COMPLETION OF THE FLIGHT? If a flight plan has been filed, FAR requires that an arrival notice be filed at the completion of the flight or when the flight terminates at other than the planned destination. This may be done (even if no tower or FSS is located at the destination) by contacting the nearest FSS by radio prior to landing, or by telephone after landing. Contrary to popular belief, the control tower at the airport of landing does not automatically close the flight plan when the landing is completed. The tower is not always aware of which aircraft were on a flight plan. It is the pilot's responsibility to file the arrival notice with the tower or FSS (simply asking by radio, telephone, or personal visit that the flight plan be closed).

WHAT ACTION IS TAKEN IF AN AIRCRAFT IS OVERDUE OR UNREPORTED? When communication cannot be established with the aircraft and it fails to arrive or report within 30 minutes of the ETA (the second of two consecutively missed flight watch stations for FVFR), a check of all local airports is made to determine if the aircraft has arrived there. If the aircraft has not been located within one and one half hours after the ETA, extensive communications inquiries are made throughout the flight range of the aircraft. If the results are negative or the aircraft has not been located within one hour after these inquiries are initiated, the full facilities of Search and Rescue (SARCC) are activated. Neglecting to close your flight plan may start an unnecessary search for you, with the consequent hazards to search pilots, and the waste of thousands of dollars.

* * * * * * * * * * * * * * *

FOR FLIGHT INSURANCE TAKE OUT "HELP ASSURANCE" -- FILE A FLIGHT PLAN
(See Exam-O-Gram No. 31 for complete information on how to fill out a flight plan)

FAA Aeronautical Center
Flight Standards Technical Division, AC-700
Operations Branch, AC-740
P. O. Box 25082
Oklahoma City, Oklahoma 73125

Exam-O-Grams available free of charge in limited quantities from this office.

FEDERAL AVIATION AGENCY
VFR PILOT EXAM-O-GRAM * NO. 31
FLIGHT PLANS (Series 2)

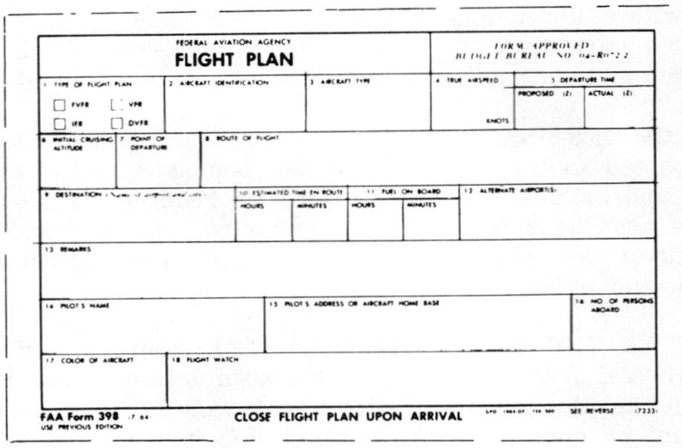


The importance and the means of filing a flight plan for a VFR flight are described in Exam-O-Gram No. 30. In order that full advantage of these services may be realized, it is fitting that the pilot be familiar with the flight plan form itself. The form is available at any FSS and at many fixed base operator's offices. Before filing the flight plan every pilot is urged to receive a preflight briefing from the FSS specialist. A Pilot's Preflight Checklist (see below) is provided on the reverse side of the form to remind the pilot of the information which he should obtain. This includes WEATHER ADVISORIES (hazardous weather that is forecast in SIGMETS and AIRMETS); EN ROUTE WEATHER (covering the stations along the proposed route); DESTINATION WEATHER; ALTERNATE WEATHER (for stations to which you may proceed if unable to reach the destination airport); FORECASTS (for stations along the route); WINDS ALOFT (for the altitudes at which the most favorable winds may be found); NOTAMS (changes or hazards appropriate to the route of flight); and adequate MAP (chart) coverage. Provided beneath the checklist is a flight log for the pilot's use if desired. Just beneath the flight log is a reminder of the information that should be given in position reports.

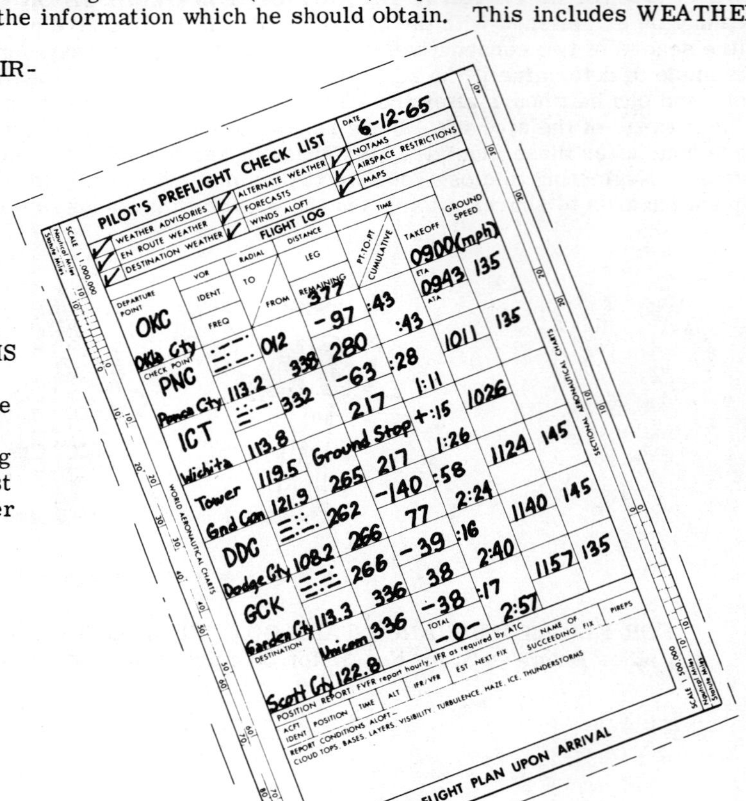


* Exam-O-Grams are non-directive in nature and are issued solely as an information service to individuals interested in Airman Written Examinations.

WHAT INFORMATION IS REQUIRED ON A FLIGHT PLAN? If you are uncertain as to the information needed or do not have the form available, the FSS specialist will assist you in completing the form or ask for specific information. However, the time spent in filing will be reduced considerably if you know ahead of time the type of information needed. Although the sample below is the most recent form, previous editions may be used since the information is generally the same.

FEDERAL AVIATION AGENCY

**FLIGHT PLAN**

*FORM APPROVED BUDGET BUREAU NO. 04-R072.2*

| 1. TYPE OF FLIGHT PLAN | 2. AIRCRAFT IDENTIFICATION | 3. AIRCRAFT TYPE | 4. TRUE AIRSPEED | 5. DEPARTURE TIME |
|---|---|---|---|---|
| ☑ FVFR ☐ VFR ☐ IFR ☐ DVFR | N 1234 B | Texaire 250 | (140 mph) 120 KNOTS | PROPOSED (Z) 1500 (0900C) / ACTUAL (Z) |

| 6. INITIAL CRUISING ALTITUDE | 7. POINT OF DEPARTURE | 8. ROUTE OF FLIGHT |
|---|---|---|
| 7500 | Will Rogers OKla City | Via V-77 PNC/Direct ICT/ Direct DDC/ Via V-10 GCK/ Direct Scott City |

| 9. DESTINATION (Name of airport and city) | 10. ESTIMATED TIME EN ROUTE HOURS | MINUTES | 11. FUEL ON BOARD HOURS | MINUTES | 12. ALTERNATE AIRPORT(S) |
|---|---|---|---|---|---|
| Scott City Muni. Scott City, Kans. | 3 | 00 | 4 | 30 | NR |

13. REMARKS :15 min. stop at ICT. Will close flight plan with GCK FSS.

| 14. PILOT'S NAME | 15. PILOT'S ADDRESS OR AIRCRAFT HOME BASE | 16. NO. OF PERSONS ABOARD |
|---|---|---|
| U.R. Wright | On file at OKC. | 3 |

| 17. COLOR OF AIRCRAFT | 18. FLIGHT WATCH |
|---|---|
| White — Red Trim | PNC — ICT - GCK |

FAA Form 398 (7-64) USE PREVIOUS EDITION    CLOSE FLIGHT PLAN UPON ARRIVAL    GPO 1964 OF—739-500    SEE REVERSE    (7233)

TYPE OF FLIGHT PLAN - Check the appropriate square for the type of flight. VFR is for a flight conducted in accordance with visual flight rules. DVFR is for a VFR flight which penetrates an ADIZ. FVFR is for a VFR flight with a request for Flight Following Service. IFR is to be checked for flights to be conducted under instrument flight rules.

AIRCRAFT IDENTIFICATION - Enter the full registration number of the aircraft (N1234B).

AIRCRAFT TYPE - Enter the aircraft manufacturer and the model designation; i.e. Cessna 140 (use of the model designation alone may cause confusion with a similar designation of a different manufacturer).

ESTIMATED TRUE AIRSPEED - Enter the estimated true airspeed at your cruising altitude and not the indicated airspeed. Although speed in knots is preferred, miles per hour is acceptable if you indicate mph.

DEPARTURE TIME - Enter the time at which you expect to take off. Greenwich Meridian Time (Z) is desired although local time may be used if you so indicate. The actual takeoff time will be recorded by the FSS after you report your actual takeoff time.

INITIAL CRUISING ALTITUDE - Enter the altitude above sea level (MSL) at which you plan to fly the first leg of the route. (Based on the magnetic course to be flown if 3,000 feet or more above the surface.)

POINT OF DEPARTURE - Enter the name of the airport and the city from which you are departing. Remember, many airports have the same name and many cities have several airports.

ROUTE OF FLIGHT - Enter the route to be flown; if a straight course from departure to destination, enter DIRECT; if other courses, describe the courses by airway number, fixes, etc., (via V-14 AEX, 180° radial ALI VOR, DIRECT KGS). A careful and exact description is important in the event search and rescue or in-flight assistance becomes necessary.

DESTINATION - Enter name of destination airport and city at which the flight is to terminate.

ESTIMATED TIME EN ROUTE - Enter estimated elapsed time from takeoff to landing at destination. This should include ground time during any stops while en route to destination.

FUEL ON BOARD - Enter the flying time available at cruise power with the amount of usable fuel. (Do not include the unusable fuel.) Remember to subtract the fuel already used since the last re-fueling.

ALTERNATE AIRPORT - No entry is required for VFR flights.

REMARKS - Enter any remarks of importance to the flight such as special radio frequencies, or that no radio is available. Any stops en route and the ground time at each stop should also be entered.

FLIGHT WATCH STATIONS - For entering the stations which the FSS may instruct you to contact while en route if you have requested Flight Following Service (FVFR flight plan).

* * * * * * *

A FLIGHT PLAN PROTECTS YOU WITH A "HELP ASSURANCE" PLAN.
FILE IT BEFORE DEPARTURE - CLOSE IT UPON ARRIVAL.

FAA Aeronautical Center
Flight Standards Technical Division, AC-700
Operations Branch, AC-740
P. O. Box 25082
Oklahoma City, Oklahoma 73125

Exam-O-Grams available free of charge in limited quantities from this office.

Permission is hereby granted to reproduce this material.

FAA AC 65-4003

FEDERAL AVIATION AGENCY
VFR EXAM-O-GRAM *NO. 34

## HOW TO OBTAIN PROPER WEATHER BRIEFING

The Civil Aeronautics Board Statistical Review of General Aviation accidents for calendar year 1963 indicates that about one in every four fatal accidents in General Aviation is the result of pilots continuing flight in adverse weather. This report lists weather as a causal factor in 1, 039 accidents of which 296 were classed as fatal accidents and 743 were non-fatal.

Most FAA General Aviation Written Examinations contain test items which deal with proper "Preflight Action" as required by Federal Aviation Regulations. Obtaining sufficient information concerning enroute and destination weather before beginning a flight is one of the most important parts of preflight action. The illustration and telephone conversation below is an example of an individual who "thinks" he has checked the weather.

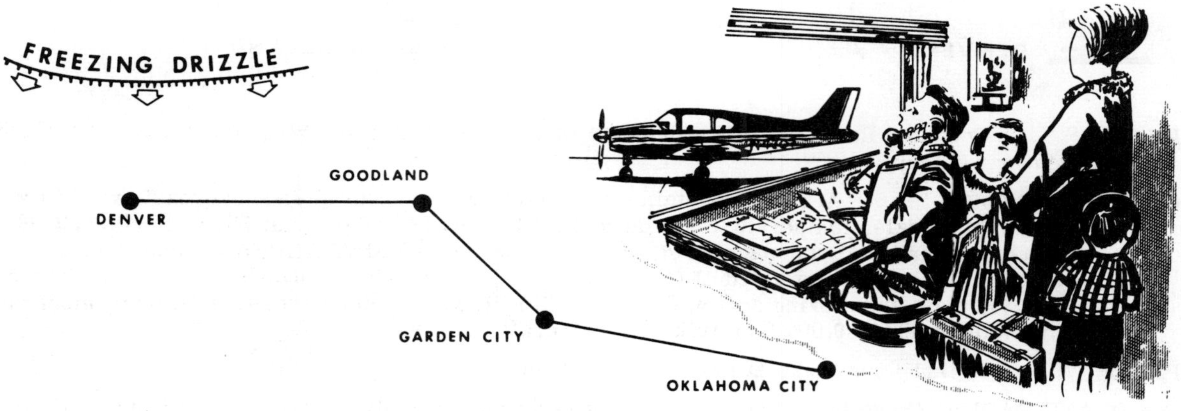


PILOT: "This is Tom Jones. Will you give me the latest Garden City, Goodland and Denver sequence reports and the winds at eighty-five hundred feet?"

WEATHER BRIEFER: "Yes sir, at 1100 Central time, Garden City was reporting clear skies, visibility more than 15, surface wind 140 degrees at 20 knots. At Goodland the visibility is still good and they report scattered cirriform clouds. . . their wind is 150 degrees at 20 knots. At 1000 Mountain time, Denver is clear, visibility 50 miles, surface wind is from 040 degrees at 10 knots. Winds at eight thousand five hundred feet will average 220 degrees at 30 knots. Sir, would you like a weather briefing for your flight? A front lies. . . " (the pilot interrupts at this point.)

PILOT: "No, that's all right, thank you. I'm in a hurry. I'll check the weather along the way."

★★★★★

**The above illustration portrays a pilot who is in a hurry to fly his family on an extended cross-country flight of approximately 463 miles. The weather looks good at his departure point and the hourly sequence reports indicate that present weather conditions are favorable along the route. However, this pilot is too anxious to get into the air -- too anxious to be on his way. He probably doesn't realize how rapidly the weather can deteriorate in the 3 hours that will be required to reach his destination--or--he is a careless or inexperienced pilot whose flying is characterized by poor judgment. If Mr. Jones had stayed on the phone for perhaps another minute, the briefer would have given him information (Terminal and Area Forecasts, AIRMETS, etc.) which should have changed his mind about attempting the flight. He would have learned that the proposed flight would take him into rapidly deteriorating ceilings and visibilities as well as freezing drizzle. The briefer offered the pilot more information and it may have appeared to him that the briefer was questioning his competence. Nevertheless, the pilot should have listened to what the briefer had to offer. Sometimes it may be superfluous--often it is vitally important.**

★★★★★

ENROUTE WEATHER BRIEFING . . .

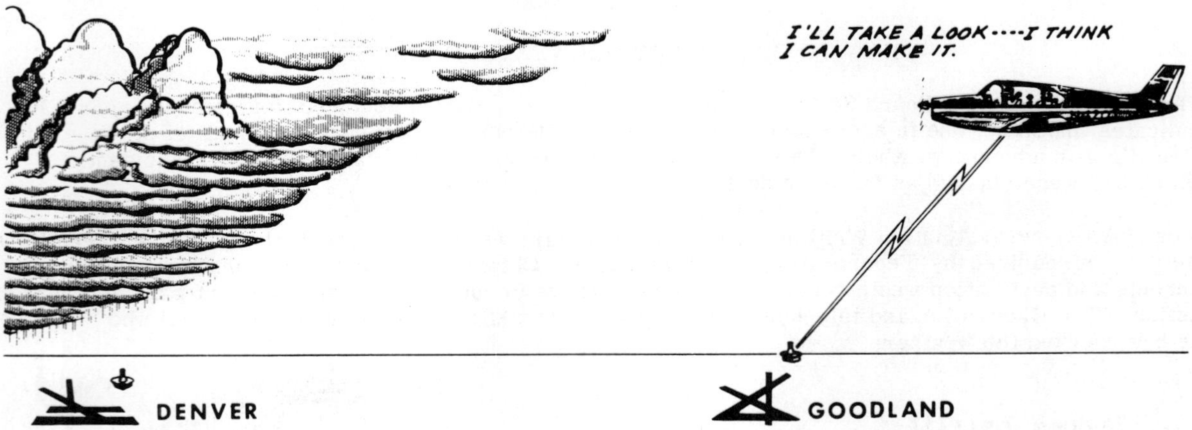


PILOT: ". . . Goodland Radio, I am 5 miles east of Goodland at 6,500. Will you give me the latest Denver weather . . . ?"

GOODLAND RADIO: ". . . At 1200 Mountain time Denver was measured 1500 overcast, visibility 10, very light drizzle, temperature 38, dewpoint 34, wind 040 degrees at 15, altimeter 29.98, low clouds northeast approaching station. . . Denver AIRMET ALPHA 2, moderate icing in precipitation in northern third of Colorado east of the Rockies, conditions continuing beyond 1500 Mountain and moving southward. . . Pilot Report at 50 northeast of Denver, moderate rime icing 6,000 to 9,000 feet MSL, type aircraft unknown. . . "

PILOT: "I believe that I can make Denver okay VFR, don't you?"

GOODLAND RADIO: "Negative. The conditions have dropped rapidly during the past hour. I'll give you the Denver forecast."

PILOT: "That's okay, I'll take a look and I'll turn around if it gets too bad. It's 1500 and 10 at Denver--I think I can make it."

○○○

(Epilogue - Unfortunately, he didn't make it!
The pilot encountered freezing drizzle, crashed,
and all occupants aboard were killed.)

So often pilots are given SIGMETS, AIRMETS, or Pilot Reports, but they do not realize the significanc of the information--nor have they gained a lasting respect for the forces of weather. Even though this pilot was made aware of the potential weather hazards on the flight from Goodland to Denver, he did no heed the warning received. It is not unusual for a pilot to be motivated to continue flight into deteriorating weather in order to: Keep a speaking engagement, attend a party with friends, get home to the family, or a multiplicity of other seemingly important reasons. Those oft spoken words "I think I can make it," are too frequently the pilot's last transmission.

★★★★★

WAS THE "PREFLIGHT ACTION" BY THE PILOT IN ACCORDANCE WITH REGULATIONS? No! FAR Part 91 states in part: "Each pilot in command shall, before beginning a flight, familiarize himself with all available information concerning that flight. This information must include, for a flight under IFR or a flight not in the vicinity of an airport, available weather reports and forecasts, fuel requirements, alternatives available if the planned flight cannot be completed . . . ."

★★★★★

This Exam-O-Gram covers several of the most common methods by which a pilot can obtain weather briefings. In certain localities, there are other ways that a pilot can receive weather information.

WHAT IS THE PROPER WAY TO REQUEST A WEATHER BRIEFING BY TELEPHONE? You get faster service and greatly assist the weather briefer by telling him:

1. That you are a pilot. (If you are a student, private, or commercial pilot--say so. The weather briefer needs to know that you are a pilot, not someone who calls just to find out the general weather picture.
2. The type of airplane you are planning to fly. (light single engine, high performance multi-engine, and jets all present different briefing problems.)
3. Your destination. (If you plan to stop somewhere enroute or deviate from the normal course, you should tell the briefer your intentions.)
4. Your estimated departure time.
5. Whether or not you can go IFR. (Instrument rated? . . . Aircraft equipped?)

WHAT PERTINENT INFORMATION SHOULD A WEATHER BRIEFING INCLUDE? A preflight weather briefing will be incomplete unless it includes:

1. Weather synopsis (position of lows, fronts, ridges, etc.).
2. Current weather conditions.
3. Forecast weather conditions.
4. Alternate routes (if necessary).
5. Hazardous weather.
6. Forecast winds aloft.

NOTE: All of these items are on the weather briefer's checklist!

★ ★ ★ ★ ★

WHERE CAN YOU OBTAIN A WEATHER BRIEFING? Flight Service Stations (FSS) provide aviation weather briefing service, and at major strategically located cities weather briefings are also available through both the FSS and Weather Bureau Airport Station (WBAS). At many locations, the Weather Bureau furnishes Pilot's Automatic Telephone Weather Answering Service (PATWAS), which means that the weatherman records a briefing that is available to the pilot over his local telephone. Pilots may receive continuous broadcasts of weather information over many of the low and medium frequency navigational aids - known as Recorded Weather Briefings (TWEB). TWEB and PATWAS are similar in that they provide weather information for a radius of 250 miles.

WHERE CAN YOU FIND A LISTING OF FSS AND WEATHER BUREAU TELEPHONE NUMBERS? The Weather Bureau and Flight Service Stations provide weather briefings through both listed and unlisted telephone numbers. The unlisted numbers for all Weather Bureau offices and FSS's providing this service are published in the back pages (Section IVA) of the Airman's Information Manual. See excerpt.

AIRMAN'S INFORMATION MANUAL Excerpt

FSS AND WEATHER BUREAU TELEPHONE NUMBERS

OKLAHOMA

| Location | | Area Code | Telephone |
|---|---|---|---|
| Ardmore (AAS) | FSS | (405) | CA 3-6772 |
| | FSS | (405) | DU 9-5441 |
| Gage (AAS) | FSS | (405) | 923-2601 |
| Hobart (AAS) | FSS | (405) | GE 6-4234 |
| McAlester (AAS) | FSS | (918) | GA 3-4091 |
| Oklahoma City (Will Rogers) | FSS | (405) | SU 9-7452 |
| | WB | (405) | 685-4433 ★ |
| | WB | (405) | 685-6350 ◆ |
| (Answered personally after PATWAS message.) | | | |
| Ponca City (AAS) | FSS | (405) | RO 5-5485 |
| Tulsa | FSS | (918) | TE 8-3316 |
| | WB | (918) | 835-4775 ◆ |
| | | | 835-2364 ★ |

★ Indicates Pilot's Automatic Telephone Weather Answering Service (PATWAS) available, providing transcribed aviation weather information.

◆ Indicates a restricted number, use for aviation weather information

★ ★ ★ ★ ★

For long cross-country flights or flights in marginal weather, the pilot may choose a face-to-face briefing by FSS or Weather Bureau personnel. If a planned flight is short (250 miles or less) a visit with a weather briefer may be unnecessary except in marginal or poor weather situations. Often, a briefing by telephone or the information contained in recorded weather briefings (PATWAS or TWEB) will fill a pilot's needs. NOTE: At the conclusion of a PATWAS recording, you may get additional information by holding the phone and waiting for the weather briefer to answer.

## HOW CAN YOU DETERMINE WHERE RECORDED WEATHER BRIEFINGS (TWEB) ARE AVAILABLE?

By referring to the Airport/Facility Directory section of Airman's Information Manual (AIM). In the excerpt to the right, notice the black dot which appears after the frequency and indicates that recorded weather briefings are available. A pilot on the ground at a small airport 50 miles from Will Rogers World Airport might utilize this service by tuning his ADF (or low frequency) receiver to 350 kcs, and thereby receive continuous transcribed weather briefings of the area around Will Rogers within a 250 mile radius.

**OKLAHOMA—*Continued***

AIRPORT/FACILITY DIRECTORY Excerpt

**WILL ROGERS WORLD** ***IFR*** 6 SW **FSS:** OKLAHOMA CITY (DL)
1284 H98/17R-35L(4) (S-120, T-160, TT- ) BL4, 6.
S5 F5, JP1, JP4 U2 **VASI:** Rnwy 17
**Remarks:** Rnwy 3-21 and txwy (C- d
to 15000 lbs or less. N ngt;
3800' N dalgt NM NE. 1563'
(2726' MSL) 49' MSL) twr 11 NM NNE.
Due non-vsby ctc twr.
**Tower** 118.3 122.7R **Gnd Con** 121.9
**Radar Services:** (BCN)
**Oklahoma City App Con** 119.3 134.1x 122.5R 120.5
**Oklahoma City Dep Con** 121.1
**ASR** Rnwy 17, 35 Ceil 400 Vsby 1 mi Min Alt 1684
**ILS**[1] 109.9 I-OKC Apch Brg 351° **LOM:** 219/OK
**Oklahoma City (H) BVORTAC** 115/OKC 098° 8.1 NM to rnwy 12
**Oklahoma City RBn H-SAB** 350●/OKC 092° 1.6 NM to fld
**Remarks:** [1]Glide slope unusable below 1417' MSL. **VOT:** 108.8

○ ○ ○

**AIRPORT DIRECTORY** Excerpt

**CHICKASHA**
**CHICKASHA MUNI** 3 NW
1146 H38 (1) BL4 S5 F4 U-1
FSS: OKLAHOMA CITY (LC CA 4-6440)
**Remarks:** Attended dalgt hrs.

WHAT IS THE MEANING OF FOREIGN EXCHANGE TELEPHONE SERVICE AS PROVIDED BY FAA? Pilots departing from many of those airports having neither a Weather Bureau Station nor a FSS may call a nearby FSS on foreign exchange telephone service provided by FAA at no cost to the pilot. For example: In the Oklahoma City area there are 3 surrounding towns that have been provided this service. These towns are Norman, Chickasha, and Stillwater, and the distances from Oklahoma City are 22, 27, and 62 miles respectively. Calls can be placed to Oklahoma City FSS from any telephone located in these towns by dialing the local number listed in Airport Directory section of AIM. See excerpt for Chickasha, which lists the local call number CA 4-6440.

FAA Aeronautical Center
Flight Standards Technical Division, AC-700
Operations Branch, AC-740
P. O. Box 25082
Oklahoma City, Oklahoma 73125

Exam-O-Grams available free of charge in limited quantities from this office.



☆ ☆ ☆ ☆ ☆

PROPER AND COMPLETE WEATHER INFORMATION IS A PILOT'S BEST LIFE INSURANCE - IT PAYS OFF IN LIFE INSTEAD OF DEATH.

## CHECK YOUR READING

1. When should flight planning begin?
2. List the information which should be given to the weather briefer prior to receiving a weather briefing.
3. List the information contained in a weather briefing.
4. Why is the filing of a flight plan good practice although not required for VFR flight?
5. How may the flight plan be filed?
6. How may the flight plan be activated?
7. How may the flight plan be closed?

# SUMMARY

## Summary

If you have diligently studied the material in this guide; if you have given your very best in an unrelenting quest for knowledge; if you have conscientiously applied your knowledge to the practical aspects of actual flight, you have built a firm foundation in flight training.

The end of this guide book certainly should not signify the end of your learning, for the surface has just been scratched. Mountains of knowledge lie before you. The fascinating truth about flight training is that it never ends. The more you learn, the more you will want to learn. The limit to your learning lies among the stars and beyond. For, as our astronauts delve deeper and deeper into space, they find new questions which must be answered.

> "A little learning is a dangerous thing;
> Drink deep, or taste not the Pierian spring;
> There shallow draughts intoxicate the brain,
> And drinking largely sobers us again."
> Alexander Pope

# ANSWERS TO "CHECK YOUR READING"

## CHAPTER 2

1. The four forces of flight are: Lift, Gravity, Thrust, Drag.
2. Drag is the natural force which prevents an aircraft from accelerating to an infinite speed when thrust is increased.
3. Lift is produced by an airfoil in two ways. At least fifty per cent of the lift is produced by the formation of a low pressure area above the lifting surface. The pressure differential between the air under the lifting surface and the air above the lifting surface forces it towards the area of lower pressure. The remaining lift is produced by ram air packing under the wing.
4. Bernoulli's theorem states that the pressure exerted by a moving fluid decreases as the speed of the fluid increases.
5. The amount of lift produced by an airfoil is determined by the shape and area of the airfoil, the speed of the air moving by it, the density of that air, and the angle at which the air strikes the airfoil.
6. Relative wind is the air which passes the aircraft due to the relative motion between the aircraft and the air. Relative wind is not in any way affected by the natural wind when the aircraft is in flight.
7. The natural wind has no effect on the aircraft in flight. It does however, affect groundspeed and track.
8. The angle of attack is the angle between the relative wind and the chord of the lifting airfoil.

9. A stall is a condition of an airfoil in which lift is destroyed due to the burbling or tearing away of the relative wind.
10. The stall is caused by an excessive angle of attack.
11. The angle of incidence is the angle between the chord of the wing and the longitudinal axis of the aircraft. The slight positive angle of incidence produces a small angle of attack when the longitudinal axis is parallel to the path of flight. The angle of incidence is normally greater on the left wing than on the right wing to diminish the rolling effect of torque.
12. Any aircraft will stall at any airspeed at or above the minimum stall speed.
13. Any aircraft may stall from any attitude. The stall is a function of angle of attack, not aircraft attitude.

## CHAPTER 3

1. The three axes of the aircraft and the movement of the aircraft about each axis are:
   a. Longitudinal—roll.
   b. Lateral—pitch.
   c. Vertical—yaw.
2. The primary purpose of the aileron is to cause the aircraft to roll about its longitudinal axis.
3. The force which makes an aircraft turn is the force of lift which, when the aircraft is banked, is diverted towards the direction of the turn.
4. More lift must be produced in a level turn than in straight and level flight because a portion of the total lift is being used to *pull* the aircraft into a curved path. The other portion of the total lift is used to counteract gravity.
5. Aircraft pitch attitude must be controlled because pitch may control aircraft speed and lift production.
6. Six examples of torque use in everyday life are:
   a. The twisting of a doorknob.
   b. The opening of a water faucet.
   c. The turning of an automobile steering wheel.
   d. The use of a screwdriver to drive home a wood screw.
   e. The dialing of a radio frequency selector.
   f. The removal of a screw-off lid on a jar.
7. Under normal circumstances right rudder pressure must be applied when using high power and in a high pitch attitude (low airspeed) condition, due to torque which acts to yaw the aircraft nose to the left.
8. Left rudder pressure must normally be used when the aircraft is developing cruise power or less (but high airspeed), due to the absence of torque. In these conditions, the built-in torque corrections on the aircraft overcorrect for the amount of torque being produced, and left rudder must be applied to counteract these built-in corrections.
9. The aircraft manufacturer attaches the vertical stabilizer at a point slightly to the left of the longitudinal axis in order to counteract torque.
10. The angle of incidence is greater on the left wing than it is on the right wing in order to counteract torque.

11. If an aircraft tends to fly left wing low, the aileron trim tab on the right wing must be bent down. The aileron trim tab on the left wing would be bent upward.
12. The pilot should keep the aircraft in trim to reduce pilot fatigue and to insure constant flight attitude in the event his attention should be momentarily diverted to other cockpit duties.

## CHAPTER 4

1. When flaps are lowered, the camber and the chord line of the lifting airfoil are changed.
2. Drag which is produced when lift is produced is called *induced drag*.
3. Specific amounts of flaps to be used in specific flight conditions can be found in the Aircraft Operations Manual or the Aircraft Owners Manual.
4. Normally, less than full flap is used for takeoff because drag increases with higher flap settings to such an extent as to delay the takeoff.
5. The primary use for flaps on approach for landing is to increase the angle of descent without increasing the airspeed.
6. Flaps normally lower the minimum stall speed of an aircraft.
7. When using flaps we should exercise the following cautions:
   a. Airspeed must be within flap operating range. (This speed range is shown on the airspeed indicator as a white arc.)
   b. Flap deflection should be tested while on the ground to determine that both flaps actuate in the same direction and to the same degree.
   c. We must remain alert for rapid pitch changes which may be extreme on some aircraft when the flaps are lowered or raised.
   d. Power changes with flaps extended may also cause rapid pitch changes.
   e. Certain aircraft designs are such that full flap deflection may blanket out the smooth flow of air passing the tail surface. We must be prepared for reduced rudder and elevator efficiency when the flaps are lowered.
   f. Due to increased drag when the flaps are lowered, the aircraft will accelerate slowly and may climb with less than normal performance.
   g. When flaps are retracted in flight, we must expect a serious loss of lift, therefore, pilots prefer to "milk" or slowly raise the flaps when near the surface.
8. The normal procedure for raising flaps on a balked landing or *go-around* is to "milk" or slowly raise the flaps only after determining positively that the aircraft is under control and in a climbing attitude.
9. The primary difference between a plain flap and a Fowler flap is that the Fowler flap changes the camber of the wing, the angle of attack, and the area of the wing.

## CHAPTER 5

1. The weight being supported by the wings of an aircraft will be greater than the total weight of the aircraft whenever a load factor greater than 1*g* is affecting the aircraft.
2. The load factor on an aircraft in straight and level unaccelerated flight is 1*g*.

3. The load factor of an aircraft in a 60° banked level turn is $2g$.
4. The load factor in a 70° banked level turn is approximately $2.9g$.
5. The effective weight of a 6,500 pound aircraft in a 70° banked level turn is approximately 18,850 pounds.
6. The effective weight of 150 pounds of extra baggage when the aircraft is in a 60° banked level turn is 300 pounds.
7. Aircraft speed has no effect on load factor in a level turn.
8. "Negative *g*" is a resultant force which opposes natural gravity.
9. The maximum allowable load factors for most light aircraft are: 4.4 positive *g* and 1.76 negative *g* (Specific Owner's Manual and/or Aircraft Operations Limitations must be relied upon for specific aircraft).
10. Some of the instances in which a load factor of more than $1g$ may be encountered are:
    a. Level turns.
    b. Pullout from descending flight.
    c. Turbulence.

## CHAPTER 6

1. Included in the maximum allowable gross weight of an aircraft is the weight of the aircraft and everything in it or attached to it.
2. The empty weight of an aircraft is the weight of the aircraft and attached equipment plus unusable liquids.
3. The useful load of an aircraft includes passengers and crew, baggage, useful liquids, and all unattached equipment.
4. Hazards to flight in an overweight aircraft include: structural damage, either localized or massive; instability, slow control response, and/or a decrease in performance up to the inability to fly.
5. Aviation gasoline weighs 6 pounds per gallon.
6. Oil weighs $7\frac{1}{2}$ pounds per gallon.
7. When computing a weight and balance problem, weight of the unusable fuel is included in the aircraft empty weight.
8. The definition of *arm* is the distance from the fulcrum to the center of gravity of a weighted object.
9. The definition of *moment* is the product of the weight of an object multiplied by the arm of the object.
10. The definition of the *datum line* is an arbitrary reference line on or near the aircraft from which arms may be measured.
11. In light aircraft, if the final computed center of gravity is within the moment envelope at takeoff, it will remain within specified limits for the duration of the flight no matter how much fuel is burned off provided other weights on board are not shifted.
12. A weight and balance computation should be completed prior to takeoff.

## CHAPTER 7

1. A diagram of vertical gusts striking the airfoil, thereby changing the angle of attack of the relative wind.

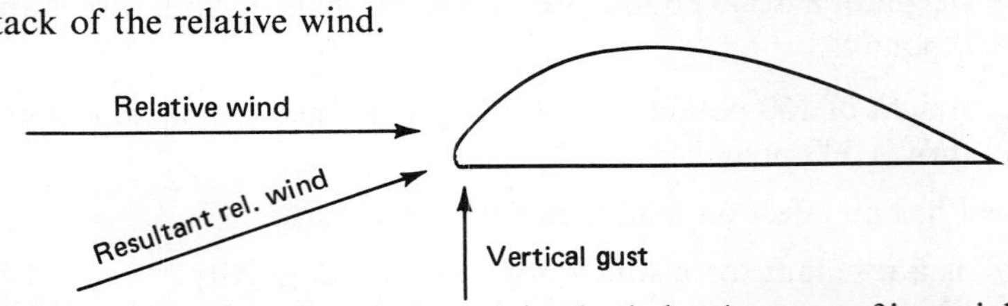


2. Momentum is the force which a moving body has because of its weight and motion.
3. The approximate maneuvering speed of an aircraft in which the minimum power off stall speed is 62 miles per hour is 105 miles per hour.
4. When encountering turbulence, airspeed should not be reduced to less than maneuvering speed due to a loss of control efficiency at slower speeds.
5. Maneuvering speed is normally found in the Aircraft Operations Limitations Manual. It may be in the form of a placard in the cockpit. Maneuvering speed is not noted on the airspeed indicator.
6. Cruising speed is a compromise between the speed which would result from maximum allowable continuous power and a much reduced power which would cause the aircraft to *mush.*
7. The cruising speed range is denoted on the airspeed indicator by a green arc.
8. The never-exceed speed is marked on the airspeed indicator by a red line.
9. We should not plan to fly an aircraft at speeds within the caution range because at those speeds any imposed severe load factor will structurally damage the aircraft and/or its components. Since turbulence may impose a load factor and since turbulence is invisible, the pilot has no way to determine when flight within the caution range is safe.
10. An aircraft will normally fly at a slower speed when the engine is producing thrust than it will when the engine is not producing thrust, first, because the slipstream from the engine is directed over a portion of the lifting surfaces which increases the relative wind, and second, because thrust tends to reduce mushing of the aircraft, which allows flight at lower speed before reaching the stall angle of attack.

## CHAPTER 8

1. We may expect to find engine wash on the surface when taxiing behind other operating aircraft, particularly in run-up areas, or near an active runway.
2. Wing-tip vortices are caused by high-pressure air under the wing spilling up and over the wing tip in an effort to fill the lower pressure area above the wing.
3. Wing-tip vortex intensity increases directly with the weight of the aircraft and inversely with the speed of the aircraft. Hence, a heavy slow aircraft will produce a more severe vortex than a light, fast aircraft.
4. When approaching an airport for landing behind a heavy aircraft, our approach path should be higher than the preceding aircraft, and the touchdown should be

planned at a point beyond the point of touchdown of the preceding aircraft (if runway length permits). This flight path will keep the aircraft above the wake turbulence of the preceding aircraft.

5. When taking off behind another aircraft, the upwind side of the runway should be used.
6. When intercepting the flight path of another aircraft, we should alter our altitude to one higher than the other aircraft to avoid the wake turbulence, which normally sags below the altitude of the producing aircraft.

## CHAPTER 9

1. *Static stability* is the initial tendency of any body to remain in its original position until disturbed.
2. *Dynamic stability* determines the action of a body after it has been disturbed.
3. *Positive dynamic stability* means a force is necessary to move an object, and once the force is removed, the object will return to its original point, but will oscillate before coming to rest.
4. An aircraft is longitudinally stable because of the system of "down loading" on the tail lifting surfaces. When aircraft speed changes the amount of lift produced by the wing and the tail decreases, thus forcing the nose down. As the nose drops, speed increases, thus increasing lift again.
5. An aircraft is laterally stable primarily due to *dihedral* which causes the low wing to produce lift which more nearly counteracts gravity than the lift on the high wing.
6. An aircraft is directionally stable due to the fact that the side area of the aircraft aft of the vertical axis is greater than the side area forward of the vertical axis. *Sweep back* in wing design also may be used to increase directional stability.
7. Improper aircraft loading may destroy aircraft stability.

## CHAPTER 10

1. The four strokes of the internal combustion reciprocating engine are:
   a. Intake stroke: the piston moves downward in the cylinder sucking the fuel-air mixture into the cylinder, through the intake valve.
   b. Compression stroke: the piston moves upward in the cylinder while both the intake and the exhaust valves are closed. The fuel-air mixture is compressed to about one-seventh of its original volume and ready for ignition.
   c. Power stroke: the fuel–air mixture is ignited, and the expansion of the rapidly burning fuel–air mixture forces the piston downward. The intake and exhaust valves are closed.
   d. Exhaust stroke: the exhaust valve is opened while the piston moves upward, pushing the burned gases out of the cylinder.
2. The aircraft ignition system is totally independent of the electrical system. The engine is equipped with a magneto which produces electricity for the spark plugs.
3. Two spark plugs are used in each cylinder of an aircraft engine for two reasons.

In case one spark plug fails, the other will ignite the fuel-air mixture. Two spark plugs provide more even, faster burning of the fuel–air mixture, which produces the best power.

4. Detonation is the explosive burning of the fuel-air mixture. Less power is developed by explosive burning in the cylinder than by rapid but even burning.
5. Preignition is the ignition of the fuel-air mixture prior to the time of spark ignition. The overheated cylinder, or glowing carbon particles in the cylinder may act as an ignitor. Preignition causes a severe power loss and places a strain on the engine.
6. If the recommended octane fuel is not available, a ***higher*** octane fuel must be used. (Fuel which is of a lower octane than is recommended must never be used.)
7. The results of preignition and detonation include:
   a. Increased engine temperature.
   b. Decreased power output.
   c. Rough running engine.
   d. Possible engine failure.
8. The supercharged engine is capable of producing high power at low altitude, and will maintain high power at higher altitudes than an unsupercharged engine because of the packing of the fuel-air mixture into the cylinders.
9. The primary indication of power output of an engine equipped with a fixed-pitch propeller is the tachometer.
10. The primary indication of power output of an engine equipped with a variable pitch propeller are the tachometer and the manifold pressure gauge.
11. On an engine equipped with a variable pitch propeller, the rpm must be increased prior to increasing the manifold pressure. Rpm must be decreased after decreasing the manifold pressure. The use of this procedure will prevent destructive pressures from forming in the cylinders at the moment of combustion.
12. The primary differences between turbine and reciprocating engines include:
    a. In the turbine engine, intake, compression, combustion, and exhaust all occur simultaneously and continuously.
    b. Thrust in the turbine engine is produced because of a reaction to escaping high-speed gases.
    c. Turbine engines develop more power for their size and weight than do reciprocating engines.
    d. The useful life of a turbine engine may be five or ten times that of a reciprocating engine.

## CHAPTER 11

1. If the recommended octane fuel is not available, a higher octane fuel must be used (never lower octane than recommended).
2. The most common fuel contaminant is water.
3. In order to reduce the amount of condensation in the fuel tanks, we should fill all tanks prior to overnight storage of the aircraft.
4. A generous sample of fuel should be drained from all tank sumps as well as from the gasculator bowl, to insure the removal of all contaminants from the fuel.

5. Fuel samples from sumps should be collected in a transparent container so that water or other contaminants will be visible. A generous sample from each sump drain will be required.
6. The only reliable method to check fuel quantity prior to flight is visually.
7. The carburetor is a fuel metering and mixing device.
8. Carburetor ice is more apt to form when the outside air is warm (temperature between 35 degrees Fahrenheit and 80 degrees Fahrenheit) and humid.
9. The primary distinction between fuel injection and carburetion is that the fuel is vaporized directly into each cylinder on most fuel injected engines, instead of being vaporized outside the cylinders and being routed through intake manifolds.
10. The amount of carburetor heat available is dependent upon the heat produced by the engine. Hence, to increase carburetor heat we must increase engine heat, by increasing power, increasing drag (lower gear and flaps) and decreasing airspeed (climb).
11. The power indications during carburetor ice formation, heat application, ice removal, and heat removal are shown in these charts.

**FIXED PITCH PROPELLER**

| CARBURETOR CONDITION | POWER INDICATION | HEAT CONTROL |
|---|---|---|
| No Ice | 2,400 RPM | Off |
| Ice Forming | RPM Slowly Decreasing | Off |
| Heavy Ice | RPM 2,100—Decreasing | On |

NOTE: When heat is applied a further power decrease will be noted due to warm air.

| | | |
|---|---|---|
| Ice Melting | RPM 2,100—Increasing | On |
| Ice Removed | RPM 2,350—Constant | On |
| Ice Removed | RPM 2,400—Constant | Off |

NOTE: When heat is removed, a power increase will be noted due to the removal of warm air.

**VARIABLE PITCH PROPELLER**

NOTE: Variable pitch propeller will maintain constant rpm. Manifold pressure will indicate power fluctuation.

| CARBURETOR CONDITION | POWER INDICATION | HEAT CONTROL |
|---|---|---|
| No Ice | M.P. 24 inches | Off |
| Ice Forming | M.P. Slowly Decreasing | Off |
| Heavy Ice | M.P. 20 inches—Decreasing | On |

NOTE: When heat is applied, a further power decrease will be noted due to warm air.

| | | |
|---|---|---|
| Ice Melting | M.P. 20 inches—Increasing | On |
| Ice Removed | M.P. 23½ inches—Constant | On |
| Ice Removed | M.P. 24 inches—Constant | Off |

NOTE: When heat is removed, a power increase will be noted due to the removal of warm air.

12. Carburetor heat should not be used indiscriminately because:
    a. Engine power output is decreased with heat applied.
    b. Fuel consumption increases with heat applied.
    c. Engine operating temperature increases with heat applied.
13. The mixture control changes the quantity of ***fuel*** which is metered to the carburetor.
14. Normally, light aircraft engines are not leaned until above 5,000 feet.
15. One commonly recommended method used when leaning a light aircraft engine includes the following steps:
    a. Adjust throttle and propeller controls for desired power output.
    b. Carefully begin leaning the mixture. This will cause a slight increase in power.
    c. At the first sign of power decrease, richen mixture slightly to return to the "rich side" of the best power indications.
16. When using the fuel consumption chart in Fig. 51, we find the following fuel consumption rates, endurance hours, and range.

**GALLONS PER HOUR**

| FLIGHT CONDITIONS | FUEL CONSUMPTION | END. HOURS | RANGE |
|---|---|---|---|
| 5,000 ft 59% bhp | 4.5 | 4.9 | 529 |
| 2,500 ft 46% bhp | 3.6 | 6.1 | 567 |
| 7,500 ft 2450 rpm | 4.2 | 5.2 | 556 |
| 10,000 ft 2730 rpm | 5.0 | 4.4 | 524 |
| 10,000 ft 2150 rpm | 3.3 | 6.7 | 630 |

## CHAPTER 12

1. The magnetic float assembly in the magnetic compass is surrounded by kerosene which dampens oscillations and lubricates the pivot point.
2. Variation is the angle between the True North Pole and the Magnetic North Pole as viewed from any specific geographic position.
3. Deviation is caused by electronic equipment, electrical wiring, or magnetized metal in the aircraft. Deviation can be cut to a minimum by properly "swinging" the compass (using the compensating magnets).
4. Oscillation error is the result of the swinging compass due to poor pilot technique or turbulence.
5. When turning from a northerly heading, the magnetic compass will show a turn in the opposite direction.
6. Assuming 10 degrees will be required to roll-out of our turn from East to South, we should begin our roll-out when the magnetic compass reads 200 degrees (in an area of 030 degrees North latitude).
7. Since northerly or southerly turning errors are nonexistent on headings of East or West, we will initiate our roll-out when the magnetic compass indicates 275 degrees, assuming 5 degrees will be required for roll-out.
8. When using the magnetic compass for heading information in a southerly direc-

tion, the slightest turn will be magnified by the southerly turning error. Therefore, the aircraft must not be turning during the period when we refer to the compass.

9. We must initiate the roll-out when the magnetic compass reads 305 degrees.
10. When accelerating on headings of East or West, we may expect the magnetic compass to indicate a turn towards the North.

## CHAPTER 13

1. The airspeed indicator, altimeter, and vertical velocity indicator are dependent upon the pitot-static system.
2. The altimeter and vertical velocity indicator depend only on the static portion of the pitot-static system.
3. When the alternate source of static pressure is used, the altimeter will read higher than normal and the vertical velocity indicator will momentarily show a climb.
4. Errors in the airspeed indicator and associated plumbing include:
   a. Installation error: caused by friction within the plumbing, and the angle with which the relative wind strikes the pitot head.
   b. Density error: caused by varying air density for which the airspeed indicator cannot adjust.
   c. Compressibility error: caused by the packing and compressing of ram air within the pitot head, particularly at high airspeeds and high altitudes.
5. Indicated airspeed is the speed with which the aircraft passes through the air, as indicated by the instrument. True airspeed is the true, actual speed with which the aircraft passes through the air. True airspeed is computed by applying corrections for air density errors to the indicated airspeed.
6. Calibrated airspeed is indicated airspeed corrected for installation error. It is sometimes called true indicated airspeed and is the airspeed to which aircraft operating data is referred.
7. The indicated airspeed which is recommended for any flight maneuver does not vary with altitude. Therefore, if final approach speed at sea level is 75 knots indicated airspeed, it will be 75 knots at any altitude. True airspeed however, will increase with altitude if a constant indicated airspeed is maintained.
8. The vertical velocity indicator lags behind the aircraft by about six to nine seconds due to a calibrated leak in the instrument which is designed to reduce oscillation in the instrument reading.
9. When reading the altimeter, we should read the small hand first, then the medium size hand, and then the longest hand.
10. *Barometric pressure reduced to sea level* is the calculated pressure at a specific weather station which represents the atmospheric pressure of the station if it were at sea level. The outside atmospheric pressure is adjusted to what it would be if the station were in the same geographic location, but at sea level.
11. Aircraft performance is based on standard density altitude.
12. Pressure altitude can be determined by turning the adjusting knob of the altimeter until 29.92 inches of mercury (standard pressure) is indicated on the setting dial. Pressure altitude can then be read directly from the altimeter.

13. Indicated altitude must be converted to pressure altitude, which in turn must be adjusted for nonstandard temperature in order to find density altitude.

14. a. Altimeter setting at departure airport = 30.12 inches
Altimeter setting at destination airport = 30.00 inches
Amount of pressure change = 0.12 inches
0.12 inches of mercury = 120 feet
Since we flew from high to low pressure, the altimeter will read higher than actual altitude (assuming new altimeter setting is not set on the altimeter). When we land at the airport, which is 220 feet above mean sea level the altimeter will indicate 220 feet plus 120 feet, or 340 feet.
b. Altimeter setting at departure airport = 29.95 inches
Altimeter setting at destination airport = 30.05 inches
Amount of pressure change = 0.10 inches
0.10 inches of mercury = 100 feet
Since we flew from low to high pressure, the altimeter will read lower than actual altitude (assuming new altimeter setting is not set on the altimeter). When we land at the airport which is 3,700 feet above mean sea level, the altimeter will indicate 3,700 feet minus 100 feet, or 3,600 feet.
c. Altimeter setting at departure airport = 29.90 inches
Altimeter setting at destination airport = 29.75 inches
Amount of pressure change = 0.15 inches
0.15 inches of mercury = 150 feet
Since we flew from high to low pressure, the altimeter will read higher than actual (assuming the new altimeter setting is not set on the altimeter). Therefore, when the altimeter reads 2,550 feet on downwind leg, we will be only 650 feet above the surface (800 feet minus 150 feet).

## CHAPTER 14

1. The three gyro-operated flight instruments are:
   a. The attitude indicator.
   b. The turn indicator (the needle portion of the turn and slip indicator).
   c. The directional gyro (heading indicator).
2. Gyroscopic inertia is determined by:
   a. Weight: a heavier mass will be deflected to a lesser degree than a lighter mass.
   b. Angular velocity: the rigidity of the gyroscope increases directly with the speed of rotation of the rotor.
   c. Radius at which the weight is concentrated: the rigidity of the gyroscope increases directly with the distance of the weight of the mass from the axis.
   d. Friction: minimum friction on the rotor bearings decreases the deflection of the gyroscope.
3. Aircraft attitude may be controlled by reference to the turn and slip indicator and the airspeed indicator. Therefore, the turn and slip indicator is normally powered by a different source than is used for the attitude indicator and the directional gyro. In the event of a vacuum system failure, the turn and slip indicator will be available if it is electrically operated.
4. A venturi is a restricted pipe. As air passes through, a low pressure area is formed at the point of restriction. A tube may be fitted to the venturi at the point of re-

striction, and air rushing through the tube (in an effort to fill the low pressure area) causes suction.

5. The attitude indicator depicts the attitude of the aircraft in respect to the artificial horizon within the instrument which matches the natural horizon.
6. The direction of bank is depicted accurately on the attitude indicator by the position of the wings of the small aircraft in relation to the horizon bar. (The bank index shows the amount of bank, but in an opposite direction.)
7. The quality of a turn is depicted by the ball portion of the turn and slip indicator.
8. The directional gyro must be reset every 15 minutes (to match the magnetic compass), because it precesses slowly from its original position in space.
9. Prior to resetting the directional gyro to match the magnetic compass, we must be sure that the aircraft is in straight and level, unaccelerated flight, because that is the only time the magnetic compass is not subjected to acceleration and turning errors.
10. The needle portion of the turn and bank indicator depicts the direction and rate of turn.
11. The rudder is commonly used to change the position of the ball in the turn and slip indicator.
12. The operating limits of the attitude indicator are 100 to 110 degrees in bank, and 60 to 70 degrees in pitch.
13. Acceleration error within the attitude indicator causes an erroneous nose-high pitch attitude reading during periods of acceleration and an erroneous nose-low pitch attitude reading during periods of deceleration.

## CHAPTER 15

1. The first item on the preflight inspection of every light aircraft must be the ignition and electrical switches to determine that they are in the "off" position.
2. Federal law requires three papers to be onboard an aircraft prior to flight.
   a. Registration Certificate.
   b. Airworthiness Certificate.
   c. Operations Limitations. (A current FAA approved Aircraft Flight Manual, placards, listings, instrument markings, or any combination thereof containing each operating limitation for the aircraft.)
3. Critical check points on the engine cowl are the fastenings, baffles, openings, and slots. The critical check points on the propeller are leading edge cracks, nicks, or dents, and propeller hub fasteners.
4. When inspecting an aircraft wheel and tire assembly which is equipped with wheel pants, we must roll the aircraft so that the entire wheel becomes visible.
5. When checking the oil quantity in the engine, we must not misplace the oil filler cap, or avoid replacing it tightly.
6. When inspecting any moveable control surface, the following items are of particular importance:
   a. Hinges and hinge pins.
   b. Push–pull rods.

c. Full and free movement of control surface.
d. Proper direction of movement, to full allowable deflection.

7. The airspeed indicator may be severely damaged by blowing into the pitot head in an effort to clean it.
8. Fuel quantity must be visually checked on a preflight inspection because of the unreliability of fuel gauges, and to determine the color (denoting octane) of the fuel.

## CHAPTER 16

1. A checklist should be used during the prestart procedure to insure against the inadvertent deletion of any required item.
2. The propeller may be damaged, as well as persons or property aft of the airplane, if the engine is started over loose debris.
3. Before engaging the starter, we must shout "clear" to warn persons in the area that the propeller is about to rotate.
4. Closely cowled engines are apt to overheat rapidly if required to deliver high power for long periods of time while on the surface.
5. A quartering tailwind condition, particularly while turning, is the most dangerous of wind conditions on the surface.
6. While taxiing in a quartering tailwind from the right, the right aileron is down, the left aileron is up, and the elevator is forward.

## CHAPTER 17

1. The takeoff should be planned into the natural wind because the length of runway required to accelerate to takeoff airspeed will be shortened by the headwind.
2. Landing should be planned into the wind because the wind shortens the landing roll.
3. Takeoff performance is affected by temperature, humidity, and altitude.
4. The worst atmospheric conditions in which to take off are hot, humid, and high altitude.
5. Altitude has the greatest effect of all atmospheric conditions on the takeoff performance of aircraft.
6. The takeoff distance required to clear a fifty foot object (according to the takeoff data chart on page 141) is 882 feet.
7. The throttle, engine operation, and flap setting are mechanical items affecting takeoff performance.
8. The best rate of climb speed should be used to climb to a specific altitude in as short a time as possible.
9. The best angle of climb speed should be used to clear obstacles.
10. The best angle and best rate of climb speeds are published in the Aircraft Owners Handbook or Operations Manual.

## CHAPTER 18

1. The airport traffic pattern is designed to keep all traffic moving in an orderly fashion.
2. A detailed description of the traffic pattern may normally be found on the bulletin board at the airport operations office.
3. The commonly used surface wind indicators are:
   a. The wind sock.
   b. The wind tee.
   c. The tetrahedron.
4. The magnetic heading on final approach to runway 33 (no crosswind) will be 330 degrees. (Runway magnetic course, rounded off to the nearest ten degrees is used to name the runway.)
5. While in a standard left hand traffic pattern for runway 03, with a strong wind blowing down the runway, we must adjust base leg heading toward the airport, to avoid drifting while on base leg. Therefore, the aircraft magnetic heading on a left base leg will be about 110 degrees (allowing 10 degrees left drift correction).
6. Blue lights are used on the airport to indicate taxiways.
7. Runways are lighted with white lights along the edge of the landing area.
8. A right-hand traffic pattern will be indicated by the segmented circle, or by a flashing amber light on the airport.
9. Normally, after takeoff, the first turn is not made until at least 400 feet above the surface.
10. Long, low final approaches, particularly in single-engine aircraft may prove dangerous in the event of a power failure. Final approach should be planned so as to allow for a power-off descent to a safe landing area.
11. Airspeed control on final approach is important for the following reasons:
    a. If airspeed is much below that specified by the aircraft manufacturer, the efficiency of aircraft controls will be less than desirable.
    b. If airspeed is much above that specified by the aircraft manufacturer, the aircraft will use unnecessary runway during the flare out and landing.
12. When our approach altitude is lower than the VASI glide slope, all bars of lights will appear red.
13. When on the VASI glide path, the downwind (nearest) light bars will be white and the upwind (farthest) light bars will be red.
14. According to Federal Air Regulation 91.87 "an airplane approaching to land on a runway served by a VASI shall maintain an altitude at or above the glide slope until a lower altitude is necessary for a safe landing."
15. When in or near the traffic pattern, we must exercise the following precautions:
    a. Diligent watchfulness for other aircraft.
    b. Strict maintenance of a rectangular course or other prescribed pattern.
    c. Courtesy towards other pilots.
    d. Final approach planning which will allow for an emergency landing.

## CHAPTER 19

1. Special preflight precautions which are necessary prior to night flight include:
   a. Operating flashlight with extra batteries.
   b. Spare fuses.
   c. Cockpit familiarization.
   d. Aircraft lighting, including position lights, instrument panel lights, anticollision lights, and landing lights.
   e. Detailed weather briefing.
2. When operating at night we must avoid showing taxi and landing lights when pointing directly at other aircraft.
3. Spacial disorientation is a potential hazard after takeoff at night, particularly in sparsely populated areas.
4. Steady red lights on or near an airport indicate obstructions to flight.
5. While enroute at night we should expect:
   a. The shape of cities to be different than in daytime.
   b. Pilotage navigation less reliable than during daytime.
   c. Cloud formations will be difficult to see.
   d. Visibility may change rapidly.
   e. The path of aircraft producing wing-tip vortices or jet-wash may be difficult to determine.
6. A green and white rotating beacon indicates a lighted land civilian airport.
7. A military airport will display a split white and green beacon.
8. An airport rotating beacon operating in daylight indicates that the weather at the airport is below VFR minimums.
9. Flashing lights outlining the wind tee or tetrahedron indicate that the weather at the airport is below VFR minimums.
10. Night flight instruction is necessary to insure safe flight in order to familiarize the pilot with the changes in depth perception experienced and different flying techniques which should be used.

## CHAPTER 20

1. Most weather exists in the layer of the atmosphere called the troposphere.
2. The atmosphere is composed of a mixture of gases, in the approximate amounts shown here:
   a. Nitrogen: 78%
   b. Oxygen: 21%
   c. Other: 1%
3. Heat is a form of energy which causes molecular activity.
4. The formula to convert Celsius temperature from Fahrenheit is:

$$\text{Celsius} = \tfrac{5}{9}(\text{Fahrenheit} + 40) - 40$$
$$\text{Celsius} = \tfrac{5}{9}(78 + 40) - 40$$
$$\text{Celsius} = \tfrac{5}{9}(118) - 40$$
$$\text{Celsius} = 65.5 - 40$$
$$\text{Celsius} = 25.5 \text{ degrees}$$

5. The air near the earth's surface changes in temperature from daytime to nighttime because of the rapid heating of the surface during the day (which in time radiates heat to the air above the surface) and the rapid cooling of the surface at night (which cools the air above it).

6. An inversion is the name of the atmospheric phenomenon in which the temperature of the air increases with altitude. The inversion is caused by rapid cooling of the earth's surface which cools the air near the surface to a temperature lower than the air aloft.

7. The inversion is caused by the rapid cooling of the earth's surface, the insurgence of cold air moving under warm air, or the wide spread settling of air aloft.

8. Standard atmospheric pressure is:
   a. 29.92 inches of mercury.
   b. 1013.2 millibars.
   c. 14.7 pounds per square inch.
   d. 760 millimeters.

9. The amount of atmospheric pressure decrease per 1,000 feet of altitude is approximately one inch of mercury.

10. Station barometric pressure is the atmospheric pressure measured at the weather station elevation.

11. Station pressure reduced to sea level is the calculated atmospheric pressure of a specific geographic location, if it were at sea level.

12. The station barometric pressure must be reduced to sea level before publication so that all published barometric pressures will be at the same base; mean sea level. This procedure makes it possible to compare the atmospheric pressure at different stations, without having to know the altitude of the reporting station.

13. Atmospheric circulation is caused by the uneven heating of the earth's surface.

14. Factors influencing atmospheric circulation include:
    a. Rotation of the earth.
    b. Pressure gradient force.
    c. Friction.
    d. Centrifugal force.
    e. Gravity.
    f. Seasonal changes.
    g. Uneven distribution of oceans.

15. The wind blows counterclockwise around a low pressure area because coriolis force deflects the air to the right as it rushes to fill the low. Centrifugal force, coriolis force, and centripetal force are balanced to create a counterclockwise movement. Air rushing out and away from a high pressure area is affected by the identical forces which cause a clockwise rotation about the high pressure area.

## CHAPTER 21

1. Isobars are lines on a weather map which connect points of equal barometric pressure reduced to sea level.

2. As the pressure gradient increases, so does the wind.

3. Sea breezes are caused by the cool air over the water rushing towards the shore to fill the void left by rising air currents over the heated land.
4. Since land masses radiate heat more rapidly than water masses, the land may cool to temperatures lower than the nearby water after sunset. The air over the warmer water rises, and cool air over the land rushes to fill the void left by the rising air.
5. Mountain winds are caused by the rapid cooling of mountain slopes after sunset. The air above the slopes cools, and, being heavier than the warmer air surrounding it, settles toward the valley.
6. Turbulence is caused by air in motion, either vertically or horizontally.
7. Aircraft reaction to turbulence varies with the intensity of the turbulence, the speed of the aircraft, the wing loading of the aircraft, and aircraft attitude.
8. Dangerous down drafts may be found on the lee side of mountains.
9. Shearing or tearing of the wind is caused by a change in wind speed and/or direction.
10. Turbulence may be classified as:
    a. Light: occupants of the aircraft may be required to wear seat belts, although unsecured objects will remain at rest.
    b. Moderate: occupants of the aircraft may be thrown against the seat belt, and unsecured objects in the aircraft may move about.
    c. Severe: Aircraft may be momentarily out of control, and occupants are thrown about violently.
    d. Extreme: aircraft may be violently tossed about, and is practically impossible to control. Structural damage may result.

## CHAPTER 22

1. Most weather occurs in the troposphere because most of the atmospheric moisture is in the troposphere, and moisture is required to produce weather.
2. Water vapor in the atmosphere comes from evaporating water in the oceans, lakes, wetlands, snow fields and ice caps.
3. Evaporation is the movement of water molecules away from a body of water, snow, or ice, and into the atmosphere.
4. Sublimation is the evaporation of frozen liquid without first changing to the liquid state.
5. Condensation is the change of state from invisible water vapor to visible moisture.
6. Evaporation of liquid requires heat. Therefore heat is taken from the liquid and surrounding area when evaporation occurs.
7. Heat which is lost from a liquid when evaporation occurs is stored in the invisible vapor, to be released into the liquid when condensation occurs.
8. The amount of water vapor a parcel of air can contain before becoming saturated depends on the temperature of the air; the higher the temperature of the air, the more water vapor it can hold.
9. The water vapor may be condensed from saturated air by reducing the temperature of the air or by adding water vapor by evaporation.

10. Dew point is the temperature at which the air will become saturated.
11. The spread between the temperature and the dew point is important to the pilot because it indicates the relative humidity of the air. As the spread decreases, the chance of low visibility increases.
12. Condensation nuclei are tiny particles of dust, salt, soot, etc., in the air on which water vapor molecules collect.
13. The form of the precipitation is dependent upon the following factors:
    a. The intensity of vertical currents determines the size of the precipitation drops or crystals.
    b. The temperature of the air determines whether the precipitation will be liquid or solid.
14. Hail is formed by vertical currents carrying liquid precipitation up and down through the layers of above and below freezing temperature.

## CHAPTER 23

1. Atmospheric stability depends on the weight distribution (vertically) of the air.
2. Atmospheric stability is important to the pilot because it determines the type of weather to be encountered; i.e. turbulence, cloud types, precipitation type, and visibility.
3. The dry adiabatic lapse rate is $5\frac{1}{2}$ degrees Fahrenheit per 1,000 feet of altitude.
4. The moist adiabatic lapse rate is 2 to 5 degrees Fahrenheit per 1,000 feet of altitude.
5. The moist adiabatic lapse rate is less than the dry adiabatic lapse rate because, as the moist air rises and water vapor in it is condensed, heat is released to the atmosphere.
6. The dew point lapse rate is one degree Fahrenheit per 1,000 feet of altitude.
7. If the surface temperature is 80 degrees Fahrenheit and the surface dew point is 55 degrees Fahrenheit, clouds may be expected to form at about 5,500 feet above the surface. (The temperature decreases $5\frac{1}{2}$ degrees Fahrenheit per 1,000 feet, while the dew point decreases one degree Fahrenheit per 1,000 feet. Temperature and dew point will become equal at about 5,500 feet above the surface.)
8. Air at high temperature is either stable or unstable depending on its moisture content.
9. Stratus type clouds form in stable air.
10. Cumulus type clouds form in unstable air.

## CHAPTER 24

1. Stratus type clouds form in stable air.
2. Cumulus (towering, cauliflower-like) clouds form in unstable air.
3. Cumulus type clouds may be accompanied by heavy local precipitation.
4. Stratus type clouds may be accompanied by steady, light precipitation.

5. The only difference between fog and clouds is the height at which the bases are formed. Fog is a cloud which is less than 50 feet above the surface.
6. High clouds have bases at 16,500 to 45,000 feet.
7. The prefix denoting high clouds is *cirrus*.
8. Middle clouds have bases at 6,500 to 16,500 feet.
9. The prefix denoting middle clouds is *alto*.
10. The word "nimbo" denotes precipitation is falling from the cloud.

## CHAPTER 25

1. Fog is a cloud formation within 50 feet of the surface.
2. In order for fog to form, the air must be moist, there must be condensation nucleii in the air, and a light wind (three to seven mph) must exist.
3. A coastal station at which the temperature/dew point spread is within three degrees, with an onshore wind, is apt to be fogged-in shortly.
4. Radiation fog is most usually formed on clear, calm nights, or during the early morning hours.
5. Advection fog may form if warm moist air moves over a cooler surface, and the wind speed is less than 15 mph.
6. The air around steam fog tends to be turbulent due to the fact that it is heated from below and rises.
7. Precipitation induced fog often forms near a weather front.

## CHAPTER 26

1. The intensity of air mass characteristics is determined by the length of time the air mass remains in the source region and the difference between the temperature of the air and the surface.
2. More complete modification of an air mass occurs when the air is colder than the surface over which it passes because the cold air rises after being heated from below spreading modifying factors to great heights.
3. The characteristics of a cold air mass include:
   a. Turbulence up to 10–12,000 feet.
   b. Good visibility.
   c. Clouds (if any) of vertical development.
   d. Precipitation (if any) scattered shower-like.
4. The characteristics of a warm air mass include:
   a. Smooth air.
   b. Poor visibility.
   c. Stratus clouds.
   d. Possible wide spread precipitation (steady, light).
5. Passengers will appreciate their first flights during warm air mass conditions because the air will be smooth.

6. The water vapor content of the air may be increased by evaporation of surface water, oceans, snow or ice fields, and precipitation.
7. The water vapor content of the air may be decreased by condensation or precipitation.
8. Orographic lifting is the term used to describe the lifting of air up a mountain slope.
9. Air may be caused to sink by being forced down a mountain slope or by being forced down from above a colder air mass. Air may also sink from its own weight (subsidence).

## CHAPTER 27

1. A front is a line of discontinuity between air masses.
2. The greatest contrast between air masses will be in the lower levels of the atmosphere.
3. Cloud types, moisture content, wind, temperature, and atmospheric pressure will normally change during frontal penetration.
4. A trough is an area where air is converging from different directions.
5. Not all fronts are associated with a major trough. Sometimes the primary trough becomes stationary while the front moves on within a secondary (minor) trough.
6. Factors affecting frontal zone weather include:
   a. Moisture content of the air.
   b. Stability of the air masses.
   c. The steepness of the frontal zone.
   d. The speed with which the front moves.
7. A cold front is the leading edge of a mass of air which is colder than the surface over which it passes.
8. A warm front is the leading edge of a mass of air which is warmer than the surface over which it passes.
9. While flying in a warm air mass, heading northwestward towards a cold front, the following weather will be observed.
   a. Altocumulus clouds become visible on the horizon.
   b. Our flight will normally be influenced by southwest winds.
   c. Barometric pressure will decrease.
   d. As we near the front, cumulus and cumulonimbus clouds will appear.
   e. Precipitation, heavy at times, will appear below the clouds.
10. While flying in a cold air mass, heading southwestward toward a warm front, the following weather will be observed.
    a. High cirrus-type clouds will become visible on the horizon.
    b. Our flight will normally be influenced by winds from the west through north.
    c. Barometric pressure will decrease.
    d. Cirrostratus and altostratus clouds will appear as we reach a point about 300 to 400 miles ahead of the front.
    e. Precipitation will begin about 200 miles in advance of the front.
    f. Low stratus clouds and fog will appear near the front.

11. An occluded front is designated warm or cold depending on whether the warm air remains on the surface or the cold air remains on the surface.

12. A low pressure area may form on a stationary front in the following manner.
    a. A small disturbance caused by uneven surface heating or irregular terrain starts a wave or bend in the frontal line.
    b. If the disturbance persists, the wave increases and cyclonic circulation begins.
    c. One section of the front begins to move as a cold front, while another section moves as a warm front.

13. Frontolysis is the diminishing or fading away of a frontal zone.

14. Frontogenesis is the forming or intensification of a frontal zone.

## CHAPTER 28

1. The atmospheric requirements necessary for the formation of thunderstorms are:
    a. Unstable air.
    b. Lifting action.
    c. Moisture.

2. The two main types of thunderstorms are *frontal* and *air mass.*

3. Frontal thunderstorms are normally found in long, narrow bands, while air mass thunderstorms are normally scattered over wide areas.

4. Common air mass thunderstorms are:
    a. Convective
    b. Orographic
    c. Nocturnal

5. The three stages in the life cycle of a thunderstorm are:
    a. Cumulus or building stage.
    b. Mature stage.
    c. Dissipating stage.

6. The most severe weather is found in a thunderstorm during the mature stage.

7. The dissipating stage of a thunderstorm is the longest of the three stages.

8. Weather hazards normally associated with thunderstorms may include:
    a. Turbulence.
    b. Strong and/or gusty surface winds.
    c. Hail.
    d. Lightning.
    e. Precipitation static.

9. Saint Elmo's fire is a blueish, brush-like flame which appears on the leading edges and components of the aircraft. It is a discharge of static electricity which causes no serious damage to the aircraft.

10. The penetration of a thunderstorm, or flight in the proximity of thunderstorms may be extremely dangerous, particularly in a light aircraft. We should never plan a flight through or near a thunderstorm and should take every precaution to avoid being trapped by thunderstorm activity.

## CHAPTER 29

1. Visible moisture and freezing temperatures are both required before structural ice will form on an aircraft.
2. The two main types of structural ice are *rime* and *clear.*
3. Rime ice is opaque, rough, and flaky. It forms in irregular shapes and is normally easy to remove. Clear ice is transparent, smooth, and conforms to the shape of the surface on which it forms. It is difficult to remove.
4. Clear ice may be expected to form in turbulent air when visible moisture and below freezing temperatures exist.
5. Rime ice may be expected to form in smooth air when visible moisture and below freezing temperatures exist.
6. The effects of structural ice on an aircraft include:
   a. Loss of lift.
   b. Increased drag.
   c. Increased weight.
   d. Unbalanced propeller.
   e. Clogged pitot head.
   f. Reduced visibility through windshield.
   g. Increased stall speeds.
7. The best method of avoiding structural ice is to execute a 180-degree turn when encountering atmospheric conditions which are conducive to its formation. If a reversal of direction is not possible, a climb should be immediately initiated, into the warmer air aloft. **Caution:** Pilots not instrument-rated and properly qualified must avoid instrument flight.
8. The most common method used to deice an aircraft is the use of inflatable rubber boots on the leading edge of the lifting surfaces.
9. Common anti-ice equipment includes:
   a. Heated wings.
   b. Alcohol spray.
   c. Pitot head heaters (electrical).
10. Frost may occur if the air temperature is below freezing and condensation occurs.
11. All frost must be removed from the aircraft prior to flight. Particular attention should be directed to hinges, static ports, and moveable control surfaces, to make sure that all traces of frost are removed.

## CHAPTER 30

1. The U.S. Weather Bureau, Federal Aviation Administration, and the military services are all concerned with the observation and dissemination of aviation weather.
2. Weather observations may be made visually, by radar, RAWINSONDE, PIBALS, rockets, PIREPS, satellites, and reconnaisance flights.
3. As pilots we will consult regularly with the local weatherman in the airport weather station.

4. Weather data may be distributed by teletypewriter, facsimile machine, long-line, or TelAutograph.
5. Weather data may be presented by recorded broadcast, in person, on closed circuit television, or by request from a Flight Service Station.
6. The aviation industry is a prime user of weather data because flight operations depend heavily upon detailed weather reporting and forecasting.

## CHAPTER 31

1. The aviation weather report provides the most up to date weather information available.
2. The time and day of the month of the aviation weather reports following the heading 023SA23182300 is 2300 hours Greenwich Mean Time on the eighteenth day of the month.
3. Sample station identifiers are:
   Kansas City: MKC    Boston: BOS    Jacksonville: JAX
4. A ceiling is indicated on the aviation weather report by the following symbols.
   a. ⦶ Broken clouds.
   b. ⊕ Overcast.
   c. ✕ Obscuration.
5. A report of scattered clouds indicates that one-tenth to five-tenths of the sky is covered.
6. The method of determining cloud heights will be included in the report only if the cloud cover constitutes a ceiling.
7. The following symbols are used to designate the method used to determine the cloud height.

| | |
|---|---|
| E = Estimated | A = Reported by aircraft |
| M = Measured | R = Radiosonde or Radar |
| W = Indefinite | D = Estimated height of cirriform clouds |
| B = Balloon | U = Cirriform ceiling, height unknown |

8. Cloud heights reported in the main body of the aviation weather report are given in reference to the surface. That is, cloud heights are given as height above the ground.
9. The symbols used for restrictions to visibility and precipitation are:

| | |
|---|---|
| A = Hail | E = Sleet |
| L = Drizzle | S = Snow |
| R = Rain | W = Showers |
| AP = Small hail | IC = Ice crystals |
| SG = Snow grains | SP = Snow pellets |
| T = Thunderstorms | Z = Freezing |
| TORNADO (Always spelled out in full) | |
| D = Dust | F = Fog |
| K = Smoke | H = Haze |
| GF = Ground Fog | B = Blowing (snow or dust) |

10. The barometric pressure in millibars for the code numbers given are:

| | | |
|---|---|---|
| 520 = 952.0 | 483 = 1048.3 | 234 = 1023.4 |
| 765 = 976.5 | 222 = 1022.2 | 123 = 1012.3 |

11. The wind direction as indicated by the code numbers given are:

| | |
|---|---|
| 22 = 220 degrees | 34 = 340 degrees |
| 35 = 350 degrees | 03 = 030 degrees |
| 10 = 100 degrees | 01 = 010 degrees |

12. The interpretation of NOTAMS is found in the Airman's Information Manual.

## CHAPTER 32

1. The data included on the Weather Depiction Chart are:
   a. Precipitation and/or other important weather phenomena.
   b. Sky cover.
   c. Visibility (when 6 statute miles or less).
   d. Cloud base (in hundreds of feet, up to 20,000 feet, in reference to height above the surface).
   e. Areas in which the ceilings are below 1,000 feet and/or visibilities are below 3 statute miles are enclosed by solid black lines and may be shaded red.
   f. Areas in which the ceilings are below 5,000 feet but not below 1,000 feet and visibilities are above 3 statute miles are enclosed by a black scalloped line and may be shaded blue.
2. Weather depiction charts are printed every three hours.
3. Pertinent data included on the Surface Weather Chart are:
   a. The station location.
   b. The total amount of cloud cover.
   c. Surface wind direction and speed.
   d. Barometric pressure in millibars.
   e. The amount of pressure change within the preceding three hours.
   f. The direction and manner of pressure change.
   g. The type of low clouds.
   h. The temperature in degrees Fahrenheit.
   i. The dew point in degrees Fahrenheit.
   j. Precipitation.
   k. Fronts.
   l. Pressure centers.
   m. Isobars.
   n. Large precipitation areas.
4. The Surface Weather Chart is printed every three hours.
5. The symbol used in the station model for a surface wind of 270 degrees at 45 knots is
6. The following symbols are used on the Surface Weather Chart to depict fronts.
   a. Cold front 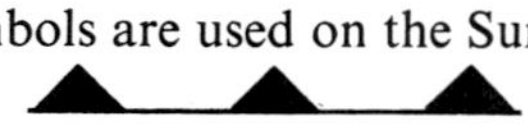
   b. Warm front 
   c. Occluded front 
7. An isobar is a line on the weather chart which connects points of equal barometric pressure (reduced to sea level).
8. Three Winds Aloft Charts are printed.
   a. Low level winds: the second level above the surface, 5,000 feet, 8,000 feet, and 10,000 feet.

b. Intermediate winds: 14,000 feet, 20,000 feet, 25,000 feet, and 30,000 feet.
c. Upper level winds: 35,000 feet, 40,000 feet, 50,000 feet, and 60,000 feet.

9. The Winds Aloft Chart depicts observed winds.
10. The Surface Weather Chart should be updated with the information presented on the Aviation Weather Report.

## CHAPTER 33

Note: The following answers were found in the sample area forecast in this chapter.

1. The weather over northern Alabama will be clear at 1900 hours Eastern Standard time.
2. When flying through the mountain valley in western North Carolina prior to 1200 hours Eastern Standard time, we can expect ceilings of 1,000 feet overcast, seven miles visibility, and occasional light snow, with visibilities reduced to two miles.
3. A few snow flurries will persist in the mountains until 2000 hours Eastern Standard time.
4. The freezing level will be at the surface over the western Carolinas.
5. Severe turbulence will occasionally be present along the east slopes of the mountains prior to late afternoon.

## CHAPTER 34

1. The 12-hour terminal forecast is issued every six hours.
2. If a ceiling is forecast in the 12-hour terminal forecast, the cloud height digit will be preceded by the letter "C."
3. Only when the visibility is expected to be 8 statute miles or less will it be included in the 12-hour terminal forecast.
4. If the wind speed is forecasted to be less than 10 knots, it will be omitted from the 12-hour terminal forecast.
5. The Winds Aloft forecast will include a forecast wind for the following altitudes.
   a. 3,000 feet (if the station elevation is 2,000 feet or lower).
   b. 5,000 feet (if the station elevation is 4,000 feet or lower).
   c. 10,000 feet.
   d. 15,000 feet.
   e. 25,000 feet.
6. The coded report for a wind from 300 degrees at 140 knots is "8040."
7. Temperatures given in the Winds Aloft forecast are in Celsius degrees.
8. The AIRMET includes weather which is considered potentially hazardous for light aircraft. The SIGMET includes weather which is significant to all aircraft.

## CHAPTER 35

1. The four basic methods of navigation are:
   a. Pilotage: referring to a chart and matching its pictorial display with the surface.
   b. Dead reckoning: mathematical computation of speed, time, distance, and drift.

   c. Radio navigation: following a broadcast signal, either aural or visual, to a predetermined location.
   d. Celestial navigation: the computation of our present position with reference to the heavenly bodies.
2. The most popular method of navigation is radio navigation.
3. Aids to navigation include:
   a. Aeronautical charts: detailed drawings of the earth's surface, or of radio stations and airways.
   b. Radio stations: a broad network of transmitters which cover the nation.
   c. Federal airways: a system of "highways of the air" which are defined by radio signals.
   d. Airman's Information Manual: a manual of information of concern to the pilot.
4. Samples of navigation which everyone uses prior to flight training include:
   a. The infant child finds his way from room to room as he learns to crawl by using pilotage. He remembers the landmarks along the way, and simply follows them.
   b. When we issue directions to a lost driver, and instruct him to "drive East for ten minutes until you see the fork in the road, then turn South," we have used a combination of dead reckoning and pilotage.

## CHAPTER 36

1. The scale of the Local Chart is 1: 250,000 or one inch equals four statute miles.
2. The scale of the Sectional Chart is 1: 500,000 or one inch equals eight statute miles.
3. The Sectional Chart is the most popular for use in light aircraft.
4. The World Aeronautical Chart is identical to the Operational Navigation Chart, except for the geographical areas which are depicted on them. The World Aeronautical Charts cover the United States except Hawaii, and the Operational Navigation Charts cover Hawaii and all the other land areas of the world.
5. The scale of the World Aeronautical Chart is 1: 1,000,000 or one inch equals 16 statute miles.
6. The Aeronautical Planning Chart is ideally suited for planning long distance flights.
7. The date of publication of each chart is listed in the Airman's Information Manual.
8. Very-high frequency facilities are printed on the navigation charts in blue ink.
9. The topography, or height of the terrain, is shown on navigation charts by relief shading, contour lines, elevation tinting, and spot elevations.
10. All flight is prohibited (without prior permission) in a prohibited area. Flight is possible in a restricted area, but is subject to the specific restriction for the area.

## CHAPTER 37

1. The grid of longitude and latitude lines which covers the globe provides a standard means of referring to any spot on the earth's surface.
2. Longitude lines converge at both poles, and therefore are not parallel lines. Latitude lines are parallel.

3. The distance between each minute of latitude is one nautical mile.
4. Each 15 degrees of longitude represents one hour of time.
5. A *minute* of longitude is one-sixtieth of a degree of longitude.
6. The prime meridian is located in Greenwich, England.
7. The coordinates of Chicago, Illinois are:
   Latitude: 41°47′ N
   Longitude: 87°45′ W
8. The *True North Pole* is used as a reference by all map makers for the purpose of standardization. The Magnetic North Pole cannot be used because its location is not constant.
9. The numerical equivalents of the four cardinal points of the compass are: North; 000 or 360 degrees. East; 090 degrees. South; 180 degrees. West; 270 degrees.
10. Direction is always measured clockwise for standardization purposes.

## CHAPTER 38

1. A *course* is a straight line. A *true course* is an angle. That is, a true course has direction.
2. The mid-longitude line should be used when measuring a true course, because the longitude lines are not parallel.
3. If we measure the true course in reference to the latitude lines on the chart, we must add or subtract 90 degrees to the resulting angle in order to find the correct true course.
4. The Magnetic North Pole cannot be used as the standard reference on charts because its location is not constant.
5. The True North Pole is an arbitrary location which is used as a standard reference point from which to measure direction.
6. Variation is the angle between the True North Pole and the Magnetic North Pole as viewed from a specific location.
7. Variation is shown in aeronautical charts in the form of a dashed line which connects points of equal variation. These lines are called *isogonic lines*.
8. An isogonic line is a line which connects points of equal variation.
9. When variation is applied to the true course, the result is called a magnetic course.
10. Deviation is the error in the compass caused by magnetic lines of force within the aircraft.
11. When deviation is applied to the magnetic course, the result is called the *compass course*.
12. The compass course is 322°.

## CHAPTER 39

1. The heading is the direction in which the nose of the aircraft points.
2. The difference between a course and a heading is *drift*, which is caused by the natural wind.

3. True heading is the angle between the heading and True North.
4. Wind effect is applied to the true course (not the magnetic course) because the wind aloft is always given in reference to True North.
5. Drift is the effect of a cross wind on the path of the aircraft in reference to the surface.
6. Magnetic heading is the angle between the heading and Magnetic North.
7. Deviation is printed on the compass correction card in the aircraft.
8. Compass heading is the angle between the heading and Compass North.
9. To solve for compass heading, the following steps must be followed.
   a. Draw the course line.
   b. Measure the true course.
   c. Apply the wind drift correction angle to the true course to find the true heading.
   d. Apply the variation to the true heading to find the magnetic heading.
   e. Apply the deviation to the magnetic heading to find the compass heading.
10. If the compass heading, variation, and deviation are known, we may compute the true heading by following these steps.
    a. Apply the deviation (in reverse) to the compass heading, to find the magnetic heading.
    b. Apply the variation (in reverse) to the magnetic heading to find the true heading.

## CHAPTER 40

1. The wind line is drawn first when drawing the basic wind vector diagram because both the length of the wind line (wind speed) and the direction of the wind line (wind direction) are known.
2. The wind vector line points *toward* the grommet on the computer so that, when the drawing wheel is rotated to the true course, the wind will appear to blow *toward* the true course.
3. True heading = 321 degrees, Groundspeed = 197 knots
   True heading = 205 degrees, Groundspeed = 81 knots
4. True heading = 257 degrees, Groundspeed = 191 knots
   True heading = 054 degrees, Groundspeed = 75 knots
5. Compass course = 073 degrees
   True heading = 068 degrees
   Magnetic course = 075 degrees
   Magnetic heading = 083 degrees
   Groundspeed = 116 knots
   Time enroute = 3 hours, 48 minutes
   Compass heading = 081 degrees

## CHAPTER 41

MULTIPLICATION

1. 544
2. 1,278
3. 12,740

4. 95.1
5. 264.6
6. 9,837

DIVISION
1. 15
2. 14
3. 5.4
4. 5.5
5. 1.34
6. 20.8

PERCENTAGES
1. 47.1%
2. 20.0%
3. 16.6%
4. 26.6
5. 366
6. 686

TIME—SPEED—DISTANCE (solve for time)
1. 131 minutes (2 hours, 11 minutes)
2. 172 minutes (2 hours, 52 minutes)
3. 108 minutes (1 hour, 48 minutes)
4. 27 minutes
5. 530 minutes (8 hours, 50 minutes)
6. 22 minutes

TIME—SPEED—DISTANCE (solve for speed)
1. 187 knots
2. 94 knots
3. 79.5 knots
4. 156 knots
5. 136 knots
6. 210 knots

FUEL CONSUMPTION
1. 27.1 gallons
2. 53.0 gallons
3. 24.0 gallons
4. 7.7 gallons
5. 23.9 gallons
6. 155 gallons

MILEAGE CONVERSION

| | NAUTICAL | STATUTE | KILOMETERS |
|---|---|---|---|
| 1. | 80 | 92 | 148 |
| 2. | 151 | 174 | 280 |
| 3. | 1890 | 2180 | 3500 |
| 4. | 20 | 23 | 37 |
| 5. | 704 | 810 | 1302 |
| 6. | 10.85 | 12.50 | 20.50 |

TEMPERATURE CONVERSIONS
1. −2 degrees Celsius
2. −22 degrees Fahrenheit

3. $+26\frac{1}{2}$ degrees Celsius
4. +59 degrees Fahrenheit
5. +77 degrees Fahrenheit
6. −18 degrees Celsius

DENSITY ALTITUDE

1. 4,400 feet
2. 12,000 feet
3. 7,450 feet
4. 13,000 feet
5. 3,800 feet
6. 5,800 feet

TRUE AIRSPEED

1. 160 knots
2. 117.5 knots
3. 228 knots
4. 92 knots
5. 183 knots
6. 210 knots

OFF COURSE CORRECTION

| | Deg. to parallel | Deg. to converge | Tot. Degrees |
|---|---|---|---|
| 1. | 3.6 | 3.1 | 6.7 |
| 2. | 3.3 | 3.6 | 6.9 |
| 3. | 10 | 5 | 15 |
| 4. | 5.7 | 2.7 | 8.4 |
| 5. | 7.1 | 9 | 16.1 |
| 6. | 3 | 1.5 | 4.5 |

## CHAPTER 42

1. The VOR operates on the frequency band between 108.0 MHz and 118.0 MHz.
2. Advantages to very-high frequency transmission include:
   a. static-free reception;
   b. reliability in most atmospheric conditions.
   Disadvantages include:
   a. reception is line-of-sight;
   b. certain propeller rpm may cause minor fluctuations in the readout.
3. After selecting the published VOR frequency, the transmission must be orally identified.
4. The outbound course from a VOR is called a radial. The inbound course may be called a radial inbound, or a track. Radials and tracks are relative to Magnetic North.
5. OMNI stations are identified with three Morse code letters which are broadcast every 15 seconds. (Most OMNI stations are also identified by recorded voice every 15 seconds.)
6. Flight Service Stations are capable of voice broadcast over the VOR for scheduled aviation weather broadcasts and answers to specific requests.
7. The VOR may be used for:
   a. navigating between VOR stations;

b. navigating from one VOR to a destination;
c. navigating from a departure point to a VOR;
d. position fixing while enroute;
e. aligning the aircraft to the final approach to a terminal;
f. determining the time and/or distance from a VOR.

8. The course deviation indicator represents the magnetic course to or from the VOR which is selected on the omni bearing selector.

9. The OMNI bearing selector is used in conjunction with the course deviation indicator and the To-From flag to indicate our position relative to the VOR.

10. At an altitude of 5,000 feet, we will normally receive a usable VOR signal up to 100 miles from the transmitter.

11. The To–From flag tells us that the course deviation indicator represents a magnetic course to or from the station.

12. To find our position relative to one VOR we simply follow these steps.
    a. Tune and identify the nearest VOR.
    b. Center the course deviation indicator with a from indication on the To-From flag, by rotating the OMNI bearing selector.
    c. The OMNI bearing selector will indicate our position from the VOR.

13. Since the aircraft heading has no effect on the VOR cockpit indications, nothing on the VOR will change when the aircraft heading is changed.

14. When using one VOR receiver to plot a two-station fix, the station from which our relative position is changing the slowest should be tuned first.

15. VOR station passage will be recognized by watching for a reversal of the To-From flag, which is the only reliable indicator of station passage.

16. The VOR receiver will indicate reverse sensing when the aircraft heading is in conflict with the OMNI bearing selector by 90 degrees or more.

## CHAPTER 43

1. The automatic direction finder operates on the low frequency band (190–1750 Kc).

2. Low frequency transmission offers the following advantages.
    a. Reception is not line-of-sight; that is, the signal may be received when on the surface, or behind obstructions.
    b. Depending on transmitter power, low frequency signals may be received from great distances.

3. Low frequency transmissions are susceptible to static produced by thunderstorms and precipitation.

4. The loop antenna measures the strength of the signal which is received on each half of the antenna. These signals are compared, and the loop rotates until the strength on each half of the antenna is equal. Therefore, the loop will automatically rotate until it is perpendicular to the line of transmission.

5. The function switch on the airborne ADF receiver can be placed on the "receive" position while tuning the proper frequency, so that the antenna does not "home" on every station which the frequency selector passes. When the correct frequency is properly tuned, the function switch is placed on the "ADF" or "automatic" position, to provide homing.

6. *Bearing* means the direction of a straight line drawn between the nose of the aircraft and a transmitter.
7. *Relative bearing* is the angle between the nose of the aircraft and a line drawn from the nose to the station.
8. *True bearing* is the angle between True North and a line drawn from the present position to the station.
9. *Magnetic bearing* is the angle between Magnetic North and a line drawn from the present position to the station.
10. Homing may be inefficient if a crosswind condition exists, causing a curved path of flight over the ground to the station.
11. *Tracking* with an ADF differs from "homing" in that drift correction is applied when tracking in order to maintain a straight line course to or from the station.
12. When tracking inbound on the 170-degree magnetic bearing from an ADF station, with no wind, the magnetic heading will be 350 degrees.
13. True bearing = true heading + relative bearing
    True bearing = 045 degrees + 270 degrees = 315 degrees
14. Magnetic bearing = magnetic heading + relative bearing
    Magnetic bearing = 240 degrees + 240 degrees = 480 degrees
    Subtract 360 degrees from 480 degrees = 120 degrees

## CHAPTER 44

1. The transponder is used to positively segregate and identify air traffic on the surface radar scope.
2. Transponder code 7700 is used to indicate an emergency condition.
3. Air Traffic Control will specify the transponder code to be used in specific conditions.
4. "Squawk Ident" is the phrase used to request the pilot to positively identify himself, by pressing the identification button on the transponder.
5. Transponder code 0000 is forbidden for use by civilian pilots.
6. DME equipment is normally located at a VOR site.
7. The mileage indicated by DME is nautical.
8. If we fly 9 miles in 3 minutes, our groundspeed is 180 knots.

## CHAPTER 45

1. The problems of speech associated with articulation include:
   a. the omission of sounds from words;
   b. the distortion of words or sounds;
   c. the substitution of one sound for another.
2. Anger must not be allowed to show in the voice because it is contagious; that is, the receiver tends to become angry also.

3. We should not broadcast while smoking or eating so that the mouth, tongue, and breath may be put to efficient use in the forming of sounds.
4. We can improve our speech habits by conscientiously studying and practicing correct methods and by listening to and analyzing the speech of others.
5. Some common examples of telescoping include:
   a. "Whereja go?" instead of "where did you go?"
   b. "Whatja know?" instead of "what do you know?"
   c. "How're you?" instead of "how are you?"
6. Common errors associated with the use of the microphone include:
   a. failure to speak directly into the microphone;
   b. holding the microphone too far from the lips;
   c. holding the microphone too close to the lips;
   d. poor timing; that is, keying the microphone after the first word or words have been spoken; releasing the microphone button before the message is completed.
7. We must use the full aircraft identification number when broadcasting until the surface communicator initiates a shortened identification number.
8. When initially contacting a ground station, the call sign of the station should be used before the aircraft call sign. This alerts the ground station to the fact that the following message is for his attention.
9. *Wilco* means, "I have received your message; I understand it; I will comply with it."
   *Roger* means, "I have received all of your last transmission."
   *Over* means, "My transmission is ended and I expect a reply from you."
   *Out* means, "This conversation has ended. No response is expected."
10. We should refrain from contacting a Flight Service Station at 15 minutes past the hour because the ground communicator will normally be broadcasting the Aviation Weather Reports at that time.
11.

| LOCAL TIME | GREENWICH MEAN TIME |
|---|---|
| 0200 EST | 0700 Z |
| 0620 PDT | 1320 Z |
| 1414 CDT | 1914 Z |

12. 17,500 feet would be broadcast "one seven thousand, five hundred."
13. Use the phonetic alphabet to spell your full name.
14. Use the latest Airman's Information Manual to answer this question.
15. If, after making three transmissions to a surface station, no reply is received, the following items should be checked:
    a. transmitter and receiver turned on;
    b. volume high;
    c. proper frequency selected for transmission and reception;
    d. microphone plugged into mike jack;
    e. fuse secure in fuse holder and intact.
16. If cleared to "taxi to" a runway, we may cross all other runways while enroute without further clearance.
17. We may not taxi on a runway unless specifically cleared. "Taxi to" a runway does not imply clearance to taxi on that runway.

18. A flashing red signal light received in flight means "airport unsafe; do not land."
19. The pilot in command is solely responsible for the safety of his flight.
20. If we do not reject an air traffic control clearance, we will be expected to comply with the clearance.
21. If we cannot comply with a clearance, we must inform the clearing agency.
22. If, after requesting "Lake Reporting Service", we do not contact a Flight Service Station every ten minutes, search and rescue procedures will be initiated.
23. The pilot in command is responsible for maintaining VFR (if the flight is not IFR) even though being vectored by Air Traffic Control.
24. The pilot in command is responsible for maintaining traffic separation, even though the flight is being advised by Radar Traffic Information Service.
25. Automatic Terminal Information Service is a continuous recorded message broadcast over VOR stations. The ceiling, visibility, surface wind, altimeter setting, and runways in use at the associated terminal airport will be supplied.

## CHAPTER 46

1. The five senses are: taste, smell, touch, hearing, and sight.
2. The only sense which is reliable during flight by reference to instruments is sight.
3. Three common problems concerning "looking at the instruments" are:
   a. Fixation: the tendency to stare at one particular instrument.
   b. Omission: the tendency to leave one instrument out of the cross check.
   c. Emphasis on one instrument, usually the most familiar.
4. Control response when flying by reference to instruments is identical to control response when flying visually.
5. An untrimmed aircraft increases pilot fatigue and tension. A properly trimmed aircraft will maintain the attitude for which it was trimmed.
6. Without proper training and constant practice, the pilot will succumb to the erroneous signals received from the four useless senses.

## CHAPTER 47

1. Physiological altitude is determined by the amount of oxygen in the blood. Oxygen content is determined by the amount and pressure of the oxygen delivered to the lungs and the ability of the blood to absorb and transport oxygen to the brain and other organs.
2. The amount of oxygen entering the bloodstream decreases with altitude because less oxygen is available and the partial pressure of oxygen decreases as altitude increases.
3. Toxicity is the effect of a chemical change in the blood, which reduces the amount of oxygen the blood is able to absorb.
4. The middle ear is vented to atmospheric pressure via the Eustachian tube.

5. Sinus barotrama is caused by the entrapment of gas in the sinus cavities, due to the swelling of mucous membranes. Severe pain may result.
6. Vision is the first bodily function to suffer from hypoxia.
7. Euphoria is a feeling of well-being, usually accompanied by giddiness.
8. Since vision is the first bodily function affected by hypoxia, most pilots attempt to detect any decrease in visual acuity to receive a warning of hypoxia. A fine-print label or instrument marking may be used as a reference to periodically determine visual acuity.
9. Smoking increases the physiological altitude of the body.
10. Prior to using drugs of any kind, pilots must consult a flight surgeon or aviation-oriented physician.
11. Supplemental oxygen should be used at or above 5,000 feet at night and at or above 10,000 feet during the day.
12. Hypoxia must be detected early in its inception before euphoria commences, making the pilot unable to realize the danger.
13. The symptoms of hyperventilation are quite similar to that of hypoxia; that is:
    a. light-headedness
    b. euphoria
    c. blue tinting of fingernails and lips
    d. "pins and needles"
    e. muscle cramps
    f. collapse

    However, symptoms vary from day to day and from person to person.
14. Hyperventilation is caused by a deficiency of carbon dioxide in the blood.
15. Hyperventilation may be prevented by a conscious effort to maintain a normal rate of breathing. Rebreathing air, in a paper bag or other container, or holding the breath will usually increase the carbon dioxide content of the blood rapidly, thus diminishing the effects of hyperventilation.
16. The rods, around the periphery of the eye, "see" best at night.
17. The rods are most susceptible to hypoxia. Hence, night vision suffers first from oxygen lack.
18. Common nighttime illusions include:
    a. Stationary lights appear to move.
    b. Reflecting lights may give the impression of another aircraft in close proximity.
    c. Lights on the horizon may suddenly appear at a higher altitude.
19. Inertia is the tendency of a mass at rest to remain at rest, or of a moving mass to continue movement in a straight line.
20. Inertia affects man in flight as it does any other mass. If inertia is to be overcome to change direction or speed, a force must be exerted on the mass.
21. "Positive *g*" is a resultant force acting in the same direction as normal gravity.
22. "Negative *g*" is a resultant force acting in a direction opposite to normal gravity.
23. Excess positive *g* results in the following symptoms:
    a. A feeling of heaviness in the seat.
    b. Sagging limbs and facial tissue.
    c. Difficult or impossible movement.

d. Dimming of vision.
e. Black out.

24. The amount of positive *g* to cause blackout is not the same for every pilot, nor is it the same for one pilot from day to day. It normally takes 2 to 4 *g*'s to cause blackout.

25. Vision is the only reliable sense when flying without outside references.

26. The vestibular mechanism cannot determine the difference between centrifugal force and the force of gravity. Only the beginning of motion is perceived, not continuous motion.

27. Fatigue and boredom in flight may be prevented by:
a. Proper aircraft design.
b. Proper air conditioning and/or pressurization.
c. Stimulating interest in the landscape, navigation, or communication.
d. Proper diet and rest habits.

28. Symptoms of fear include:
a. Pounding heart.
b. Muscle tenseness.
c. Rapid pulse.
d. Dry throat and mouth.
e. Sweating.
f. Frequent urination and loose bowels.

## CHAPTER 48

1. The Airman's Information Manual may be purchased from the Superintendent of Documents, General Printing Office, Washington, D.C.

2. The following information may be found in the specific part of the Airman's Information Manual as listed.
a. Oil burner routes: Part III (revised in part III A).
b. Communications procedures prior to departure: Part I.
c. Medical facts for pilots: Part I.
d. Notices to airmen: Part III A.
e. NOTAM codes: Part I.
f. Airport lighting: Part I.
g. VOR test stations and frequencies: Part III.
h. VOR test procedures: Part I.
i. Fuel and services available at a specific airport: Part II.

3. According to the excerpt from the Airport Directory found in Chapter 48, the following information is given concerning the Atlanta Municipal Airport, in Atlanta, Georgia.
Atlanta Municipal Airport. 7 miles south, southwest of Atlanta. Instrument airport. Flight Service Station associated with the airport, called Atlanta F.S.S. Airport elevation: 1,024 feet above mean sea level. Longest hard surfaced runway is 10,000 feet long, runway 9L-27R. Four runways available.
Runway weight bearing capacity:
a. Single wheel type gear: 120,000 pounds.
b. Twin wheel type gear: 170,000 pounds.
c. Twin tandem type gear: 310,000 pounds.

Rotating light on airport. Other lighting available:
a. Medium intensity runway.
b. High-intensity runway.
c. High-intensity instrument approach.
d. Sequence flashing lights.
e. Visual approach slope indicator.
f. Runway end identifier lights.

Storage, major airframe, and major powerplant repairs.
Fuel: 115/145 octane (and lower), Kerosene.
Unicom: 123.0 MHz.
VASI on runway 27R and 27L, runway 33.
Runway end identifier lights for runway 27R.
Remarks: an air traffic control tower and/or an instrument landing system are associated with the airport.
Television towers:
a. 1,042-foot tower (2,049 feet above MSL) located 8.5 nautical miles northeast of the airport.
b. 1,076-foot tower (2,049 feet above MSL) located 7.5 nautical miles from the airport, northeast.
c. 1,100-foot tower (2,049 feet above MSL) located 10 nautical miles northeast of the airport.

Landing fee for commercial flights only.
For additional information, see Part III.

4. According to the Airport Facility Directory found in Chapter 48, the following information is given concerning Lovell Field, Chattanooga, Tennessee.
   a. Tower frequencies are: 118.3 MHz, 119.2 MHz, 122.5 MHz (receive only).
   b. Ground Control frequency is: 121.9 MHz.
   c. Field elevation: 682 feet above MSL.
   d. VFR traffic information available. Contact approach control when 25 nautical miles out, on 119.2 MHz.

## CHAPTER 49

1. The two-way radio is a most effective tool to request aid in combatting emergencies.
2. The Four C's are: Confess, Communicate, Climb, and Comply.
3. The standard emergency frequency is 121.5 MHz.
4. "Mayday" is the standard word used to indicate an emergency transmission.
5. We should state fuel remaining in hours and minutes during an emergency.
6. Except for fire in flight, we should climb during most emergencies in order to gain time (due to higher altitude) and to become more visible to more radar stations.
7. If unable to comply with emergency instructions, we should advise the ground station communicator.
8. To receive a D.F. steer we should:
   a. Contact any Flight Service Station or control tower on 121.5 MHz.
   b. Request an immediate D.F. steer.
   c. When requested, we should key the microphone and count slowly to ten.

9. If the engine fails on takeoff, the best glide angle must be immediately established.
10. If the engine fails at altitude, we should check all engine controls and fuel selectors while gliding to the landing area.
11. The electrical system, including the master switch, has no effect on engine operation.
12. In the event of a generator failure, all unnecessary equipment should be shut off to conserve battery power.
13. In the event of engine overheating, we should reduce power, increase airspeed, open cowl flaps, richen the mixture, and plan on a landing at the nearest suitable airport.
14. In the event of low oil pressure indications, we should reduce the power to that required to maintain maneuvering speed, plan on an immediate landing, and make as few power changes as possible.
15. Military airports may be used by civilian pilots in an emergency.

## CHAPTER 50

1. According to Federal Air Regulations, an aircraft is any device which is used or intended to be used for flight in the air, while an airplane is an engine-driven, fixed-wing, heavier-than-air device which is supported in flight by the dynamic reaction of the air against its wings.
2. An airport traffic area is that airspace within a horizontal radius of 5 statute miles from the center of any airport at which a control tower is operating, extending from the surface up to, but not including, 2,000 feet above the surface. The airport traffic area is in force whenever the control tower is operating.
3. Air Traffic Control promotes the safe, orderly and expeditious flow of traffic.
4. When used in respect to *airman* certification, *category* means a broad classification of aircraft, such as airplane, watercraft, glider, and lighter-than-air.
5. When used in respect to an *airman*, *class* means a classification of aircraft within a category, having similar operating characteristics, such as single-engine, multi-engine, land, water, gyroplane, helicopter, airship, and free balloon.
6. When used in respect to an *airman*, *type* means a specific make and basic model of aircraft, including modifications thereto that do not change its handling or flight characteristics, such as the DC-9, DC-8, B-707, Cessna 150, Cessna 182, etc.
7. Cloud heights are reported by surface observers as heights above the surface. Pilot reports of cloud heights or cloud tops are normally given in reference to mean sea level.
8. A ceiling is the height above the surface of the lowest layer of clouds or obscuring phenomenon that is reported as "broken", "overcast", or "obscuration", and not classified as "thin" or "partial."
9. Flight time is "block to block" time, or the time from the moment the aircraft first moves under its own power for the purpose of flight until the moment it comes to rest at the next point of landing.
10. Flight in a prohibited area is *prohibited*, but flight in a restricted area is subject only to specific restrictions.

## CHAPTER 51

1. A person may not act as a pilot in command or in any other capacity as a required pilot flight crewmember of an aircraft under a certificate issued to him under Part 61.3 unless he has in his *personal possession* an appropriate current medical certificate.

2. A pilot must present his pilot certificate for inspection upon the request of the Administrator or an authorized representative of the National Transportation Safety Board or of any Federal, state, or local law enforcement offices.

3. A student pilot certificate expires at the end of the 24th month after the month it was issued.

4. A student pilot certificate issued on April 1, 1970 will expire on April 30 of 1972.

5. A private pilot license is issued without expiration date.

6. The category ratings to be placed on a private pilot certificate are:
   a. Airplane.
   b. Rotorcraft.
   c. Glider.
   d. Lighter-Than-Air.

7. When applicable, the airplane class ratings to be placed on a private pilot certificate are:
   a. Single-engine land.
   b. Multiengine land.
   c. Single-engine sea.
   d. Multiengine sea.

8. A private pilot must hold a category and class ratings for the aircraft in which he acts as pilot in command while carrying another person.

9. A private pilot may not act as pilot in command for compensation.

10. It is wise to keep an accurate logbook because the flight time used to meet the experience requirements for any pilot certificate or rating or the recent flight experience requirements of FAR 61.47 must be shown by a reliable record.

11. A private pilot may log as second in command time that flight time during which he is performing the duties of a second in command. However, a private pilot may not be credited with more than 50 hours of flight time as second in command and only 50 percent of that kind of flight time may be credited toward the total flight time required for a higher certificate or rating.

12. A private pilot needs only a third class physical. Therefore, a first class medical certificate may be used by a private pilot for the same duration as a third class medical certificate; that is, it expires at the end of the last day of the 24th month after the month in which it was issued. A first class medical certificate, if received on June 20, 1970 and used for operations requiring only a private pilot or student certificate, would expire on June 30, 1972.

13. A third class medical certificate expires at the end of the last day of the 24th month after the month in which it is issued.

14. No person may act as pilot in command while he has a known physical deficiency that would make him unable to meet the physical requirements for his current medical certificate.

15. Prior to carrying passengers at night as pilot in command, a properly certificated

pilot must have made at least five takeoffs and five landings to a full stop during nighttime, within the preceding 90 days.

16. *Night*, when concerned with pilot experience, is the period beginning one hour after sunset and ending one hour before sunrise.
17. A student pilot must be 14 years of age to pilot a glider and 16 years of age to pilot any other aircraft.
18. A student pilot may not operate an airplane in solo flight outside of a local area designated by his flight instructor until his pilot certificate is endorsed by a rated flight instructor who finds that the student has complied with FAR 61.65.
19. A student pilot may never carry passengers.
20. To be eligible for a private pilot certificate, a person must be at least 16 years of age for a glider rating and at least 17 years of age for any other rating.
21. A private pilot may hold a first, second, or third class medical certificate.
22. A private pilot may share the operating expense of an airplane with his passengers.
23. A private pilot who is an aircraft salesman and who has at least 200 hours of logged flight time may demonstrate an aircraft in flight to a prospective buyer.
24. A private pilot may not fly an aircraft in instrument conditions as pilot in command unless he holds an instrument rating.
25. A private pilot may, for compensation or hire, act as pilot in command of an aircraft in connection with any business or employment if the aircraft does not carry passengers or property for compensation or hire.

## CHAPTER 52

1. Each Federal airway includes the airspace within parallel boundary lines 4 nautical miles each side of the centerline.
2. Unless otherwise specified, each Federal airway includes that airspace extending from 700 feet above the surface of the earth.
3. Unless otherwise specified, each Federal airway extends upward to, but not including, 18,000 feet M.S.L.
4. A Federal airway does not include the airspace of a prohibited area.
5. The control zone extends upward from the surface of the earth.
6. The Airman's Information Manual lists specific information about special use airspace.
7. No person may operate an aircraft within a restricted area between the designated altitudes and during the time of designation unless he has the permission of the using agency or the controlling agency.
8. No person may operate an aircraft within a prohibited area unless authorization has been granted by the using agency.
9. The continental control area begins at 14,500 feet M.S.L, but does not include the airspace less than 1,500 feet above the surface of the earth.
10. A control zone extends upward to infinity.

## CHAPTER 53

1. The pilot in command of an aircraft is directly responsible for, and is the final authority as to, the operation of that aircraft.
2. In an emergency requiring immediate action, the pilot in command may deviate from any rule of subpart A or B of Part 91 to the extent required to meet that emergency.
3. Except in an emergency, no pilot of a civil aircraft may allow a person who is obviously under the influence of intoxicating liquors or drugs (except a medical patient under proper care) to be carried in that aircraft.
4. No person may operate a civil aircraft unless it has within it an appropriate and current airworthiness certificate, a registration certificate issued to its owner, and operations limitations.
5. The pilot in command is responsible for determining that the aircraft is in condition for safe flight.
6. The colored speed range arcs on the airspeed indicator in many light aircraft are placards which are part of the operations limitations, and therefore are required for flight.
7. An approved seat belt must be provided for each passenger.
8. No person may operate an aircraft in formation flight except by arrangement with the pilot in command of each aircraft in the formation. No person may operate an aircraft carrying passengers for hire in formation flight.
9. When two *airplanes* (all *airplanes* are of the same category) are converging at approximately the same altitude (except head-on or nearly so) the aircraft to the other's right has the right of way. However, an aircraft towing or refueling other aircraft has the right of way over all engine-driven aircraft.
10. When two *airplanes* (or any aircraft) are approaching each other head-on or nearly so, each pilot of each airplane shall alter course to the right.
11. Each aircraft that is being overtaken has the right of way, and each pilot of an overtaking aircraft shall alter course to the right to pass well clear.
12. Aircraft, while on final approach to land or while landing, have the right of way over all other aircraft in flight or operating on the surface.
13. When below 10,000 feet MSL, no person may operate an aircraft at an indicated airspeed of more than 250 knots.
14. Unless otherwise authorized by ATC, no person may operate an aircraft within an airport traffic area at an indicated airspeed of more than:
    a. 156 knots: reciprocating engine aircraft.
    b. 200 knots: turbine-powered aircraft.
15. No person may, during the period from sunset to sunrise, operate an aircraft unless it has lighted position lights.
16. The minimum safe altitude *anywhere* is an altitude allowing, if a power unit fails, an emergency landing without undue hazard to persons or property on the surface.
17. The minimum safe altitude over a congested area is defined as an altitude of 1,000 feet above the highest obstacle within a horizontal radius of 2,000 feet of the aircraft.

18. The minimum altitude at which flight may be made over a ship in the high seas is an altitude which is not closer than 500 feet from the vessel.
19. A flight plan is not required for VFR flight.
20. Unless otherwise authorized by ATC, no person may operate an aircraft within an airport traffic area except for the purpose of landing at, or taking off from, an airport within that area.
21. Since no person may operate an aircraft to, from, or on an airport with an operating control tower, or operate within an airport traffic area, without having two-way radio contact with the tower, we must maintain an altitude above the airport traffic area if we do not contact the tower.
22. An airplane approaching to land on a runway served by a VASI shall maintain an altitude at or above the glide slope until a lower altitude is necessary for a safe landing.
23. A clearance to "taxi to" the runway is a clearance to cross all intersecting runways, but is not a clearance to "taxi-on" the assigned runway.
24. No person may operate an aircraft within a positive control area unless that aircraft is operated IFR and is flown by a pilot rated for instrument flight.
25. The VFR minimums for flight within controlled airspace, 1,200 feet or less above the surface, are three statute miles visibility and 500 feet clearance below clouds, 1,000 feet clearance above clouds, and 2,000 feet horizontal clearance from clouds.
26. The VFR minimums outside controlled airspace between 1,200 feet above the surface and 10,000 feet MSL are three statute miles visibility and 500 foot clearance below clouds, 1,000 feet clearance above clouds, and 2,000 feet horizontal clearance from clouds.
27. The VFR minimums outside controlled airspace when 1,200 feet or less above the surface are one statute mile visibility, and clear of clouds.
28. On a magnetic course of 050 degrees and when more than 3,000 feet above the surface, we must use an odd thousand foot altitude plus 500 feet for VFR flight. Between 4,000 and 7,000 feet, therefore, we must maintain 5,500 feet.
29. On a magnetic course of 005 degrees, over terrain which is at or near sea level, we must use an odd thousand foot altitude plus 500 feet, for VFR flight. Therefore, between 4,000 and 6,000 feet, 5,500 feet must be maintained.
30. The owner or operator of an aircraft is primarily responsible for maintaining that aircraft in an airworthy condition.
31. An airworthiness certificate must be renewed every 12 calendar months.

## CHAPTER 54

1. ADIZ and DEWIZ procedures define the special precautions which have been taken to identify aircraft operating across the borders of the United States.
2. With certain specific exceptions, flight in an ADIZ requires a two-way radio.
3. The following information must be included in a report prior to penetrating a domestic ADIZ.
   a. Time, position and altitude at which the aircraft passed the last reporting point before penetration.

b. The estimated time of arrival over the next appropriate reporting point along the flight route.

4. If the pilot operating an aircraft under DVFR in an ADIZ cannot maintain two-way radio communications, he may proceed in accordance with his original DVFR flight plan or land as soon as practicable. The pilot shall report the radio failure to an appropriate aeronautical facility as soon as possible.
5. Any FAA Flight Service Station will aid the pilot concerning questions of ADIZ or DEWIZ operations.

## CHAPTER 55

1. An aircraft accident is an occurrence associated with the operation of an aircraft which takes place between the time any person boards the aircraft with the intention of flight, until such time as all such persons have disembarked, in which any person suffers death, or serious injury as a result of being in or upon the aircraft or by direct contact with the aircraft or anything attached thereto, or the aircraft receives substantial damage.
2. A fatal injury means any injury which results in death within 7 days.
3. A serious injury means any injury which:
   a. Requires hospitalization for more than 48 hours, commencing within seven days from the date the injury was received;
   b. Results in a fracture of any bone (except simple fractures of fingers, toes, or nose);
   c. Involves lacerations which cause severe hemorrhages, nerve, muscle, or tendon damage;
   d. Involves injury to any internal organ; or
   e. Involves second or third degree burns, or any burns affecting more than five percent of the body surface.
4. Occurrences which must be immediately reported to the National Transportation Safety Board are:
   a. An aircraft accident or any of the following listed incidents:
      (1) Flight control system malfunction or failure;
      (2) Inability of any required flight crewmember to perform his normal flight duties as a result of injury or illness;
      (3) Turbine engine rotor failures excluding compressor blades and turbine buckets;
      (4) Inflight fire;
      (5) Aircraft collide in flight.
   b. An aircraft is overdue and is believed to have been involved in an accident.
5. The holder of a pilot certificate issued under Part 61 may perform preventive maintenance on any aircraft owned and operated by him that is not used in air carrier service.

## CHAPTER 56

1. Flight planning should begin well in advance of the flight. Diet and rest patterns for days before the flight should be consistent with good health care.

2. When contacting the weather briefer, we should supply the following information:
   a. Type of pilot.
   b. Type of aircraft.
   c. Destination.
   d. Estimated time of departure and time enroute.
   e. Whether VFR or IFR.
3. A weather briefing should contain the following:
   a. General synopsis.
   b. Current conditions.
   c. Forecast conditions.
   d. Alternate routes.
   e. Hazardous weather.
   f. Forecast winds aloft.
4. Even though not required to do so by law, we should file a VFR flight plan so as to aid in search and rescue procedures should the need arise.
5. The flight plan may be filed in person, by telephone, or by two-way radio with any Flight Service Station or other FAA communications facility.
6. The flight plan may be activated by reporting the time off the ground by two-way radio or by telephone to any Flight Service Station or FAA communications facility.
7. The flight plan may be closed by telephone, by two-way radio, or in person at any Flight Service Station or FAA communications stations.

# INDEX

**G**



**H**



**I**

**Q**



**R**



**S**

**Y**